The ROV Manual

A User Guide for Remotely Operated Vehicles

The ROV Manual
A User Guide for Remotely Operated Vehicles

Second Edition

Robert D. Christ

Robert L. Wernli, Sr.

AMSTERDAM • BOSTON • HEIDELBERG • LONDON
NEW YORK • OXFORD • PARIS • SAN DIEGO
SAN FRANCISCO • SINGAPORE • SYDNEY • TOKYO
Butterworth-Heinemann is an imprint of Elsevier

Butterworth-Heinemann is an imprint of Elsevier
225 Wyman Street, Waltham, MA 02451, USA
The Boulevard, Langford Lane, Kidlington, Oxford, OX5 1 GB, UK

First edition 2007
Second edition 2014

Library of Congress Cataloging-in-Publication Data
A catalog record for this book is available from the Library of Congress

British Library Cataloguing-in-Publication Data
A catalogue record for this book is available from the British Library

ISBN: 978-0-08-098288-5

For information on all Butterworth–Heinemann publications visit
our website at http://store.elsevier.com

14 15 16 17 18   10 9 8 7 6 5 4 3 2 1

Contents

PART 2   VEHICLE

CHAPTER 3   Design Theory and Standards 55

PART 3   PAYLOAD SENSORS

Foreword

The world watched in horror in September 2005 as the video feeds from the hurricane-ravaged city of New Orleans were broadcast worldwide. As the city was struggling to recover from Hurricane Katrina, Hurricane Rita delivered the final blow to both the people on the coast of the Northern Gulf of Mexico as well as to the oilfield infrastructure. In the aftermath, close to 200 oil and gas production structures lay on the sea floor. The remaining structures fortunate enough to still stand incurred heavy damage.

In the midst of this crisis, we were busily trying to complete the first edition of this manual to meet an April 2006 publishing deadline. But the world changed for us on January 1, 2006 when the phone rang demanding Bob Christ's immediate travel to survey an oil barge that struck one of those 200 unmarked submerged platforms, which resulted in it spilling 72,000 barrels of fuel oil on the sea floor. Then a platform damage inspection was needed... dive support for decommissioning... structural repairs due to wind and wave stresses from hurricane force winds and seas... the publication deadline passed and yet Bob was still in the field. Bob Wernli was juggling his consulting, at-sea test support and a publication deadline for his second novel. Time was running out for both of us so we quickly buttoned up the first edition, although it was not as complete as we had originally envisioned.

In this second edition, we have come closer to our goal of producing a broad overview of ROV technology. Through the help of leaders and companies from throughout the industry, we have produced a solid survey of the current state of this capability. Our sincere gratitude and thanks go out to those who contributed to our quest. These contributors are recognized in the Acknowledgements section. What we envisioned from the beginning for this manual is a basic **How To** for ROV technology. The US Military has this type of top-level technology manual in their "Dash 10" series. The aviation industry has what is commonly known as the "Jeppesen Manual" (named for the company that wrote the original manual). We hope that we have achieved this goal through this edition of *The ROV Manual*.

This manual is a living breathing entity. Every book is a piece of history upon the publication date; therefore, we welcome comments on this edition. We hope to revise this manual in the future as the technology evolves and would like your comments for further refining this text if/when we put forth another edition. Each subject within this manual could fill an entire book in and of itself. We struggled with editing this manual (with a nominal word and text cap) to include all subjects in as short and succinct a manner as possible while still getting the point across. Although, due to the size constraints, we could not address the larger work-class ROVs as much as we desired, the technology, sensors, tools, manipulators, and related equipment apply across the board to all systems. We hope that this text will whet the reader's appetite for further research into the technologies, equipment, and systems discussed.

The entire body of knowledge encompassing ROV technology is evolving rapidly and the lines between ROV and AUV are quickly occluding as the field of robotics morphs from space to land to sea. The subsea oilfield is firmly embracing the land-based model of network interconnectivity bringing man (remotely) back into the harsh environment of the subsea world. The easy *finds* for the world's minerals have already been achieved and exploited. The frontier has moved from the

29% of the Earth covered by land to the 71% of the Earth covered by sea. The mineral riches of the world are hidden beneath those waves. The only way to get to them is with robotics. And that is where the fun starts!

Bob Christ
SeaTrepid
Covington, LA
www.SeaTrepid.com

Robert Wernli Sr
First Centurion Enterprises
San Diego, CA
www.wernlibooks.com

Acknowledgments

In the first edition, the authors performed all of the basic research with some participation from industry in the form of recommendations and figure submissions. However, in this edition we elicited considerable industry participation—and we certainly received it!

Arnt Olsen of Kongsberg again participated with the contribution of Kongsberg's seminal work on acoustic theory. Stephen Dodd of GRI Simulations contributed to Chapter 4. Tyler Schilling along with Peter MacInnes, Steve Barrow, and Matt Whitworth of Schilling Robotics were instrumental in the production of both the manipulator as well as the tooling chapters. Alasdair Murray and Steve Stepinoff of Sub-Atlantic and Chris Roper of Saab SeaEye supplied a plethora of materials for mid-sized vehicles and Jim Teague, J. Teague Enterprises, contributed to the chapter on floatation.

And a special thanks to pioneers of the cables and connector industry, Cal Peters, Kevin Hardy, and Brock Rosenthal, who completely drafted Chapter 8. Kevin Hardy is President of Global Ocean Design, specializing in free vehicle component technologies. He retired from the Scripps Institution of Oceanography/UCSD after 36 years in ocean engineering. Hardy's instruments have successfully operated from the Arctic Circle to the southern oceans, and from the Arctic surface to the deepest ocean trenches, including James Cameron's dive to the bottom of the Mariana Trench. Cal Peters is the Director of Engineering for Falmat (San Marcos, CA), a manufacturer of custom cables for the offshore industry and other markets. He has 32 years experience in the design and manufacture of EM, signal, power, faired, and neutrally buoyant underwater cables for diverse applications, including towed instruments, moorings, ROVs, and manned vehicles. Brock Rosenthal is the President and founder of Ocean Innovations (La Jolla, CA), a distributor of underwater connectors, cables, and other quality oceanographic hardware. Rosenthal has helped numerous end users clearly define their operational requirements before selecting their underwater equipment. Their support in this endeavor is truly appreciated.

But it did not stop there. Practically all of the titans of this industry volunteered their time and resources to this project. It is difficult to express our thanks in broad enough terms. All contributors are listed alphabetically by chapter below.

General

- Johnny Johnson   Oceaneering
- Noel Nelson   Remote Inspection Technologies
- Steve Stepinoff   Forum Energy Technologies

Chapter 1: The ROV Business

- Brian Luzzi   VideoRay
- Drew Michel   Marine Technology Society
- Matt Whitworth   FMC Technologies
- Steve Walsh   SeaTrepid International

Chapter 2: The Ocean Environment

- Steve Fondriest   Fondriest Environmental

Chapter 3: Theory and Standards

- Alasdair Murrie   Sub-Atlantic
- Brock Rosenthal   Ocean Innovations
- Chris Roper   Saab SeaEye
- Gordon Durward   Forum Energy Technologies
- Peter MacInnes   Schilling Robotics

Chapter 4: Vehicle Control and Simulation

- Ioseba Tena   SeeByte
- Neil Noseworthy   GRI Simulations
- Rick Cisneros   CA Richards
- Steve Dodd   GRI Simulations

Chapter 5: Vehicle Design and Stability

- Amanda Ellis   Deep Sea Power & Light
- Jim Teague   J Teague Enterprises
- Robert Kelly   Trelleborg

Chapter 6: Thrusters

- Andrew Bazeley   Tecnadyne
- Bob Gongwer   Innerspace
- Chris Gibson   VideoRay
- Dick Frisbie   Oceaneering
- Ian Griffiths   Soil Machine Dynamics
- Jesse Rodocker   SeaBotix
- Scott Bentley   VideoRay
- Steve Barrow   Schilling Robotics
- Tyler Schilling   Schilling Robotics

Chapter 7: Power and Telemetry

- George "Buddy"
  Mayfield   Outland Technology
- Scott Allen   MacArtney

Chapter 8: Cables and Connectors

- Brad Fisher   SeaCon
- Brock Rosenthal   Ocean Innovations

- Cal Peters   Falmat
- Glenn Pollock   SeaCon
- Kevin Hardy   Global Ocean Design
- Shawn Amirehsani   Falmat

Chapter 9: LARS and TMS

- Bill Kirkwood   Monterey Bay Aquarium Research Institute
- David Owen   Caley Ocean Systems
- Jeff Conger   MacArtney
- Jesse Rodocker   SeaBotix
- Jim Honey   Shark Marine
- Joan Gravengaard   MacArtney
- Joe Caba   SeaTrepid International
- Kim Fulton-Bennett   Monterey Bay Aquarium Research Institute
- Mike Obrien   SeaTrepid International
- Rasmus Frøkjær Bonde   MacArtney
- Steve Barrow   Schilling Robotics
- Vergne Caldwell   SeaTrepid International
- Wes Keepers   Dynacon

Chapter 10: Video

- Pat McCallan   Pat McCallan Graphics
- Rick Cisneros   CA Richards

Chapter 11: Vehicle Sensors and Lighting

- Ronan Gray   SubAqua Imaging Systems
- Cyril Poissonnet   Remote Ocean Systems

Chapter 12: Sensor Theory

- Chris Roper   Saab SeaEye
- Keith Pope   Teledyne TSS
- Keith Wittie   Entergy

Chapter 13: Communications

- Carl Schneider   Grey Insurance
- Vergne Caldwell   SeaTrepid International

Chapter 14: Underwater Acoustics

- Arnt-Helge Olsen   Kongsberg Maritime

Chapter 15: Sonar

- Blair Cunningham   Coda Octopus
- Helmut Lanziner   Imagenex Technology
- Jan Trienekens   Remote Inspection Technologies
- Jim O'Neill   Coda Octopus
- Joe Burch   Sound Metrics
- Joe Caba   SeaTrepid International
- John Beekman   FugroChance
- Maurice Fraser   Tritech International
- Willy Wilhelmsen   Imagenex Technology

Chapter 16: Acoustic Positioning

- Arnt-Helge Olsen   Kongsberg Maritime
- Julian Rickards   Sonardyne

Chapter 17: Navigational Sensors

- Margo Newcombe   Teledyne RDI
- Paul Watson   CDL
- Rick Cisneros   CA Richards

Chapter 18: Ancillary Sensors

- Alistair Coutts   Seatronics Group
- Chris Roper   Saab SeaEye
- Euan Mackay   Seatronics Group
- Jan Trienekens   Remote Inspection Technologies
- Keith Pope   Teledyne TSS
- Lori Strosnider   Chevron
- Margo Newcombe   RD Instruments (Maybe this is Margo in C17?)
- Rod Sanders   Cygnus Instruments
- Scott Vidrine   Tracerco
- Tony Landry   SeaTrepid International

Chapter 19: Manipulators and Tooling

- Brett Kraft   Kraft Telerobotics
- Chris Gibson   VideoRay
- Don Rodocker   SeaBotix
- Jesse Rodocker   SeaBotix
- Jessica Montoya   Schilling Robotics
- Matt Whitworth   Schilling Robotics
- Megan Anderson   Schilling Robotics
- Peter MacInnes   Schilling Robotics

- Steve Stepinoff   Forum Energy Technologies
- Tyler Schilling   Schilling Robotics
- Wendy Glover   Hydro-Lek

Chapter 20: Tooling and Sensor Deployment

- Andrew Bazeley   Tecnadyne
- Antone Forneris   Cavidyne
- Deborah Geisler   Stanley Infrastructure Solutions
- Graham Sloane   Planet Ocean
- John Clark   Wachs Subsea
- John Merrifield   Stanley Infrastructure Solutions
- Jonathan Bochner   WEBTOOL—Variators Ltd.
- Mike Cardinal   Stanley Infrastructure Solutions
- Noel Nelson   Remote Inspection Technologies
- Scott Crosier   Crosier Design
- Tom Ayars   Seanic Ocean Systems
- Steve Barrow   Schilling Robotics
- Susan Bazeley   Tecnadyne
- Tim Sheehan   Wachs Subsea

Chapter 21: Practical Applications

- Blades Robinson   Dive Rescue International
- Brett Seymour   US National Park Service
- Jack Fisher   JW Fishers
- Julian Rickards   Sonardyne
- Ken McDaniel   US Department of Homeland Security
- Marco Flagg   Desert Star Systems
- Willy Wilhemsen   Imagenex Technology

Chapter 22: The Little Things That Matter

- Joe Caba   SeaTrepid International
- Vergne Caldwell   SeaTrepid International

Chapter 23: The Future of ROV Technology

- Alain FIDANI   Cybernetix
- Andy Bowen   Woods Hole Oceanographic Institution
- Art Schroeder   Energy Valley
- Chris Nicholson   Deep Sea Systems International
- Chris Roper   Saab SeaEye
- Dan McLeod   Lockheed Martin
- Deanna Talbot   Bluefin Robotics
- Drew Michel   Marine Technology Society

- Erika Fitzpatrick   Woods Hole Oceanographic Institution
- Ian Crowther   WFS Technologies
- Jake Klara   C&C Technologies
- John Westwood   Douglas-Westwood
- Lou Dennis   Lockheed Martin
- Steve Cowen   SPAWAR Systems Center Pacific

Introduction

This edition of *The ROV Manual* substantially expands upon the previous edition. This text is divided into five logical parts covering the industry and environment, the basics of ROV technology, payload sensors, intervention tooling as well as practical field applications. In the last chapter of the book, we look into the future in order to examine what industry analysts feel is the direction subsea technology is heading with a specific focus on the field of subsea robotics.

It is often said that *for every mathematical formula within a book, the population of book purchasers is halved.* As authors we certainly appreciate that thought, but seek to go from general terms to specific (as well as from simple to complex) toward reaching a broad readership for this subject. Some of the chapters are heavily focused on theory (e.g., Chapter 14 on underwater acoustics is heavily math-based) while others (e.g., Chapters 21 and 22, which focus on field applications and procedures) contain little or no mathematics. The general technology user should feel free to skip over the math-based sections, while those with a more academic bent or specific application should delve into the technical aspects of theory.

Chapters 1 and 2 (Part 1) seek to paint a background picture of the industry, as well as the environment, where ROVs operate and this technology applies. Chapters 3−11 (Part 2) drill down to the actual vehicle in a good bit of detail while Chapters 12−18 (Part 3) branch out into the broad subject of payload sensors. It is often said in this industry, "It is not about the vehicle, it is about the sensors and tooling." Therefore, we additionally break out Part 4 (Chapters 19 and 20) to address the manipulators and tooling aspects of ROV technology while the final section (Part 5— Chapters 21−23) focuses on both practical applications and standard operating procedures, then closes with the authors' take on the future of this technology.

While this text casts a wide net over the entire field of Remotely Operated Vehicle technology, we focus specifically on the classes of vehicle (narrowly defined in Chapter 3) in the observation class ROV (OCROV) and mid-sized ROV (MSROV) categories. The only subject missing to cover the full gamut of vehicles (i.e., the Work Class ROV, WCROV) is high-pressure hydraulics. And that subject will be left for a future iteration of this manual as we continue to refine this work while the industry (and technology) continues to evolve.

The divisions (parts) of this manual each address a separate readership. Part 1 is geared towards the business side and should be applicable to project managers making use of this technology while Part 2 focuses specifically on the ROV technician. Part 3 is addressed to the project manager but should also be of interest to the survey team as well as the ROV technician for gaining a general understanding of deployed sensor technologies. Part 4 is directed toward intervention technicians over a broad range of users (from Project Manager to Corporate Executive to Regulatory Officials to, of course, ROV Technicians). And Part 5 wraps it all up with both practical considerations and a look into the future.

No text of this size can do any measure of justice to the field of ROV technology. The authors have carefully carved out individual subjects in order to form an introduction into each field. We hope that you, the reader, come away with a general understanding of this industry and its advanced technologies, thus encouraging further investigation into your specific field of interest.

Welcome to the exciting field of subsea robotics!

Industry and Environment

The ROV Business

CHAPTER CONTENTS

1.1 The ROV

1.1.1 What is an ROV?

Currently, underwater vehicles fall into two basic categories (Figure 1.1): manned underwater vehicles and unmanned underwater vehicles (UUVs). The US Navy often uses the definition of UUV as synonymous with autonomous underwater vehicles (AUVs), although that definition is not a standard across the industry.

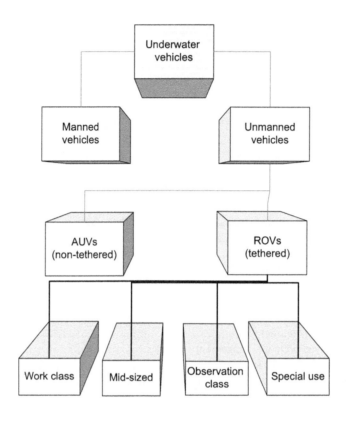

FIGURE 1.1

Underwater vehicles to ROVs.

According to the US Navy's UUV Master Plan (2004 edition, section 1.3), an "unmanned undersea vehicle" is defined as a:

Self-propelled submersible whose operation is either fully autonomous (preprogrammed or real-time adaptive mission control) or under minimal supervisory control and is untethered except, possibly, for data links such as a fiber-optic cable.

The civilian moniker for an untethered underwater vehicle is the AUV, which is free from a tether and can run either a preprogrammed or logic-driven course. The difference between the AUV and the remotely operated vehicle (ROV) is the presence (or absence) of a direct hardwire (for communication and/or power) between the vehicle and the surface. However, AUVs can also be (figuratively) linked to the surface for direct communication through an acoustic modem, or (while on the surface) via an RF (radio frequency) and/or an optical link. But in this book, we are concerned with the surface-directed, hard-wired (tethered) ROV.

The ROV falls within a broad range of mobile robotic vehicles generally termed "remotely con-trolled mobile robots." The motion of the vehicle can be via autonomous logic direction or remote operator control depending upon the vehicle's capability and the operator's degree of input. The

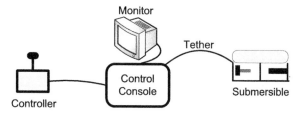

FIGURE 1.2

Basic ROV system components.

power of the vehicle can be onboard (i.e., battery or engine powered), offboard (i.e., power delivered through conductors within the tether), or a hybrid of both (e.g., onboard battery powered with a power recharge transmitted remotely through the tether).

Simplistically, an ROV is a camera mounted in a waterproof enclosure, with thrusters for maneuvering, attached to a cable to the surface over which a video signal and telemetry are transmitted (Figure 1.2). Practically all of today's vehicles use common industry standards for commercial off-the-shelf (COTS) components.

As the ROV goes from its simplest shallow water form toward the more complex deep water work vehicles, the required degree of sophistication of its operators as well as the depth of its support network climbs substantially in a similar fashion to that of aircraft or large surface vehicles.

The modern ROV is a mature technology with established standards of operator qualifications, safe operations, and a proven history of getting work done in the "dull, dirty and dangerous" work environments of the world's waters.

The following sections will provide a better understanding of the scope of this definition. Subsequent chapters will provide in-depth discussions of the design of ROV systems along with operational considerations.

1.1.2 ROV size classifications

ROVs can be more specifically described as teleoperated free-swimming robotic unmanned (or "uninhabited" for the more modern term) underwater vehicles. These are used in a variety of applications from diver support to heavy marine subsea construction. The market is substantially segmented into four broad categories based upon vehicle size and capabilities:

1. *Observation class ROVs (OCROV)*: These vehicles go from the smallest micro-ROVs to a vehicle weight of 200 pounds (100 kg). They (Figure 1.3) are generally smaller, DC-powered, inexpensive electrical vehicles used as either backup to divers or as a diver substitution for general shallow water inspection tasks. Vehicles in this classification are generally limited to depth ratings of less than 1000 ft (300 m) of seawater (fsw/msw) due to the weight of the power delivery components and one atmosphere pressure housings—which imposes limitations upon the vehicle size (i.e., neutral buoyancy must be maintained if the vehicle is to have the ability to swim). The vehicles within this class are typically hand launched and are free flown from the

FIGURE 1.3

Examples of OCROVs (Note: shoes atop for scaling).

(Courtesy SeaTrepid.)

FIGURE 1.4

Example of MSROV.

(Courtesy Forum Energy Technologies.)

surface with hand tending of the tether. The older/antiquated term for vehicles of this classification is "low-cost ROVs."

2. *Mid-sized ROVs (MSROV)*: These vehicles weigh from 200 pounds (100 kg) to up to 2000 pounds (1000 kg). They (Figure 1.4) are generally a deeper-rated version of the OCROVs with

FIGURE 1.5

Example of WCROV.

(Courtesy FMC Technologies.)

sufficient AC power delivery components and pressure housings capable of achieving deeper depths over longer tether/umbilical lengths. These also are generally all-electric vehicles (powering prime movers (thrusters) and camera movement controls) with some hydraulic power for the operation of manipulators and small tooling package options. The vehicle electrical power is stepped down to a manageable voltage for operation of the various components and can be either AC or DC power. Vehicles in this classification are sometimes termed "light work class" vehicles to fully differentiate them from OCROVs. Due to the weight of these vehicles, a launch and recovery system (LARS) as well as a tether management system (TMS) is often needed.

3. *Work class ROVs (WCROV)*: Vehicles in this category are generally heavy electromechanical vehicles running on high-voltage (>3000 V) AC circuits from the surface to the vehicle (Figure 1.5). The power delivered to the vehicle generally is changed immediately to mechanical (hydraulic) power at the vehicle for locomotion as well as all manipulation and tooling functions.

4. *Special-use vehicles*: Vehicles not falling under the main categories of ROVs due to their non-swimming nature such as crawling underwater vehicles, towed vehicles, or structurally compliant vehicles (i.e., non-free-swimming). The special-use vehicle coverage is outside the purview of this text.

The general difference between the OCROV and the MSROV is the power transmission and depth rating. The general difference between the MSROV and the WCROV is the size of the hydraulic power pack and the horsepower rating for the operation of manipulators and tooling. Both the MSROV and the WCROV are deep-rated vehicles and both can be delivered to deep work sites. The WCROV, however, can perform heavier tasks than the MSROV is capable of achieving due to the added muscle of hydraulic actuation of its components (versus the electrical actuation of the MSROV). There are a few "electrical WCROVs" on the market, but the vast majority of the

Table 1.1 ROV Classifications

Classification	General Capability
OCROV	Flying eyeball with limited intervention
MSROV	Full gigabit data throughput with light intervention
WCROV	Full gigabit data throughput with heavy intervention

Table 1.2 Representative Vehicle Characteristics

Size Category	Input Power	Vehicle Power	Telemetry Type	Depth Rating	Launch Method	TMS	Thruster/ Tooling	Tooling Fluid Flow
OCROV	110/220 VAC 1Φ	Low-voltage DC	Copper only	+/− 300 m (1000 ft)	Hand deploy	No	Electric/ electric	Electric only
MSROV	440/480 VAC 3Φ	Medium-voltage DC or AC	Copper or fiber	>1000 m (3000 ft)	Crane or A-frame	Optional	Electric/ hydraulic	15 lpm (4 gpm)
WCROV	440/480 VAC 3Φ	High-voltage AC	Fiber only	>3000 m (10,000 ft.)	A-frame	Yes	Hydraulic/ hydraulic	70 lpm (18 gpm)

worldwide population of WCROVs is hydraulic due to the power requirement and the inherent reliability of hydraulic systems over their electric counterparts within the seawater environment. OCROVs are generally called "flying eyeballs" as their main job is to function as a shallow water video platform. The MSROV has additional deepwater capabilities along with fiber optic telemetry for full gigabit sensor throughput. The WCROV possesses all of the attributes of both the OCROV and the MSROV along with high-powered hydraulic manipulators and tooling capabilities (Table 1.1).

Table 1.2 depicts representative vehicle configurations and power/telemetry requirements. Configurations vary within each category from vehicle to vehicle, but these represent the general characteristics of vehicles within the specified size category. For specific vehicle parameters, please consult the vehicle manufacturer's technical specifications.

1.1.3 Vehicle shape versus mission

The length to width (termed "aspect") ratio of the vehicle directly affects its hydrodynamics. Long slender vehicles generally have lower drag characteristics at higher speeds but exhibit poor station-holding capabilities. Short vehicles have much better station-keeping capabilities along with higher maneuverability in all three axes of travel (x/y/z) but have much higher drag profiles as the speed ramps up. AUVs typically exhibit the classical torpedo shape with a high aspect ratio and minimal number of thrusters coupled with control fins for long-distance travel at higher speeds. The ROV is typically used in station-keeping (or slow speed observation) tasks for close inspection and/or sensor and tooling delivery.

AUVs typically have a closed frame for uninterrupted fluid flow about the vehicle (for minimal drag at higher speeds) while ROVs typically have an open frame to allow for fluid flow through the frame due to a higher number of internally mounted thrusters.

Within the ROV group, vehicles with a higher aspect ratio are used for longer travel distances (e.g., pipeline surveys or regional transects) while lower aspect ratio vehicles are used for close inspections or tasks requiring station-keeping capabilities.

1.1.4 ROV depth rating by classification

General man and machine capabilities for working within the marine environment fall within the depth limitations provided in Table 1.3.

At the current state of technology for telepresence with mobile robotic systems, man (a diver) in the environment will be much more capable than a machine. This is due to the *in situ* situational awareness provided by man as well as the ultimate dexterity of the end effector (a hand for a man and a manipulator claw for the machine). But the costs and danger of placing man in the high-pressure work environment of the deep sea are considerable. Within the commercial, scientific, and governmental industries, more and more customers are looking for a robotic solution to the ever-increasing requirements for subsea sensing and intervention; their goal is to limit the cost and liability exposure brought on from possibly harming "the man." Look for vast future improvements in both the technology and demands for this category of robotics.

1.1.5 Size versus ability to deploy sensors/tooling

It is best to view any underwater vehicle as simply a "bus ride to the work site." The purpose of the vehicle will always be the delivery of the sensor and/or tooling package to the work site in order to accomplish the mission. The question then becomes: "How large does the vehicle need to be in order to effectively accommodate the sensors and tooling package?"

In general, any ROV can float any sensor or tooling package—all that is necessary is to place ample flotation aboard the vehicle to offset the in-water weight of the payload. The size of the vehicle then needs to vary to optimally accommodate the prime mover (thrusters), tooling, and sensors while still powering these items and thrusting/moving the package. Also, communication capabilities must match the sensor bandwidth requirements to ensure sufficient data throughput exists to transmit the telemetry to the operator station for logging sensor data and controlling the tooling package.

Table 1.3 Capabilities Versus Depth

Category	Depth Limitation
Air diving	190 ft (60 m)
Mixed gas diving	300 ft (100 m)
Saturation diving	1000 ft (300 m)
Atmospheric diving system	2300 ft (700 m)
OCROV	1000 ft (300 m)
MSROV	>3000 ft (>1000 m)
WCROV	>10,000 ft (>3000 m)

Table 1.4 Sensors and Data Transmission Requirements

Sensor Type	Bandwidth	Protocol	Telemetry Type
Single beam sonar	Low	RS-232/422/485	Copper or optical fiber
2D multibeam sonar	Medium to high	Ethernet	Copper or optical fiber
3D multibeam sonar	High	Ethernet	Optical fiber
UT metal thickness	Low	RS-232/422/485	Copper or optical fiber
Standard-def. video	Medium	Composite	Copper or optical fiber
High-def. video	Medium to high	Ethernet	Optical fiber
Radiation sensor	Low	RS-232/422/485	Copper or optical fiber
Pipe tracker	Low	RS-232/422/485	Copper or optical fiber

Table 1.5 Tool Type Versus Vehicle Characteristics

Tool Type	Weight in-Water	Power	Minimum Vehicle Size
Single-function manipulator	Low	DC electrical	OCROV
Light duty 4-function manipulator	Medium	Low pressure/flow hydraulic	MSROV
Heavy-duty 7-function hydraulic manipulator	High	High-pressure/low-flow hydraulic	WCROV
Wire rope cutter	Medium	High-pressure/low-flow hydraulic	MSROV and WCROV
Hydraulic grinder	Medium	Low-pressure/high-flow hydraulic	MSROV and WCROV
Diamond wire saw	Low	High-pressure/high-flow hydraulic	WCROV

With the advent of fiber optic communication capabilities on smaller mobile platforms, sensor throughput capabilities have taken on a new dimension for the ROV as a sensor delivery platform. While data over copper is still the prevalent setup for OCROVs (due to its inherent short-distance transmission requirements and low-cost structure), fiber optics is the predominant telemetry conductor for MSROVs and WCROVs. A short sample of sensors along with their data transmission requirements and likely telemetry type is provided in Table 1.4.

For tooling, the main difficulties are (i) the vehicle's ability to carry the tool as payload (i.e., the size of the vehicle vis-à-vis the tool itself) with a large enough vehicle-to-tool ratio so that the tool does not "wag the dog"—or provide so much drag that it renders the vehicle uncontrollable with the tool attached, (ii) the vehicle's power delivery capability is sufficient to power the tool while maintaining station, and (iii) the vehicle's ability to physically reach the work site. A short sample of tooling along with the typical payload and vehicle size requirements is provided in Table 1.5.

1.2 Types of ROV services

1.2.1 Call-out versus contract work

ROV services are subdivided based upon a number of factors including whether the job is a primary versus supporting role, equipment footprint, duration of assignment, and other factors.

- *Primary versus supporting role*: An example of a primary function for an ROV is any diverless operation such as subsea structural inspection, deepwater remote intervention, subsea pipeline survey, and any other function where a diver is not present in the water or at the job site. A supporting role would be as diver backup, such as a tool delivery vehicle, diver support as backup, and for diver monitoring or monitoring of diver-mounted tooling.
- *Larger ROV spreads* require a larger footprint and higher support requirements. Whereas the OCROV can simply plug into the vessel's standard 110/220 VAC consumer-grade single-phase power source, the MSROV and larger WCROV spreads require independent high-voltage/high-service three-phase power sources. The OCROVs can be hand launched over the vessel bulwarks while larger ROVs require a LARS. Many LARS require special configuration in order to get the vehicle from the deck, into the water and back again in all sea states without damage to equipment and danger to personnel. Further, in order to deliver the vehicle to deep work sites, a TMS is required to manage the soft tether from the depressor weight (i.e., heavy weight holding the vehicle steady at the work site) to the vehicle and to protect the vehicle from damage during transit to the work site. This involves a separate electrical or hydraulic system, essentially requiring two control systems in the water (doubling complexity).
- *The two types of ROV assignments* are broadly defined as "contract" and "call-out" work.
  - *Contract work* involves long-term (greater than 6 months, duration) assignments that generally involve (and cost-justify) integrating the vehicle into the work platform with the requisite detailed and complicated mobilization. An example of a contract job would be a drill support assignment whereby a complete section of the rig is dedicated to the ROV spread. Another would be a dedicated ROV vessel with the LARS and control system integrated into the superstructure of the vehicle. Integrating an ROV spread into a vessel of opportunity can be a very expensive and time-consuming proposition undertaken only for jobs that will allow for mobilization cost amortization over a long period (lowering per work day mobilization costs).
  - *Call-out work* involves short-term (less than 6 months) assignments whereby minimal integration work (termed "bolt on" integration) is performed into the vessel of opportunity due to its limited duration. The exception to this would be a fully integrated ROV vessel performing short-term work.

The cost of mobilizing a WCROV spread can be substantial. The LARS must be trucked over roads with a special wide-load road permit; the work package (vehicle, TMS, LARS, and winch with umbilical) weighs in excess of 100,000 pounds (50,000 kg) on deck, and the spread is up to four vans (control, work, generator, and survey—in addition to the work package spread). The cost of mobilizing an OCROV spread is simply hand-carrying the cases onto the vessel of opportunity and plugging equipment into the vessel's readily available consumer power.

Most mobilizations require several days of working out equipment issues in the field (normally described as "tweaking the spread") in order to achieve the optimum equipment configuration. This problem is multiplied as the complexity of the equipment increases. In short, mobilizing a WCROV spread on a new vessel of opportunity for a short-term assignment is seldom worth the trouble or the expense. The only way to justify a WCROV spread for call-out work is to have a dedicated vessel with the ROV spread permanently integrated. But considering the relatively low vessel/vehicle utilization of call-out work, it is an expensive asset to have idle, moored at the dock.

The inland work is the realm of the OCROV and MSROV, but unless it is a construction project (requiring heavy-lift WCROV capability) it is generally call-out work. The offshore marketplace is generally populated with larger exploration and production (E&P) companies performing construction projects requiring heavy WCROV equipment. In most cases, a smaller ROV would be more than adequate to perform the scope of work for the project assignment but the engineering section of these E&P companies are used to paying more for their ROV services and do not object to the higher cost structure.

The market draw for OCROV and the MSROV services encompasses "the 80% solution" at a substantially lower cost structure than the larger ROVs. The nature of the call-out business is short-term and very profitable work. The upside to the call-out business is there are not many players to dilute the already industry-wide low utilization inherent in call-out work. The downside of the call-out business is the lack of any predictably sustainable work levels. It is, for the most part, either feast or famine.

1.2.2 Day rate versus project management

This discussion is more applicable to the offshore hydrocarbon mining industry since it dominates deepwater construction but also crosses to any deepwater construction industry (e.g., wind farms, and seafloor mining).

The deepwater offshore oil and gas (O&G) service industry begins with the seismic survey and ends with the gas pumping into the passenger car. The need for waterborne robotic services begins with the precasing survey and ends when the pipeline crosses the preselected 1000 fsw (300 msw) curve. The following is a sample list of services/tasks needed:

- Pipeline/umbilical/flowline pre-lay survey
- Subsea site survey
- Drill support
- Installation of wellhead valve trees
- Installation of subsea tie-back
- Inspection, repair, and maintenance (IRM) of subsea facilities
- Fabrication and installation of pull-tubes
- Subsea coiled tubing flowline procurement and installation
- Subsea wellhead control umbilical procurement and installation
- Subsea pipeline end terminations procurement and installation
- Spool piece measurements, fabrications, and installation
- Subsea umbilical termination assemblies procurement and installation
- Planning, management, and execution of pigging and testing flowline programs
- Conduct deepwater pipeline hot taps and deep tap projects
- Conduct deepwater diverless pipeline repairs utilizing ROVs onboard vessel

- Plug and abandonment (P&A) of subsea valve trees and templates
- Umbilical and pipeline system recovery and decommissioning
- Subsea development survey, salvage, installation of long baseline (LBL) acoustic arrays
- Jumper and flying lead installations
- Strake installations to tension leg platforms
- Hydrotesting and commissioning subsea field developments
- Steel catenary riser tie-in, inspection, repair, and maintenance
- Mattress installations and crossing of pipelines, flowlines, and umbilical
- Platform and pipeline inspections to regulatory standards

The clear-bottom "survey" functions can be accomplished with logic-driven equipment and fall more within the survey function than within the construction function. The logic-driven vehicles are the AUVs and fall more within the traditional services of the survey company (i.e., outside of the traditional service offerings of ROV service companies). Construction/IRM-type robotic services require a more active approach to subsea vehicle control—which requires a teleoperated vehicle (i.e., ROV) for active "man-in-the-loop" control.

As a day rate player, the ROV service company's primary (some would say, "only") function is to have "uptime." As a project management company, the primary function is to get tasks accomplished. As a "day rater" (DR), the goal is to have sufficient equipment to get the job and stay on the job. As a "project manager" (PM), the goal is to have access to a large menu of services so as to have them available when needed in order to fulfill the end goal (a completed task). The PM company is farther up the proverbial "food chain" as it manages its own projects (as opposed to taking directions from a PM). The deepwater intervention market is the realm of the ROV. Required of the PM or DR company is ROVs with sufficient capability to deliver sensors and tooling for the deepwater environment.

The end product for a DR company is the delivery of either sensors or tooling to the work site with maximum uptime. The end product of a PM company is to accomplish an assigned project (as outlined by the customer). In the risk/reward curve, the PM company stands a much higher operational and financial risk with a corresponding potential for huge profits. The DR player has a much lower risk profile with a correspondingly lower profit upside.

The major players within the industry fall out to either day rate players or project management players. The smaller OCROV and MSROV service companies generally own and operate their equipment aboard the client's vessel of opportunity. The typical day rate players in the WCROV space are boat companies who use the ROV to sell the boat due to the substantial cost of mobilizing the WCROV spread onto the vessel (Section 1.2.1). The typical project management company has the ROV services as one of the many tools within the tool chest.

On the day rate front, the investment (and hence the reward) is in the boat—not (significantly) in the ROV. The boat companies are willing to go to a very low (or zero!) day rate on the ROV by hiding the ROV's day rate in the vessel day rate. This squeezes out any would-be purely day rate WCROV-only companies.

"Is it an ROV vessel or is it a boat with ROV capability?" This is an important question. It decides on the risk for downtime (the main purpose of a day rate company) should (when) the ROV encounter operational or maintenance issues. Most DR companies resolve this issue by splitting the customer contracts between the ROV company and the vessel company, thereby compartmentalizing the downtime risk (to the customer's detriment). A project management company does not have that luxury since if the ROV is down the entire mission is compromised.

In a post-Macondo offshore minerals extraction world, there is a new concept within the field support vessel service that has arisen to address newer and more stringent regulations. The traditional platform supply vessel (PSV) is changing from a simple mud/chemical/fluids/bulk transfer vehicle to a fully integrated deepwater support vessel. The new paradigm reasons that all future field support vessels will require an ROV capability.

The move to deepwater O&G exploration and production has seen the movement of the wellhead of the production platform from the surface to the seafloor requiring all IRM tasks to be accomplished robotically. All field support vessels will naturally be combined DR ROV companies as well as vessel companies. And the demand will rise rapidly in the oilfields of the developed world for the integrated deepwater support vessel.

So, the two choices are to be a DR player or a PM player. The equipment requirements for each follow:

DR player: The trend within the deepwater IRM market is for "dynamically positioned" (DP—in this case DP-2 for DP with redundancy) vessel capability in order for oil companies to allow PSVs to approach deepwater production platforms. At a minimum, all vessels from which ROVs operate in close proximity to deepwater production platforms will require a DP-2 capability. The typical DP-1 (dynamical positioning with no redundancy) vessel (for production support) remains within the 170 ft (55 m) length overall (LoA) range with a high of 205 ft (65 m) and a low of about 140 ft (45 m). The typical deepwater DP-2 platform supply vessel (PSV) is in the 300 ft (95 m) LoA range. For the ROV spread, a name-brand hydraulic vehicle (i.e., one of the major international WCROV manufacturers) with at least 150 hydraulic horsepower is required as most contracts specify a minimum horsepower rating (and many uninformed customers equate ROV horsepower with ROV capability).

PM player: This requirement is more task oriented as the risk of task completion borne by the PM company is much more logic driven (as opposed to market driven). And it all boils down to either a fiber (for sensor throughput and/or deepwater vehicle telemetry) or a pump (for remotely powering subsea tooling). For sensor delivery, the vehicle simply needs to be big enough to deliver the sensor while having a fiber for full gigabit data throughput. For tooling, the vehicle's auxiliary pump needs to be sized to drive the highest requirement tool anticipated.

Sensor delivery vehicle: This vehicle can be a small MSROV with electric thrusters and minimal tooling capability. The vehicle must have a fiber optic capability for data transmission along with some type of depressor to keep it at depth (most likely a cage or top-hat deployment system), pressure-compensated components and a high-voltage power system to handle the tether lengths/depths. Deepwater fiber optic-based electric MSROVs are certainly sufficient for practically all sensor delivery tasking. Hydraulic WCROVs certainly have the sensor delivery capability but are seen as expensive overkill for the sensor delivery task.

Tooling delivery vehicle: "It is all about the pump!" The highest anticipated tooling need for IRM and light intervention tasks is exemplified by the hydraulic flow requirement of a diamond wire saw as the hydraulic motor of the mid-sized saw requires both high pressure and high flow rate. Polling the various tooling manufacturers of diamond wire saws finds that the mid-sized diamond wire saw requires a 30 gpm (115 lpm) flow rate at 2000 psi (140 bar). The typical

45 cc auxiliary pump on a 150 hp WCROV provides 20 gpm (75 lpm) at 3000 psi (200 bar), which is the same hydraulic horsepower as 30 gpm (115 lpm) at 2000 psi (140 bar). The typical 150 hp WCROV should be sufficient to cover the highest draw tool initially anticipated for IRM (and most construction) tasks. If the ROV company is a DR company, the company has little choice other than to buy a name-brand system from a major manufacturer. If the ROV company is a PM company, the choice is open as to whether to buy the WCROVs or to manufacture the company's own proprietary design.

1.2.3 Strategy for service package deployment

There will be two separate strategies with regard to deployment of services for a DR company versus a PM company. Both will have some common requirements, but the PM company will have additional requirements above the DR company in order to field a "one stop shop" concept to its customer base. Further, the customer base for a PM company will be much more sophisticated with a much longer lead time for ordering services.

DR company: As explained above, it is useless to be a full-service DR WCROV company without full ownership of the deployment vessel as the boat makes the lion's share of the revenues/profits. For the MSROV and OCROV company (due to the call-out nature of the business), vessel ownership (while certainly an option) would probably be an expensive luxury as opposed to a necessity. At best, the ROV company could purchase vessels and dig into an unserviced (or underserviced) niche left open by the void between the larger international PSV companies (who offer ROV services as an added option to the vessel platform) and the smaller PSV companies (who only offer vessel-only charters).

PM company: The strategy for a PM company is the same as for a DR company with the addition of a project management function. It would be preferable to keep the DR and PM companies in relatively close geographical proximity to one another so that the engineering-to-deployment process is done with a teamwork approach allowing for maximum face time with minimal travel time.

1.3 ROV economics

1.3.1 Capital expenditure (CAPEX) versus day rate

Investment in ROV equipment can be quite expensive and financially risky. The cost of capital involves daily interest charges as well as financial carrying costs. In addition to the cost of capital, the equipment must be insured, maintained in an operationally ready status (when not in the field earning revenues), housed in a secure location, staffed by qualified and trained personnel, and operated periodically (the "use it or lose it" proposition) to verify full readiness. Further, anything that is put in the water has the chance of breaking free, becoming entangled, and/or becoming unrecoverable (equating to a full loss scenario). In short, not only does the ROV company require recapture of its cost of capital, it also requires sufficient revenues to counter the carrying costs of maintaining the vehicle during both revenue and nonrevenue days as well as compensation for risk involved

Table 1.6 CAPEX Versus Vehicle Sizes

Category	Typical Depth Rating	Typical Cost	Typical Day Rate	Type Work	Payback Period
OCROV	300 m (1000 ft)	$100,000	$750	Call out	133 days
MSROV	2000 m (6600 ft)	$1.5 million	$3000	Call out	500 days
WCROV	>3000 m (>10,000 ft)	$5 million	$5000	Contract	1000 days

Table 1.7 Utilization Rate Versus Recapture Time

Category	Typical Cost	Typical Day Rate	Payback Period	Utilization Per Year	Revenue Days Per Year	Years to Recapture
OCROV	$100,000	$750	133 days	10%	36 days	3.65
MSROV	$1.5 million	$3000	500 days	40%	146 days	3.42
WCROV	$5 million	$5000	1000 days	75%	274 days	3.65

along with some measure of profit. Table 1.6 describes a representative sample of day rates, costs, and recapture on the various size categories of vehicles.

There is a trade-off between utilization and day rate in order to achieve an acceptable annual return on investment (ROI) so as to substantiate the investment. Contract work typically enjoys a higher utilization percentage over its call-out counterpart. But a contract ROV with a 1000-day payback period (days, revenue to full recapture of the system cost) with a 75% utilization will take the same time period to recapture its investment as a call-out system with a 10% utilization depending upon the combined factors. Table 1.7 provides a sample of how the same recapture time is achieved between a high recapture percentage coupled with a low utilization as opposed to a high utilization coupled with a longer payback period. As demonstrated in Table 1.7, in the day rate world of ROV services, it is all about utilization!

1.4 ROV services by industry

In this section, the various typical missions by industry will be explored with a conclusion as to the typical vehicle type and sensor/tooling configurations for accomplishing those mission-related tasks.

1.4.1 Science

Industry description: The general need for governmental/university/industrial research organizations involves the gathering of sensor data and the taking of samples for the understanding of the subject environment. Often, scientific manned diving or manned submersibles are used for shallow and deeper water efforts. ROVs predominate as the robotic vehicle of choice for teleoperation due to the lower cost structure over manned submersibles as well as the lower logistical requirements of the ROV versus manned submersibles.

Typical mission: ROV equipment is typically used in this application to take physical *in situ* samples and to deliver sensors for gathering data from the operational environment.

Typical vehicle type and configuration: As no significant heavy work is performed during science missions (other than geological sampling), minimal intervention is needed requiring small electric actuators, manipulators, and end effectors. Typical vehicles for this mission are the OCROV and/or MSROV with high data-throughput capabilities and small electrical manipulators/ end effectors.

1.4.2 **Fisheries and aquaculture**

Industry description: As the world becomes increasingly more populous and the world's oceans remain over-exploited, fish farming has become much more pronounced in the production of foodstuff for the world's consumers. As a result, various usages of ROV equipment have become prevalent for both the production support at fish farms as well as use in regulatory compliance assurances by policing authorities in open water fisheries locations.

Typical mission: The typical service provided by ROVs in this industry is the inspection of fish cages within a fish farm and for various usages including checking nets for holes, assuring the integrity of moorings for the farm, and the retrieving of "morts" (dead fish) from the cage for health/sanitation purposes.

Typical vehicle type and configuration: As the intervention needs of this mission are minimal and the operational environment is predominantly shallow water, the "flying eyeball" OCROV with a simple video camera and a small manipulator is the vehicle/configuration of choice.

1.4.3 **Military**

Industry description: The predominant need and usage of ROVs in a military application involve the three basic functions: mine countermeasures (MCM), object retrieval/recovery, and inspection/ security tasks.

Typical mission: For the MCM mission, the ROV is sent to a location of targets identified through other sensors or means (mine-hunting sonar, laser line scanners, intelligence, etc.). Once the mine is located, some form of end effector is required in order to neutralize the mine for final disposition. For the object retrieval function, much heavier vehicles are needed in order to rig heavy-lifting gear for retrieval to the surface. And for the inspection/security function, a simple video camera along with basic sensors is needed.

Typical vehicle type and configuration: For the MCM mission, the predominant vehicle is a special-use explosives delivery platform whereby the vehicle delivers a charge and then egresses the area for detonation or (for the more poignant "suicidal" approach) the vehicle carries the charge within close proximity of the mine and then detonates itself and (hopefully) the mine simultaneously. This single-shot MCM vehicle is clearly an OCROV (hopefully of minimal cost) while the charge delivery-then-evacuation vehicle is typically an MSROV with a dexterous electric manipulator capability. For the object retrieval ROV, a heavy-duty WCROV is needed along with hydraulic manipulators and deepwater capabilities. The inspection/security vehicle is clearly an OCROV with minimal sensor and tooling requirements.

1.4.4 Homeland security

Industry description: Homeland security needs involve the periodic inspection of various vulnerable locations for structural integrity and presence (hopefully, absence) of security threats.

   Typical mission: ROVs are typically used in this application for periodic ship hull and pier security inspections and sweeps.

   Typical vehicle type and configuration: This function is clearly the realm of the OCROV with minimal sensor and tooling capabilities. The cheaper the per inspection cost the more likely and often the inspection will take place (thus increasing the security proportionately).

1.4.5 Public safety

Industry description: The public safety industry is typically the realm of the police and fire department. Every year, recreational boaters drown, often requiring first responders to perform search and rescue (more often "search and recovery") in response to public needs.

   Typical mission: Many public safety diving (PSD) teams are attached to various municipalities and/or regional governmental authorities. In many cases, the PSD team has an ROV capability assigned to a team member to augment the team's capability. By the time the ROV is typically called in, however, the team is in full recovery mode (as opposed to rescue mode). The typical mission of a PSD team is search and gathering/recovery of crime scene evidence or recovery of inaccessible items (e.g., drowning victim).

   Typical vehicle type and configuration: The budgets of most municipalities obviate the funding for anything other than OCROVs with minimal tooling and sensor capabilities; therefore, the OCROV dominates this function.

1.4.6 O&G drill support

Industry description: Support of drilling operations has become a requirement as the search for hydrocarbons has pushed into deeper waters off the coasts of the world's continents. In waters deeper than 1000 fsw (300 msw), the wellhead and blowout preventer stack have moved from the surface to the seafloor, requiring all intervention tasks be performed robotically. The requirement for ROV support during all drilling functions in deepwater has become the industry standard.

   Typical mission: ROVs for drill support are used from the first spud-in (initial drill bit penetration into the ocean bottom) through to well completion. Missions include observation of the seafloor environment, mounting of well casing seals and guides, guiding of tooling and drill equipment into the well along with various other operations. Recent regulations, in the wake of the April 2010 Macondo oil spill in the northern Gulf of Mexico, have required a second standby ROV, operated from a vessel separate from the drilling rig, to manually operate the BOP should there be a service interruption on the main drilling rig.

   Typical vehicle type and configuration: The typical ROV size and configuration for drill support are a larger MSROV or a light WCROV. A drill support operation typically requires a drill-rig-located ROV with a 7-function hydraulic manipulator along with a second 5-function manipulator/grabber for steadying the vehicle during work. As the heavy-lifting functions are mostly handled from the surface, the vehicle does not require the muscle of a construction project, thus allowing the vehicle to be in the 50–100 hp range.

1.4.7 Inspection, repair, and maintenance

Industry description: The inspection, repair, and maintenance (IRM) market (also referred to as "IMR," or inspection, maintenance, and repair) broadly comprises the IRM of subsea fixed or floating structures for various industries. These industries include offshore wind farms, fish farms (although this is considered under Section 1.4.2), civil engineering projects (marine deepwater intakes, outfalls, tidal energy production structures, etc.), ship husbandry, and IMR of various offshore O&G industry support structures.

Typical mission: The typical mission for inspection of subsea structures involves the visual and nondestructive testing/evaluation of various man-made items for safety, security, structural integrity, and functionality of the fixture as well as primary and supporting systems. The repair and maintenance functions are carried out during the course of the asset's life through various techniques for supporting the life of the project.

Typical vehicle type and configuration: The ROV need during the inspection phase varies depending upon the operating environment (depth, currents, surface conditions, etc.) and the type of inspection being conducted (e.g., high-bandwidth acoustic mapping, structural flooded member detection, and cathodic potential measurement). For basic visual shallow water inspections, a small OCROV will be sufficient, but for higher-bandwidth sensor delivery and/or deepwater operations in harsh conditions, an MSROV will be required. For the repair and maintenance functions, most operations can be accomplished with an MSROV with light intervention and tooling capabilities. As the need for further mechanical tasks becomes heavier, the project may have periodic need for a WCROV, but in most cases the MSROV will suffice.

1.4.8 Construction (O&G as well as civil)

Industry description: The marine subsea construction industry encompasses the full gamut of heavy-lift ROV needs for the setting and assembling of subsea structures.

Typical mission: ROVs for use in this mission typically are tasked with setting and pulling rigging, guiding large construction pieces into place, moving heavy pieces from location to location, laying and burying cables and pipelines and setting mattresses as well as the various tasks outlined in Section 1.2.2.

Typical vehicle type and configuration: Vehicles in use for the subsea construction tasking are typically higher powered (>150 hydraulic hp) and specification WCROVs with dual 7-function manipulators and high pressure/flow remote tooling delivery capabilities.

1.5 CONCLUSIONS

The business aspect of the ROV service industry can be just as challenging as the technical aspect. It is quite daunting to the un-indoctrinated. Like the aviation and maritime counterparts to the ROV business, the support structure requires not only a deep equipment and spares pool, it also requires a thorough training program, procedures, controls, and a commitment to service quality. It is a financially risky business rife with pitfalls. But the rewards are clear and evident for the entrepreneur willing to navigate the regulatory and economic minefields.

Field personnel must be properly trained with a positive "can do" attitude with thorough troubleshooting skills and a trained eye toward accomplishing the task in a safe and cost-effective manner. In this business, it is all about the people!

What follows in the remainder of the manual is the fun part—the technical aspects of the ROV system and its use.

CHAPTER CONTENTS

The ROV Manual.
© 2014 Robert D Christ and Robert L Wernli. Published by Elsevier Ltd. All rights reserved.

In order to comprehend the concepts of operating in the ocean world, an understanding of the details of this environment is needed. The content of this chapter explores the makeup of fresh water and seawater and then goes into the interaction of this substance with the world of robotics. We will explore the basic concepts of water density, ocean circulation, currents and tides and how each of these affects the operation of ROV equipment. Armed with a general knowledge of ocean-ography, work site predictions may be made on such variables as turbidity (affecting camera optics), temperature/salinity (affecting acoustic equipment and vehicle buoyancy), tide and current flows (affecting drag computations on the submersible/tether combination), and dissolved gases (affecting biological population). This section condenses information from complete college curric-ulums; therefore, for further details, please see the references in the bibliography. Special thanks go to Steve Fondriest of Fondriest Environmental, Inc. for his contribution to the fundamentals of environmental monitoring and data collection instrumentation.

2.1 Physical oceanography

2.1.1 Distribution of water on earth

Earth is the only planet known to have water resident in all three states (solid, liquid, and gas). It is also the only planet to have known liquid water currently at its surface. Distribution of the earth's water supply is given in Table 2.1.

As shown in Table 2.1, most (97%) of the world's water supply is in the oceans. Water can dis-solve more substances (and in greater quantities) than any other liquid. It is essential to sustain life and is a moderator of our planet's temperature, a major contributor to global weather patterns, and, of course, essential for operation of an ROV.

The oceans cover 70.8% of the earth's surface, far overreaching earth's land mass. Of the ocean coverage, the Atlantic covers 16.2%, the Pacific 32.4%, the Indian Ocean 14.4%, and the margin and adjacent areas the balance of 7.8%. It is also interesting to note that the Pacific Ocean alone covers 3.2% more surface area on earth than all of the land masses combined.

2.1.2 Coastal zone classifications and bottom types

General coastal characteristics tend to be similar for thousands of kilometers. Most coasts can be classified as either erosional or depositional depending upon whether their primary features were

Table 2.1 Earth's Water Supply

Water Source	Water Volume, in Cubic Miles	Water Volume, in Cubic Kilometers	Percent of Fresh water	Percent of Total Water
Oceans, seas, and bays	321,000,000	1,338,000,000	–	96.5
Ice caps, glaciers, and permanent snow	5,773,000	24,064,000	68.7	1.74
Groundwater				
Fresh	2,526,000	10,530,000	30.1	0.76
Saline	3,088,000	12,870,000	–	0.94
Soil moisture	3959	16,500	0.05	0.001
Ground ice and permafrost	71,970	300,000	0.86	0.022
Lakes				
Fresh	21,830	91,000	0.26	0.007
Saline	20,490	85,400	–	0.006
Atmosphere	3095	12,900	0.04	0.001
Swamp water	2752	11,470	0.03	0.0008
Rivers	509	2120	0.006	0.0002
Biological water	269	1120	0.003	0.0001

Source: US Geological Survey.

created by erosion of land or deposition of eroded materials. Erosional coasts have developed where the shore is actively eroded by wave action or where rivers or glaciers caused erosion when the sea level was lower than its present level. Depositional coasts have developed where sediments accumulate either from a local source or after being transported to the area in rivers and glaciers or by ocean currents and waves.

Of primary interest to the ROV pilot, with regard to coastal zones, is the general classification of these zones and its effect upon general water turbidity in the operational area. Depositional coasts tend to have a higher quantity of suspended solids in the water column, thus a higher turbidity and degraded camera performance. Erosional coasts tend to possess fewer suspended particles, thus featuring better camera optics. Further, the depositional source will greatly affect the level of turbidity since mud deposited from a river estuary will have a higher turbidity than a rock and sand drainage area.

As stated earlier, oceans cover 70.8% of the earth's surface. Of that composition, the distribution between continental margins and deep-sea basins is provided in Table 2.2.

A substantial amount of scientific and oil exploration/production work is done in the continental margins with ROVs. The continental margins are, in large part, depositional features. Their characteristics are driven by runoff deposited from the adjacent continent.

Sediments are carried from the marine estuaries and then deposited onto the continental shelf. As the seafloor spreads due to tectonic forces, the sediments fall down the continental slope and come to rest on the abyssal plain. Substantial amounts of oil and gas deposits are locked in these

Table 2.2 Ocean Coverage Distribution

	10^6 km^2	Percentage of Earth's Surface Area
Continental margin	93	18.2
Deep basins	268	52.6
Total	361	70.8

sediments and are the focus of exploration and production efforts. The general bottom characteristics of this shelf are mud and sediment.

2.2 Chemical oceanography

Water is known as the "universal solvent." While pure water is the basis for life on earth, as more impurities are added to that fluid the physical and chemical properties change drastically. The chemical makeup of the water mixture in which the ROV operates will directly dictate operational procedures and parameters if a successful operation is to be achieved.

Two everyday examples of water's physical properties and their effect on our lives are (i) ice floats in water and (ii) we salt our roads in wintertime to "melt" snow on the road. Clearly, it is important to understand the operating environment and its effect on ROV operations. To accomplish this, the properties and chemical aspects of water and how they are measured will be addressed to determine their overall effect on the ROV.

The early method of obtaining environmental information was by gathering water samples for later analysis in a laboratory. Today, the basic parameters of water are measured with a common instrument named the "CTD sonde." Some of the newest sensors can analyze a host of parameters logged on a single compact sensing unit.

Fresh water is an insulator, with the degree of electrical conductivity increasing as more salts are added to the solution. By measuring the water's degree of electrical conductivity, a highly accurate measure of salinity can be derived. Temperature is measured via electronic methods, and depth is measured with a simple water pressure transducer. The CTD probe measures "conductivity/temperature/depth," which are the basic parameters in the sonic velocity equation. Newer environmental probes are available for measuring any number of water quality parameters such as pH, dissolved oxygen and CO_2, turbidity, and other parameters.

The measurable parameters of water are needed for various reasons. A discussion of the most common measurement variables the commercial or scientific ROV pilot will encounter, the information those parameters provide, and the tools/techniques to measure them follows.

2.2.1 Salinity

2.2.1.1 Salt water

The world's water supply consists of everything from pure water to water plus any number of dissolved substances due to water's soluble nature. Water quality researchers measure salinity to assess the purity of drinking water, monitor salt water intrusion into fresh water marshes and groundwater aquifers, and research how the salinity will affect the ecosystem.

Table 2.3 Dissolved Salts in Water

Component	Weight in Grams
Pure water	965.31
Major constituents	
Chlorine	19.10
Sodium	10.62
Magnesium	1.28
Sulfur	2.66
Calcium	0.40
Potassium	0.38
Minor constituents	0.24
Trace constituents	0.01
Total (in grams)	1000.00

The two largest dissolved components of typical seawater are chlorine (56% of total) and sodium (31% of total), with the total of all lumped under the designation of "salts." Components of typical ocean water dissolved salts are comprised of major constituents, minor constituents, and trace constituents. An analysis of 1 kg of seawater (detailing only the major constituents of dissolved salts) is provided in Table 2.3.

The total quantity of dissolved salts in seawater is expressed as salinity, which can be calculated from conductivity and temperature readings. Salinity was historically expressed quantitatively as grams of dissolved salts per kilogram of water (expressed as percentage) or, more commonly, in parts per thousand (ppt). To improve the precision of salinity measurements, salinity is now defined as a ratio of the electrical conductivity of the seawater to the electrical conductivity of a standard concentration of potassium chloride solution. Thus, salinity is now defined in practical salinity units (PSU), although one may still find the older measure of salt concentration in a solution as parts per thousand or percentage used in the field.

Ocean water has a fairly consistent makeup, with 99% having between 33 and 37 PSU in dissolved salts. Generally, rain enters the water cycle as pure water and then gains various dissolved minerals as it travels toward the ocean. Water enters the cycle with a salt content of 0 PSU, mixes with various salts to form brackish water (in the range of 0.5–30 PSU) as it blends with rivers and estuaries, homogenizes with the ocean water (75% of the ocean's waters have between 33 and 34 PSU of dissolved salts) as the cycle ends, and then renews with evaporation.

Just as a layer of rapid change in temperature (the thermocline) traps sound and other energy, so does an area of rapid change in salinity, known as a "halocline." These haloclines are present both horizontally (see cenote example later in this chapter) and vertically (e.g., rip tides at river estuaries).

As the salinity of water increases, the freezing point decreases. As an anecdote to salinity, there are brine pools under the Antarctic ice amid the glaciers in the many lakes of Antarctica's McMurdo Dry Valleys. A team recently found a liquid lake of super-concentrated salt water, seven times saltier than normal seawater, locked beneath 62 ft (19 m) of lake ice—a record for lake ice cover on earth.

Salts dissolved in water change the density of the resultant seawater for these reasons:

- The ions and molecules of the dissolved substances are of a higher density than water.
- Dissolved substances inhibit the clustering of water molecules (particularly near the freezing point), thus increasing the density and lowering the freezing point.

Unlike fresh water, ocean water continues to increase in density up to its freezing point of approximately $-2°C$. At 0 PSU (i.e., fresh water), maximum density is approximately 4°C with a freezing point of 0°C. At 24.7 PSU and above, ocean water has a freezing point of its maximum density; therefore, there is no maximum density temperature above the freezing point. The maximum density point scales in a linear fashion between 0 and 24.7 PSU (Figure 2.1).

Thus, ocean water continues to increase in density as it cools and sinks in the open ocean. As a result, the deepwaters of the world's oceans are uniformly cold.

Comparisons of salinity and temperature effects upon water density yield the following:

- At a constant temperature, variation of the salinity from 0 to 40 PSU changes the density by about 0.035 specific gravity (or about 3.5% of the density for 0 PSU water at 4°C).
- At a constant salinity (i.e., 0 PSU), raising the temperature from 4°C (maximum density) to 30°C (highest temperature generally found in surface water) yields a decrease in density of 0.0043 (1.000−0.9957) for a change of 0.4%.

Clearly, salinity has a much higher effect upon water density than does temperature.

As a practical example, suppose a 220 pound (100 kg) ROV is ballasted for 4°C fresh water at exactly neutral buoyancy. If that same submersible were to be transferred to salt water, an additional ballasting weight of approximately 7.7 pounds (3.5 kg) would be required to maintain that vehicle at neutral buoyancy.

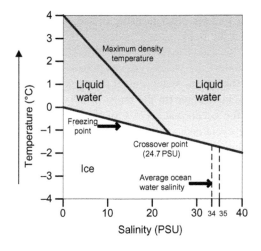

FIGURE 2.1

Salt water density, salinity versus temperature.

Ice at 0°C has a density of 57.25 lb/ft^3 (0.917 g/cm^3), which is about 8% less than that of water at the same temperature. Obviously, water expands when it freezes, bursting pipes and breaking apart water-encrusted rocks, thus producing revenues for marine and land plumbing contractors.

2.2.1.2 Fresh water

A vast majority of the world's fresh water supply is locked within the ice caps and glaciers of the high Arctic and Antarctic regions. Fresh water is vital to man's survival, and thus most human endeavors surround areas of fresh water. Due to the shallow water nature of the fresh water collection points, man has placed various items of machinery, structures, and tooling in and around these locations. The ROV pilot will, in all likelihood, have plenty of opportunity to operate within the fresh water environment.

The properties of water directly affect the operation of ROV equipment in the form of temperature (affecting components and electronics), chemistry (affecting seals, incurring oxidation, and degrading machinery operation), and specific gravity (buoyancy and performance). These parameters will determine the buoyancy of vehicles, the efficiency of thrusters, the numbers and types of biological specimens encountered, as well as the freezing and boiling points of the operational environment. The water density will further affect sound propagation characteristics, directly impacting the operation of sonar and acoustic positioning equipment.

Fresh water has a specific gravity of 1.000 at its maximum density. As water temperature rises, molecular agitation increases the water's volume, thus lowering the density of the fluid per unit volume. In the range between 3.98°C and the freezing point of water, the molecular lattice structure (in the form of ice crystals) again increases the overall volume, thus lowering its density per unit volume (remember, ice floats). The point of maximum density for fresh water occurs at 3.98°C (the point just before the formation of ice crystals). At the freezing point of water, the lattice structure rapidly completes, thus significantly expanding the volume per unit mass and lowering the density at that temperature point. A graph describing the temperature/density relationship of pure water is shown in Figure 2.2.

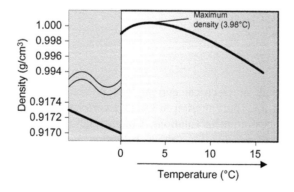

FIGURE 2.2

Water density versus temperature.

Fresh water has a maximum density at approximately 4°C (Figure 2.1) yet ocean water has no maximum density above the freezing point. As a result, lakes and rivers behave differently at the freezing point than ocean water. As the weather cools with the approach of winter, the surface water of a fresh water lake is cooled and its density is increased. Surface water sinks and displaces bottom water upward to be cooled in turn. This convection process is called "overturning." This overturning process continues until the maximum density is achieved, thus stopping the convection churning process at 4°C. As the lake continues to cool, the crystal structure in the water forms, thus allowing the cooler water at the surface to decrease in density, driving still further the cooler (less dense) surface water upward, allowing ice to form at the surface.

2.2.2 Pressure

The SI unit for pressure is the kilopascal (expressed as kPa). One pascal is equal to one newton per square meter. However, oceanographers normally use ocean pressure with reference to sea-level atmospheric pressure. The imperial unit is one atmosphere. The SI unit is the bar. The decibar is a useful measure of water pressure and is equal to 1/10 bar. Seawater generally increases by one atmosphere of pressure for every 33 ft of depth (approximately equaling 10 m). Therefore, one decibar is approximately equal to one meter of depth in seawater (Figure 2.3).

As an ROV pilot, seawater pressure directly affects all aspects of submersible operation. The design of the submersible's air-filled components must withstand the pressures of depth, the flotation must stand up to the pressure without significant deformation (thus losing buoyancy and sinking the vehicle), and tethers must be sturdy enough to withstand the depth while maintaining their neutral buoyancy.

2.2.3 Compressibility

For the purposes of ROV operations, seawater is essentially incompressible. There is a slight compressibility factor, however, that does directly affect the propagation of sound through water. This compressibility factor will affect the sonic velocity computations at varying depths (see Section 2.2.8 later in this chapter).

2.2.4 Conductivity

Conductivity is the measure of electrical current flow potential through a solution. In addition, because conductivity and ion concentration are highly correlated, conductivity measurements are used to calculate ion concentration in solutions. Commercial and military operators observe conductivity for gauging water density (for vehicle ballasting and such) and for determining sonic velocity profiles (for acoustic positioning and sonar use). Water quality researchers take conductivity readings to determine the purity of water, to watch for sudden changes in natural or wastewater, and to determine how the water sample will react in other chemical analyses. In each of these applications, water quality researchers count on conductivity sensors and computer software to sense environmental waters, log and analyze data, and present this data.

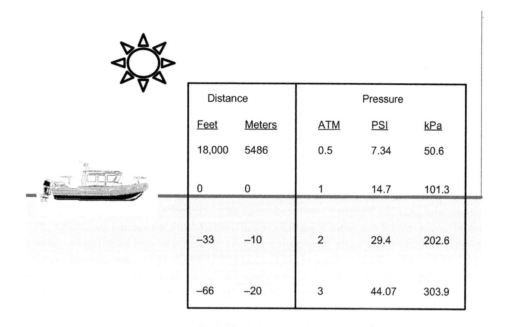

Distance		Pressure		
Feet	Meters	ATM	PSI	kPa
18,000	5486	0.5	7.34	50.6
0	0	1	14.7	101.3
−33	−10	2	29.4	202.6
−66	−20	3	44.07	303.9

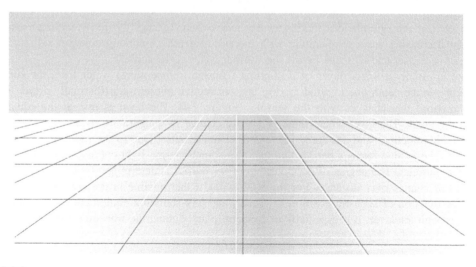

FIGURE 2.3

Pressure in atmospheres from various levels.

2.2.5 **Water temperature**

Water temperature is a measure of the kinetic energy of water and is expressed in degrees Fahrenheit (F) or Celsius (C). Water temperature varies according to season, depth, and, in some cases, time of day. Because most aquatic organisms are cold blooded, they require a certain temperature range to survive. Some organisms prefer colder temperatures and others prefer warmer temperatures. Temperature also affects the water's ability to dissolve gases, including oxygen. The lower the temperature, the higher the solubility. Thermal pollution, the artificial warming of a body of water because of industrial waste or runoff from streets and parking lots, is becoming a common threat to the environment. This artificially heated water decreases the amount of dissolved oxygen and can be harmful to cold water organisms.

In limnological research, water temperature measurements as a function of depth are often required. Many reservoirs are controlled by selective withdrawal dams, and temperature monitoring at various depths provides operators with information to control gate positions. Power utility and industrial effluents may have significant ecological impact with elevated temperature discharges. Industrial plants often require water temperature data for process, use, and heat transfer calculations.

Pure water freezes at 32°F (0°C) and boils at 212°F (100°C). ROV operations do not normally function in boiling water environments; therefore, the focus here will be upon the temperature ranges in which most ROV systems operate (i.e., 32°F to 86°F or 0°C to 30°C). The examination of salinity will be in the range from fresh water to the upper limit of seawater.

Temperature in the oceans varies widely both horizontally and vertically. On the high temperature side, the Persian Gulf region during summertime will achieve a maximum of approximately 90°F (32°C). The lowest possible value is at the freezing point of 28.4°F (−2°C) experienced in polar region(s).

The vertical temperature distribution nearly everywhere (except the polar regions) displays a profile of decreasing water temperature with increasing depth. Assuming constant salinity, colder water will be denser and will sink below the warmer water at the surface.

There is usually a mixed layer of isothermal (constant temperature) water from the surface to some near-surface depth due to wind mixing and convective overturning (thermally driven vertical density mixing) that changes with the seasons (Figure 2.4). The layer is thin at the equator and thick at the poles. The layer where there is a rapid change in temperature over a short distance is termed a "thermocline" and has some interesting characteristics. Due to the rapid temperature gradient, this thermocline forms a barrier that can trap sound energy, light energy, and any number of suspended particles. The degree of perviousness of the barrier is determined by the relative strength or degree of change over distance. For the ROV pilot, a thermocline in the area of operation can hinder the function of acoustic positioning, sonar, and any sounding equipment aboard attempting to burn through the layer. It is especially of concern to anti-submarine warfare technicians.

2.2.6 **Density**

Density is mass per unit volume measured in SI units in kilograms per cubic meter (or, on a smaller scale, grams per cubic centimeter). The density of seawater depends upon salinity, temperature, and pressure. At a constant temperature and pressure, density varies with water salinity. This measure is of particular importance to the ROV pilot for the determination of neutral buoyancy for the vehicle.

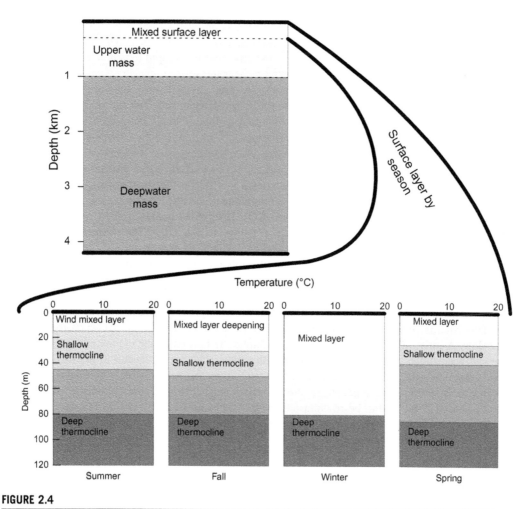

FIGURE 2.4

Surface layer mixing by season.

The density range for seawater is from 1.02200 to 1.030000 g/cm^3 (Thurman, 1994). In an idealized stable system, the higher density water sinks to the bottom while the lower density water floats to the surface. Water under the extreme pressure of depth will naturally be denser than surface water, with the change in pressure (through motion between depths) being realized as heat. Just as the balance of pressure/volume/temperature is prevalent in the atmosphere, so is the temperature/salinity/pressure model in the oceans.

A rapid change in density over a short distance is termed a "pycnocline" and can trap any number of energy sources from crossing this barrier, including sound (sonar and acoustic positioning systems), current, and neutrally buoyant objects in the water column (underwater vehicles). Changing operational area from a lower latitude to a higher latitude produces a mean temperature change in the surface layer. As stated previously, the deep ocean is uniformly cold due to the higher density cold water sinking to the bottom of the world's oceans. The temperature change from the warm surface at the tropics to the lower cold water can be extreme, causing a rapid temperature swing within a few meters of the surface. This surface layer remains near the surface, causing a small tight "surface duct" in the lower latitudes. In the higher latitudes, however, the difference between ambient surface temperature and the temperature of the cold depth is less pronounced. The thermal mixing layer at these latitudes, as a result, is much larger (over a broader range of depth between the surface and the isothermal lower depths) and less pronounced (Figure 2.5). In Figure 2.5(a), density profiles by latitude and depth are examined to display the varying effects of deepwater temperature/density profiles versus ambient surface temperatures. Figure 2.5(b) and (c) looks at the same profiles only focusing on temperature and salinity. Figure 2.5(d) demonstrates a general profile for density at low to midlatitudes (the mixed layer is water of constant temperature due to the effects of wave mechanics/mixing).

A good example of the effect of density on ROV operations comes from a scientific mission conducted in 2003 in conjunction with *National Geographic* magazine. The mission was to the cenotes (sinkholes) of the Northern Yucatan in Mexico. Cenotes are a series of pressure holes in a circular arrangement, centered around Chicxulub (the theoretical meteor impact point), purportedly left over from the K−T event from 65 million years ago that killed the dinosaurs.

The top water in the cenote is fresh water from rain runoff, with the bottom of the cenote becoming salt water due to communication (via an underground cave network) with the open ocean. This column of still water is a near perfect unmixed column of fresh water on top with salt water below. A micro-ROV was being used to examine the bottom of the cenote as well as to sample the salt water/fresh water (halocline) layer. The submersible was ballasted to the fresh water on the top layer. When the vehicle came to the salt water layer, the submersible's vertical thruster had insufficient downward thrust to penetrate into the salt water below and kept "bouncing" off the halocline. The submersible had to be re-ballasted for salt water in order to get into that layer and take the measurements, but the vehicle was useless on the way down due to its being too heavily ballasted to operate in fresh water.

2.2.7 Depth

Depth sensors, discussed below, measure the distance from the surface of a body of water to a given point beneath the surface, either the bottom or any intermediate level. Depth readings are used by researchers and engineers in coastal and ocean profiling, dredging, erosion and flood monitoring, and construction applications.

Bathymetry is the measurement of depth in bodies of water. Further, it is the underwater version of topography in geography. Bottom contour mapping details the shape of the seafloor, showing the features, characteristics, and general outlay. Tools for bathymetry and sea bottom characterization are as follows.

2.2.7.1 Echo sounder

An echo sounder measures the round trip time it takes for a pulse of sound to travel from the source at the measuring platform (surface vessel or on the bottom of the submersible) to the sea bottom and return. When mounted on a vessel, this device is generally termed a "fathometer" and when mounted on a submersible it is termed an "altimeter."

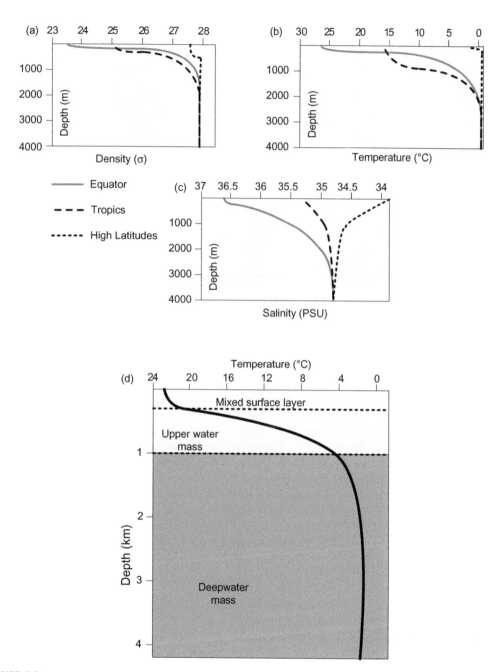

FIGURE 2.5

Individual sound velocity components (a) density, (b) temperature, and (c) salinity combined into a composite (d) sound velocity curve by depth.

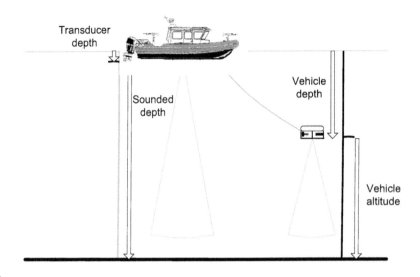

FIGURE 2.6

Vessel-mounted and sub-mounted sounders.

According to Bowditch (2002), "the major difference between various types of echo sounders is in the frequency they use. Transducers can be classified according to their beam width, frequency, and power rating. The sound radiates from the transducer in a cone, with about 50 percent actually reaching the sea bottom. Beam width is determined by the frequency of the pulse and the size of the transducer. In general, lower frequencies produce a wider beam, and at a given frequency, a smaller transducer will produce a wider beam. Lower frequencies penetrate deeper into the water, but have less resolution in depth. Higher frequencies have a greater resolution in depth, but less range, so the choice is a trade-off. Higher frequencies also require a smaller transducer. A typical low-frequency transducer operates at 12 kHz and a high-frequency one at 200 kHz." Many smaller ROV systems have altimeters, such as the Imagenex 852, on the same frequency as their imaging system for easier software integration (the same software can be used for processing both signals) and reduced cost (Figure 2.6).

Computation of depth as determined by an echo sounder is determined via the following formula:

$$D = (V \times T/2) + K + D_r$$

where D is depth below the surface (or from the measuring platform), V is the mean velocity of sound in the water column, T is time for the round trip pulse, K is the system index constant, and D_r is the depth of the transducer below the surface.

2.2.7.2 Optic–acoustic seabed classification

Traditional seafloor classification methods have, until recently, relied upon the use of mechanical sampling, aerial photography, or multiband sensors (such as Landsat™) for major bottom-type

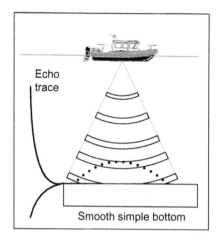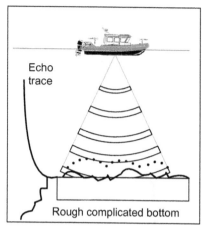

FIGURE 2.7

Acoustic seabed classification.

discrimination (e.g., mud, sand, rock, sea grass, and corals). Newer acoustic techniques for collecting hyperspectral imagery are now available through processing of acoustic backscatter.

Acoustic seabed classification analyzes the amplitude and shape of acoustic backscatter echoed from the sea bottom for the determination of bottom texture and makeup (Figure 2.7). Seafloor roughness causes impedance mismatch between the water and the sediment. Further, reverberation within the substrate can be analyzed in determining the overall composition of the bottom being insonified. Acoustic data acquisition systems and a set of algorithms that analyze the data allow for determining the seabed acoustic class.

2.2.8 Sonic velocity and sound channels

Sound propagation (vector and intensity) in water is a function of its velocity. And velocity is a function of water density and compressibility. As such, sound velocity is dependent upon temperature, salinity, and pressure and is normally derived expressing these three variables (Figure 2.8). The speed of sound in water changes by 3−5 m/s/°C, by approximately 1.3 m/s/PSU salinity change, and by about 1.7 m/s/100 m change in depth (compression). The speed of sound in seawater increases with increasing pressure, temperature, and salinity (and vice versa).

The generally accepted underwater sonic velocity model was derived by W. D. Wilson in 1960. A simplified version of Wilson's (1960) formula on the speed of sound in water follows:

$$c = 1449 + 4.6T - 0.055T^2 + 0.0003T^3 + 1.39(S - 35) + 0.017D$$

where c is the speed of sound in meters per second, T is the temperature in degrees Celsius, S is the salinity in PSU, and D is the depth in meters.

Temperature/salinity/density profiles are important measurements for sensor operations in many underwater environments, and they have a dramatic effect on the transmission of sound in the

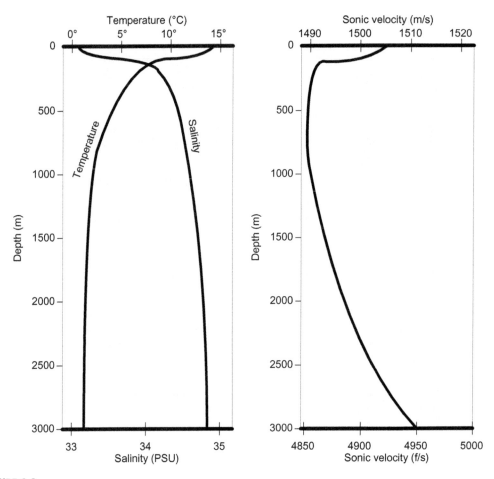

FIGURE 2.8

Sonic velocity profiles with varying temperature/salinity/depth.

ocean. A change in overall water density over short range due to any of these three variables (or in combination) is termed a "pycnocline." Overall variations of pressure and temperature are depicted graphically in Figure 2.9.

This layering within the ocean, due to relatively impervious density barriers, causes the formation of sound channels within bodies of water. These "channels" trap sound, thus directing it over possibly long ranges. Sound will refract based upon its travel across varying density layers, bending toward the denser water and affecting both range and bearing computations for acoustics (Figure 2.10). Over short ranges (tens or hundreds of meters) this may not be a substantial number and can possibly be disregarded, but for the longer distances of some larger ROV systems, this becomes increasingly a factor to be considered.

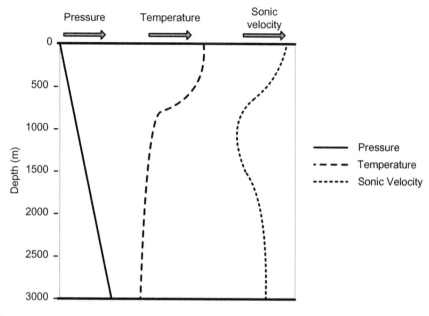

FIGURE 2.9

Variations of pressure and temperature with depth producing sound velocity changes.

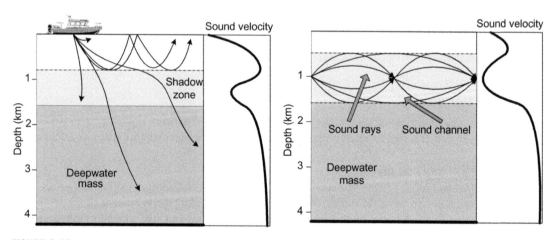

FIGURE 2.10

Sound velocity profiles and channeling at various depths and layers.

2.2.9 Viscosity

Viscosity is a liquid's measure of internal resistance to flow or resistance of objects to movement within the fluid. Viscosity varies with changes in temperature/salinity, as does density. Seawater is more viscous than fresh water, which will slightly affect the computations of vehicle drag.

2.2.10 Water flow

Water flow is the rate at which a volume of water moves or flows across a certain cross-sectional area during a specified time and is typically measured in cubic feet per second/cubic meters per second (cfs/cms). The flow rate changes based upon the amount of water and the size of the river or stream being monitored. Environmental researchers monitor water flow in order to estimate pollutant spread, to monitor groundwater flow, to measure river discharge, to manage water resources, and to evaluate the effects of flooding.

2.2.11 Turbidity

Turbidity (which causes light scattering—see Section 2.2.17.3), the measure of the content of suspended solids in water, is also referred to as the "cloudiness" of the water. Turbidity is measured by shining a beam of light into the solution. The light scattered off the particles suspended in the solution is then measured, and the turbidity reading is given in nephelometric turbidity units. Water quality researchers take turbidity readings to monitor dredging and construction projects, examine microscopic aquatic plant life, and monitor surface, storm, and wastewater.

2.2.12 Chlorophyll

In various forms, chlorophyll is bound within the living cells of algae and other phytoplankton found in surface water. Chlorophyll is a key biochemical component in the molecular apparatus that is responsible for photosynthesis, the critical process in which the energy from sunlight is used to produce life-sustaining oxygen. In the photosynthetic reaction, carbon dioxide is reduced by water, and chlorophyll assists this transfer.

2.2.13 Water quality

Water quality researchers count on sensors and computer software to sense environmental waters and log and analyze data. Factors to be considered in water quality measurement are discussed below.

2.2.13.1 Alkalinity and pH

The acidity or alkalinity of water is expressed as pH (potential of hydrogen). This is a measure of the concentration of hydrogen (H^+) ions. Water's pH is expressed as the logarithm of the reciprocal of the hydrogen ion concentration, which increases as the hydrogen ion concentration decreases (and vice versa). When measured on a logarithmic scale of 0–14, a pH of 0 is the highest acidity, a pH of 14 is the highest alkalinity, and a pH of 7 is neutral (Figure 2.11). Pure water is pH neutral, with seawater normally at a pH of 8 (mildly alkaline).

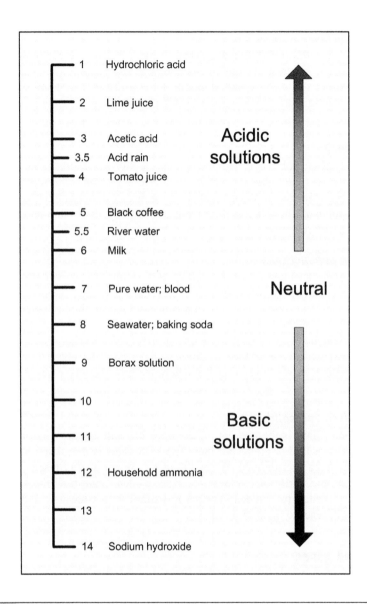

FIGURE 2.11

Graphical presentation of common items with their accompanying pH.

pH measurements help determine the safety of water. The sample must be between a certain pH to be considered drinkable, and a rise or fall in pH may indicate a chemical pollutant. Changes in pH affect all life in the oceans; therefore, it is most important to aquatic biology to maintain a near-neutral pH. As an example, shellfish cannot develop calcium carbonate hard shells in an acidic environment.

2.2.13.2 Oxidation reduction potential

Oxidation reduction potential (ORP) is the measure of the difference in electrical potential between a relatively chemically inert electrode and an electrode placed in a solution. Water quality researchers use ORP to measure the activity and strength of oxidizers (those chemicals that accept electrons) and reducers (those that lose electrons) in order to monitor the reactivity of drinking water and groundwater.

2.2.13.3 Rhodamine

Rhodamine, a highly fluorescent dye, has the unique quality to absorb green light and emit red light. Very few substances have this capability, so interference from other compounds is unlikely, making it a highly specific tracer. Water quality researchers use rhodamine to investigate surface water, wastewater, pollutant time of travel, groundwater tracing, dispersion and mixing, circulation in lakes, and storm water retention.

2.2.13.4 Specific conductance

Specific conductance is the measure of the ability of a solution to conduct an electrical current. However, unlike the conductivity value, specific conductance readings compensate for temperature. In addition, because specific conductance and ion concentration are highly correlated, specific conductance measurements are used to calculate ion concentration in solutions. Specific conductance readings give the researcher an idea of the amount of dissolved material in the sample. Water quality researchers take specific conductance readings to determine the purity of water, to watch for sudden changes in natural or wastewater, and to determine how the water sample will react in other chemical analyses.

2.2.13.5 Total dissolved solids

Total dissolved solids (TDS) is the measure of the mass of solid material dissolved in a given volume of water and is measured in grams per liter. The TDS value is calculated based on the specific conductance reading and a user-defined conversion factor. Water quality researchers use TDS measurements to evaluate the purity or hardness of water, to determine how the sample will react in chemical analyses, to watch for sudden changes in natural or wastewater, and to determine how aquatic organisms will react to their environment.

2.2.14 Dissolved gases

Just as a soda dissolves CO_2 under the pressure of the soda bottle, so does the entire ocean dissolve varying degrees of gases used to sustain the life and function of the aquatic environment. The soda bottle remains at gas–fluid equilibrium under the higher-than-atmospheric pressure condition until the bottle is opened and the pressure within the canister is lowered. At that point, the gas bubbles out of solution until the gas/air mixture comes back into balance. If, however, that same bottle were opened in the high-pressure condition of a saturation diving bell deep within the ocean, that soda would (instead of bubbling) absorb CO_2 into solution until again saturated with that gas.

The degree of dissolved gases within a given area of ocean is dependent upon the balance of all gases within the area. The exchange of gases between the atmosphere and ocean can only occur at the air–ocean interface, i.e., the surface. Gases are dissolved within the ocean and cross the air/water

Table 2.4 Distribution of Gases in the Atmosphere and Dissolved in Seawater

	Percentage of Gas Phase by Volume		
Gas	Atmosphere	Surface Oceans	Total Oceans
Nitrogen (N_2)	78	48	11
Oxygen (O_2)	21	36	6
Carbon dioxide (CO_2)	0.04	15	83

Source: Segar (1998).

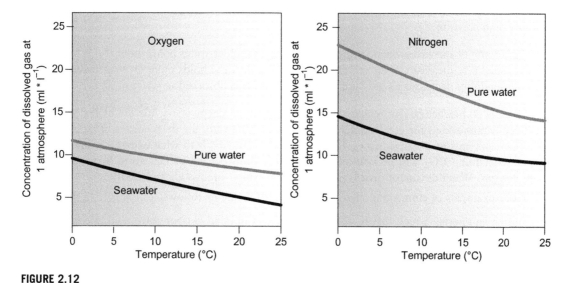

FIGURE 2.12

Solubility of two sea-level pressure gases based upon temperature.

interface based upon the balance of gases between the two substances (Table 2.4). A certain gas is said to be at local saturation if its distribution within the area is balanced given the local environmental conditions. The degree upon which a gas is able to dissolve within the substance is termed its "solubility" and in water is dependent upon the temperature, salinity, and pressure of the surrounding fluid (Figure 2.12). Once a substance is at its maximum gas content (given the local environmental conditions), the substance is saturated with that gas. The degree to which a substance can either accept or reject transfer of gas into a substance is deemed its saturation solubility.

The net direction of transfer of gases between the atmosphere and ocean is dependent upon the saturation solubility of the water. If the sea is oversaturated with a certain gas, it will off-gas back into the atmosphere (and vice versa). Once equilibrium is again reached, net gas transfer ceases until some environmental variable changes.

Oxygen and CO_2 are dissolved in varying degrees within the world's oceans. Most marine life requires some degree of either of these dissolved gases in order to survive. In shallow water, where photosynthesis takes place, plant life consumes CO_2 and produces O_2. In the deepwaters of the oceans, decomposition and animal respiration consume O_2 while producing CO_2.

Dissolved oxygen is the amount of oxygen that is dissolved in water. It is normally expressed in milligrams per liter (mg/l = ppm) or percent air saturation. Dissolved oxygen can also enter the water by direct absorption from the atmosphere (transfer across the air/water interface).

Aquatic organisms (plants and animals) need dissolved oxygen to live. As water moves past the gills of a fish, microscopic bubbles of oxygen gas, dissolved oxygen, are transferred from the water to the bloodstream. Dissolved oxygen is consumed by plants and bacteria during respiration and decomposition. A certain level of dissolved oxygen is required to maintain aquatic life. Although dissolved oxygen may be present in the water, it could be at too low a level to maintain life.

The amount of dissolved oxygen that can be held by water depends on the water temperature, salinity, and pressure.

- Gas solubility increases when temperature decreases (colder water holds more O_2).
- Gas solubility increases when salinity decreases (fresh water holds more O_2 than seawater).
- Gas solubility decreases as pressure decreases (less O_2 is absorbed in water at higher altitudes).

There is no mechanism for replenishing the O_2 supply in deepwater while the higher pressure of the deep depths allows for greater solubility of gases. As a result, the deep oceans of the world contain huge amounts of dissolved CO_2. The question that remains is to what extent the industrial pollutants containing CO_2 will be controlled before the deep oceans of the world become saturated with this gas. Water devoid of dissolved oxygen will exhibit lifeless characterization.

Three examples of contrasting dissolved oxygen levels are as follows:

1. During an internal wreck survey of the USS *Arizona* in Pearl Harbor performed in conjunction with the US National Park Service, it was noted that the upper decks of the wreck were in a fairly high state of metal oxidation. However, as the investigation moved into the lower spaces (less and stable oxygenation due to little or no water circulation), the degree of preservation took a significant turn. On the upper decks, there were vast amounts of marine growth that had decayed the artifacts and encrusted the metal. However, on the lower living quarters, fully preserved uniforms were found where they were left on that morning over 60 years previous, still neatly pressed in closets and on hangers.

2. During an internal wreck survey of a B-29 in Lake Mead, Nevada (again done with the US National Park Service), the wreck area at the bottom of the reservoir was anaerobic (lacking significant levels of dissolved oxygen). The level of preservation of the wreck was amazing, with instrument readings still clearly visible, aluminum skin and structural members still in a fully preserved state, and the shiny metal data plate on the engine still readable.

3. During the cenote project in the Northern Yucatan (mentioned earlier in this chapter), the fresh water above the halocline was aerobic and alive with all matter of fish, plant, and insect life flourishing. However, once the vehicle descended into the anaerobic salt waters below the halocline, the rocks were bleached, the leaves dropped into the pit were fully preserved, and nothing lived.

2.2.15 Ionic concentration

There are four environmentally important ions: nitrate (NO_3^-), chloride (Cl^-), calcium (Ca^{2+}), and ammonium (NH_4^+). Ion-selective electrodes used for monitoring these parameters are described below.

- *Nitrate* (NO_3^-): Nitrate ion concentration is an important parameter in nearly all water quality studies. Nitrates can be introduced by acidic rainfall, fertilizer runoff from fields, and plant or animal decay or waste.
- *Chloride* (Cl^-): This ion gives a quick measurement of the salinity of water samples. It can even measure chloride levels in ocean salt water or salt in food samples.
- *Calcium* (Ca^{2+}): This electrode gives a good indication of the hardness of water (as Ca^{2+}). It is also used as an end point indicator in EDTA-Ca/Mg hard water titrations.
- *Ammonium* (NH_4^+): This electrode measures levels of ammonium ions introduced from fertilizers. It can also indicate aqueous ammonia levels if sample solutions are acidified to convert NH_3 to (NH_4^+).

2.2.16 Solar radiation

Solar radiation is the electromagnetic radiation emitted by the sun and is measured in some underwater scientific applications.

2.2.17 Light and other electromagnetic transmissions through water

Light and other electromagnetic transmissions through water are affected by the following three factors:

1. Absorption
2. Refraction
3. Scattering

All of these factors, which can be measured by the ROV using light sensors, are normally lumped under the general category of attenuation.

2.2.17.1 Absorption

Electromagnetic energy transmission capability through water varies with wavelength. The best penetration is gained in the visible light spectrum (Figure 2.13). Other wavelengths of electromagnetic energy (radar, very low-frequency RF, etc.) are able to marginally penetrate the water column (in practically all cases only a few wavelengths), but even with very high intensity transmissions only very limited transmission rates/depths are possible under current technologies. Submerged submarines are able to get RF communications in deepwater with very low-frequency RF, but at that frequency it may take literally minutes to get through only two alphanumeric characters.

In the ultraviolet range, as well as in the infrared wavelengths, electromagnetic energy is highly attenuated by seawater. Within the visible wavelengths, the blue/green spectrum has the greatest energy transparency, with other wavelengths having differing levels of energy transmission.

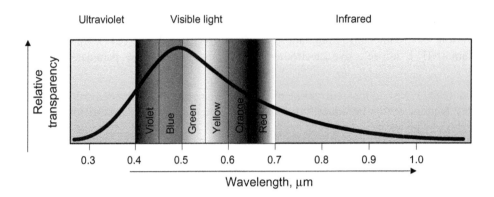

FIGURE 2.13

Light transparency through water (by wavelength).

Disregarding scattering (which will be considered below), within 1 m of the surface, fully 60% of the visible light energy is absorbed, leaving only 40% of original surface levels available for lighting and photosynthesis. By the 10-m depth range, only 20% of the total energy remains from that of the surface. By 100 m, fully 99% of the light energy is absorbed, leaving only 1% visible light penetration—practically all in the blue/green regime (Duxbury and Alison, 1997). Beginning with the first meter of depth, artificial lighting becomes increasingly necessary to bring out the true color of objects of interest below the surface.

Why not put infrared (IR) cameras on ROVs? The answer is simple—the visible light spectrum penetration in water favors the use of optical systems (Figure 2.13). IR cameras can certainly be mounted on ROV vehicles, but the effective range of the sensor suffers significantly due to absorption. The sensor may be effective at determining reflective characteristics, but the sensor would be required to be placed at an extremely close range, negating practically all benefits from non-optical IR reflectance.

2.2.17.2 Refraction

Light travels at a much slower speed through water, effectively bending (refracting) the light energy as it passes through the medium. This phenomenon is apparent not only with the surface interaction of the seawater, but also with the air/water interface of the ROV's camera system.

2.2.17.3 Scattering

Light bounces off water molecules and suspended particles in the water (scattering), further degrading the light transmission capability (in addition to absorption) by blocking the light path. The scattering agents (other than water molecules) are termed "suspended solids" (e.g., silt, single-cell organisms, salt molecules, etc.) and are measured in milligrams per liter on an absolute scale. Modern electronic instruments have been developed that allow real-time measurement of water turbidity from the ROV submersible or other underwater platform. The traditional physical measure of turbidity, however, is a simple measure of the focal length of a reflective object as it is lost from sight. Termed "Secchi depth," a simple reflective Secchi disk (coated with differing colors and textures) is lowered into the water until it just disappears from view.

All of the above issues and parameters will aid in determining the submersible's capability to perform the assigned task within a reasonable time frame.

2.3 Ocean dynamics
2.3.1 Circulation

The circulation of the world's water is controlled by a combination of gravity, friction, and inertia. Winds push water, ice, and water vapor around due to friction. Water vapor rises. Fresh water and hot water rise. Salt water and cold water sink. Ice floats. Water flows downhill. The high-inertia water at the equator zooms eastward as it travels toward the slower-moving areas near the poles (Coriolis effect—an excellent example of this is the Gulf Stream off the East Coast of the United States). The waters of the world intensify on the Western boundary of oceans due to the earth's rotational mechanics (the so-called Western intensification effect). Add into this mix the gravitational pull of the moon, other planets, and the sun, and one has a very complex circulation model for the water flowing around our planet.

In order to break this complex model into its component parts for analysis, oceanographers generally separate these circulation factions into two broad categories, "currents" and "tides." Currents are broadly defined as any horizontal movement of water along the earth's surface. Tides, on the other hand, are water movement in the vertical plane due to periodic rising and falling of the ocean surface (and connecting bodies of water) resulting from unequal gravitational pull from the moon and sun on different parts of the earth. Tides will cause currents, but tides are generally defined as the diurnal and semidiurnal movement of water from the sun/moon pull.

A basic understanding of these processes will arm the ROV pilot with the ability to predict conditions at the work site, thus assisting in accomplishing the work task.

According to Bowditch (2002), "currents may be referred to according to their forcing mechanism as either wind-driven or thermohaline. Alternatively, they may be classified according to their depth (surface, intermediate, deep, or bottom). The surface circulation of the world's oceans is mostly wind driven. Thermohaline currents are driven by differences in heat and salt. The currents driven by thermohaline forces are typically subsurface." If performing a deep dive with an ROV, count on having a surface current driven by wind action and a subsurface current driven by thermohaline forces—plan for it and it will not ruin the day.

An example of the basic differences between tides and currents is as follows:

- In the Bay of Fundy's Minas Basin (Nova Scotia, Canada), the highest tides on planet Earth occur near Wolfville. The water level at high tide can be as much as 45 ft (16 m) higher than at low tide! Small Atlantic tides drive the Bay of Fundy/Gulf of Maine system near resonance to produce the huge tides. High tides happen every 12 h and 25 min (or nearly an hour later each day) because of the changing position of the moon in its orbit around the earth. Twice a day at this location, large ships are alternatively grounded and floating. This is an extreme example of tides in action.

- There is a vertical density current through the Straits of Gibraltar. The evaporation of water over the Mediterranean drives the salinity of the water in that sea slightly higher than that of the Atlantic Ocean. The relatively denser high-salinity waters in the Mediterranean flow out of the bottom of the Straits while the relatively lower (less dense) salinity waters from the Atlantic flow in on the surface. This is known as a "density current." Trying to conduct an ROV operation there will probably result in a very bad day.
- Currents flow from areas of higher elevation to lower elevation. By figuring the elevation change of water over the area, while computing the water distribution in the area, one can find the volume of water that flows in currents past a given point (volume flow) in the stream, river, or body of water. However, the wise operator will find it much easier to just look it up in the local current/tide tables. There are people who are paid to make these computations on a daily basis, which is great as an intellectual exercise but is not recommended to "recreate the wheel."

2.3.1.1 Currents

The primary generating forces for ocean currents are wind and differences in water density caused by variations in heat and salinity. These factors are further affected by the depth of the water, underwater topography, shape of the basin in which the current is running, extent and location of land, and the earth's rotational deflection. The effect of the tides on currents is addressed in the next section.

Each body of water has its peculiar general horizontal circulation and flow patterns based upon a number of factors. Given water flowing in a stream or river, water accelerates at choke points and slows in wider basins per the equations of Bernoulli. Due to the momentum of the water at a river bend, the higher volume of water (and probably the channel) will be on the outside of the turn. Vertical flow patterns are even more predictable with upwelling and downwelling patterns generally attached to the continental margins.

Just as there are landslides on land, so are there mudslides under the ocean. Mud and sediment detach from a subsea ledge and flow downhill in the oceans, bringing along with it a friction water flow known as a turbidity current. Locked in the turbidity current are suspended sediments. This increase in turbidity can degrade camera optics if operating in an area of turbidity currents—take account of this during project planning.

Currents remain generally constant over the course of days or weeks, affected mostly by the changes in temperature and salinity profiles caused by the changing seasons.

Of particular interest to ROV operators is the wind-driven currents culled into the so-called Ekman spiral (Figure 2.14). The model was developed by physicist V. Walfrid Ekman from data collected by arctic exploration legend Fridtjof Nansen during the voyage of the *Fram*. From this model, wind drives idealized homogeneous surface currents in a motion 45° from the wind line to the right in the Northern Hemisphere and to the left in the Southern Hemisphere. Due to the friction of the surface water's movement, the subsurface water moves in an ever-decreasing velocity (and ever-increasing vector) until the momentum imparted by the surface lamina is lost (termed the "depth of frictional influence"). Although the depth of frictional influence is variable depending upon the latitude and wind velocity, the Ekman frictional transfer generally ceases at approximately 100 m depth. The net water transfer is at a right angle to the wind.

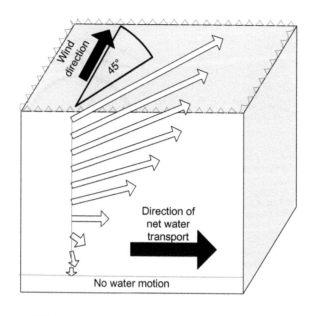

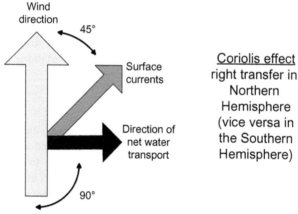

FIGURE 2.14

Ekman spiral.

2.3.1.2 Tides

Tides are generally defined as the vertical rise and fall of the water due to the gravitational effects of the moon, sun, and planets upon the local body of water (Figure 2.15). Tidal currents are horizontal flows caused by the tides. Tides rise and fall. Tidal currents flood and ebb. The ROV pilot is concerned with the amount and time of the tide, as it affects the drag velocity and vector computations on water flow across the work site. Tidal currents are superimposed upon non-tidal currents such as normal river flows, floods, and freshets. Put all of these factors together to find what the actual current will be for the job site (Figure 2.16).

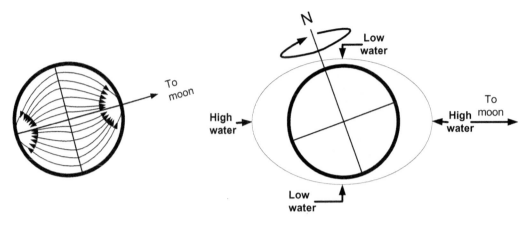

FIGURE 2.15

Tidal movement in conjunction with planets.

On a vertical profile, the tide may interact with the general flow pattern from a river or estuary—the warm fresh water may flow from a river on top of the cold salt water (a freshet, as mentioned above) as the salt water creeps for a distance up the river. According to Van Dorn (1993), fresh water has been reported over 300 miles at sea off the Amazon. If a brackish water estuary is the operating area, problems to be faced will include variations in water density, water flow vector/speed, and acoustic/turbidity properties.

2.3.1.3 Water velocity

Water velocity is the measure of the speed at which water travels, or the distance it travels over a given time, and is measured in meters per second. Hydrologists and other researchers measure water velocity for monitoring current in rivers, channels, and streams, to measure the effect of vessel traffic in harbors and ports, and to calculate water flow. To account for drift, water velocity readings are key factors in knowing where and when to deploy buoys and other environmental devices to ensure their correct location.

2.3.1.4 Waves and the Beaufort scale

Most operations manuals will designate launch, recovery, and operational parameters, as they relate to sea state, which is measured in the Beaufort scale (Table 2.5). Waves are measured with several metrics, including wave height, wavelength, and wave period. Wave height, -length, and period depend upon a number of factors, such as the wind speed, the length of time it has blown, and its fetch (the straight distance it has traveled over the surface).

Energy is transmitted through matter in the form of waves. Waves come in several forms, including longitudinal, transverse, and orbital waves. The best example of a longitudinal wave is a sound wave propagating through a medium in a simple back and forth (compression and rarefaction) motion. A transverse wave moves at right angles to the direction of travel, such as does a

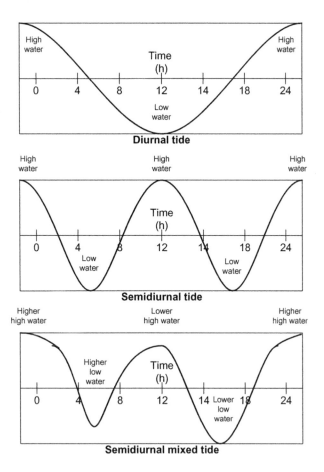

FIGURE 2.16

Tide periods over a 24-hour span.

guitar string. An orbital wave, however, moves in an orbital path as between fluids of differing densities.

A visual example of all three waves can be gained with a simple spring anchored between two points demonstrated as follows:

- Longitudinal wave—Hit one end with your hand and watch the longitudinal wave travel between end points.
- Transverse wave—Take an end point and quickly move it at a 90° axis to the spring and watch the transverse wave travel between end points.
- Orbital wave—Now rotate the end points in a circular fashion to propel the circular wave toward its end point. Orbital waves have characteristics of both longitudinal and transverse waves.

Table 2.5 Beaufort Scale (Thurman 1994)

Beaufort No.	Surface Winds in m/s (mph)	Seaman's Description	Effect at Sea
0	<1 (<2)	Calm	Mirror-like sea
1	0.3–1.5 (1–3)	Light air	Ripples with appearances of scales; no foam crests
2	1.6–3.3 (4–7)	Light breeze	Small wavelets; crests of glassy appearance, no breaking
3	3.4–5.4 (8–12)	Gentle breeze	Large wavelets; crests beginning to break; scattered whitecaps
4	5.5–7.9 (13–18)	Moderate breeze	Small waves, becoming longer; numerous whitecaps
5	8.0–10.7 (19–24)	Fresh breeze	Moderate waves, taking longer to form; many whitecaps; some spray
6	10.8–13.8 (25–31)	Strong breeze	Large waves begin to form; whitecaps everywhere; more spray
7	13.9–17.1 (32–38)	Near gale	Sea heaps up and white foam from breaking waves begins to be blown in streaks
8	17.2–20.7 (39–46)	Gale	Moderately high waves of greater length; edges of crests begin to break into spindrift; foam is blown in well-marked streaks
9	20.8–24.4 (47–54)	Strong gale	High waves; dense streaks of foam and sea begins to roll; spray may affect visibility
10	24.5–28.4 (55–63)	Storm	Very high waves with overhanging crests; foam is blown in dense white streaks, causing the sea to appear white; the rolling of the sea becomes heavy; visibility reduced
11	28.5–32.6 (64–72)	Violent storm	Exceptionally high waves (small- and medium-sized ships might be for a time lost to view behind the waves); the sea is covered with white patches of foam; everywhere the edges of the wave crests are blown into froth; visibility further reduced
12	32.7–36.9 (73–82)	Hurricane	The air is filled with foam and spray; sea completely white with driving spray; visibility greatly reduced

Described here are the basic components of ocean waves in an idealistic form to allow the analysis of each component individually. As described in Thurman (1994):

> As an idealized progressive wave passes a permanent marker, such as a pier piling, a succession of high parts of the wave, crests, will be encountered, separated by low parts, troughs. If the water level on the piling were marked when the trough passes, and the same for the crests, the vertical distance between the marks would be the wave height (H). The horizontal distance between corresponding points on successive waveforms, such as from crest to crest, is the wavelength (L). The ratio of H/L is wave steepness. The time that elapses during the passing of one

wavelength is the period (T). Because the period is the time required for the passing of one wave-
length, if either the wavelength or period of a wave is known, the other can be calculated, since
L (m) = 1.56 (m/s)T².

Speed (S) = L/T.

For example: Speed (S) = L/T = 156 m/10 s = 15.6 m/s.

Another characteristic related to wavelength and speed is frequency. Frequency (f) is the
number of wavelengths that pass a fixed point per unit of time and is equal to 1/T. If six wave-
lengths pass a point in one minute, using the same wave system in the previous example, then:

Speed (S) = Lf = 156 m × 6/min = 936 m/min × (1 min/60s) = 15.6 m/s.

Because the speed and wavelength of ocean waves are such that less than one wavelength
passes a point per second, the preferred unit of time for scientific measurements, period (rather
than frequency), is the more practical measurement to use when calculating speed.

The complexities of air/wave as well as air/land and fixed structure interaction, while quite important to the operation of ROV equipment in a realistic environment, are beyond the scope of this text. Please refer to the bibliography for a more in-depth study of the subject.

A more practical system of gauging the overall sea state/wind combination was developed in 1806 by Admiral Sir Francis Beaufort of the British Navy. The scale runs from 0 to 12 (Table 2.5), calm to hurricane, with typical wave descriptions for each level of wind speed. The Beaufort scale reduces the wind/wave combination into category ranges based upon the energy of the combined forces acting upon the sea surface.

2.3.2 Effects of wave pattern upon ROV operation

There is a plethora of information on wave propagation and its effect upon vessel management. Please refer to the bibliography for more detailed information. This section will address the effect of waves upon ROV operations.

The energy to produce a sea wave comes principally from wind but can also be generated through some lesser factors, such as submarine earthquakes (which happen less often but are possibly devastating when they occur), volcanic eruptions, and, of course, the tide. The biggest concern for wave pattern management in ROV operations is during the launch and recovery of the vehicle. During vessel operations, the hull is subject to motions broken down into six components—pitch, roll, and yaw for rotational degrees of freedom, and heave, surge, and sway for translational motion. The distance from the vessel's pivot point on any axis, combined with the translational motion from waves hitting the hull, will translate into the total swinging moment affecting the suspended weight—the ROV.

The closer the launch platform is located to the center of the vessel's pivotal point, the smaller will be the arm for upsetting the suspended weight. The longer the line is from the hard point on the launch winch, the longer the arm for upsetting the suspended weight. Add a long transit distance from the hard point to the water surface, with an over-the-side launch from a larger beam vessel, and the situation is one just waiting to get out of control (Figure 2.17).

Regardless of the sea state, it is best to complete this vulnerable launch operation (from the time the vehicle is lifted from the deck to the vehicle's submergence) as quickly as possible, given operational and safety constraints. This concept will be addressed further in Chapter 9.

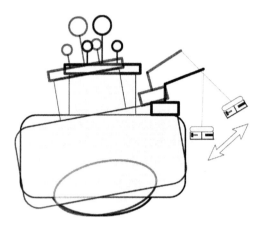

FIGURE 2.17

Vehicle suspended from vessel over the side.

ROV operations manuals should set the maximum sea state in which to deploy the submersible. In a high sea state on a rocking vessel, the submersible becomes a wrecker ball suspended in mid-air, waiting to destroy the vessel structure as well as the vehicle, deployment system, and any bodies unfortunate enough to be in the wrecker ball's path. If the sea state ramps up unexpectedly while the vehicle is still in the water, most operations manuals specify to leave the vehicle submerged until a sea state that allows safe recovery.

Another important wave propagation factor affecting ROV operations includes the heave component of the vessel while the vehicle is in the water. Particularly affected is the umbilical or tether length between the hard point of the launching platform and the clump weight, cage deployment platform, tether management system, or the vehicle itself. Of particular concern is the snap loading of the tether due to tether/umbilical pull with the vessel heaving. This rapid loading of the tether/umbilical can easily exceed the structural limitation of the tether and/or part conductors and communications components within the line. If this snap load parts the line, the vehicle could be lost. Larger and more sophisticated launch systems have a stress limit adjustable to the tether or umbilical loading limit. The system will slack or pull based upon the given parameters to maintain tension while avoiding overstressing the line.

The Vehicle

This part of the manual discusses the major components of a typical ROV system along with every-day underwater tasks ROVs perform.

As discussed earlier, the design team must consider the overall system. To reinforce the importance of this point, a few additional words of wisdom about the design process are warranted.

An ROV is essentially a robot. What differentiates a robot from its immovable counterparts is its ability to move under its own power. Along with that power of locomotion comes the ability to navigate the robot, with ever increasing levels of autonomy to achieve some set goal. While the ROV system, by its nature, is one of the simplest robotic designs, complex assignments can be accomplished with a variety of closed-loop aids to navigation. Some ROV manufacturers are aggressively embracing the open-source computer-based control models, allowing users to design their own navigation and control matrix. This is an exciting development in the field of subsea robotics and will allow development of new techniques, which will only be limited by the user's imagination. This concept takes the control of the development of navigation capabilities (which is the mission) from the hands of the design engineer (who may or may not understand the user's needs) into the hands of the end user (who does understand the needs). Designing efficient and cost-effective systems with the user in mind is critical to the success of the product and ultimately the mission.

Do not over design the system. The old saying goes that a chain is only as strong as its weakest link. Accordingly, all components of an ROV system should be rated to the maximum operating depth of the underwater environment anticipated, including safety factors. However, they should not be overdesigned. As the operating depth proceeds into deeper water, larger component wall thicknesses will be required for the air-filled spaces (pressure-resistant housings) on the vehicle. This increased wall thickness results in an increased vehicle weight, which requires a larger floatation system to counter the additional weight. This causes an increase in drag due to a larger cross section, which requires more power. More power drives the cable to become larger, which increases drag, etc. It quickly becomes a vicious design spiral.

Careful consideration should be given during the design phase of any ROV system to avoid overengineering the vehicle. By saving weight and cost during the design process, the user will receive an ROV that has the capability of providing a cost-effective operation. This is easier said than done, as "bells and whistles" are often added during the process or the "latest and greatest" components are chosen without regard to the impact on the overall system. Keep these ideas in mind as the various component choices are presented in the following chapters.

Design Theory and Standards

3

CHAPTER CONTENTS

3.1 A bit of history

3.1.1 Introduction

The strange thing about history is that it never ends. In the case of remotely operated vehicles (ROVs), the history is a short one, but very important nonetheless, especially for the observation-class ROVs.

Two critical groups of people have driven ROV history: (i) dedicated visionaries and (ii) exploiters of technology. Those who drove the development of ROVs had a problem to solve and a vision, and they did not give up the quest until success was achieved. There were observation-class vehicles early in this history, but they were far from efficient. In time, however, the technology caught up with the smaller vehicles, and those who waited to exploit this technology have led the pack in fielding smaller, state-of-the-art ROVs.

This section will discuss what an ROV is, address some of the key events in the development of ROV technology, and address the breakthroughs that brought ROVs to maturity.

3.1.2 In the beginning

One way to discuss the historical development of ROVs is to consider them in terms of the cycle of life—from infancy to maturity. Anyone who has raised a child will quickly understand such a categorization. In the beginning the ROV child was "nothing but a problem: Their bottles leaked, their hydraulics failed, sunlight damaged them, they were too noisy and unreliable, were hard to control and needed constant maintenance. Beginning to sound familiar?" (Wernli, 1998).

Some have credited Dimitri Rebikoff with developing the first ROV—the *POODLE*—in 1953. However, the vehicle was used primarily for archeological research and its impact on ROV history was minimal—but it was a start.

Although entrepreneurs like Rebikoff were making technology breakthroughs, it took the US Navy to take the first real step toward an operational system. The Navy's problem was the recovery of torpedoes that were lost on the seafloor. Replacing a system that essentially grappled for the torpedo, the Navy (under a contract awarded to VARE Industries, Roselle, New Jersey) developed a maneuverable underwater camera system—a mobile underwater vehicle system. The original VARE vehicle, the *XN-3*, was delivered to the Naval Ordnance Test Station in Pasadena, California, in 1961. This design eventually became the Cable-controlled Underwater Research Vehicle (CURV).

The Navy's *CURV* (and its successor—*CURV III*) made national headlines twice:

- The *CURV* retrieved a lost atomic bomb off the coast of Palomares, Spain, in 1966, from 2850 ft (869 m) of water, even though working beyond its maximum depth. The *CURV*'s sister vehicle, *CURV II*, is shown in Figure 3.1.
- *CURV III*, which had become a "flyaway" system, was sent on an emergency recovery mission from San Diego to an offshore point, near Cork, Ireland, in 1973. With little air left for the two pilots of the *PISCES III* manned submersible, which was trapped on the bottom in 1575 ft (480 m) of water, the *CURV III* attached a recovery line that successfully pulled the doomed crew to safety.

FIGURE 3.1

The Navy's *CURV II* vehicle.

FIGURE 3.2

Co-author R. Wernli (right) directs the launch of the US Navy's WSP/PIV.

With such successes under its belt, the Navy expanded into more complex vehicles, such as the massive Pontoon Implacement Vehicle (PIV), which was developed to aid in the recovery of sunken submarines, shown with the integrated Work Systems Package (WSP) (Figure 3.2).

At the other end of the scale, the US Navy developed one of the very first small-size observation ROVs. The *SNOOPY* vehicle, which was hydraulically operated from the surface, was one of the first portable vehicles (Figure 3.3).

This version was followed by the *Electric SNOOPY*, which extended the vehicle's reach by going with a fully electric vehicle. Eventually sonars and other sensors were added and the childhood of the small vehicles had begun.

Navy-funded programs helped Hydro Products (San Diego, CA) get a jump on the ROV field through the development of the *TORTUGA*, a system dedicated to investigating the utility of a submarine-deployed ROV. These developments led to Hydro Products' RCV line of "flying eyeball" vehicles (Figure 3.4).

FIGURE 3.3

US Navy's hydraulic *SNOOPY.*

FIGURE 3.4

Hydro Products' *RCV 225* and *RCV 150* vehicles.

These new intruders, albeit successful in their design goals, still could not shake that lock on the market by the manned submersibles and saturation divers. In 1974, only 20 vehicles had been constructed, with 17 of those funded by various governments. Some of those included are as follows:

- France—*ERIC* and Telenaute and ECA with their *PAP* mine countermeasure vehicles
- Finland—*PHOCAS* and Norway—the *SNURRE*
- UK—British Aircraft Corporation (*BAC-1*) soon to be the *CONSUB 01; SUB-2, CUTLET*
- Heriot-Watt University, Edinburgh—*ANGUS* (001, 002, and 003)
- Soviet Union—*CRAB-4000* and *MANTA* vehicles

It could be said that ROVs reached adolescence, which is generally tied to a growth spurt, accented by bouts of unexplained or irrational behavior, around 1975. With an exponential upturn, the number of vehicles grew to 500 by the end of 1982. And the funding line also changed during this period. From 1953 to 1974, 85% of the vehicles built were government funded. From 1974 to 1982, 96% of the 350 vehicles produced were funded, constructed, and/or bought by private industry.

The technological advancements necessary to take ROVs from adolescence to maturity had begun. This was especially true in the electronics industry, with the miniaturization of the onboard systems and their increased reliability. With the ROV beginning to be accepted by the offshore industry, other developers and vehicles began to emerge:

- USA—Hydro Products—the *RCV 125, TORTUGA, ANTHRO*, and *AMUVS* were soon followed by the *RCV 225*, and eventually the *RCV 150*; AMETEK, Straza Division, San Diego—turned their Navy-funded *Deep Drone* into their *SCORPIO* line; Perry Offshore, Florida—started their RECON line of vehicles based on the US Navy's *NAVFAC SNOOPY* design.
- Canada—International Submarine Engineering (ISE) started in Canada (*DART, TREC*, and *TROV*).
- France—Comex Industries added the *TOM-300*, C. G. Doris produced the *OBSERVER* and *DL-1*.
- Italy—Gay Underwater Instruments unveiled their spherical *FILIPPO*.
- The Netherlands—Skadoc Submersible Systems' *SMIT SUB* and *SOP*.
- Norway—Myers Verksted's *SPIDER*.
- Sweden added SUTEC's *SEA OWL* and Saab-Scandia's *SAAB-SUB*.
- UK—Design Diving Systems' *SEA-VEYOR*, Sub Sea Offshore's *MMIM*, Underwater Maintenance Co.'s *SCAN*, Underwater and Marine Equipment Ltd.'s *SEA SPY, AMPHORA*, and *SEA PUP*, Sub Sea Surveys Ltd.'s *IZE*, and Winn Technology Ltd.'s *UFO-300, BOCTOPUS, SMARTIE*, and *CETUS*.
- Japan—Mitsui Ocean Development and Engineering Co., Ltd., had the *MURS-100, MURS-300*, and *ROV*.
- Germany—Preussag Meerestechnik's *FUGE*, and VFW-Fokker GmbH's *PINGUIN B3* and *B6*.
- Other US—Kraft Tank Co. (*EV-1*), Rebikoff Underwater Products (*SEA INSPECTOR*), Remote Ocean Systems (*TELESUB-1000*), Exxon Production Research Co. (*TMV*), and Harbor Branch Foundation (*CORD*).

From 1982 to 1989 the ROV industry grew rapidly. The first ROV conference, ROV'83, was held with the theme "A Technology Whose Time Has Come!" Things had moved rapidly from 1970, when there was only one commercial ROV manufacturer. By 1984 there were 27. North American firms (Hydro Products, AMETEK, and Perry Offshore) accounted for 229 of the 340 industrial vehicles produced since 1975.

Not to be outdone, Canadian entrepreneur Jim McFarlane bought into the business with a series of low-cost vehicles—*DART, TREC*, and *TROV*—developed by ISE in Vancouver, British Columbia. But the market was cutthroat; the dollar to pound exchange rate caused the ROV technology base to transfer to the United Kingdom in support of the oil and gas operations in the North Sea. Once the dollar/pound exchange rate reached parity, it was cheaper to manufacture vehicles in the United Kingdom. Slingsby Engineering, Sub Sea Offshore, and the OSEL Group cornered the North Sea market, and the once dominant North American ROV industry was soon decimated. The only North American survivors were ISE (due to their diverse line of systems and the can-do attitude of their owner) and Perry, which wisely teamed with their European competitors to get a foothold in the North Sea.

However, as the oil patch companies were fighting for their share of the market, a few companies took the advancements in technology and used them to shrink the ROV to a new class of

FIGURE 3.5

Kaiko—the world's deepest diving ROV.

small, reliable, observation-class vehicles. These vehicles, which were easily portable when compared to their larger offshore ancestors, were produced at a cost that civil organizations and academic institutions could afford.

The *MiniRover*, developed by Chris Nicholson, was the first real low-cost, observation-type ROV. This was soon followed by Deep Ocean Engineering's (DOE's) Phantom vehicles. Benthos (now Teledyne Benthos) eventually picked up the *MiniRover* line and, along with DOE, cornered the market in areas that included civil engineering, dam and tunnel inspection, police and security operations, fisheries, oceanography, nuclear plant inspection, and many others.

The 1990s saw the ROV industry reach maturity. Testosterone-filled ROVs worked the world's oceans; no job was too hard or too deep to be completed. The US Navy, now able to buy vehicles off the shelf as needed, turned its eyes toward the next milestone—reaching the 20,000 ft (6279 m) barrier. This was accomplished in 1990, not once, but twice:

- *CURV III*, operated by Eastport International for the US Navy's Supervisor of Salvage, reached a depth of 20,105 ft (6128 m).
- The Advanced Tethered Vehicle (ATV), developed by the Space and Naval Warfare Systems Center, San Diego, broke the record less than a week later with a record dive to 20,600 ft (6279 m). The ATV was the first deep ocean ROV to incorporate three multi-mode optical fibers and a Kevlar strength member into its 23,000 foot (7012 m) cable.

It did not take long for this record to be not only beaten by Japan but obliterated. Using JAMSTEC's *Kaiko* ROV (Figure 3.5), Japan reached the deepest point in the Mariana Trench—35,791 ft (10,909 m)—a record that can be tied but never exceeded.

The upturn in the offshore oil industry is increasing the requirement for advanced undersea vehicles. Underwater drilling and subsea complexes are now well beyond diver depth, some

FIGURE 3.6

Schilling Robotics UHD.

exceeding 3000 m deep. Due to necessity, the offshore industry has teamed with the ROV developers to ensure integrated systems are being designed that can be installed, operated, and maintained through the use of ROVs. ROVs, such as Schilling Robotics UHD ROV (Figure 3.6), are taking underwater intervention to a higher technological level.

In the late 1990s, a new entrant to market by the name of VideoRay brought micro-ROV technology to common use. Followed shortly thereafter, SeaBotix entered the market with the Little Benthic Vehicle (LBV) and legitimized the micro-ROV as an affordable and useful ROV platform. Soon, sensor manufacturers began to tailor their products to fit in the smaller form factor, thus making the micro-ROV a fully industrial system.

By the end of the first decade of the twenty-first century, many manufacturers were entering into the market for OCROV and MSROV. On the educational front, several ROV-specific training programs were forming all over Europe and North America to cater to the ever-expanding fleet of operational industrial ROV systems. The Marine Advanced Technology Education (MATE) Center was formed in Monterey, California, in order to encourage high school and college-age enthusiasts into the field of subsea robotics. Today the MATE program sponsors international ROV competitions worldwide.

Also in the late 1990s (Wernli, 1998), it was estimated that there were over 100 vehicle manufacturers, and over 100 operators using approximately 3000 vehicles of various sizes and capabilities. We will not even try to determine the OCROVs in the field today as the number of operational systems is growing almost exponentially. According to *The World ROV Market Forecast 2011−2015*, there were 747 WCROVs being operated by 21 major companies, not counting the systems involved in noncommercial operations. As far as the number of actual vehicles in the field today ... well, tracking that number will be left to the statisticians.

3.2 Underwater vehicles to ROVs

Previous sections provided an introduction to what an ROV is and is not, along with a brief history of how these underwater robots arrived at their present level of worldwide usage. Since these vehicles are an extension of the operator's senses, the communication with them is probably the most critical aspect of vehicle design.

The communication and control of underwater vehicles is a complex issue and sometimes occludes the lines between the ROV and the autonomous underwater vehicle (AUV). Before addressing the focus of this text, the issues involved will be investigated further.

The basic issues involved with underwater vehicle power and control can be divided into the following categories:

- Power source for the vehicle
- Degree of autonomy (operator controlled or program controlled)
- Communications linkage to the vehicle

3.2.1 Power source for the vehicle

Vehicles can be powered in any of the following three categories: surface-powered, vehicle-powered, or hybrid system.

- *Surface-powered* vehicles must, by practicality, be tethered, since the power source is from the surface to the vehicle. The actual power protocol is discussed more fully later in this text, but no vehicle-based power storage is defined within this power category.
- *Vehicle-powered* vehicles store all of their power-producing capacity on the vehicle in the form of a battery, fuel cell, or some other means of power storage needed for vehicle propulsion and operation.
- A *hybrid system* involves a mixture of surface and submersible supplied power. Examples of the hybrid system include the battery-powered submersible with a surface-supplied charger (through a tether) for recharging during times of less-than-maximum power draw; a surface-powered vehicle with an onboard power source for a transition from ROV to AUV (some advanced capability torpedo designs allow for swim-out under ship's power to transition to vehicle power after clearing the area) and other variations to this mix. (Chapter 23 discusses the more advanced hybrid vehicles that will appear offshore in the future.)

3.2.2 Degree of autonomy

According to the National Institute of Standards and Technology (Huang, 2004), unmanned vehicles may be operated under several modes of operation, including fully autonomous, semi-autonomous, teleoperation, and remote control (RC).

- *Fully autonomous*: A mode of operation of an unmanned system (UMS) wherein the UMS is expected to accomplish its mission, within a defined scope, without human intervention. Note that a team of UMSs may be fully autonomous while the individual team members may not be, due to the need to coordinate during the execution of team missions.
- *Semi-autonomous*: A mode of operation of a UMS wherein the human operator and/or the UMS plan(s) and conduct(s) a mission and requires various levels of human−robot interaction.
- *Teleoperation*: A mode of operation of a UMS wherein the human operator, using video feedback and/or other sensory feedback, either directly controls the motors/actuators or assigns

incremental goals, waypoints in mobility situations, on a continuous basis, from off the vehicle and via a tethered or radio/acoustic/optic/other linked control device. In this mode, the UMS may take limited initiative in reaching the assigned incremental goals.

- *RC*: A mode of operation of a UMS wherein the human operator, without benefit of video or other sensory feedback, directly controls the actuators of the UMS on a continuous basis, from off the vehicle and via a tethered or radio-linked control device using visual line-of-sight cues. In this mode, the UMS takes no initiative and relies on continuous or nearly continuous input from the user.

3.2.3 Communications linkage to the vehicle

The linkage to the vehicle can come in several forms or methods depending upon the distance and medium through which the communication must take place. Such linkages include:

- Hard-wire communication (either electrical or fiber optic)
- Acoustic communication (via underwater analog or digital modem)
- Optical communication (while on the surface)
- Radio frequency (RF) communication (while on or near the surface)

What is communicated between the vehicle and the operator can be any of the following:

- *Telemetry*: The measurement and transmission of data or video through the vehicle via tether, RF, optical, acoustic, or other means.
- *Tele-presence*: The capability of a UMS to provide the human operator with some amount of sensory feedback similar to that which the operator would receive if inside the vehicle.
- *Control*: The upload/download of operational instructions (for autonomous operations) or full teleoperation.
- *Records*: The upload/download of mission records and files.

ROVs receive their power, their data transmission, or their control (or all three) directly from the surface through direct hard-wire communication (i.e., the tether). In short, the difference between an ROV and an AUV is the tether (although some would argue that the divide is not that simple).

3.2.4 Special-use ROVs

Some of the special-use ROVs come in even more discriminating packages:

- *Rail cameras*: Work on the drilling string of oil and gas platforms/drilling rigs (a pan and tilt camera moving up and down a leg of the platform to observe operations at the drill head with or without intervention tooling).
- *Bottom crawlers*: Lay pipe as well as communications cables and such while heavily weighted in the water column and being either towed or on tracks for locomotion
- *Towed cameras*: Can have movable fins that allow "sailing" up or down (or side to side) in the water column behind the towing vehicle.

- *Swim-out ROVs*: Smaller free-swimming systems that launch from a larger ROV, AUV, or manned vehicle system.

Although this text covers many of the technologies associated with all underwater vehicles, the subject matter will focus on the free-swimming, surface-powered, teleoperated (or semi-autonomous) observation-class to mid-sized ROVs with submersible weights from the smallest of sizes to 2000 lb (907 kg).

3.3 Autonomy plus: "why the tether?"

In order to illustrate where ROVs fit into the world of technology, an aircraft analogy will be discussed first and then the vehicle in its water environment.

Autonomy with regard to aerial vehicles runs the full gamut from man occupying the vehicle while operating it (e.g., a pilot sitting in the aircraft manipulating the controls for positive navigation) to artificial intelligence on an unmanned aerial vehicle making unsupervised decisions on navigation and operation from start to finish (Figure 3.7). However, where the human sits (in the vehicle or on a separate platform) is irrelevant to the autonomy discussion, since it does not affect how the artificial brain (i.e., the controller) thinks and controls.

3.3.1 An aircraft analogy

To set the stage with an area most are familiar with, the control variations of an aircraft will be defined as follows:

- *Man in vehicle*: Pilot sitting aboard the aircraft in seat manning controls.
- *Man in vehicle with AutoPilot*: Pilot sitting aboard aircraft in seat with AutoPilot controlling the aircraft's navigation (pilot supervising the systems).
- *Man in remote location with teleoperation*: Technician sitting in front of control console on the ground (or another aerial platform) with RF link to the unmanned aircraft while the technician is manipulating the controls remotely.

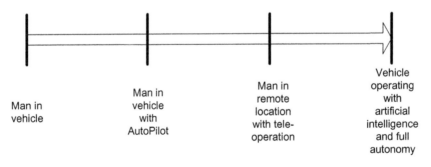

FIGURE 3.7

Degrees of autonomy.

- *Vehicle operating with artificial intelligence and full autonomy*: No human supervisor directly controlling the vehicle. The vehicle controls are preprogrammed with the vehicle making objective decisions as to the conduct of that flight from inception to termination based upon the "Sense/Plan/Act" paradigm.

Predator UAVs have recently been retrofitted with weapons, producing the new designation unmanned combat aerial vehicle (UCAV). The most efficient technology would allow that UCAV (without human intervention) to find, detect, classify, and deliver a lethal weapon upon the target, thus eliminating the threat. And here is the crux—is a responsible commander in the field comfortable enough with the technology to allow a machine the decision of life and death? This may be an extreme example, but (for now) a human must remain in the decision loop. To continue this example, would any passenger (as a passenger) fly in a commercial airliner without a pilot physically present in the cockpit? This may not happen soon, but one can safely predict that unpiloted airliners are in our future. Unattended trolleys are currently used in many airports worldwide.

3.3.2 Underwater vehicle variations

Now, the aircraft analogy will be reconsidered with underwater vehicle control in mind.

- *Man in vehicle*: Manned submersible pilot sitting aboard the vehicle underwater in the pilot's seat manning the controls and directly commanding the vehicle.
- *Man in vehicle with AutoPilot*: Same situation with AutoPilot controlling the submersible's navigation (pilot supervising the systems).
- *Man in remote location with teleoperation*: Technician sitting in front of control console on the surface (or other submerged platform) with tether or other data link to the submersible while the technician is manipulating the controls remotely.
- *Vehicle operating with artificial intelligence and full autonomy*: No human supervisor directly controlling the vehicle. The vehicle controls of the AUV are preprogrammed, with the vehicle making objective decisions as to the conduct of that dive from inception to termination.

During operation Iraqi Freedom, mine countermeasure AUVs were used for mine clearance operations. The AUV swam a preprogrammed course over a designated area to search and detect mine-like objects on the bottom. Other vehicles (or marine mammals or divers) were then sent to these locations to classify and (if necessary) neutralize the targets.

The new small UUVs are going through a two-stage process where they Search (or Survey), Classify, and Map. The Explosive Ordinance Disposal personnel then return with another vehicle (or marine mammal or human divers) to Reacquire, Identify, and Neutralize the target. Essentially, the process is to locate mine-like targets, classify them as mines if applicable and then neutralize them. What if the whole process can be done with one autonomous vehicle? And again the crux—is the field commander comfortable enough with the vehicle's programming to allow it to distinguish between a Russian KMD-1000 Bottom-type influence mine and a manned undersea laboratory before destroying the target? For the near term, man will remain in the decision loop for the important operational decisions. But again, one can safely predict that full autonomy is in our future.

3.3.3 Why the tether?

RF waves penetrate only a few wavelengths into water due to water's high attenuation of its energy. If the RF is of a low frequency, the waves will penetrate farther into water due to longer wavelengths. But with decreasing RF frequencies, data transmission rates suffer. In order to perform remote inspection tasks, live video is needed at the surface so that decisions by humans can be made on navigating the vehicle and inspecting the target. Full teleoperation (under current technology) is possible only through a high-bandwidth data link.

With the UAV example above, full teleoperation was available via the RF link (through air) between the vehicle and the remote operator. In water, this full telemetry is not possible (with current technologies) through an RF link. Acoustic in-water data transmission (as of 2013) is limited to less than 100 kB/s (insufficient for high-resolution video images). A hard-wire link to the operating platform is needed to have a full teleoperational in-water link to the vehicle. Thus, the need exists for a hard-wire link of some type, for the foreseeable future, for real-time underwater inspection tasks.

3.3.4 Teleoperation versus remote control

An ROV pilot will often operate a vehicle remotely with his/her eyes directly viewing the vehicle while guiding the vehicle on the surface to the inspection target. This navigation of the vehicle through line of sight (as with the RC airplane) is termed "RC (remote control) mode." Once the inspection target is observed through the vehicle's camera or sensors, the transition is made from RC operational mode to teleoperation mode. This transition is important because it changes navigation and operation of the vehicle from the operator's point of view to the vehicle's point of view. Successful management of the transition between these modes of operation during field tasks will certainly assist in obtaining a positive mission completion.

Going back to the UAV analogy, many kids have built and used RC model aircraft. The difference between an RC aircraft and a UAV is the ability to navigate solely by use of onboard sensors. A UAV can certainly be operated in an RC mode while the vehicle is within line-of-sight of the operator's platform, but once line-of-sight is lost, navigation and control are only available through teleoperation or preprogramming.

The following is an example of this transition while performing a typical observation-class ROV inspection of a ship's hull: The operator swims the vehicle on the surface (Figure 3.8) via RC to the hull of the vessel until the inspection starting point is gained with the vehicle's camera and then transitions to navigation via the vehicle's camera.

3.3.5 Degrees of autonomy

An open-loop control system is simply a condition on a functioning machine whereby the system has two basic states: "On" or "Off." The machine will stay On/Off for as long as the operator leaves it in that mode. The term "open-loop" (or essentially "no loop") refers to the lack of sensor feedback to control the operation of the machine. An example of an open-loop feedback would be a simple light switch that, upon activation, remains in the "On" or "Off" condition until manually changed.

FIGURE 3.8

Surface swim in RC mode.

Beginning with pure teleoperation (which is no autonomy), the first step toward full autonomy is the point at which the vehicle begins navigation autonomously within given parameters. This is navigation through "closed-loop feedback."

Closed-loop feedback is simply control of an operation through sensor feedback to the controller. A simple example of a closed-loop feedback system is the home air-conditioning thermostat. At a given temperature, the air conditioner turns on, thus lowering the temperature of the air surrounding the thermostat (if the air-conditioning is ducted into that room). Once the air temperature reaches a certain preset value, the thermostat sends a signal to the air conditioner (closing the control signal and response loop) to "turn off," completing this simple closed-loop feedback system.

The most common first step along this line for the ROV system is the auto heading and auto depth functions. Any closed-loop feedback control system can operate on an ROV system, manipulating control functions based upon sensor output. Operation of the vertical thruster as a function of constant depth (as measured by the variable water pressure transducer) is easily accomplished in software to provide auto depth capability. For example, consider an auto depth activation system on an ROV at 100 ft (30 m) of seawater. The approximate (gauge) pressure is 3 atmospheres or 45 psig (3 bar). As the submersible sinks below that pressure (as read by the pressure transducer on the submersible), the controller switches on the vertical thruster to propel the vehicle back toward the surface until the 45 psig (3 bar) reading is reacquired (the reverse is also applicable).

The same applies to auto altitude, where variation of the vertical thruster maintains a constant height off the bottom based upon echo soundings from the vehicle's altimeter. Similarly, auto standoff from the side of a ship for hull inspections can be based upon a side-looking acoustic sensor, where variation of the sounder timing can be used to vary the function of the lateral thruster.

Any number of closed-loop variables can be programmed. The submersible can then be given a set of operating instructions based upon a matrix of "if/then" commands to accomplish a given mission. The autonomy function is a separate issue from communications. A tethered ROV can be operated in full autonomy mode just as an untethered AUV may be operated in full autonomy mode. The only difference between a fully autonomous ROV and a fully autonomous AUV is generally considered to be the presence/absence of a hard-wire communications link, i.e., a tethered AUV is actually an ROV.

Such design issues will be addressed in Chapter 23.

3.4 Vehicle classifications

Vehicle classification is a hotly debated issue due to the lack of any widely accepted standards on the subject. In the early days of OCROVs, the (rather condescending) moniker "low-cost ROV" drew some rather interesting discussions due to its inference to a child's toy. Later, as the technology matured, other lines of demarcation between vehicle classes evolved starting with "hydraulic" and "electric" for the emphasis on drive force for the prime mover (i.e., thruster). Now in the second decade of the twenty-first century, ROVs range from less than 10 lb (5 kg) for the smallest to multiton hydraulic vehicles topping 250 hp in drive motors. But the classification debate continues.

As the typical client for ROV services generally has limited knowledge of vehicle performance, the larger operators have tended to link vehicle capabilities directly to horsepower. But where is the horsepower measured (even if it is a ridiculous performance metric)?

Many commercial tenders for WCROV services will specify a minimum horsepower rating for the ROV system but will not define the term "horsepower." For instance, on a recent tender for ROV services an oil company specified a 150 hp minimum rating for a drill support contract. A large electric vehicle was certainly fully capable of fulfilling all of the task requirements specified in the tender. One ROV operator solved the problem very competitively (some would say "aggressively") by offering a 100 hp vehicle with a 50 hp pump located on the TMS (the TMS is certainly part of the system ...). Voilà—they were awarded the contract! Problem solved. But the naming convention occlusion remains. Expect some standards on this front soon from the oil industry standards organizations such as American Petroleum Industry (API) or International Standards Organization (ISO).

In this section, we will examine an older classification example (in this case, the IMCA classification system). Then we will expand upon the older classification system to better stratify the modern vehicle systems currently available on the open market.

3.4.1 Size classifications of ROVs

As an initial note, the International Maritime Contractors Association (IMCA—http://www.imca-int.com/) lists its own classification of ROVs as follows:

a. Class I—observation ROVs (small vehicles fitted with camera/lights and sonar only)
b. Class II—observation ROVs with Payload Option (vehicles fitted with two simultaneously viewable cameras/sonar as standard and capable of handling additional sensors as well as a basic manipulative capability)

c. Class III—work-class vehicles (vehicles large enough to carry additional sensors and/or manipulators)

d. Class IV—towed and bottom-crawling vehicles (vehicles pulled through the water by a surface craft or winch, and bottom-crawling vehicles using a wheel or track system to move across the seafloor)

e. Class V—prototype or development vehicles (those still being developed and those regarded as prototypes).

This naming convention has been in effect for quite some time (mostly with the European operators, but also with the US-based ADCI (Association of Diving Contractors International)). However, due to the rapid development of the technology, a much more robust naming convention is necessary to fully define the full range of vehicle options.

As discussed in Chapter 1, ROVs break into three broad categories based upon their vehicle weight:

1. *OCROV*—from the smallest vehicles to submersible weights up to 200 lb (91 kg)
2. *MSROV*—submersible weights from 200 lb (91 kg) to 2000 lb (907 kg)
3. *WCROV*—submersible weights in excess of 2000 lb (907 kg)

With the dynamic expansion of this technology, further delineation is required to fully define each category.

OCROV categories

Within the OCROV category fall three subcategories based upon vehicle weight:

a. *Micro* (or small) OCROVs—those vehicles with a basic weight of less than 10 lb (4.5 kg) (e.g., VideoRay, GNOM, and AC-ROV)

b. *Mini* (or medium) OCROVs—vehicles with submersible weight between 10 lb (4.5 kg) and 70 lb (32 kg), i.e., the limit of single-person hand deployment (e.g., SeaBotix LBV, Outland 1000, JW Fisher SeaLion/SeaOtter, and Seamor)

c. *Large* OCROV—vehicles with weights between 70 lb (32 kg) and 200 lb (90 kg) (e.g., Benthos StingRay, SeaEye Falcon, Sub-Atlantic Mohave, and Seatronics Predator).

MSROV categories

Within the MSROV category fall three subcategories based upon vehicle performance and depth capability:

a. *Shallow* MSROV: These vehicles are typically low-power vehicles with copper (or fiber) telemetry and <3300 ft (1000 m) depth capability (e.g., Benthos SeaRover, Sub-Atlantic Mohawk, DOE S5N, and SeaEye Falcon DR).

b. *Deepwater* MSROV: These vehicles are typically deepwater versions of the shallow vehicles and may run single or dual light manipulator systems along with high-voltage power, light-duty electric and hydraulic manipulator systems (Hydro-Lek or similar), and fiber-optic telemetry (e.g., SeaEye Tiger or Cougar, Sub-Atlantic Super Mohawk or Mohican, and Argus Rover).

c. *Heavy* MSROV: These are often named "light work class" and typically have electric thrusters, dual medium-duty hydraulic manipulator systems (Schilling Orion or similar), and a hydraulic power unit for operation of medium-duty hydraulic tooling (e.g., Sub-Atlantic Comanche and Seaeye Jaguar).

WCROV categories

For the largest class of vehicle, the focus shifts from depth capability (it is assumed that these are fully capable of deepwater operation) to the size of the drive pump. WCROVs are measured in terms of horsepower of the primary motor *on the vehicle* (emphasis added):

a. *Standard work class*: These vehicles are in the 100−200 hp range typically used in drill support or light construction (e.g., Argus Worker, Perry XLR/XLS/XLX, Schilling HD, SMD Atom or Quasar, and Oceaneering Magnum).

b. *Heavy work class*: These vehicles are very large and heavy work vehicles of 200 hp or greater for heavy construction work (e.g., Perry XLX 200, Schilling UHD, Oceaneering Millennium Plus or Maxximum).

3.4.2 Today's observation-class vehicles

One can spend a lifetime running numbers and statistics; however, it can be easily said that the OCROV technology is here to reach into the shallower depths in a cost-effective manner and complete a series of critical missions. Whether performing dam inspections, body recoveries, fish assessment, or treasure hunting, technology has allowed the development of the advanced systems necessary to complete the job—and stay dry at the same time.

Technology has moved from vacuum tubes, gear trains, and copper/steel cables to microprocessors, magnetic drives, and fiber-optic/Kevlar cables. That droopy-drawered infant has now graduated from college and can work reliably without constant maintenance. As computers have moved from trunk-sized "portable" systems to those that fit in a pants pocket, observation-class ROVs have moved from portable, i.e., a team of divers can carry one, to handheld vehicles that can complete the same task. Examples of observation-class vehicles in use today are shown in Figure 3.9. Table 3.1 provides a summary of those vehicles that weigh less than 91 kg (200 lb) and have 25 or more in the field.

FIGURE 3.9

Right side (a) and left side (b) view of various OCROV systems.

Table 3.1 Observation-Class Vehicles Weighing Less Than Approximately 91 kg (200 lb)

Name	Company	Weight (kg) in Air	Depth (m)
AC-ROV	AC-CESS CO, UK	3	75
Firefly	Deep Ocean Engineering, USA	5.4	46
H300	ECA Hytec, France	65	300
Hyball	SMD Hydrovision Ltd., UK	41	300
LBV	SeaBotix, Inc., USA	10–15	150–1500
Navaho	Sub-Atlantic (SSA Alliance), UK	42	300
Offshore Hyball	SMD Hydrovision Ltd., UK	60	300
Outland 1000	Outland Technology Inc., USA	17.7	152
Phantom 150	Deep Ocean Engineering, USA	14	46
Phantom XTL	Deep Ocean Engineering, USA	50	150
Prometeo	Elettronica Enne, Italy	48–55	–
RTVD-100MKIIEX	Mitsui, Japan	42	150
Predator	Seatronics Group, UK	65	300
Seaeye 600 DT	Seaeye Marine Ltd., UK	65	300
Seaeye Falcon	Seaeye Marine Ltd., UK	50	300–1000
Stealth	Shark Marine Technologies Inc., Canada	40	300
VideoRay	VideoRay LLC, USA	4–4.85	0–305

The following chapters will provide an overview of various classes of ROVs, related technologies, the environment they will work in, and words to the wise about how to—and how not to—use them to complete an underwater task. How well this is done will show up in the History chapter of future publications.

3.4.3 Today's mid-sized vehicles

The MSROV (vehicles weighing between 200 and 2000 lb (91–907 kg) technology is rapidly catching up to the larger WCROVs with deeper depth capabilities and more efficient power delivery. Several manufacturers (Table 3.2) are popping up worldwide, but the current predominant players are Saab SeaEye (near Southampton, UK) and Sub-Atlantic (Aberdeen, Scotland). Both companies manufacture a full line of deep-rated electric vehicles that extend up to the light work class. Most of the heavier (>2000 lb (1000 kg)) systems are standard rated to 6500 ft (2000 m) depth with expansion to 10,000 ft (3000 m) depth ratings (or higher). The lines between the OCROV, MSROV, and WCROV classes continue to blur with the smaller vehicles gaining further capabilities of the next higher size.

3.4.4 The ROV "spread"

An ROV spread is defined as all items needed to deploy and support the ROV system in the field. They range from one simple suitcase for the micro-ROV all the way to a fully integrated heavy WCROV system integrated into the large DP-2 vessel with internal control station, hangar-housed

Table 3.2 Sampling of Mid-Size Vehicles Weighing Over 91 kg (200 lb) and Less Than 907 kg (2000 lb)

Name	Company	Weight (kg) in Air	Depth (m) Standard
S5N	Deep Ocean Engineering, USA	114	1000
Lynx	Saab SeaEye, UK	200	1500
Mohawk	Sub-Atlantic, UK	165	1000
Tiger	Saab SeaEye, UK	150	1000
Rover	Argus Remote Systems, Norway	450	500
Mohican	Sub-Atlantic, UK	230	2000
Cougar XT	Saab SeaEye, UK	344	2000
Super Mohawk	Sub-Atlantic, UK	290	2000
Panther XT	Saab SeaEye, UK	500	1500
Tomahawk	Sub-Atlantic, UK	1075	2000
Jaguar	Saab SeaEye, UK	1500	3000
Comanche	Sub-Atlantic, UK	1130	2000

vehicle storage (with accompanying large work area for fabrication as well as on-site repairs) and internally mounted high-powered generator.

Spreads are broadly classified as:

1. *Carry-on*: OCROV spreads hand-carried aboard and powered off of vessel power source.
2. *Bolt-on*: OCROV, MSROV, or WCROV spreads that are self-contained in standardized marine containers capable of quickly bolting onto (and removal from) a vessel of opportunity.
3. *Integrated*: All classes of ROV that are semipermanently mounted onto the vessel and function as part of the vessel's primary mission (e.g., an ROV support or deepwater construction vessel with the ROV as primary mission support).

3.5 Design theory

This chapter describes the different types of underwater systems, the basic theory behind vehicle design/communication/propulsion/integration, and the means by which a typical ROV gets every-day underwater tasks performed.

3.5.1 Unmanned underwater vehicle objectives

Underwater vehicles perform a wide range of tasks depending upon the application. The type of vehicle required will depend upon the intended use—specifically, the vehicle will be optimized for movement over either long or short ranges. This will dictate the shape of the vehicle, the method of power delivery—onboard (e.g., onboard battery powered), offboard (e.g., surface powered or powered from some remote location through a tether), or hybrid (e.g., onboard power with constant charge from some offboard hard-wired power cable).

Table 3.3 Mission Caparisons for Moving Versus Stationary Vehicles

Stationary Vehicle (ROV)	Rapidly Moving Vehicle (AUV)
Close visual inspection	Wide area sensor deployment
Localized NDT or NDE sensor deployment	General visual inspection
Physical manipulation or intervention of stationary structures	High-speed weapons delivery
Localized sensor deployment	

Table 3.4 Design Attributes of a Stationary Versus Moving Vehicle

Stationary Efficiency (ROV)	Moving Efficiency (AUV)
Low vehicle aspect ratio (length to width)	Optimized for travel over one axis only
Similar drag profiles about all axes of sway	Closed frame for drag minimization
Easy thrust and movement about all planes of motion	Lowest possible frontal profile
Open frame design	Onboard power
Drag as secondary design consideration	Tetherless for drag minimization
Typically offboard power through hard-wired umbilical/tether	

The mission of the AUV (untethered) is typically in the survey function where it efficiently delivers a sensor suite over a wide area. The typical mission of an ROV, on the other hand, is for localized stationary (or close area) work. Therefore, the ROV will be optimized for station keeping in various environmental conditions along all axes of travel while the AUV will be optimized for a single axis of travel. A comparison of missions is listed at Table 3.3.

The rapidly moving platform has the complex task of managing a hard-wired tether from the command platform to the vehicle (especially over long distances) while the stationary vehicle minimizes this problem. The highest bandwidth data rate through water will be achieved via a hard-wired communications link between the vehicle and the command platform. This is due to the high-attenuation rate through water with many of the communications technologies deployed versus through air or vacuum (e.g., RF, optics, acoustics). Therefore, the AUV is powered onboard with logic-directed operation while the ROV is typically powered via offboard with man-in-the-loop command and control. Further, the need for intervention along with close-in maneuvering often requires a human (with real-time control and feedback) for direct operation in a nonstructured environment.

3.5.2 Designing for mission efficiency

The design efficiency will dictate the overall design philosophy. Table 3.4 lists some of the various design attributes based upon the stationary or moving platform objective.

3.5.3 Drag discussion

The subject of this text is ROVs; therefore, we will concentrate this discussion on the ROV question. But elements of this discussion pervade all aspects of underwater vehicles (manned and unmanned).

The function of an ROV is to act as a delivery platform (for sensors and tooling) to a remote work site. All items and subsystems of the vehicle support this function. The vehicle must have some type of locomotion to take it to the work site and perform the work. In order to achieve the locomotion objective, the vehicle must power itself and overcome the fluid drag of the vehicle/tether combination to travel to and remain at the work site. This sounds simple, but the devil is in the details.

3.5.3.1 The drag equation

Nestled in the Appalachian Mountains of central Pennsylvania is the Applied Research Lab at Penn State University. The facility houses a closed-circuit water tunnel for inducing all of the characteristics for identifying a theoretical fluid flow equation, namely:

1. a fully enclosed fluid
2. a blunt form factor
3. a large enough Reynolds number to induce turbulence downstream from the item.

There is no better place to isolate the components of the drag equation than in this controlled environment. Imagine the item (let us call it a cube for now) stationary in the tunnel with no water flow. There would be no fluid drag. However, turn on the pump and make the water flow. As the water flow through the tunnel is slowly increased (with the cube still stationary in the tunnel), the drag will increase in an amount proportional to the density of the fluid (in this case either 62.43 lb/ft^3 (1000 kg/m^3) for fresh water or 64.62 lb/ft^3 (1035 kg/m^3) for seawater at maximum density) as well as the square of the object's speed relative to the fluid. This may sound complicated, but it is quite simple (double the speed/quadruple the drag). Also, with a constant object volume, the shape of the object will directly affect its drag force—this factor is referred to as the "coefficient of drag" (C_d).

Let us first examine C_d. In Figure 3.10, various shapes are depicted along with their value of C_d (often called "shape or form drag"). The value of C_d, as will be shown shortly, will play a significant role in determining the overall drag of the ROV system.

3.5.3.2 Vehicle stability

As with a child's seesaw, the further a weight is placed from the fulcrum point, the higher the mechanical force, or moment, needed to "upset" that weight (the term "moment" is computed by the product of the weight times the arm or distance from the fulcrum). It is called "positive stability" when an upset object inherently rights itself to a steady state. When adapting this to a submersible, positive longitudinal and lateral stability can be readily achieved by having weight low and buoyancy high on the vehicle. This technique produces an intrinsically stable vehicle on the pitch and roll axis. In most observation-class ROV systems, the higher the stability the easier it is to control the vehicle. With lower static stability, expect control problems (Figure 3.11).

External forces, however, do act upon the vehicle when it is in the water, which can produce apparent reductions in stability. For example, the force of the vertical thruster when thrusting down appears to the vehicle as an added weight high on the vehicle and, in turn, makes the center of gravity appear to rise, which destabilizes the vehicle in pitch and roll. The center of buoyancy and center of gravity can be calculated by taking moments about some arbitrarily selected point.

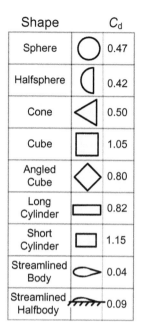

Shape		C_d
Sphere		0.47
Halfsphere		0.42
Cone		0.50
Cube		1.05
Angled Cube		0.80
Long Cylinder		0.82
Short Cylinder		1.15
Streamlined Body		0.04
Streamlined Halfbody		0.09

FIGURE 3.10

Measured drag coefficients by shape.

(*NASA.*)

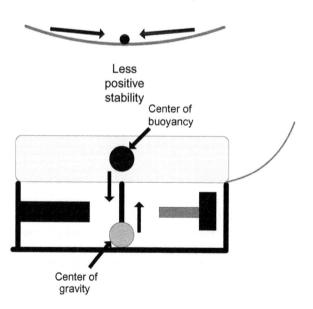

Less positive stability

Center of buoyancy

Center of gravity

FIGURE 3.11

ROV with ballast moved up.

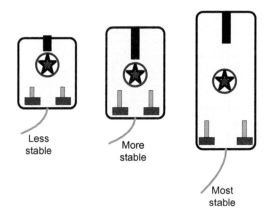

FIGURE 3.12

Vehicle geometry and stability.

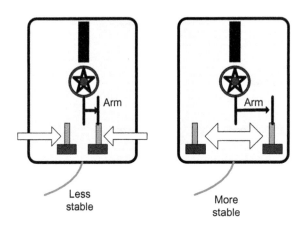

FIGURE 3.13

Thruster placement and stability.

Other design characteristics also affect the stability of the vehicle along the varying axes. The so-called aspect ratio (total mean length of the vehicle versus total mean width of the vehicle) will determine the vehicle's hull stability (Figure 3.12), as will thruster placement (Figure 3.13).

Most attack submarine designers specify a 7:1 aspect ratio as the optimum for the maneuvering-to-stability ratio (Moore and Compton-Hall, 1987). For ROVs, the optimal aspect ratio and thruster placement will be dependent upon the anticipated top speed of the vehicle, along with the need to maneuver in confined spaces.

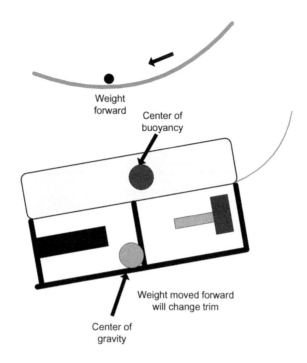

Weight
forward

Center of
buoyancy

Weight moved forward
will change trim

Center of
gravity

FIGURE 3.14

Vehicle trim with weight forward.

3.5.3.2.1 Mission-related vehicle trim

Two examples of operational situations where ROV trim could be adjusted to assist in the completion of the mission are as follows:

1. If an ROV pilot requires the vertical viewing of a standpipe with a camera tilt that will not rotate through 90°, the vehicle may be trimmed to counter the lack in camera mobility (Figure 3.14).
2. If the vehicle is trimmed in a bow-low condition while performing a transect or a pipeline survey (Figure 3.15), when the thrusters are operated to move forward, the vehicle will tend to drive into the bottom, requiring vertical thrust (and stirring up silt in the process). The vehicle ballast could be moved aft to counter this condition.

3.5.3.2.2 Point of thrust/drag

Another critical variable in the vehicle control equation is the joint effect of both the point of net thrust (about the various axes) and the point of effective total drag.

The drag perspective will be considered first. One can start with the perfect drag for a hydrodynamic body (like an attack submarine) and then work toward some practical issue of manufacturing an ROV.

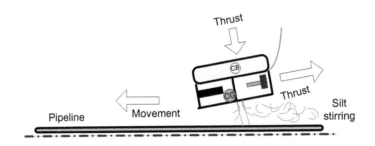

FIGURE 3.15

Movement down a pipeline with vehicle out of trim.

As stated best in Burcher and Rydill (1994), there are two basic types of drag with regard to all bodies:

1. *Skin friction drag*: Friction drag is created by the frictional forces acting between the skin and the water. The viscous shear drag of water flowing tangentially over the surface of the skin contributes to the resistance of the vehicle. Essentially this is related to the exposed surface area and the velocities over the skin. Hence, for a given volume of vehicle hull, it is desirable to reduce the surface area as much as possible. However, it is also important to retain a smooth surface, to avoid roughness and sharp discontinuities, and to have a slowly varying form so that no adverse pressure gradients are built up, which cause increased drag through separation of the flow from the vehicle's hull.

2. *Form drag*: A second effect of the viscous action of the vehicle's hull is to reduce the pressure recovery associated with nonviscous flow over a body in motion. Form drag is created as the water is moved outward to make room for the body and is a function of cross-sectional area and shape. In an ideal nonviscous flow, there is no resistance since, although there are pressure differences between the bow and stern of the vehicle, the net result is a zero force in the direction of motion. Due to the action of viscosity, there is reduction in the momentum of the flow and, while there is a pressure buildup over the bow of the submersible, the corresponding pressure recovery at the stern is reduced, resulting in a net resistance in the direction of motion. This form of drag can be minimized by slowly varying the sections over a long body, i.e., tending toward a needle-shaped body even though it would have a high surface-to-volume ratio.

As shown in Figure 3.16, there is an optimum aspect ratio whereby the total drag formed from both form drag and skin friction is minimized. Assuming a smoothly shaped contour forming a cylindrical hull, that aspect is somewhere in the range of a 6:1 aspect ratio (length-to-diameter ratio). The practicalities of building a cost-effective underwater vehicle (including the engineering headaches of procuring and forming constantly changing form factors) always get in the way of obtaining the perfect underwater design. Figure 3.17 shows the ideal submarine form and a slightly modified form factor popular in the defense industry. From this perfect form, the various aspects of the drag computation can be analyzed.

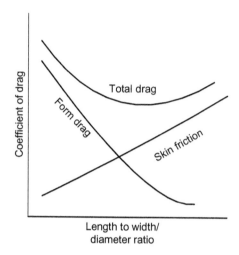

FIGURE 3.16

Vehicle drag curves.

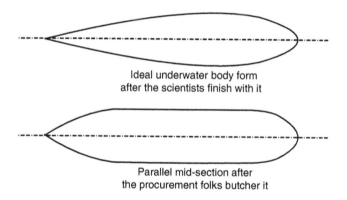

Ideal underwater body form
after the scientists finish with it

Parallel mid-section after
the procurement folks butcher it

FIGURE 3.17

Underwater vehicle body forms.

Skin friction drag

When considering the effect of the skin on the drag of the vehicle, the Reynolds number comes into play. Instead of starting with the technical description, let us take an example from the master himself—Theodore von Kármán:

Suppose that a book containing many pages is placed on a desk and the upper cover is slowly pushed parallel to the surface of the desk. The pages slide over each other, but the lower cover sticks to the desk. Similarly, fluid particles stick to the surface of a body, so that there is no slip between fluid and solid surface. Near the surface, however, the fluid velocity increases with the

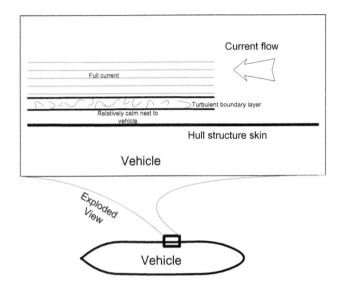

FIGURE 3.18

Ideal form with skin surface detail.

distance from the surface, i.e., it exhibits a certain gradient. The velocity gradient across the flow
produces friction between successive fluid layers which we call vicious friction.

The Reynolds number is a dynamic factor for fluid flow and comes into place for determining the flow characteristics around the vehicle (which directly affects the drag equation). The three modes of flow around a body are as follows:

1. laminar—smooth flow over the body
2. transient—approaching the critical Reynolds number where laminar transitions to turbulent
3. turbulent—disorganized flow over the body

With that simple introduction in place, a more analytical discussion can now be considered.

The drag dynamics of submerged vehicles were worked out during manned submersible research done in the 1970s by the office of the Oceanographer of the Navy. According to Busby (1976):

Skin friction is a function of the viscosity of the water. Its effects are exhibited in the adjacent,
thin layers of fluid in contact with the vehicle's surface, i.e., the boundary layer (Figure 3.18).
The boundary layer begins at the surface of the submersible where the water is at zero velocity
relative to the surface. The outer edge of the boundary layer is at water stream velocity.
Consequently, within this layer is the velocity gradient and shearing stresses produced between
the thin layers adjacent to each other. The skin friction drag is the result of stresses produced
within the boundary layer. Initial flow within the boundary layer is laminar (regular, continuous
movement of individual water particles in a specific direction) and then abruptly terminates into

a transition region where the flow is turbulent and the layer increases in thickness. To obtain high vehicle speed, the design must be toward retaining laminar flow as long as possible, for the drag in the laminar layer is much less than that within the turbulent layer.

An important factor determining the condition of flow about a body and the relative effect of fluid viscosity is the "Reynolds number." This number was evolved from the work of Englishman Osborne Reynolds in the 1880s. Reynolds observed laminar flow become abruptly turbulent when a particular value of the product of the distance along a tube and the velocity, divided by the viscosity, was reached. The Reynolds number expresses in nondimensional form a ratio between inertia forces and viscous forces on a particle, and the transition from the laminar to the turbulent area occurs at a certain critical Reynolds number value. This critical Reynolds number value is lowered by the effects of surface imperfections and regions of increasing pressure. In some circumstances, sufficient kinetic energy of the flow may be lost from the boundary layer such that the flow separates from the body and produces large pressure or form drag.

The Reynolds number effect of a fluid flowing around a cylinder (such as a tether being pulled against an oncoming current) is depicted at Figure 3.19 and can be calculated by the following formula:

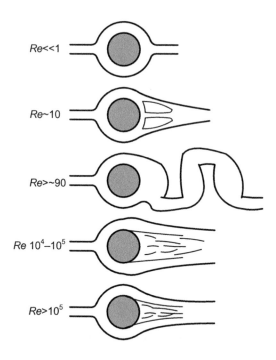

$Re \ll 1$

$Re \sim 10$

$Re > \sim 90$

$Re \ 10^4 - 10^5$

$Re > 10^5$

FIGURE 3.19

Water flow around a cylinder by Reynolds number as the flow increases.

$$Re = pVl/m = Vl/v,$$

where:

p = density of fluid (slugs/ft^3),
V = velocity of flow (ft/s),
m = coefficient of viscosity (lb-s/ft^2),
$v = m/p$ = kinematic viscosity (ft^2/s),
l = a characteristic length of the body (ft).

An additional factor is roughness of the body surface, which will increase frictional drag. Naval architects generally add a roughness-drag coefficient to the friction-drag coefficient value for average conditions.

Form drag

A variation on a standard dynamics equation can be used for ROV drag curve simulation. With an ROV, the two components causing typical drag to counter the vehicle's thruster output are the tether drag and the vehicle drag (Figure 3.20). The function of an ROV submersible is to push its hull and pull its tether to the work site in order to deliver whatever payload may be required at the work site. The only significant metric that matters in the motive performance of an ROV is the net thrust to net drag ratio. If that ratio is positive (i.e., net thrust exceeds net drag), the vehicle will make headway to the work site. If that ratio is negative, the vehicle becomes a very high-tech and very expensive boat anchor.

ROV thrusters must produce enough thrust to overcome the drag produced by the tether and the vehicle. The drag on the ROV system is a measurable quantity derived by hydrodynamic factors that include both vehicle and tether drag. The drag produced by the ROV is based upon the following formula:

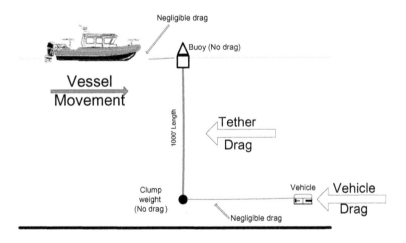

FIGURE 3.20

System drag components.

$$\text{Vehicle drag} = 1/2 \times \sigma A V^2 C_{\text{d}},$$

where

σ = density of seawater/gravitational acceleration, where density of seawater is 64 lb/ft^3 (1035 kg/m^3) and gravitational acceleration is 32.2 ft/s^2 (9.8 m/s^2);

A = characteristic area on which C_{d} (the drag coefficient) is non-dimensionalized. For an ROV, A is defined as the cross-sectional area of the front or the vehicle. In some cases, the ROV volume raised to the 2/3 power is used;

V = velocity in feet per second—1 knot = 1.689 ft/s = 0.51 m/s;

C_{d} = nondimensional drag coefficient. This ranges from 0.8 to 1 when based on the cross-sectional area of the vehicle.

Total drag of the system is equal to the vehicle drag plus the tether drag (Figure 3.21).

In the case of cables, the characteristic area, A, is the cable diameter in inches divided by 12, times the length perpendicular to the flow.

The C_{d} for cables ranges from 1.2 for unfaired cables, 0.5–0.6 for hair-faired cable, and 0.1–0.2 for faired cables (although ROV cables are not typically faired).

Since the ROV's tether is typically the highest drag item on the ROV system, an understanding of the concept of vortex shedding (Figure 3.22) around a cylindrical item (e.g., a tether

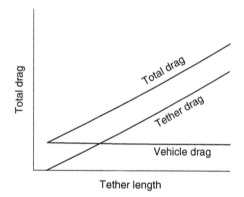

FIGURE 3.21

Component drag at constant speed.

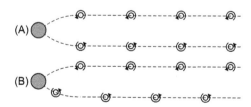

FIGURE 3.22

Vortex shedding (a) symmetrical and (b) asymmetrical.

perpendicular to the water flow) is warranted. This so-called Von Kármán shedding begins to appear with Reynolds numbers approaching 90 (Figure 3.19) and disappear once the Reynolds number approaches 10^4. Vortex shedding is a real problem as the flow speed increases. The turbulent vortex cells on either side of the cylinder can be either symmetrical (Figure 3.22(a)) or (much more often) asymmetrical (Figure 3.22(b)). As the flow increases, and the asymmetrical oscillations begin, the cylinder is pulled back and forth with the oscillations further exacerbating the drag and causing structural stresses. Engineers for large offshore structures mount strakes on the skin to inhibit the formation of these vortexes. ROV operators have the choice of either fairing the tether cable or accepting the problem.

Based on the above discussions, the total drag of the system is defined as:

$$\text{Total drag} = 1/2\sigma A_v V^2 C_{dv} + 1/2\sigma A_u V^2{}_u C_{du} \text{ (where v = vehicle, u = umbilical)}$$

A simple calculation can be performed if it is assumed that the umbilical cable is hanging straight down and that the tether from the end of the umbilical (via a clump or TMS) to the vehicle is horizontal with little drag (Figure 3.20). For this calculation, it will be assumed that the ship is stationary keeping in a 1 knot current (1.9 km/h) and the vehicle is working at a depth of 500 ft (152 m). The following system parameters will be used:

Unfaired umbilical diameter = 0.75 in. (1.9 cm)
A, the characteristic area of the vehicle = 10 ft^2 (0.93 m^2)

Based on the above, the following is obtained:

Vehicle drag = $1/2 \times 64/32.2 \times 10 \times (1.689)^2 \times 0.9 = 25.5$ lb (11.6 kg)
Umbilical drag = $1/2 \times 64/32.2 \times (0.75/12 \times 500) \times (1.689)^2 \times 1.2 = 106.3$ lb (48.2 kg)

Note: Computations will be the same in both imperial and metric if the units are kept consistent. This simple example shows why improvements in vehicle geometry do not make significant changes to system performance. *The highest factor affecting ROV performance is tether drag.* The following discussion will consider the drag of individual components.

Drag computations for the vehicle assume a perfectly closed frame box. Drag computations for the tether are in the range of a cross-section of ROV systems sampled during recent field trials of small observation-class systems.

By varying the tether diameter, the relationships in Figure 3.23 can be developed. Figure 3.24 shows that by varying the speed with a constant length of tether, the vehicle will display a similar curve, producing a drag curve that is proportional to velocity squared.

The power required to propel an ROV is calculated by multiplying the drag and the velocity as follows:

$$\text{Power} = \text{Drag} \times V/550$$

The constant 550 is a conversion factor that changes foot-pounds/second to horsepower. As discussed previously, the drag of a vehicle is proportional to the velocity of the vehicle squared.

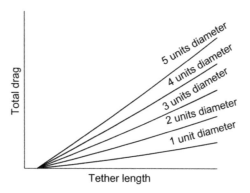

FIGURE 3.23

Linear tether drag at constant speed with varying diameter.

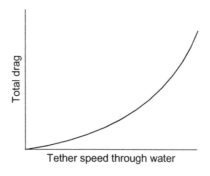

FIGURE 3.24

Linear tether drag at varying speed with constant diameter.

Accordingly, the propulsion power used is proportional to the velocity cubed. To increase the forward velocity by 50%, for example, from 2 knots to 3 knots, the power increases by $(3/2)^3$, or $(1.5)^3$, which is 3.4 times more power. To double the speed, the power increases by $(2)^3$ or eight times. Increased speed requirements have a severe impact on vehicle design.

Table 3.5 lists some observation-class systems tested during US Coast Guard procedure trials (without specific names and using figures within each vehicle manufacturer's sales literature) with their accompanying dimensions.

At a given current velocity (i.e., 1 knot), the drag can be varied (by increasing the tether length) until the maximum thrust is equal to the total system drag. That point is the maximum tether length for that speed that the vehicle will remain on station in the current. Any more tether in the water (i.e., more drag) will result in the vehicle losing way against the current. Eventually (when the end

Table 3.5 Specifications of ROVs Evaluated

System and Parameter	Large ROV A	Small ROV A	Small ROV B	Large ROV B	Small ROV C	Medium ROV A
Depth rating (ft)	500	330	500	1150	500	1000
Length (in.)	24	10	14	39	21	18.6
Width (in.)	15	7	9	18	9.65	14
Height (in.)	10	6	8	18	10	14
Weight in air (lb)	39	4	8	70	24	40
Number of thrusters	4	3	3	4	4	4
Lateral thruster	Yes	No	No	Yes	Yes	No[a]
Approximate thrust (lb)	25	2	5	23	9	12
Tether diameter (in.)	0.52	0.12	0.44	0.65	0.30	0.35
Rear camera	No	No	Yes	No	No	No
Side camera	No	No	No	Yes	No	No
Generator req. (kW)	3	1	1	3	1	3

[a]*Medium ROV A possesses lateral thrusting capabilities due to offset of vertical thrusters.*

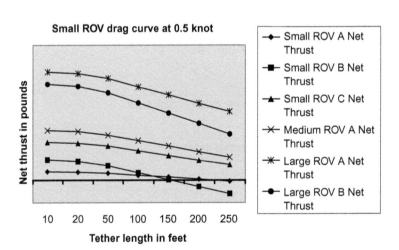

FIGURE 3.25

Drag curves of systems tested at 0.5 knot.

of the tether is reached), the form drag will turn the vehicle around, causing the submersible to become the high-tech equivalent of a sea anchor.

The charts in Figures 3.25–3.28 show the approximate net thrust (positive forward thrust versus total system drag) curves at 0.5, 1, 1.5, and 2.0 knots for the ROVs described in Table 3.5.

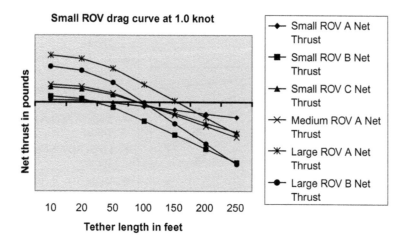

FIGURE 3.26

Drag curves of systems tested at 1.0 knot.

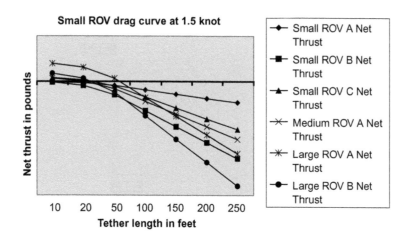

FIGURE 3.27

Drag curves of systems tested at 1.5 knots.

The net thrust is shown with the horizontal line representing zero net thrust. All points below the zero thrust line are negative net thrust, causing the vehicle to lose headway against an oncoming current. Note: These tether lengths represent theoretical cross-section drag for a length of tether perfectly perpendicular to the oncoming water (Figure 3.20). Vehicle drag assumes a perfectly closed box frame with the dimensions from Table 3.5 for the respective system.

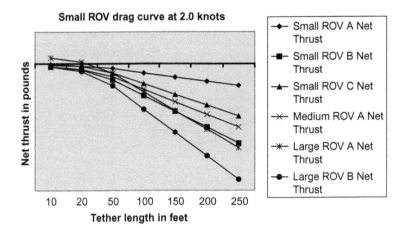

FIGURE 3.28

Drag curves of systems tested at 2.0 knots.

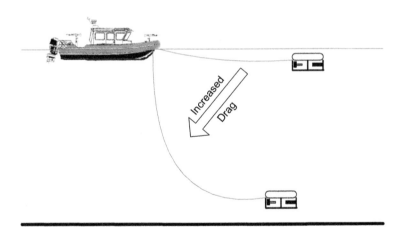

FIGURE 3.29

Tether profile drag increases to the perpendicular point.

With a tether in a perfectly streamlined configuration (i.e., tether following directly behind the vehicle), the tether drag profile changes and is significantly reduced (Figure 3.29).

The obvious message from this data is that the tether drag on the vehicle is the largest factor in ROV deployment and usage. The higher the thrust-to-drag ratio and power available, the better the submersible pulls its tether to the work site.

3.6 **Standards and specifications**

"And God said, Let there be light: and there was light" (Book of Genesis). "And light became a standard" (Book of Bob). Once light was chosen as a standard, sensors (e.g., eyes) could be designed to a common protocol (e.g., polarity, focus, orientation) and tuned to a common wavelength spectrum (for humans, 380−740 nm). Other beings could similarly make use of the electromagnetic spectrum wavelengths for other sensors (IR/UV) as well as other electromagnetic measures for use in their survival. The ones that effectively made use of the proper wavelengths and protocols survived and thrived while the beings using the wrong parameters either died or were consumed by the entities more suited for survival. It is the *exact* same measure for a standard—and evolution of standards lives on (nods to Mr. Charles Darwin . . .).

Standards are necessary in practically all cooperative fields so that everyone knows the rules and can project the future for investment of time and resources with the knowledge that the other participants in the field will play along. The industry players cooperate in an additive fashion so as to achieve synergy in the development of any technology or endeavor. Standards are imperative—imagine a design team for an aircraft where one half of the engineers used the metric system and the other imperial units. "Let's play football!" One team arrives with a soccer ball, the second for a rugby game, while the third shows up in helmet and pads (US gridiron style).

The word "standard" is described in the Oxford dictionary as, "used or accepted as normal or average; (of a size, measure, design, etc.) such as is regularly used or produced (not special or exceptional); (of a work, repertoire, or writer) viewed as authoritative or of permanent value and so widely read or performed." Further, a *per se* standard is only as good as the standards organization holding it. A standard held by a small local club or company is unlikely to gain wide regional acceptance while some governmental standards (e.g., military specifications or "MILSPEC") are unlikely to be adapted by civilian organizations (although some MILSPEC standards are widely used in civilian applications).

In the subsea industry, there are few organizations that are large enough to drive a standard. On the military side, those services that concentrate in the subsea world typically do not share their secrets (thus killing any would-be standards). On the civilian side, the largest application is with minerals extraction—currently dominated by oil and gas exploration and production. The two oil and gas industry-specific organizations promulgating standards are the American Petroleum Industry (API) and the International Standards Organization (ISO).

There is an interesting anecdote on standards evolution. In 2005, the Program Executive Officer of Littoral and Mine Warfare of the US Navy had a problem. He was tasked with bringing about the development of the Littoral Combat Ship's new integrated robotic fighting platform, but there were few established standards with which all of the disparate systems could communicate (one specifically was the Mission Reconfigurable Unmanned Undersea Vehicle, or "MRUUV"). He was faced with an extremely tight development timeline and realized how slowly the military develops standards. So he did a nontraditional move—he threw the standard to the civilian side to develop. He lost the benefit of control but gained the benefit of rapidity and reduced cost. As a very credible sponsoring agency, the US Navy chose the American Society for Testing and Materials (ASTM) to hold the standard for the MRUUV (ASTM Committee F41—later expanded its coverage from its roots of the UUV to Unmanned Surface Vehicle (USV)). The Navy expected other manufacturers to build to the new F41 standard (so as to reduce costs and widen availability through a larger installed user base), but the civilian world has largely passed this standard by in favor of other standards. Why? The reason this is such an interesting story is that in Chapter 23, we will examine some of the future trends in the subsea world. We may see this standard come full circle and either evolve or merge into more accepted civilian API or ISO standards as the subsea world further develops.

As with its natural counterpart, industrial standards start out with a need, then evolve, grow, change, die, are born again, then merge into other standards. A typical standards development process starts out with an industry group meeting with a common problem. The group collectively cooperates to come forth with an agreed standard adapted by the group. The standard will only survive if it is widely accepted within the group. If it favors a particular segment over another, the blighted organization will simply refuse to adapt and the whole standard will die. If the standard is provisionally accepted then it must evolve to meet the changing technology (or, again, it dies). An interesting example of this is the VHS versus Betamax videotape format war—VHS won in the short term but it eventually died in favor of DVD (which is rapidly being overtaken by Blu-Ray—which will probably be gobbled up by another standard later). There are various types of standards including:

1. *Consensus standards*—agreed within a group (e.g., "we will adapt the Roberts Rules of Order while voting"—interesting because it is adapting as standard someone else's standard)
2. De facto *standards*—accepted via fact due to market dominance (can anyone say "Windows Operating System?"); also, the *de facto* standard for subsea fiber-optic multiplexers is Focal Technologies and if one wants a manipulator system, the default is Schilling (for large hydraulic vehicles) or Hydro-Lek (for mid-sized electric vehicles)
3. *Geographical standards*—accepted within a region (e.g., (in England) "we will drive on the left-hand side of the road")
4. *Open standards*—publically available (e.g., metric/imperial)
5. *Regulatory standards*—governmental laws promulgated whereby industry is legally required to comply
6. *Standard of care*—legal standards (while intending to be objective) generally leave a considerable margin for subjective bias and interpretation.

In Chapter 13, some of the communications standards currently on the market will be examined and in Chapter 19 API/ISO common mechanical interface standards will be covered.

An ROV involves many types of technologies, including materials, data communications, video, hardware, software, and a host of other standardized items. These standards are typically listed and referenced within their category or usage. A short, yet representative, list of common ROV standards (by category) is provided in Table 3.6.

The last subject for discussion within standards is one that is definitely not germane only to the ROV industry. It is the eternal battle between the equipment manufacturer and the equipment operator.

The economic incentive of the manufacturer is to make equipment that is completely proprietary so that not only does the manufacturer make a profit once the product is sold, but continues to milk an ongoing revenue stream from that sale through replacement parts sales and services along with engineering services for any modifications needed to the system for the entire service life of the equipment. The incentive of the service provider is for a completely open architecture with freely available replacement parts at the lowest price/highest quality point resulting in many different avenues for sourcing replacement parts.

Table 3.6 Common ROV Standards (by Category)

Category	Representative Standard	Comment
General	MILSPEC (Military Specification)	US military specifications jointly adapted by the US Department of Defense (merged from the various services' standards)
	ISO (International Organization for Standards)	Internationally recognized standards organization covering a wide range of standards—including hydraulics
Hydraulic fittings	JIC (Joint Industry Council) fittings	Merged SAE J514 and MIL-F-18866 standards which evolved from AN (Army-Navy) standards
	SAE (SAE International—f/k/a Society of Automotive Engineers)	US-based professional association and international organization for promulgating standards for the automotive industry
	BSP (British Standard Pipe) fitting	UK Standard adapted internationally
	JIS (Japanese Industrial Standard)	Japanese standards organization
	Parker (Parker Hannifin Corporation)	Market-dominant hydraulic components manufacturer adapting many of the above standards (and does a few of its own)
Video	NTSC (National Television Standards Committee)	Standards organization for television formed in the 1940s in the United States
	PAL (Phase Alternating Lines)	Proprietary standard eventually adapted by CEN (European Committee for Standardization) then widely adapted internationally
Communications	Ethernet	IEEE (Institute of Electrical and Electronics Engineers) standard 802.3 (later adapted by ISO)
	RS-232	Electronic Industries Association (EIA) recommended standard RS-232 (EIA ceased operation in 2011 and the standard rolled into other organizations—it is a process...)

This is a case of opposite incentives. If the manufacturer loses the follow-on revenue stream, he or she may not recover the engineering investment for a low-quantity production run (typical of an ROV manufacturer) and thus may go out of business (it happens all too often). If the service company has no other choice for sourcing replacement parts, due to a single-source provider, the service company has the difficult choice once replacement parts are needed of either accepting whatever terms are provided from the manufacturer or to reverse engineer and manufacture the part (putting the service company into a position where it must then be a manufacturer in addition to its service offerings).

The conversation goes something like this:

Open Architecture	Closed Architecture
Manufacturer: The price for the part is $1000.	*Manufacturer:* The price for that part is $1000.
Customer: I can get it cheaper from China.	*Customer:* Whatever. I must have that part!
Manufacturer: But that is my cost.	*Manufacturer:* But you still owe for invoice x.
Customer: That is not my problem.	*Customer:* That invoice is in dispute due to quality issues and late delivery.
	Manufacturer: Well, then I won't ship you the part until you pay invoice x.
	Customer: The system you sold me is down and costing me $25,000 per day!
	Manufacturer: That is not my problem.

There is a fair balance between these two extremes. But the ROV operator is wise to choose his manufacturer carefully. The ROV manufacturer is wise to strike a balance between open and proprietary architecture—and, of course, to be fair to its customers.

CHAPTER CONTENTS

In this chapter, the overall levels of control for operated a remotely operated vehicle (ROV) from remote control (RC) will be examined, all the way through full logic-driven operations with goal orientation. The end of the chapter will address some of the newest schemes for vehicle simulation as well as scenario mimicking for rehearsing field operations before deployment.

The discussion of vehicle control levels will begin with a metaphor on neutralizing a sea mine (which both of the authors loathe) that we will generally term "Mr. Nasty."

1. *Direct control*: "I am driving my manned submersible in the harbor looking for sea mines when I spot *Mr. Nasty* protruding from the bottom within the sea lane. I directly operate the fire control button. *RIP Mr. Nasty [and hopefully <u>not</u> yourself!]*."

2. *Remote control (RC)*: "I am standing on a hill overlooking the sea lanes while operating the joystick linked to the thrusters of my ROV. I sense and control vehicle movement through line of sight (i.e., from my own point of view) as I spot *Mr. Nasty* floating near the surface. I direct my ROV [with contact explosive charge mounted to the vehicle's nose] to collide directly into the mine. *RIP Mr. Nasty [and vehicle]*."

3. *Teleoperation*: "I am sitting in my air conditioned control room in some remote location sipping a soda while operating a joystick linked to the thrusters of my ROV. I am viewing the undersea world through a camera on the ROV's nose (i.e., controlling from the vehicle's point of view). I see a sea mine and operate the vehicle's fire control system to shoot a projectile into the mine. *RIP Mr. Nasty*."

The ROV Manual.

93

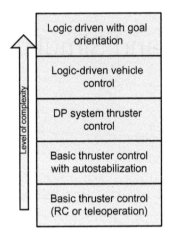

FIGURE 4.1

Vehicle control from simple thruster control to full logic control with goal orientation.

4. *Logic driven*: "I am sitting in that same control room observing the ROV automatically navigating a series of waypoints while running a search pattern when I observe *Mr. Nasty* anchored to the bottom. The ROV automatically operates [or I do the deed remotely] the fire control system. *RIP Mr. Nasty.*"

5. *Logic driven with goal orientation*: "I upload goal instructions (i.e., *Find and neutralize Mr. Nasty.*) for layering on top of the basic thruster controls, then launch my ROV [or AUV] into the morning waters of the harbor. I am at the pub in the evening sipping a beer when I receive a text from the ROV/AUV saying, *RIP Mr. Nasty.*"

As one of the basic premises of ROV control is the "R" (meaning "Remotely"), direct control will not be examined. RC mode normally occurs at the surface while the operator is directing (typically) an ROV to a location whereby visual contact is made using the vehicle's camera for transitioning from RC mode to teleoperation mode. Practically all ROV operations are performed via teleoperation (e.g., visually, acoustically, magnetically, etc.—in other words, any sensor feedback from the vehicle may be used for the *sense* portion—sense/plan/act). New vehicles, however, layer on levels of automation from basic auto-heading/depth/altitude control to full dynamic positioning (DP) with waypoint navigation. The top level of Logic Driven with Goal Orientation is addressed in more detail in Chapter 23 as this method of control is clearly where the technology is destined. Figure 4.1 depicts the levels of vehicle control from simple to complex.

4.1 Vehicle control

In this section, the levels of vehicle control will be examined, ranging from the basic joystick control of thrusters for vehicle mobility to full logic-driven vehicle management. There are many factors driving control decision making including (but hardly limited to) schedule, cost, environment, operator experience, and just plain lack of knowledge of the situation. This subject is so broad that a small chapter cannot possibly do it justice; therefore, the discussion will deal with some basic concepts and allow the reader to further research this topic separately.

4.1.1 Basic thruster control

As discussed further in Chapter 6, basic ROV mobility is achieved through control of thrusters to reactively vector fluid for vehicle movement. In the early days of ROVs, each electrical thruster motor (driving a propeller) was controlled individually via a rheostat linked directly to the thruster. This made for a dreadfully difficult control regime! As the technology evolved, control mechanisms arose allowing for a more intuitive human—machine interface. Later iterations gave rise to the joystick for scaling thruster output, thus allowing thrust vectoring and, hence, finer control of vehicle movement. Logic drive can overlie (or replace) the direct drive of a joystick (which will be examined later in this section). The basic paradigm of "sense, plan, and act" works for a human as well as a computer. As computer control of machines evolves, the "Human-in-the-Loop" control will continually be replaced with automation.

For direct human control of thrusters, a joystick is normally used. The joystick typically outputs a pulse width modulation signal (or some other electrical scaling signal) to direct motor output in some linear fashion. That signal can control power output to either an electric motor (for electrical thrusters) or a hydraulic valve pack/servo (for hydraulic thrusters).

The joystick control matrix can also significantly affect the ease of control over the smaller vehicle (Figures 4.2 and 4.3). For example, during a small turning adjustment, such as 20°, the thrusters may ramp up power so quickly that the operator cannot stop the turn until after reaching the 90° rotation point. Effective control of the vehicle would be lost. As the size of the vehicle

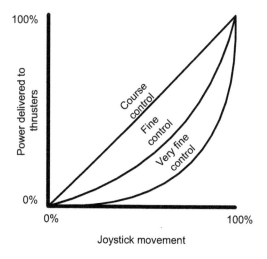

FIGURE 4.2

Joystick control matrix variation based upon varying power delivery versus joystick position.

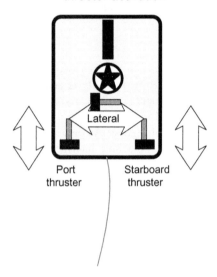

Joystick position versus thruster activation

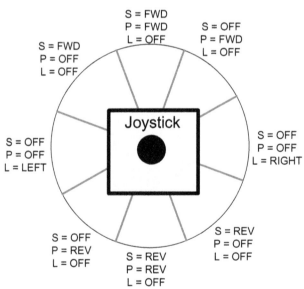

FIGURE 4.3

Joystick position versus thruster activation on a four-thruster configuration.

increases, fine thruster control becomes less critical due to the amount of mass requiring movement (larger mass vehicles react much more slowly to thruster output due to simple inertia).

Vehicle manufacturers use a variety of techniques for gross versus fine control of vehicle movement to conform to the operating environment. One vehicle manufacturer makes use of a horizontal and vertical gain setting to vary the power versus joystick position. Another vehicle manufacturer allows for variable power delivery scaling via the controller's software. The reason this power scaling is necessary is that when towing the tether and vehicle combination to the work site, the full power complement is needed for the muscling operation. Once at the work site, finer adjustments are needed to ease the ROV into and out of tight places. If the power were set to full gain in a confined area, a quarter joystick movement could over-ramp the power so quickly that the vehicle could ram into a wall, damaging the equipment and causing some embarrassing personnel reviews.

Unlike underwater vehicles built for high speed (examples of high-speed underwater vehicles are a torpedo or a nuclear attack submarine), most ROV submersibles are designed for speeds no greater than 3 knots. In fact, somewhere in the speed range of 6−8 knots for underwater vehicles, interesting hydrodynamic forces act upon the system, which require strong design and engineering considerations that address drag and control issues. At higher speeds, small imperfections in vehicle ballasting and trim propagate to larger forces that simple thruster input may not overcome. As an anecdote to unexpected consequences for high-speed underwater travel, during trials for the USS *Albacore* (AGSS 569), it was noted with some surprise (especially by the Commanding Officer) that the submarine snap rolled in the direction of the turn during high-speed maneuvering!

For the ROV at higher speeds, any thruster that thrusts on a plane perpendicular to the relative water flow will have the net vector of the thrust reaction move in the direction of the water flow (Figure 4.4). Also, one current vehicle manufacturer makes use of vectored thrust for vehicle control, which mitigates the thrust vector problem at higher speeds.

At what point is control over the vehicle lost? The answer to that is quite simple. The loss of control happens when the vehicle's thrusters can no longer counter the forces acting upon the vehicle while performing a given task. Once the hydrodynamic forces exceed the thruster's ability to counter these forces (on any given plane), control is lost. One of the variables must be changed in order to regain control.

4.1.2 Autostabilization

With sensor feedback fed into the vehicle control module, any number of parameters may be used in vehicle control through a system of closed-loop control routines. Just as dogs follow a scent to its source, ROVs can use sensor input for positive navigation. Advances are currently being made for tracking chemical plumes from environmental hazards or chemical spills (although this is considered a higher, logic-driven control level). A much simpler version of this technique is the rudimentary auto-depth/altitude/heading.

Auto-depth is easily maintained through input from the vehicle's pressure-sensitive depth transducer. Auto-altitude is equally simple, but the vehicle manufacturer is seldom the same company as the sensor manufacturer (causing some issues with communication standards and protocols between sensor and vehicle). The most common compass modules used in observation-class ROV systems are the inexpensive flux gate type. These flux gate type compasses have a sampling rate (while accurate) slower than the yaw swing rate of most small vehicles, which cause the vehicle to "chase

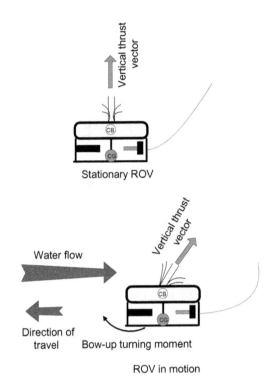

FIGURE 4.4

Apparent thrust vector change due to water flow across vehicle.

the heading." Flux gate auto-heading is better than no auto-heading, but several manufacturers of small systems have countered this "heading chase" problem by using a gyro.

As described in Chapter 11, gyros for ROVs come in two basic types, the slaved gyro and the rate gyro. The slaved gyro samples the magnetic compass to slave the gyro periodically to correspond with its magnetic counterpart. Since the auto-heading function of an ROV is simply a heading hold function, some manufacturers have gotten away with using a simple rate gyro for "heading stabilization." When the heading hold function is slaved to a gyro only, sensing a turn away from the initial setting and a rate at which the turn is progressing (i.e., the rate gyro has no reference to any magnetic heading), the vehicle is then only referenced to a given direction. Hence the term "heading stabilization" due to the lack of any reference to a specific compass direction.

4.1.3 ROV dynamic positioning

The first DP system to gain wide acceptance with ROV manufacturers and operators was produced by UK-based SeeByte Ltd. As the technology gained acceptance, the various ROV original equipment manufacturers (OEMs) have gotten into the act to produce their own proprietary systems. Now, DP systems proliferate all the way down to the smallest of vehicles (Figure 4.5).

FIGURE 4.5

MicroROV with integrated SeeByte DP system.

(Courtesy VideoRay.)

The basis for an ROV DP system involves the data fusion between various navigational sensors to hold position as well as move along some user-specified track. As more fully explained in Chapter 18, various survey-grade sensors allow for positioning relative to some nominal parameter. For instance, a Doppler Velocity Log is able to sense movement across the bottom once bottom lock is gained. Fuse this data with a motion reference unit (basically, an inertial navigation system combined with inclinometers to sense motion relative to absolute coordinates), a gyro, North-seeking compass, and acoustic positioning, and the operator is then able to precisely track the vehicle. Once the vehicle is in a known position, any variation from that position can be sensed, thus determining what thrust vector is required to maintain the desired orientation. The instructions are then sent to the thrusters to maintain that specified orientation. The latest developments in this field have also seen the introduction of DP systems that use a multibeam imaging sonar to keep station relative to objects within the environment.

This is an amazing enabling technology that relieves the operator from the stress of constant inputs to keep a vehicle in the proper location and orientation while performing a precise intervention or sensing task.

4.1.4 Logic-driven control

Layer on top of a DP system some higher level of control (e.g., waypoint navigation fused with some other operational task for sensing or physical intervention) and you have a high-level logic-driven circuit. As ROVs are typically operated with more of a "Human-in-the-Loop" paradigm, the higher-level logic-driven controls are left to its AUV brethren.

4.1.5 Logic drive with goal orientation

Take the logic-driven control and lay on top of this a goal-oriented mission planning instruction set and you have a fully autonomous control system able to make onboard decisions toward some assigned goal. This is the ultimate goal of roboticists worldwide, whereby machines will eventually be able to emulate man (or woman) toward a fully compliant robotic system. As inventor, author, and futurist Ray Kurzweil puts it, "The Singularity Is Near."

4.2 Simulation

Initially conceived purely as a pilot training tool, ROV simulator system technology has advanced and permeated offshore operations. These integrated software systems are used to great effect by militaries, ROV manufacturers, operators, and oil companies for the as-built field visualization, mission planning, dive plan development/evaluation as well as subsea equipment design and testing for ROV missions prior to field deployment. This includes ROV accessibility, engineering design and analysis, hazard identification, and clash detection. It also assists in producing interactive procedure visualizations (to complement the dive plan) and 2D field layout documentation. When field and equipment design is completed (and verified), the simulator comes full circle to be employed in equipment-specific along with mission- and/or site-specific ROV pilot training.

The initial ambition of ROV simulation is to enhance pilot training by system and procedural familiarization. The trainee gains skills for "*knobology*" (the location and purposes of the pilot interface or console controls) and in developing a sense of 3D awareness. Tether management skills, as the ROV is piloted through a virtual environment, help the student pilot learn in a controlled, structured, and low-stress environment.

Simulators were designed to efficiently assist in the creation of ROV pilot capability and confidence toward expediting the pilot's progress from trainee (company financial burden) to field technician (revenue producing). This is done by providing stick-time or hands-on experience without tying up expensive field equipment, creating (expensive) wear and tear, and/or risking ROV systems that could otherwise be employed on jobs offshore.

Starting as early as 1990, substantial development efforts in the area of ROV simulation (most notably by Imetrix) utilized first-generation graphical work stations (such as those produced by Silicon Graphics) to support the processing required for the complex hydrodynamic and collision response solutions of an ROV and its tether.

A turning point came in 1999 with the release of *VROV 1: Pilot Training*, which featured a new approach of networked PCs featuring distributed processing of simulated camera graphics, sonar data, controls interface, and a dynamics module (DM) that included hydrodynamic modeling. This package provided for multiple modeled parameters including sea state, vehicle/tether drag, and buoyancy, along with hydrodynamic friction/drag of an ROV umbilical/TMS combination (with a 100-m interactive tether). Two years later, *VROV 2: Mission Planning and Rehearsal* added to the system's capability with multiple dynamic objects (including two ROV systems) and fully interactive manipulators. This fully integrated model has defined the role of ROV simulation since that time.

Uses and capabilities of ROV simulation have been in a constant feedback loop wherein higher resolution and stabler dynamics capabilities have been required, and in turn enabled, more complicated and higher-level interactive applications in systems design and analysis.

Simulation is now a common element of the design and testing of the hardware, operations, and procedures related to complex interactive subsea structures, tooling, vehicles, and control software.

4.2.1 Enter the ROV simulator

In general, ROV simulators apply virtual reality and computer modeling technologies to generate realistic, scenario-based ROV simulations. Simulation (to a greater or lesser extent) models standard components of a real-life ROV system (e.g., acoustic imaging, obstacle avoidance, unit tracking, underwater cameras and lights, manipulator arms, and environmental factors such as currents, variable turbidity, and tether effects) in the context of a real-time interactive operation.

Most currently available commercial systems are capable of being implemented for particular ROVs, utilizing distributed processing that uses networked CPUs (central processing units) to operate a range of modules. These relate to the simulation of systems supported by the ROV system. WCROV simulators involve the simulation of a larger number of subsystems and interfaced controls, while OCROV systems can vary greatly from minimal onboard systems to such complex multiple systems that would make some WCROV manufacturers envious.

Since 1999, simulator designers have favored various modules that performed different elements of the simulation—each running on a dedicated CPU. Currently, CPUs and GPUs (graphics processing units) will run multiple modules described in the following sections from a single desktop or laptop. Not all simulators will have the same numbers and types of modules but will consist of some combination thereof.

The instructor control module

The instructor control module (ICM) acts as the simulator's "referee" computer and also as the control center for the simulator. The instructor can monitor the trainee's progress while performing a mission in real time. He or she may then initiate various commands that change the operational status of the ROV and peripheral equipment, as well as modify undersea environmental parameters.

The ICM enables the instructor to have total control of the VROV's subsystems and its environment during each training session. It will normally feature a "God's-Eye view" (Figure 4.6) of the entire scene including an external rendering of the ROV for the instructor's viewing as it proceeds through a mission. It can also provide split screens that include access to the displays that the ROV pilot and navigator would have. The ICM capabilities vary by manufacturer but would normally include (but are not limited to) the following:

1. *Selective degradation or disabling of vehicle components* to simulate subsystem failures
2. *Control over environmental parameters* such as ambient lighting, current direction/velocity, and water turbidity (muddy/murkiness)
3. *Recording and playback* training session

The console interface module

The console interface module (CIM) executes a serial communications program that interacts with an OEM ROV operator console. The CIM's data stream can mimic that of the ROV and essentially becomes the ROV during execution of ROV training simulations. Sometimes simulators use replicas of OEM console equipment where the Pilot Program and Interface are combined. Pilot screen overlays are either provided through the OEM console equipment (or graphical user interface) or can be replicated in the simulator software.

When used for prototype vehicle and control system development, the combination of precision dynamic configuration and complete control software integration provides the opportunity for verification and fine tuning of vehicle design and control system functionality in a virtual environment.

FIGURE 4.6

VROV with split screens including "God's eye" 3D external view, forward camera, and multibeam sonar displays.

(Courtesy Soil Machine Dynamics Ltd. (SMD) and GRI Simulations Inc.)

This creates tremendous opportunities for savings in iterations of hardware fabrication and wet testing. This can be as simple as premission checking for the effectiveness of different types or orientations of thrusters, tooling skid geometries, or the testing of control loops (e.g., auto-heading and auto-depth) to complete autonomous vehicle mission control.

Acoustic positioning and sonar modules

In most simulation systems, acoustic positioning and sonar modules are notified of the position and orientation of the platform (ROV/TMS). Then, based upon this simulated acoustic data, the software either sends this data to be processed by an OEM topside module or processes and then displays simulated sonar images to mimic typical ROV acoustic instrumentation. Other systems use the ROV's position to perform image manipulation to create a similar effect.

Camera modules

Similarly, camera modules generate realistic underwater imagery to simulate the video cameras mounted on an ROV and/or cage/TMS combination—again, based upon the position and orientation created by the DM and the console inputs such as pan/tilt, zoom, and focus. Each camera view can be configured to match the desired field of view, color, and distortion.

The camera module can provide adjustable levels of visibility and depth fogging, automatically (or on user-adjusted command) from the ICM, to simulate the various visibility conditions of the subsea environment. Most systems support multiple camera outputs and split screen to simulate multiple camera views that may simultaneously be available on the ROV system.

Dynamics module

A DM is dedicated to computing the motion of the ROV and other scenario objects. The vehicle's reaction to various programmed parameters will simulate motion according to operator-commanded thruster forces along with collisions with obstacles and environmental forces acting upon the vehicle and components (e.g., body, manipulators, umbilical cable, and other appendages). It broadcasts object positions and orientations to the display modules along with forces to integrated engineering applications. These outputs are recorded for use in the pilot performance evaluation. All objects are provided with a dynamics configuration file that contains all of the relevant physical parameters having impact upon the movement of (and interaction between) scenario objects.

The lack of visual cues in the camera displays due to dirt kick-up (as the result of (or punishment for) hitting the seafloor) or engaging thrusters while too close to the bottom can result in sloppy piloting. How well a simulator performs as a generic or mission-specific training (and particularly as an engineering support) tool is predominantly dependent in its dynamics engine—the heart of any simulator system. Off the shelf hardware (particularly video cards) can enable any system to provide good effects with a high refresh rate for a large number of polygons (which are the building blocks that 3D models are made of).

It is the DM that processes and translates operator control commands into forces in the context of the objects and environment around the ROV system. It is the complexity of the models (as well as the speed/accuracy with which the module processes this) that determines the realism of the simulation. Thus, the effectiveness with which the module is able to provide positive training and reliable verification of operator performance may be measured for employee evaluation and mission achievability. The DM performs its task behind a cloak of secrecy masked by the video (and sonar) displays. The next section provides a look behind the curtain at this most critical of elements.

4.2.2 Physics simulation

A core requirement of an ROV simulation provides that after every step of the simulation, all of the components of that simulation (i.e., the objects being simulated) are in plausible locations and orientations. To maximize the possible range of scenarios developed (thus reducing developer workload), a stand-alone physics package is often utilized for handling position and orientation updates. These scenarios are then compounded into a single world transform (Figure 4.7). This package, commonly referred to as a "physics engine," is software that provides an approximate simulation of certain physical systems such as rigid body dynamics (including collision detection), soft body dynamics, and fluid dynamics. These are extensively used in the fields of computer graphics, video gaming, cinema special effects, and high-performance scientific simulation. While their main uses have been in the gaming industry (typically as middleware), they have come to be increasingly relied upon by various engineering disciplines. For example, the European Space Agency uses the PhysX library for verification of the Mars sample rover for the ExoMars Program tentatively slated to investigate the Martian environment.

From a programmer's point of view, the physics engine handles all calculations required to update every object's transform after each simulation step (Figure 4.8). This includes integrating

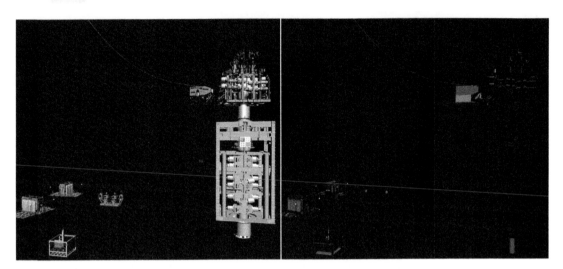

FIGURE 4.7

ICM views of high-resolution visual (left) and collision (right) models ensure high fidelity for critical planning and operations training.

(Courtesy Oil Spill Response Ltd., The Subsea Well Response Project, Trendsetter Engineering Inc., and GRI Simulations Inc.)

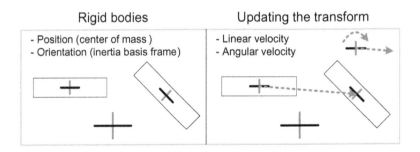

FIGURE 4.8

Computation methods for a simulated object's kinematics.

any user-supplied forces (along with gravitational forces), performing a collision detection step while generating contact points, resolving any contacts and any jointed elements with the previously predicted motion, and finally integrating the result into a final transformation for each object.

The image in Figure 4.9 goes into further details about the process, but the key point that should be stressed from an ROV simulation point of view is that the physics engine only supplies the gravitational force. All other forces must still be computed outside of the physics engine and then applied to each body as required. This means that phenomena such as buoyancy, added mass, drag, and current must still be derived and implemented by the ROV simulation programmer (though they may make use of values (such as velocity or inertial) computed by the engine tensors).

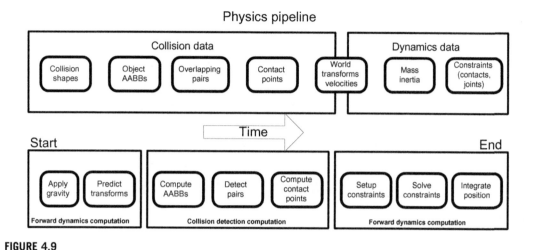

FIGURE 4.9

Physics engine logic progression for predicting, modeling, and rendering simulated objects.

Proper joint configuration and constraint solver parameters are essential as well in order to simulate believable manipulators and cables. Complex scenario logic (including triggers) is implemented in a similar fashion. Collision models that have been tagged as sensors can report when they have collided with their corresponding target sensors. But any resulting actions must be specified by the programmer or scenario developer (Figure 4.9). Also, in certain situations the developer may wish to override the physics engine and manually position objects or joints (such as in a positionally controlled mechanism).

When considered from this perspective, the physics engine is simply one of many tools that aid developers and programmers in creating realistic and plausible simulations. It is not a "magic box" that can instantly provide a robust ROV simulator without configuration or effort. Much work has to be done to determine and generate the correct forces for each object. Scenarios often must be optimized after initial design to remove excess contact points and redundant calculations. Decisions about certain trade-offs must be made such as the choice between using an iterative constraint solver (that is faster) or a direct solver (that is more accurate). Utilized properly, the physics engine can enable a faster scenario development process, thus reducing the amount of divergence from real-world behavior. This increases the resiliency of the simulation while permitting the level of flexibility developers require in order to implement convincing scenarios.

4.2.3 The future

The future of ROV simulation is here. Engineering analysis tools are bringing together various relevant technologies to enhance realism. Developers see the interactivity of critical elements as inputs to real-time human-in-the-loop systems such as ROVs. These allow for a method to visualize technologies in context (Figure 4.10).

FIGURE 4.10

Pipeline repair system simulation with real-time finite element analysis and soil modeling.

(Courtesy GRI Simulations Inc. and Dynamic Systems Analysis.)

Simulation provides a powerful tool for the ROV operator to test and train in realistic yet controlled environments at substantially lower operational costs while enhancing both operational safety and rapid trainee learning.

Vehicle Design and Stability

5

CHAPTER CONTENTS

This section, along with several others in this manual, will investigate the effect that vehicle geometry and stability have on the motive performance of an ROV, especially when taking into account the effect of the tether. Vehicle geometry is driven primarily by the vehicle's mission, e.g., pipeline survey vehicles require long straight travel distances while construction vehicles require close-quarter maneuverability. Such mission-related considerations affect the frame design and thus the accommodation of the buoyancy material required to float the vehicle. The overall shape of the vehicle determines the drag it will experience not counting the effect of the tether. Then the impact of the tether upon the vehicle's dynamic stability and the thrust to drag capability have to be taken into account. Just as one would use a leash to walk and control where his/her dog is going, so goes the tether's effect on the vehicle.

Accordingly, the perfect ROV would have the following characteristics:

- Minimal tether diameter (for instance, a single strand of unshielded optical fiber)
- Powered from the surface having unlimited endurance (as opposed to battery operated with limited power available)
- Very small in size (to work around and within structures), yet extremely stable
- Have an extremely high data pipeline for sensor throughput

Unfortunately, the perfect ROV is hard to develop, especially when considering the many tasks required by the WCROVs. ROV systems are a trade-off of a number of factors including cost, size, deployment resources/platform, and operational requirements. This chapter will address the design and stability of an ROV. Chapter 8 will address the principles of tether design, integration, and management since they can help create (or destroy) the perfect ROV.

5.1 Vehicle design

The overall mechanical design of the ROV is driven by the job it has to accomplish and the conditions it is expected to work within. It is essentially a "truck" to transport a payload to a location. This could be any number of items including a camera, sensor, manipulator, and/or tool to the worksite and be stable enough once on-site to perform the assigned task. The various payloads and onboard equipment that the ROV may use are discussed elsewhere in the manual. Needless to say, the payload must be designed to be integrated with the vehicle or attached to it (such as using a tool skid mounted below the vehicle). With that in mind, the designer's goal is to "connect the dots" between payload, thrusters, pressure vessels, tether, etc., and create a stable, maneuverable platform. The dots are connected with the vehicle's frame. Buoyancy is then added to float the hardware and stabilize the system. The tether is then integrated with the vehicle, its effect often overwhelming many of the carefully designed aspects of the vehicle (see Chapter 3). These key areas will be discussed in the following sections.

5.1.1 Frame

The frame of the ROV provides a firm platform for mounting (or attaching) the necessary mechanical, electrical, and propulsion components. This includes special tooling/instruments such as sonar, cameras, lighting, manipulator(s), vehicle/payload sensors, and sampling equipment. ROV frames have been made of materials ranging from plastic composites to aluminum tubing. In general, the materials used are chosen to give the maximum strength with the minimum weight. Since weight has to be offset with buoyancy, this is critical.

The ROV frame must also comply with regulations concerning load and lift path strength. The frame can range in size from 6 in. × 6 in. (15 cm × 15 cm) to 20 ft × 20 ft (6 m × 6 m). The size of the frame is dependent upon the following criteria:

- Weight of the complete ROV unit in air
- Volume of the onboard equipment
- Volume of the sensors and tooling
- Volume of the buoyancy
- Load-bearing criteria of the frame, in many cases requiring shock resistance

The benefit to today's engineer is that computer-aided design software makes the job of designing the frame much easier. However, ignoring critical design details can cause problems in the long run. Additional things to consider include:

- Requirement for cathodic protection systems
- Attachment points for ancillary tools or tool skid
- Effect of manipulator and/or tool movement on stability

- Thruster location to counterbalance CG (center of gravity) shifts
- Magnetism and its effect on sensors
- Through-frame lift and required certifications
- Capacity to add buoyancy should a heavier than normal payload be required
- Logical integration point for the tether
- Frame integration into a TMS and/or LARS
- Water ingress into frame (a metal frame holding salt water invites rapid corrosion)
- Anticipated frame contact with bottom sediments will hold these particles and possibly affect buoyancy
- Hydrodynamic characteristics of frame (e.g., shape vis-à-vis water flow).

Assuming the designer has the above topics under control, the types and characteristics of the buoyancy material, discussed in the next section, must be considered.

5.1.2 Buoyancy

Archimedes' principle states: "An object immersed in a fluid experiences a buoyant force that is equal in magnitude to the force of gravity on the displaced fluid." Thus, the objective of underwater vehicle flotation systems is to counteract the negative buoyancy effect of heavier than water materials on the submersible (frame, pressure housings, etc.) with lighter than water materials (i.e., buoyancy material with specific gravity less than that of the ambient water conditions). A near neutrally buoyant state is the goal. The flotation material should maintain its form and resistance to water pressure at the anticipated operating depth. The most common underwater vehicle flotation materials encompass two broad categories: rigid, lightweight materials such as polyurethane or polyvinyl chloride (PVC) foams, for shallower depths, and syntactic foams that can support full ocean depth systems. For the deepest trenches, innovative techniques for buoyancy such as ceramic spheres have also been used. All three approaches will be discussed in the following sections along with a summary of things to consider when choosing a buoyancy material.

5.1.2.1 Lightweight foam

The term "rigid polyurethane foam" comprises two polymer types: polyisocyanurate formulations and polyurethane formulas. There are distinct differences between the two, both in the manner in which they are produced and in their ultimate performance.

Polyisocyanurate foams (or "trimer foams") are generally low-density, insulation-grade foams, usually made in large blocks via a continuous extrusion process. These blocks are then put through cutting machines to make sheets and other shapes. ROV manufacturers generally cut, shape, and sand these inexpensive foams and then coat them with either a fiberglass covering or a thick layer of paint to help with abrasion and water intrusion resistance. These resilient foam blocks have been tested to depths of 1000 ft of seawater—"fsw" (330 m of sea water—"msw")—and have proven to be an inexpensive and effective flotation system for shallow water applications (Figure 5.1).

Polyisocyanurate foams have excellent insulating value, good compressive-strength properties, and temperature resistance up to 300°F (149°C). They are made in high volumes at densities between 1.8 and 6 lb/ft^3 (29–96 kg/m^3) and are reasonably inexpensive. Their stiff, brittle consistency and their propensity to shed dust (friability) when abraded can serve to identify these foams.

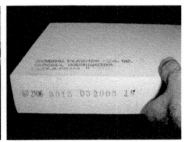

FIGURE 5.1

Polyurethane fiberglass encased and simple painted foam blocks.

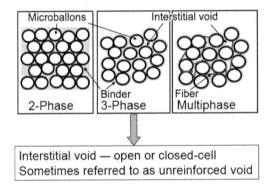

FIGURE 5.2

Types of syntactic foams.

(Courtesy Trelleborg.)

Composite tubes with closed end caps have also been used in shallow water applications, as well as a very good PVC material known as Divinycell. This material is blown foam that is quite capable for depths not exceeding 1000 m. It is an open-cell foam and should have a skin added to it for increased protection at depth. It does have some drawbacks, however, as it does not cycle very well. When buoyancy materials are developed and tested, they are subject to 1000 cycles to test depth before they are offered up for sale. Every manufacturer does this and has the data to back up the program.

5.1.2.2 *Syntactic foam*

For deepwater applications, syntactic foam has been the material of choice. "Syntactic" is a derivative of the Greek word, *syntac*, which means "to disperse in an orderly arrangement." Accordingly, syntactic foam is simply an air/microballoon structure encased within a resin body (Figure 5.2). The amount of trapped air within the resin structure will determine the density as well as the durability of the foam at deeper depths.

FIGURE 5.3

Macroscopic (a) as well as microscopic (b) views of glass microspheres.

(Courtesy Trelleborg.)

Syntactic foams have become widely accepted throughout the oceanographic and military communities, especially for the deeper diving vehicles, because of their reliability to perform over numerous excursions to their design depth. The desired density remains stable for many years of service. Syntactic foam typically exceeds a service life of 20 years. This section will address the understanding, care, and feeding of this high-performance material.

When glass microspheres (Figure 5.3) are used within the resin—some call them balloons while others call them bubbles—and the disbursement is controlled so the optimum packing factor is attained, then syntactic foam is created. This is referred to as "solid syntactic." Solid syntactics range in density from 18 lb/ft³ (0.29 g/cc) to 42 lb/ft³ (0.67 g/cc.) depending upon the service depth requirements. Simply stated, the density of seawater (64 lb/ft³, 1.0 g/cc), minus the air weight of the syntactic foam, equals the amount of buoyancy in seawater. This simple equation is used whether it is syntactic foam, blown foam, ceramic spheres, glass balls, or balsa wood. Syntactics become the work horse when going beyond depths of 2000 ft (600 m). From there to 36,000 ft (11,000 m), in most instances, they are the buoyancy material of choice.

The theoretical maximum dry packing factor for a solid material is 68%. Using special techniques, the syntactic manufacturers are able to attain 80% packing of the spheres, thereby providing the maximum amount of buoyancy, or payload, to the vehicle. This method is called "using a binary," a combination of glass spheres, typically at a 7:1 size differential, to provide the maximum amount of packing. Unfortunately, glass sphere manufacturers are not prone to manufacture these sizes as the market is limited and the costs are high. However, manufacturers that have a good R&D department can find ways around this shortcoming.

In most applications, solid syntactic foams are used due to their low moisture pickup at test depth. A well-constructed syntactic will have <3% water absorption over an extended period at depth. In fact, a well-designed system will have <1% absorption for 24 h at test depth. Solid syntactics are easily machined using the proper tools, feed rates, and speeds. (For an excellent syntactic foam machining guide, contact Engineered Syntactic Systems—www.esyntactic.com.)

Solid syntactic foams have acoustic properties that allow them to be either an absorber or a reflector in seawater, dependent upon the frequencies encountered. Various materials have also been added to the matrix to achieve the desired performance. The benefit is that the performance usually does not change throughout the water column being worked in.

Using solids has drawbacks and benefits. Typically, a solid can be used for full ocean depths while a combination, or macrofoam syntactic (Figure 5.4), has not been generally accepted beyond 10,000 fsw (>3000 msw). This is due to the large spheres used, i.e., 3/8 of an inch (0.95 cm) average diameter. The reasoning behind this is that the larger spheres may have a tendency to implode at great depths, possibly causing a sympathetic implosion on the surrounding spheres, which could lead to a catastrophic failure of the buoyancy system.

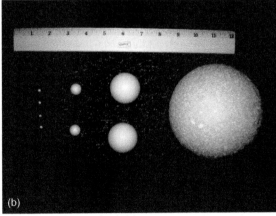

FIGURE 5.4

Macrofoam (a) and macroballoons (b).

(*Courtesy Trelleborg.*)

FIGURE 5.5

Syntactic foam filling mud tank voids on a 688I boat.

(Courtesy Newport News Shipbuilding.)

FIGURE 5.6

Ceramic buoyancy spheres.

(Courtesy Deep Sea Power & Light.)

The drawback to solid syntactics is that they cannot be cast as very large parts. They are typically 1 ft^3 (28 cm^3) or less due to the excessive heat or exothermic reaction generated during the curing process. Macrofoam, however, can be cast as very large parts; in fact, dihedrals and mud tanks with a cavity of 125 ft^3 (3.54 m^3) have been injected both in-plant and on-site at various shipyards. Figure 5.5 shows the inside of a mud tank on a 688I boat where the lines feed through foam that fills the voids in the frame bays and provides 3 long tons of buoyancy to the aft area.

5.1.2.3 Ceramic spheres

With the assault from the ocean's deepest trenches, ceramic spheres have been developed that can withstand the pressure of such locations. These spheres (Figure 5.6 sphere with logo) are typically

3.6 in. (9 cm) in diameter and capable of providing buoyancy to 36,000 ft (11,000 m). In fact, Woods Hole Oceanographic Institute's *Nereus* vehicle relies primarily on these spheres for its buoyancy. One shortcoming is the potential for sympathetic implosion. To date there is no known testing of that potential failure mode. This phenomenon is also a potential for large glass balls.

5.1.2.4 Summary

Regardless of the material chosen, the following should be taken into consideration when choosing the type of material:

- Specific gravity of the material
- Crush point and safety factor
- Shrinkage due to pressure, i.e., loss of buoyancy with depth
- Abrasion resistance, brittleness
- Potential protective coatings
- Available shapes and machinability, including hazardous material requirements
- Water absorption and thus loss of buoyancy
- Placement and stability considerations
- Ability to modify in the future

Historically, deepwater foams have been encased in a fiberglass shell. It was thought that the fiberglass would protect the syntactic from damage due to impact while being handled and transported aboard ship, which is true. What is becoming apparent, however, is that while the glass protects the foam during transit and slight impacts, in fact it may be causing further damage when the units are deployed. When the compressive modulus is considered, 2.8 million psi (2 billion kg/m^2) for fiberglass versus 400,000 psi (281 million kg/m^2) for the syntactic, the delta is extreme.

The buoyancy material needs to be able to expand and contract, that is "swim," during diving operations. Unfortunately, the modulus of the fiberglass is not compatible with the modulus of the syntactic and causes stress and strain on the foam systems. Whereas it was thought the glass was protecting the systems, it may have actually been transferring the impact load deep into the syntactic where it cannot be seen until the unit goes deep and stress relieves.

Another area of concern, when designing and machining the buoyancy materials, is to be sure to leave no sharp edges or corners. While these look great for the design, they set up stress risers (Figure 5.7) which can ultimately lead to cracking of the buoyancy system.

Recent developments in high-build urethane materials have allowed the incorporation of these materials to relieve the stress on the buoyancy system. They are flexible, easy to install, available in most colors, off the shelf items, and provide for the buoyancy material to "swim" during the dive cycle. Manufacturers of ROVs and AUVs are leaning more toward the urethane materials for coating the foam. With fiberglass weighing 128 lb/ft^3 (2.0 g/cc), using the urethane material is not only saving the syntactic but also regaining buoyancy.

Most buoyancy materials are inflammable. Due to where the vehicles operate it has not been necessary to include brominates in the resin matrix to reduce or inhibit the potential to burn. Only one instance has been documented: In the early 1980s, DSV 3 Turtle had a fire when an oil-filled line ruptured and a spark hit it.

When considering the crush depth of the buoyancy material, it is usually 1.1% times the service depth, except in the case of man-rated materials. Manned vehicles such as the *Alvin* require

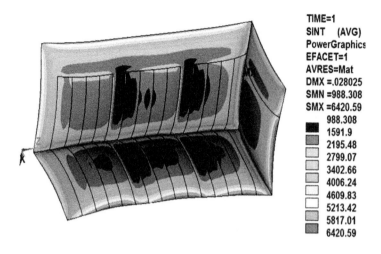

TIME=1
SINT   (AVG)
PowerGraphics
EFACET=1
AVRES=Mat
DMX =.028025
SMN =988.308
SMX =6420.59

- 988.308
- 1591.9
- 2195.48
- 2799.07
- 3402.66
- 4006.24
- 4609.83
- 5213.42
- 5817.01
- 6420.59

FIGURE 5.7

FEA analysis of fiberglass coating.

(Courtesy Engineered Syntactic Systems.)

FIGURE 5.8

Syntactic foam taken to crush depth.

(Photo courtesy Engineered Syntactic Systems.)

man-rated syntactics. To man-rate the materials an extensive amount of testing is required to insure the buoyancy will return without incident. Whereas the normal crush depth of the syntactic is 1.1%, man-rated systems are tested to 1.5% of test depth. The latest *Alvin* specification can readily be attained online. Figure 5.8 shows a typical high-performance material, rated to 20,000 ft (approximately 6000 m), that went well beyond its test depth. As can be seen, the material, designed (with safety factor) for 14.5 Kpsi ($10.2\,Mkg/m^2$), eventually crushed at a pressure of 16 Kpsi ($11.2\,Mkg/m^2$).

With the appropriate buoyancy material chosen, the designer still needs to consider the final design and its effect on the stability of the vehicle. This is considered in the next section.

5.2 Buoyancy and stability

Designing an awesome looking vehicle with an integrated frame and buoyancy system is an excellent design goal. But if that vehicle turns turtle when deploying its payload, this is clearly not the desired path. Vehicle stability, as discussed below, is critical to the successful completion of any underwater task.

As discussed further in Chapter 3, any vehicle has movement about six degrees of freedom (Figure 5.9): three translations (surge, heave, and sway along the longitudinal, vertical, and transverse (lateral) axes, respectively) and three rotations (roll, yaw, and pitch about these same respective axes). This section will address the interaction between vehicle static and dynamic stability and these degrees of freedom.

ROVs are not normally equipped to pitch and roll. The system is constructed with a high center of buoyancy and a low CG to give the camera platform maximum stability about the longitudinal and lateral axes (Figure 5.10). Most ROV systems have fixed ballast with variable positioning to allow trimming of the system nose-up/nose-down or for roll adjustment/trim. In the observation class, the lead (or heavy metal) ballast is normally located on tracks attached to the bottom frame to allow movement of ballast along the vehicle to achieve the desired trim.

5.2.1 Hydrostatic equilibrium

According to Archimedes' principle, any body partially or totally immersed in a fluid is buoyed up by a force equal to the weight of the displaced fluid. If somehow one could remove the body and instantly fill the resulting cavity with fluid identical to that surrounding it, no motion would take place: The body weight would exactly equal that of the displaced fluid.

The result of all of the weight forces on this displaced fluid (Figure 5.11) is centered at a point within the body termed the "center of gravity" (CG) (Figures 5.10 and 5.12). This is the sum of all

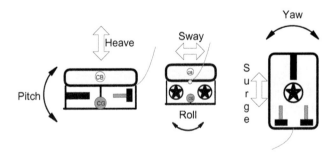

FIGURE 5.9

Vehicle degrees of freedom.

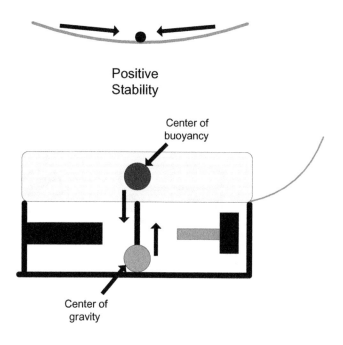

FIGURE 5.10

Vehicle with positive stability.

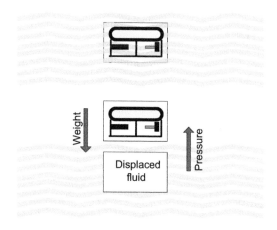

FIGURE 5.11

Hydrostatic equilibrium.

the gravitational forces acting upon the body by gravity. The resultant of the buoyant forces countering the gravitational pull acting upward through the CG of the displaced fluid is termed the "center of buoyancy" (CB). There is one variable in the stability equation that is valid for surface vessels with a nonwetted area that is not considered for submerged vehicles. The point where the

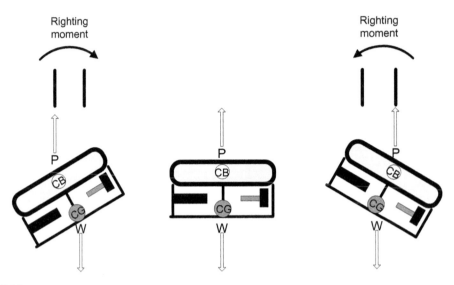

FIGURE 5.12

Vehicle's righting moment.

CB intersects the hull centerline is termed the "metacenter" and its distance from the CG is termed the "metacentric height" (GM). For ROV considerations, all operations are with the vehicle submerged and ballasted very close to neutral buoyancy, making only the separation of the CB and the CG the applicable reference metric for horizontal stability.

According to Van Dorn (1993):

> *The equilibrium attitude of the buoyant body floating in calm water is determined solely by interaction between the weight of the body, acting downward through its CG, and the resultant of the buoyant forces, which is equal in magnitude to the weight of the body and acts upward through the CB of the displaced water. If these two forces do not pass through the same vertical axis, the body is not in equilibrium, and will rotate so as to bring them into vertical alignment. The body is then said to be in static equilibrium.*

5.2.2 Transverse stability

Paraphrasing Van Dorn (1993) to account for ROVs, having located the positions of the CG and the upright CB of the vehicle, one can now investigate the transverse (lateral) stability. This is done without regard to external forces, merely by considering the hull of the vehicle to be inclined through several angles and calculating the respective moments exerted by the vertically opposing forces of gravity and buoyancy. These moments are generated by horizontal displacements (in the vehicle's reference frame) of the CB relative to the CG, as the vehicle inclines, such that these forces are no longer collinear, but are separate by same distance d, which is a function of the angle of inclination, θ. The magnitude of both forces remains always the same, and equal to the vehicle's

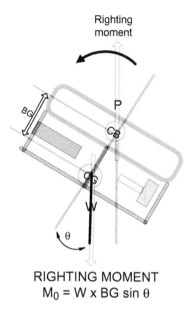

RIGHTING MOMENT
$$M_0 = W \times BG \sin \theta$$

FIGURE 5.13

Detail of righting moment.

weight W, but their moment ($W \times d$) is similarly a function of θ. If the moment of the buoyancy (or any other) force acts to rotate the vehicle about its CG opposite to the direction of inclination, it is called a "righting moment"; if in the same direction, it is called a "heeling moment."

Referring to Figure 5.13, as the BG becomes smaller, the righting moment decreases in a logarithmic fashion until static stability is lost.

5.2.3 Water density and buoyancy

It is conventional operating procedure to have vehicles positively buoyant when operating to ensure they will return to the surface if a power failure occurs. This positive buoyancy would be in the range of 1 lb (450 g) for small vehicles and 11–15 lb (5–7 kg) for larger vehicles, and in some cases, work-class vehicles will be as much as 50 lb (23 kg) positive. Another reason for this is to allow near-bottom maneuvering without thrusting up, forcing water down, thus stirring up sediment and destroying camera effectiveness through decreased visibility. It also obviates the need for continual thrust reversal. Very large vehicles with variable ballast systems that allow for subsurface buoyancy adjustments are an exception. The vehicles used by most ROV operators will predominantly have fixed ballast.

As more fully described in Chapter 2, the makeup of the water in which the submersible operates will determine the level of ballasting needed to properly operate the vehicle. The three major water variables affecting this are temperature, salinity, and pressure.

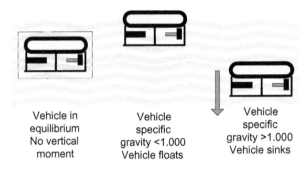

Ambient water specific
gravity = 1.000

FIGURE 5.14

Specific gravity versus vehicle buoyancy.

More than 97% of the world's water is located in the oceans. Many of the properties of water are modified by the presence of dissolved salts. The level of dissolved salts in seawater is normally expressed in grams of dissolved salts per kilogram of water (historically expressed in imperial units as parts per thousand, or "ppt," with the newer accepted unit as the practical salinity unit, or "PSU"). Open ocean seawater contains about 35 PSU of dissolved salt. In fact, 99% of all ocean water has salinity of between 33 and 37 PSU.

Pure water has a specific gravity of 1.00 at maximum density temperature of about 4°C (approximately 39°F). Above 4°C, water density decreases due to molecular agitation. Below 4°C, ice crystals begin to form in the lattice structure, thereby decreasing density until the freezing point. It is well known that ice floats, demonstrating the fact that its density is lower than water.

At a salt content of 24.7 PSU, the freezing point and the maximum density temperature of seawater coincide at −1.332°C. In other words, with salt content above 24.7 PSU, there is no maximum density of seawater above the freezing point.

Most ROVs have a fixed volume. When transferring a submersible from a fresh water environment (where the system was neutrally ballasted) to a higher density salt water environment, the ROV pilot will notice that the system demonstrates a more positive buoyancy, much like an ice cube placed into a glass of water. In order to neutralize the buoyancy of the system, ballast weights will need to be added to the submersible until neutral buoyancy is re-achieved. The converse is also true by going from salt water into fresh water or between differing temperature/salinity combinations with the submersible (Figure 5.14).

Water is effectively (for our purposes) incompressible. At deeper depths, water will be at a higher density, slightly affecting the buoyancy of the submersible. The water density buoyancy shift at the deeper operating depths is partially offset due to the compression of the air-filled spaces of the submersible. This balance is more or less dependent upon the system design and the amount of air-filled space within the submersible.

Thrusters

6

CHAPTER CONTENTS

As a part of any mobile robotic system, some means of locomotion is necessary in order to move the robotic system. In free-swimming ROV technology, thrusters rule. Thrusters are a critical design consideration for any ROV system. Without the proper thrust the vehicle can be overwhelmed by the environmental conditions and thus unable to perform the desired tasks. This chapter will address the basics of thruster design and placement on the vehicle and provide examples of some of today's leading thruster designs. The final section will discuss the over-arching question of which type of thruster to choose (electric or hydraulic), and ultimately which type of vehicle (all electric, all hydraulic, or a combination) should be chosen for the job. Examples of the full range of electric and hydraulic vehicles currently available on the open market are provided. But first, a discussion of the dreaded "design spiral" is warranted.

It is simple at first blush. The designer must determine what size thruster is needed for the vehicle. The decision process goes something like this:

- What is the task that must be done and thus the work system and tools necessary for the ROV? (There is no sense sending in a vehicle if it cannot do the task—which could be as simple as observation or as complex as changing out an AX ring or rigging a shackle.)
- What size power system is necessary to support the work system and tools (and the other electrical components: control system, lights, cameras, etc.)?
- What size frame and amount of buoyancy are necessary to support the power and work system?
- What are the physical and environmental conditions (current, depth, required operational footprint, etc.)?
- What is the drag on the vehicle and cable to meet the operating conditions and footprint?

The ROV Manual.
© 2014 Robert D Christ and Robert L Wernli. Published by Elsevier Ltd. All rights reserved.

- What size thrusters, taking into consideration the efficiency of the overall system, are needed to move the vehicle at the necessary speed above the stated environmental conditions?
- What is the total power requirement input to the electric/hydraulic power system?
- What is the size of the umbilical/tether to provide the required power?

Now, the lucky engineer will reach this point and exclaim "voilà" because the estimates for cable and vehicle drag were perfect and the size of the vehicle is sufficient to meet all of the required operational conditions. Unfortunately, this is usually not the case. The size of the cable and/or the vehicle will increase, and thus the necessary buoyancy and then drag, which drives the power required, which requires increased thrust output, which requires (get the picture?).

Once the initial design is finished, that is about the time that the sponsor or supervisor steps in and says, "I'd like to increase the operating current from one to two knots." No, problem, correct? Double the current, then double the thrust. Right? Wrong!

As discussed in Section 3.5.3, the drag (D) on the vehicle is defined by:

$$D = 1/2\sigma A V^2 C_d$$

where V is the velocity of the vehicle.

Thus, the drag (and the required thrust) is proportional to the square of the velocity. So, that simple doubling of the velocity actually requires four times the thrust! And do not forget the cable drag, which is also proportional to the square of the velocity. In addition, this change not only affects the vehicle design, but as the cable grow (obviously, the vehicle will need more power and thus a larger umbilical/tether) so grows the size, weight, and power requirements of the cable and handling system on the platform above. It is highly recommended to set the operating requirements properly up front. Do not expect minor changes in those requirements to have a minor impact on the vehicle design because the carry-over effects of simple changes typically magnify exponentially. As an example, see Figure 6.1. The Schilling HD electric motor turns a dual spline shaft driving two pumps—one for the main hydraulic system (for thrusters and manipulators) and one for the auxiliary hydraulics (for tooling). On the auxiliary side, the electric motor powers a Rexroth A10VSO45 Axial Piston Pump with the shaft turning at 1875 rpm. You cannot just switch the pump to the next higher size since the power draw could disable the main hydraulic system. Also, you cannot simply turn the motor faster as this could shear the spline shaft due to excess torque. Changes can result in a significant ripple effect.

So, in reality, the design of an ROV is a complex spiral that will eventually take the engineer to a point where a thruster that supports all requirements (above and below the water) can be chosen. The following sections will provide additional input to those decisions.

6.1 Propulsion and thrust

The propulsion system significantly impacts the vehicle design. The types of thrusters, their configuration, and the power source to drive them usually take priority over many of the other components.

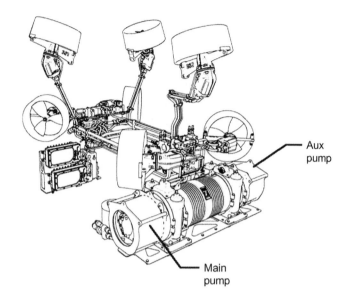

Aux
pump

Main
pump

FIGURE 6.1

HD vehicle hydraulic system.

(Courtesy Schilling Robotics.)

6.1.1 Propulsion systems

ROV propulsion systems come in three different types: Electrically driven propeller, hydraulically driven propeller, and (rarely) ducted jet propulsion (either electrically or hydraulically driven). These different types have been developed to suit the size of vehicle and anticipated type of work. In some cases, the actual location of the work task has dictated the type of propulsion used. For example, if the vehicle is operated in the vicinity of loosely consolidated debris, which could be pulled into rotating thrusters, ducted jet thruster systems could be used, albeit for the smaller sized vehicles. But more often, propeller covers are used mitigating the propeller fouling issue. If the vehicle requires heavy duty tooling for intervention, the vehicle could be operated with hydraulics (including thruster power). Hydraulic pump systems are driven by an electrical motor on the vehicle, requiring a change in energy from electrical to mechanical to hydraulic—a process that is quite energy inefficient. A definite need for high mechanical force is required to justify such an energy loss and corresponding costs. As ROVs transition from observation class to heavy duty work-class systems, in most cases (as discussed later in this chapter), the propulsion systems transition from electric to hydraulic.

The main goal for the design of ROV propulsion systems is to have high thrust-to-physical size/drag and power-input ratios. The driving force in the area of propulsion systems is the desire of ROV operators to extend the equipment's operating envelope. The more powerful the propulsion of the ROV, the stronger the sea current in which the vehicle can operate. Consequently, this extends the system's performance envelope.

Another concern is the reliability of the propulsion system and its associated subcomponents. In the early development of the ROV, a general practice was to replace and refit electric motor units every 50–100 hours of operation. This increased the inventory of parts required and the possibilities of errors by the technicians in reassembling the motors. Thus, investing in a reliable design from the beginning can save both time and money.

The propulsion system has to be a trade-off between what the ROV requires for the performance of a work task and the practical dimensions of the ROV. Typically, the more the thruster power required, the heavier the equipment on the ROV. All parts of the ROV system will grow exponentially larger with the power requirement continuing to increase. Thus, observation ROVs are normally restricted to a few minor work tasks without major modifications that would move them to the next heavier class. As the work tasks increase, so does the size of the vehicle and thus the size of the thruster system. The goal is to design an efficient thruster system with minimal (and easily repairable) components. A good example of such a system for a hydraulically operated heavy work-class vehicle is that of Schilling's HD vehicle (Figure 6.1). The hydraulic pumps are designed for easy removal and replacement on the drive motor.

6.1.2 Thruster basics

The ROV's propulsion system is made up of two or more thrusters that propel the vehicle in a manner that allows navigation to the work site. Thrusters must be positioned on the vehicle so that the moment arm of their thrust force, relative to the central mass of the vehicle, allows a proper amount of maneuverability and controllability.

Thrust vectoring is the only means of locomotion for an ROV. There are numerous placement options for thrusters to allow varying degrees of maneuverability. Maneuvering is achieved through asymmetrical thrusting based upon thruster placement as well as varying thruster output.

The three-thruster horizontal arrangement (Figure 6.2) allows only fore/aft/yaw, while the fourth thruster also allows lateral translation. The five-thruster variation allows all four horizontal thrusters to thrust in any horizontal direction simultaneously while multiple vertical thrusters further allow for pitch and roll functions via asymmetrical thrusting.

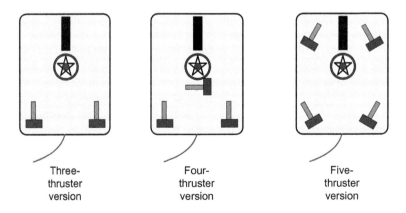

Three-
thruster
version

Four-
thruster
version

Five-
thruster
version

FIGURE 6.2

Thruster arrangement.

FIGURE 6.3

Thruster aligned off the longitudinal axis.

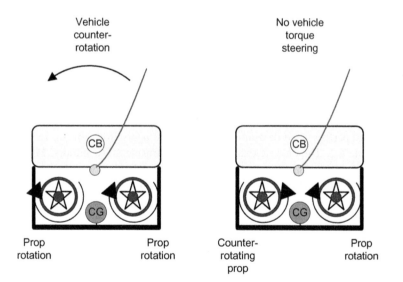

FIGURE 6.4

Thruster rotational effect on vehicle.

Also, placing the thruster off alignment from the longitudinal axis of the vehicle (Figure 6.3) will allow a better turning moment, while still providing the vehicle with strong longitudinal stability.

One problem with multiple horizontal thrusters along the same axis, without counter-rotating propellers, is the "*torque steer*" issue (Figure 6.4). With two or more thrusters operating on the

Table 6.1 Thruster Comparison of Representative COTS Thruster Systems

Thruster Comparison						
Type	Model	Manufacturer	Prop Diameter	Weight in Water	Input	Bollard Thrust
Electric	SPE-250	Sub-Atlantic	9.7 in. 246 mm	17.6 lb 8 kg	600 VDC	220 lbf 100 kgf
Electric	1002	Innerspace	9.27 in. 235 mm	41 lb 18.6 kg	600 VDC	665 lbf 302 kgf
Electric	8020	Tecnadyne	12 in. 305 mm	34–40 lb 15.5–18 kg	600 VDC	505 lbf 230 kgf
Hydraulic	SA300	Sub-Atlantic	11.8 in. 300 mm	22 lb 10 kg	26.8 hp 20 kW	700 lbf 318 kgf
Hydraulic	1002MTL	Innerspace	9.27 in. 235 mm	20 lb 9.1 kg	21 hp 15.7 kW	639 lbf 290 kgf

same plane of motion, a counter-reaction to this turning moment will result. Just as the propeller of a helicopter must be countered by the tail rotor or a counter-rotating main rotor, the ROV must have counter-rotating thruster propellers in order to avoid the torque of the thrusters rolling the vehicle, especially in the smaller observation-class vehicles, counter to the direction of propeller rotation. If this roll does occur, the resulting asymmetrical thrust and drag loading could give rise to course deviations—the effect of which is known as "torque steering."

Thruster location is critical to the performance of the vehicle as is the choice of the type of thruster. Design and performance data on similar sized thrusters from several leading manufacturers is presented in Table 6.1. The electric examples all use brushless DC motors. Most of the manufacturers make a wide range of thrusters, although the Sub-Atlantic example is their largest electric thruster. Images of the thrusters are also provided (Figures 6.5–6.7).

6.1.3 Thruster design

Underwater electrical thrusters are composed of the following major components:

- Power source
- Electric motor
- Motor controller (this may be part of the internal thruster electronics or may be part of a driver board in a separate pressure housing)
- Thruster housing and attachment to vehicle frame
- Gearing mechanism (if thruster is geared)
- Drive shafts, seals, and couplings
- Propeller
- Kort nozzle and stators

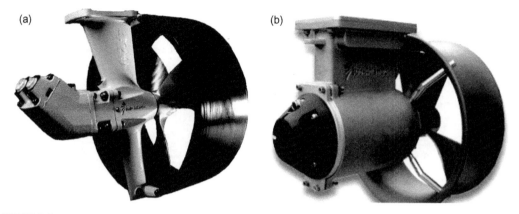

(a)   (b)

FIGURE 6.5

Sub-Atlantic Model SA300 hydraulic (a) and SPE-250 electric (b) thrusters.

(*Courtesy Forum Subsea Technologies.*)

The most critical of these components will be discussed in more detail in the following sections.

Power source

As more fully explained in Chapter 7, with a surface-powered ROV system, power arrives to the vehicle from a surface power source. The surface power can be in any form from basic shore power (e.g., 110 VAC 1Φ 60 Hz or 220 VAC 1Φ 50 Hz—which is standard for most consumer electrical power delivery worldwide) to 480 VAC 3Φ (50 or 60 Hz) to a DC battery source.

For an observation-class ROV system running on DC power at the vehicle, the AC source is first rectified to DC (either on the surface or at first arrival at the vehicle) and then sent to the submersible's components for distribution to the thrusters. The driver and distribution system location will vary between manufacturers and may be anywhere from on the surface control station, within the electronics bottle of the submersible, to within the actual thruster unit. The purpose of this power source is the delivery of sufficient power to drive the thruster through its work task.

For the mid-sized and work-class ROVs (MSROV and WCROV), power is converted/conditioned at the surface to a form appropriate for the length of umbilical/tether (typically 3000 VAC or greater) and then sent to the vehicle. Typically, the vehicle manufacturer decides upon the depth capability and designs the vehicle electrical system to conform to the tether length. Most vehicle sensors operate on low-voltage DC power; therefore, a rectifier must be in the circuit somewhere. Deepwater ROVs (MSROV or WCROV) run high-voltage AC power down the umbilical to the TMS. Most MSROVs have a step-down transformer at the TMS that then powers the components of the TMS and ROV at lower voltage. Some electrical thrusters are designed to operate on AC power while others operate on DC (which determines the location of the rectifier). From there, multiple power circuits are powered depending upon the needs of the components. The thruster circuit is typically on a separate circuit compliant with the needs of the electric thrusters. On the hydraulic vehicles, it is simply the lights, sensors/electronics, and the pump.

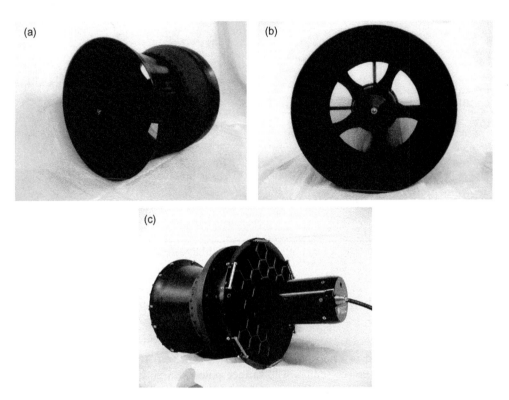

FIGURE 6.6

Innerspace Model 1002MTL hydraulic (spinner view (a) and prop view (b)) and Model 1002 electric (c) thrusters.

(Courtesy Innerspace Corp.)

Electric motor

Electric motors come in many shapes, sizes, and technologies, each designed for different functions. By far the most common thruster motor on observation-class ROV systems is the DC motor, due to its power, availability, variety, reliability, and ease of interface. The DC motor, however, has some difficult cost, design, and operational characteristics. Factors that make it less than perfect for this application include:

- The optimum motor speed is much higher than the normal in-water propeller rotation speed, thus requiring gearing to gain the most efficient speed of operation.
- DC motors consume a high amount of current.
- They require a rather complex pulse width modulation motor control scheme to obtain precise operations.

Permanent magnet DC motors

According to Clark and Owings (2003), the permanent magnet DC motor has, within the mechanism, two permanent magnets that provide a magnetic field within which the armature

FIGURE 6.7

Tecnadyne Model 8020 electric thruster.

(Courtesy Tecnadyne.)

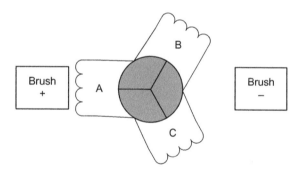

FIGURE 6.8

Commutator and brushes.

rotates. The rotating center portion of the motor (the armature) has an odd number of poles—each of which has its own winding (Figure 6.8). The winding is connected to a contact pad on the center shaft called the "commutator." Brushes attached to the (+) and (−) wires of the motor provide power to the windings in such a fashion that one pole will be repelled from the permanent magnet nearest it and another winding will be attracted to another pole. As the armature rotates, the commutator changes, determining which winding gets which polarity of the magnetic field. An armature always has an odd number of poles, and this ensures that the poles of the armature can never line up with their opposite magnet in the motor, which would stop all motion.

Near the center shaft of the armature are three plates attached to their respective windings (A, B, and C) around the poles. The brushes that feed power to the motor will be exactly opposite each other, which enables the magnetic fields in the armature to forever trail the static magnetic fields of the magnets. This causes the motor to turn. The more current that flows in the windings, the stronger the magnetic field in the armature, and the faster the motor turns.

Even as the current flowing in the windings creates an electromagnetic field that causes the motor to turn, the act of the windings moving through the static magnetic field of the motor causes a current in the windings. This current is opposite in polarity to the current the motor is drawing from the power source. The end result of this current, and the countercurrent (CEMF—counter electromotive force), is that as the motor turns faster it actually draws less current.

This is important because the armature will eventually reach a point where the CEMF and the draw current balance out at the load placed on the motor and the motor attains a steady state. The point where the motor has no load is the point where it is most efficient. It is also the point where the motor is the weakest in its working range. The point where the motor is strongest is when there is no CEMF; all current flowing is causing the motor to try to move. This state is when the armature is not turning at all. This is called the "stall" or "start" current and is when the motor's torque will be the strongest. This point is at the opposite end of the motor speed range from the steady-state velocity (Figure 6.9). The start current phenomenon is most critical during the motor startup phase on a WCROV as the motor start current can become quite excessive. There are several "soft start" methods to mitigate excess start current (thus avoiding overloading the circuits and/or the power source) including pressure relief on the system and sequentially step-up powering the motor as it achieves its fully rated speed.

While the maximum efficiency of an electric motor is at the no-load point, the purpose of having the motor in the first place is to do work (i.e., produce mechanical rotary motion that may

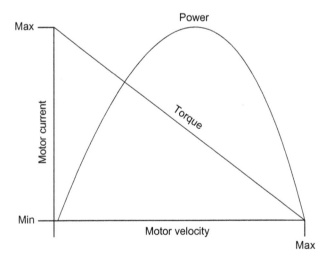

FIGURE 6.9

Motor velocity versus current.

be converted into linear motion or some other type of work). The degree of turning force delivered to the drive shaft is known as "motor torque."

Motor torque is defined as the angular force the motor can deliver at a given distance from the shaft. If a motor can lift 1 kg from a pulley with a radius of 1 m it would have a torque of 1 newton-meter (1 N m). 1 N equals 1 kg-m/s^2, which is equal to 0.225 lb (1 in. equals 2.54 cm, 100 cm are in a meter, and there are 2π radians in one revolution).

The formula for mechanical power in watts is equal to torque times the angular velocity in radians per second. This formula is used to describe the power of a motor at any point in its working range. A DC motor's maximum power is at half its maximum torque and half its maximum rotational velocity (also known as the no-load velocity). This is simple to visualize from the discussion of the motor working range: Where the maximum angular velocity (highest revolutions per minute (rpm)) has the lowest torque, and where the torque is the highest, the angular velocity is zero (i.e., motor stall or start torque).

Note that as motor velocity (rpm) increases, the torque decreases; at some point the power stops rising and starts to fall, which is the point of maximum power.

When sizing a DC motor for an ROV, the motor should be running near its highest efficiency speed, rather than its highest power, in order to get the longest running time. In most DC motors, this will be at about 10% of its stall torque, which will be less than its torque at maximum power. So, if the maximum power needed for the operation is determined, the motor can be properly sized. A measure of efficiency and operational life can then be obtained by oversizing the motor for the task at hand.

Brushless DC motors

For brushless DC motors, a sensor is used to determine the armature position. The input from the sensor triggers an external circuit that reverses the feed current polarity appropriately. Brushless motors have a number of advantages that include longer service life, less operating noise (from an electrical standpoint), and, in some cases, greater efficiency. Today's brushless DC motors, with integrated control electronics, have been used to field efficient, highly reliable thrusters.

Gearing

DC motors can run from 8000 to 20,000 rpm and higher. Clearly, this is far too fast for ROV applications if vehicle control is to be maintained. Thus, to match the efficient operational speed of the motor with the efficient speed of the thruster's propeller, the motor will require gearing. Gearing allows two distinctive benefits—the power delivered to the propeller is both slower and more powerful. Further, with the proper selection of a gearbox with a proper reduction ratio, the maximum efficiency speed of the motor can match the maximum efficiency of the thruster's propeller/kort nozzle combination. However, the efficiency of the overall system may be lowered due to the energy necessary to drive the gearbox. Further, piloting technique (smooth operation of the thrusters versus over-controlling—i. e., constantly reversing thruster direction from full power in one direction to full power in the other) will determine both thruster efficiency as well as longevity of the gearing mechanism. Over-controlling a thruster will quickly destroy the gearing mechanism, resulting in both downtime and an angry client.

Drive shafts, seals, and couplings

The shafts, seals, and couplings for an ROV thruster are much like those for a motorboat. The shaft is designed to provide torque to the propeller while the seal maintains a watertight barrier that prevents water ingress into the motor mechanism.

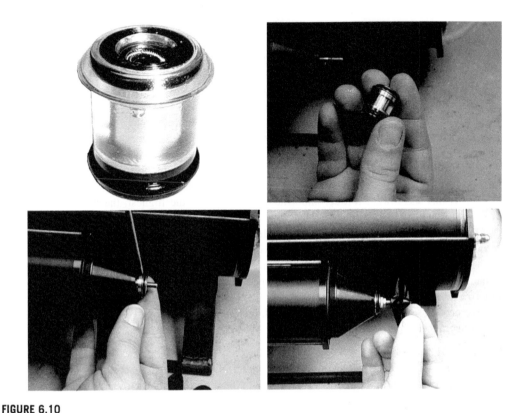

FIGURE 6.10

Fluid sealing of direct drive thruster coupling.

Drive shafts and couplings vary with the type of propeller driving mechanism. Direct drive shafts, magnetic couplings, and mechanical (i.e., geared) couplings are all used to drive the propeller. Technology advances are being exploited in attempts to miniaturize thrusters. In one case, a new type of thruster housing places the drive mechanism on the hub of the propeller instead of at the drive shaft, allowing better torque and more efficient propeller-tip flow management. Others are developing miniature electric ring thrusters, where the propeller, which can be hubless, is driven by an external "ring" motor built into the surrounding nozzle. Such a design eliminates the need for shafts, and sometimes seals, altogether.

There are various methods for sealing underwater thrusters. Some manufacturers use fluid-filled thruster housings to lower the difference in pressure between the seawater and the internal thruster housing pressure by simply matching the two pressures (internal and external). Still others use a lubricant bath between the air-filled spaces and the outside water (Figure 6.10). A common and highly reliable technique is the use of a magnetically coupled shaft, which allows the air-filled housing to remain sealed (Figure 6.11).

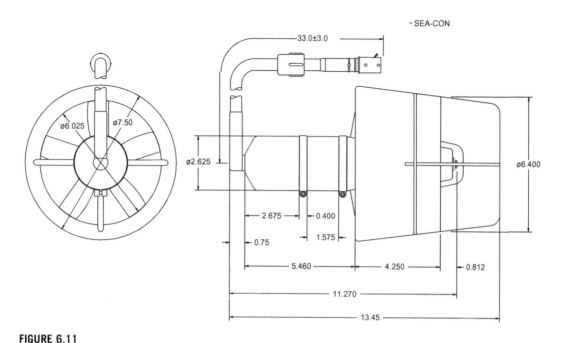

FIGURE 6.11

Magnetically coupled thruster diagram.

Hydraulic motor

The hydraulic version of the thruster motor is a much simpler design. The valve pack directs fluid to either the "A" port or "B" port to drive a motor—pump fluid to A to go forward and to B to reverse (or vice versa). The small nuances of thruster design will be left to OEM salespersons.

Propeller design

The propeller is a turning lifting body designed to move and vector water opposite to the direction of motion. Many thruster propellers are designed so their efficiency is much higher in one direction (most often in the forward and the down directions) than in the other. Propellers have a nominal speed of maximum efficiency, which is hopefully near the vehicle's normal operating speed. Some propellers are designed for speed and others are designed for power. When selecting a propeller, choose one with the ROV's operating envelope in mind. The desired operating objectives should be achieved through efficient propeller output below the normal operating envelope.

Also, if diver/ROV simultaneous operations are planned, propeller covers are generally a good idea and are sometime required by diving standards organizations (e.g., IMCA, ADCI).

Kort nozzle

A kort nozzle is common on most underwater thruster models. The efficacy of a kort nozzle is the mechanism's help in reducing the amount of propeller vortices generated as the propeller turns at high speeds. The nozzle, which surrounds the propeller blades, also helps with reducing the incidence of foreign object ingestion into the thruster propeller. Also, stators help reduce the tendency of rotating propellers' swirling discharge, which tends to lower propeller efficiency and cause unwanted thruster torque acting upon the entire vehicle.

6.2 Thrusters and speed

As the speed of the vehicle ramps up, the low speed stability due to static stability is overtaken by the placement of the thrusters with regard to the center of total drag. In short, for slow-speed vehicles the designer can get away with improper placement of thrusters. For higher speed systems, thruster placement becomes a more important consideration in vehicle control (Figure 6.12).

Propeller efficiency and placement

Propellers come in all shapes and sizes based upon the load and usage. Again using the aircraft correlation, on fixed-pitch aircraft propellers for small aircraft there are "climb props" and "cruise props." The climb prop is optimized for slower speeds, allowing better climb performance while sacrificing cruise speeds. Conversely, a cruise prop has better range and speed during cruise, but climb performance suffers. Likewise, a tugboat propeller would not be best suited for a high-speed passenger liner.

Propellers have an optimum operational speed. Some propellers are optimized for thrust in one direction over another. A common small ROV thruster on the market today uses such a propeller for forward and downward thrusting. The advantage to this propeller arrangement is better thrust performance in the forward direction while sacrificing turning reversal and upward thrust performance. Other propellers have equal thrusting capabilities in both directions. Both have their strengths and weaknesses. All ROVs are slow systems and should make use of propellers' maximizing power at slow speeds in order to counter the combined tether/vehicle drag. Remember, an ROV is a tugboat and not a speedboat.

Propellers also produce both cavitations and propeller-tip vortices, causing substantial amounts of drag. As the spinning propeller moves water across the blade, the vector/inertial force of the moving water (instead of moving aft to produce a forward thrust vector) throws the water toward the tips of the blade, spilling over the end of the blade in turbulent flow. Kort nozzles form a basic hub around the propeller to substantially reduce the instance of tip vortices. The kort nozzle then maintains the water volume within the thruster unit, allowing for more efficient

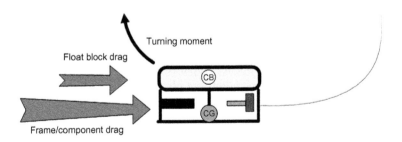

FIGURE 6.12

Bow turning moment due to asymmetrical drag as speed ramps up.

movement of the water mass in the desired vectored direction. Propeller cavitations are a lesser problem due to the speed at which the small ROV thruster propeller turns and are inconsequential to this analysis.

Thrust to drag and bollard pull

The following factors come into play when calculating vehicle speed and ability to operate in current:

- *Bollard pull* is a direct measurement of the ability of the vehicle to pull on a cable. Values provided by manufacturers can vary due to lack of standards for testing: "Actual bollard pull can only be measured in full scale, and is performed during so-called bollard pull trials. Unfortunately the test results are not only dependent on the performance of the [vehicle] itself, but also on test method and set-up, on trial site and on environmental conditions ..." (Jukola and Skogman, 2002).
- *Hydrodynamics* is another aspect of ROV design that must be considered holistically. Although a vehicle shape and size may make it very hydrodynamic (i.e., certain smaller enclosed systems), there is often a trade-off in stability. Some manufacturers seem to spend considerable effort making their ROVs more hydrodynamic in the horizontal plane, but in deep-sea operations diving to depth may consume considerable time.

It is bollard pull, vehicle hydrodynamics, and tether drag together that determine most limitations on vehicle performance. The smaller the tether cable diameter, the better—in all respects (except, of course, power delivery). Stiffer tethers can be difficult to handle, but they typically provide less drag in the water than their more flexible counterparts. Flexible tethers are much nicer for storage and handling, but they tend to get tangled or hang up more often than those that are slightly stiffer.

The use of ROVs in current is an issue that is constantly debated among users, designers, and manufacturers. This is not a topic that can be settled by comparing specifications of one vehicle to another. One of the most common misconceptions is that maximum speed equates to an ability to deal with current. When operating at depth (versus at the surface), the greatest influence of current is on the tether cable. It is the ability of the vehicle to pull this cable that allows it to operate in stronger currents. A vehicle with more power, but not necessarily more speed, will be better able to handle the tether (an example of which would be bollard pull of a tugboat versus that of a speedboat). The most effective way to determine a vehicle's ability to operate in current is to test the vehicle in current. Operator experience can have a significant effect on how the vehicle performs in higher current situations. Realistically, no small surfaced-powered ROV can be considered effective in any current over 3 knots. Even with the larger MSROV and WCROV, similar issues prevail. The only way to effectively operate any ROV in currents above 3 knots is to avoid them altogether by placing the TMS/ROV combination in a location outside of the main force of the current (e.g., the leeward side of a structure, below the surface current).

6.3 Electric versus hydraulic

Depending upon which manufacturer is contacted, it will (in all likelihood) be stated that their vehicle (whether electric or hydraulic) will definitely be considered the best vehicle for whatever task is being considered. The good news is that the vehicles developed by most major manufacturers today have become extremely reliable. Such reliability is a critical aspect of ROV operations since any vehicle downtime that prevents the user's equipment from getting back on line can be extremely costly. When the vehicle has to be repaired, the fewer parts necessary for the functioning of the vehicle, the faster they can be replaced/installed.

One of the selling points by the all-electric vehicle manufacturers is that their electric thrusters have fewer parts and a simpler overall system. A vehicle with hydraulic thrusters has substantially more parts including the hydraulic power system, compensators, valve packages, and (potentially) leaky lines that could land one in trouble with environmental regulators. With a hydraulically powered vehicle, the electricity being supplied down the umbilical has to be converted to hydraulic power with the consequent loss of energy due to the conversion efficiency. Exacerbating this are the additional losses in the hydraulic lines and components. However, the conversion at the prime mover (e.g., a hydraulic pump) is only done once and the hydraulic power is connected directly to whichever component requires it. For an all-electric vehicle, a prime mover is required at each point of use and the electric actuators are usually larger than their hydraulic counterparts. Therefore, this can also lead to a heavier and bulkier system.

On the other hand, the electrical components will not necessarily be ruined should there be a failure in the system. The high-power hydraulic systems on WCROVs are worked hard and should they not be maintained adequately could experience a pump failure—with the subsequent contamination and damage to the entire hydraulic system. Even with environmentally acceptable hydraulic fluids, a leak in the system would not be politically correct. The use of seawater hydraulics has been on the engineer's shelf for some time and may see a role in the future.

With the power required to run tools for construction, installation, and field R&M work reaching 75-plus horsepower, having a single prime mover may be a weight savings. If the electric vehicle has to have a hydraulically powered skid added to support the tooling, then a lot of the electric vehicle benefit would be lost in the heavy-duty WCROVs.

Regardless of the system trade-offs, as efficiency goes down at the working end of the umbilical, the same overall thrust and work system output is still required at the thrusters and tools. The conversion losses from electrical to mechanical must be compensated for. Therefore, the topside power system will have to increase and possibly the size of the umbilical and tether, which eventually impact the overall spread.

One may not think that the diameter of the umbilical or tether has that much of an impact on the system design. But as discussed earlier, the drag on the umbilical/tether is proportional to the projected area of that same umbilical/tether. As an example, consider the umbilical for Schilling's earlier all-electric *Quest* ROV. As a comparison, the *Quest*'s umbilical diameter is 1.06 in (27 mm) as compared to their present *UHD* ROV, with an umbilical diameter of 1.56 in. (39.7 mm). The projected area of the umbilical, and thus the drag, increases by about 50%. But the cross-sectional area of the umbilical, and thus the umbilical volume for a given length, increases by over 100%, which one can consider (assuming a similar effect on the weight) reflects the increase in the weight of the umbilical. Reportedly, this reduction in the diameter of the umbilical reduced the total *Quest* system weight by 20,000 lb (9066 kg). The ROV will not be pulling the umbilical around if it has a heavy TMS at the end. But the same factors apply to the tether between the TMS and the vehicle. For the smaller vehicles without a TMS, the entire tether in the water will see increased drag as the diameter increases.

Any increase in the size of the topside equipment can have a dramatic effect on the support platform requirements. As the offshore ROV service companies contemplate taking their operations even deeper offshore, the size of the umbilical, tether, winch, power system, vehicle, spares, control van, and launch system will all grow. There will be a point, possibly as near as 4000 m, where the heavily armored steel umbilicals will become too heavy to support their own weight *much less the TMS and vehicle on the other end.* This fact supports the arguments by the all-electric vehicle manufacturers that their systems (which are typically simpler, more efficient, and thus lighter) will be able to support future deep operations more efficiently than the heavy hydraulically powered work vehicles.

Hydraulic vehicle aficionados will argue that they will be able to use a lighter nonmetallic umbilical. Such umbilicals are made of Kevlar and used by most 20,000 ft (6000 m) depth-rated ROVs. This solves the suspension weight problem, albeit at a higher cost for the umbilical. However, the synthetic fiber materials used in such umbilicals may not withstand the wear and tear experienced during the years of service expected of them in the oil field. The steel-armored cables of today's systems have been proven to be a cost-effective solution.

It will also be argued that today's manipulators and tools require hydraulic power (which is, at this time, very true for the heavy-duty systems). There are some smaller all-electric manipulators on the market, but the size and construction (and thus the strength-to-weight ratio) of electric manipulators cannot yet match their hydraulic counterparts developed for the offshore market. Just as today's electric vehicles are leveraging the advancements in electronics and electric drive technology, manipulator manufacturers are keeping an eye on any advancements that would allow the development of a comparable electric underwater manipulator.

However, for the time being, the all-electric vehicle manufacturers will continue to supply easily integrated skid packages that can provide auxiliary hydraulic power, manipulators, and tools as necessary without modifying their basic ROV system.

So, who wins the debate? Neither side, actually. Because the electric, hydraulic, and electro-hydraulic vehicles all have their niche within the market (at this time), both types of vehicles will continue to proliferate. Table 6.2 provides a cross section of the size and capability of all three vehicle types. As can be seen, once a vehicle is big enough for a heavy-duty hydraulic system, then that is the class of system on which the manufacturers tend to concentrate. But do not discount the all-electric vehicles when manufacturers such as Seaeye and Sub-Atlantic are marketing 20,000 ft (6000 m) work vehicles. A relatively new player on the block, Elsub Technologies in Norway, is touting their 200 hp all-electric vehicle that carries an auxiliary hydraulic package for manipulator and tooling operations.

Oceaneering had also developed an all-electric vehicle—the E-Magnum. Unfortunately, after 7-plus years of reliable operation, it was lost with the *Deepwater Horizon* catastrophe. They are now looking at designs for use in the Arctic where an all-electric vehicle called *Calypso* will be operated from shore through the main field umbilical for up to 6 months between accessibility visits. *Calypso*, which will be called a resident ROV, will handle observation, maintenance, and light work tasks. On top of this, they are drawing on their experience supporting NASA and the space program to develop an electric manipulator that will match the functionality of today's hydraulic manipulators.

Table 6.2 Electric Versus Hydraulic ROVs

Name	Manufacturer	Class	Maximum Depth	Approximate Size HxLxW	Weight in Air	Electric Power	Hydraulic Power
			Electric Versus Hydraulic ROVs				
VideoRay Pro4	VideoRay	Observation	1000 ft 300 m	11 × 15 × 9 in. 29 × 38 × 22 cm	13.5 lb 6.1 kg	100 – 240 VAC	N/A
LBV300-5	Seabotix	Observation	1000 ft 300 m	10 × 21 × 18 in. 0.26 × 0.52 × 0.45 m	29 lb 13 kg	1.3 hp 1 kW	N/A
Outland 1000	Outland Technology	Observation	1000 ft 300 m	11 × 25 × 15 in.	39 lb	1.7 hp 1.3 kW	N/A
Stingray	Teledyne Benthos	Observation	1150 ft 350 m	18 × 48 × 18 in. 0.46 × 1.22 × 0.46 m	70 lb 32 kg	3 hp 2.2 kW	N/A
Falcon	Seaeye	Observation	1000 ft 300 m	20 × 39 × 24 in. 0.5 × 1 × 0.6 m	132 lb 60 kg	3.8 hp 2.8 kW	N/A
Mojave	Sub-Atlantic	Observation	1000 ft 300 m	20 × 40 × 24 in. 0.5 × 1 × 0.6 m	187 lb 85 kg	6 hp 4.4 kW	N/A
Sea-Wolf 2	Shark Marine	Observation Light work	3000 ft 900 m	24 × 36 × 27 in. 0.64 × 0.91 × 0.69 m	250 lb 113 kg	8 hp 6 kW	N/A
Mohican	Sub-Atlantic	Observation	10,000 ft 3000 m	31 × 43 × 32 in. 0.8 × 1.1 × 0.8 m	506 lb 200 kg	17 hp 13 kW	N/A
Cougar XT	Seaeye	Observation Light work	6600 ft 2000 m	2.7 × 5 × 3.3 ft 0.8 × 1.5 × 1 m	902 lb 409 kg	3 phase 380 – 480 VAC	Optional Tool skids
Comanche	Sub-Atlantic	Work/Survey	20,000 ft 6000 m	4.2 × 7 × 4.3 ft 1.3 × 2.1 × 1.3 m	2490 lb 1130 kg	47 hp 35 kW	20 hp Aux. Hyd.
Atom	SMD	Work	13,000 ft 4000 m	5 × 8 × 5 ft 1.5 × 2.5 × 1.5 m	4412 lb 2000 kg		100 hp 75 kW

						3 phase 380 – 480 VAC / 3000 VAC	Manipulator/ Tooling
Jaguar	Seaeye	Work	10–20,000 ft / 3–6000 m	5 × 7.2 × 5 ft / 1.5 × 2.2 × 1.3 m	4630 lb / 2100 kg	3 phase 380 – 480 VAC	200 hp (E)
Elsub 200	Elsub Technologies	Work	10,000 ft / 3000 m	6 × 8.2 × 5.6 ft / 1.8 × 2.5 × 1.7 m	6840 lb / 3100 kg	3000 VAC	2 × 17 hp Aux. Hyd.
Triton XLR125	Forum Energy Tech.	Work	13,000 ft / 4000 m	6.2 × 9.8 × 6.2 ft / 1.9 × 3 × 1.9 m	8000 lb / 3600 kg		125 hp / 95 kW
Millennium Plus	Oceaneering	Heavy work class	10–13,000 ft / 4000 m	6.3 × 10.8 × 5.5 ft / 1.8 × 3.3 × 1.7 m	8800 lb / 3991 kg		2 × 110 hp (E)
UHD ROV	Schilling Robotics	Heavy work class	13,124 ft / 4000 m	6.8 × 9.8 × 6.3 ft / 1.9 × 2 × 3 m	11,626 lb / 5270 kg		200 hyd hp
Maxximum	Oceaneering	Heavy work class	10–13,000 ft / 4000 m	7 × 10 × 6 ft / 2 × 135 hp(E)	10,700 lb / 4840 kg		2 × 135 hp (E)

Note: Most of the smaller observation ROVs now offer skid assemblies with hydraulics, manipulators, and tooling tailored to the ROV's size.

It was not all that long ago that offshore companies were discussing the possibility of working in 10,000 ft (3000 m) depths. Now it is commonplace and the vehicles to support that work are readily available. Operations in depths of 13,000 ft (4000 m), and much deeper, are definitely in the queue and the vehicles to work efficiently at those depths will be required. As electric actuator power densities continue to improve, the electric tools and manipulators should be coming to support the deep all-electric vehicle. Schilling Robotics is poised to offer an all-electric version of their HD ROV if the market wants it.

However, going back to the beginning of this section and the design spiral that was discussed, the choice of the type of vehicle is actually straightforward. The smallest, lightest simplest vehicle with the least impact on the support platform, and the pocketbook, that meets the operational requirements is the one to choose. It will all start with the capability of the working end of the system and end with the size of the umbilical and tether. Unfortunately, as the engineer who performs the necessary design spiral iterations will soon realize, that answer does not come easily, and the electric versus hydraulic debate will continue.

Power and Telemetry

CHAPTER CONTENTS

"Failure is not an option!"

In the 1995 Hollywood movie "*Apollo 13,*" actor Gary Sinise plays astronaut Ken Mattingly as he sits in a simulator on the ground desperately attempting to come up with a power budget to operate the stricken craft in space (commanded by astronaut Jim Lovell). The tension builds as the hours wear on and the batteries continue to deplete. The spacecraft has lost a majority of its power—and when it is gone, there will be no power to run the life support systems. There is a finite amount of power left. Each switch on the control panel energizes a system, depleting precious stored energy. On the one side of the power account is the limited power available, and on the other side is the power consumed. They must decide upon which systems are essential and which ones are not while prioritizing based upon the system's consumption versus its critical need. It is a simple math equation, but if they choose incorrectly ... *Will they make it* (see Figure 7.1)?

FIGURE 7.1

They made it—Astronaut Jim Lovell signs co-author Bob Christ's logbook in Punta Arenas, Chile, 1999.

7.1 Electrical considerations

The entire Apollo 13 team at NASA understood the power budget equation and balanced the power availability with the needs in order to achieve the objective of powering the necessary systems at the appropriate times to complete the mission. This is an excellent segue into a discussion on ROV power delivery as this is a very close metaphor for remote power and data delivery to any subsea vehicle through a tether. Note: this chapter assumes a general knowledge of electricity and electrical power delivery.

The following sections discuss specific issues and relationships regarding the tether, power, data, and connectors that bring it all together.

7.1.1 "So you wanna design an ROV—are you sure?"

Imagine a fire hose pumping water to douse a fire. With a given diameter hose, there is a nominal amount of water that can be pumped through that hose at a given pressure. To increase flow, the diameter of the hose must be increased (*lower resistance*), the pipe must be shortened (also *lower resistance*), or the water pressure at the source must be increased (*increase pressure*). The water requirement (*the requirement*), of course, depends upon the size of the fire (i.e., the mission driving *the requirement*). The direct correlation to an ROV system varies the length/size of the conductors in the umbilical/tether (*resistance*) to carry the electrical flow and the voltage (*pressure*) at the power source to push the power down the line to satisfy the vehicle power requirements (*requirement*).

As discussed in Chapter 6, an ROV system is a long series of trade-offs between cost, weight/buoyancy, energy requirements, and materials. In this chapter, the various components in the power delivery system will be covered with a focus on the power delivered to the vehicle (as that is the power requirement for the system).

7.1.2 Power systems (general)

An ROV power system is requirements driven. The general rule of thumb is to start with the end requirement, and then build back toward the generator to size the power source. The default is to always get a "bigger generator" (due to mission creep)—which typically sets the design spiral spinning out of control. The general rule is to make all components big enough (but not bigger), to high enough specifications (but not higher) while using materials of the best quality (but not better). This optimizes weight, performance, and cost (but is not an easy task).

The supply side of the requirements budget is then generated with any conversion from the resident input power (e.g., AC to DC, high-to-low/low-to-high voltage, electrical to mechanical) resulting in a loss in efficiency, thus increasing the input power requirement. Each component (e.g., cameras, lights, thrusters) has an individual power requirement. Each conductor has a resistance, as does each connection in the circuit. Each energy conversion costs efficiency. The power budget must balance in order to achieve the required operation at the end effector or sensor (as this is the mission). Supply must be greater than or equal to the demand (Figure 7.2) in order for the system to function as designed.

The power requirements progression, from the smallest of vehicles to the largest, emulates a logarithmic curve (Figure 7.3); as components are added, the vehicle size increases (and the power requirements take off!). All vehicles from OCROV to WCROV require some common components,

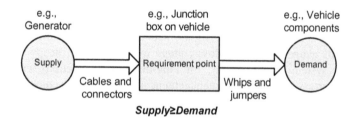

FIGURE 7.2

Top-side power budget diagram.

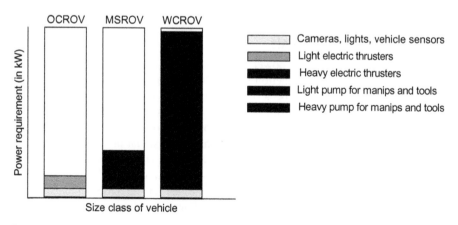

FIGURE 7.3

Power requirements (in kilowatts) by component class and vehicle category.

Table 7.1 System Components by Vehicle Size

Component	OCROV	MSROV	WCROV
Lights	X	X	X
Cameras	X	X	X
Thrusters	X	X	X
Manipulators	[optional]	[optional]	X
Transformer		[optional]	X
TMS	[optional]	[optional]	[optional]
Hydraulic pump		[optional]	X
LARS (see note)	[optional]	X	X
Electronics	X	X	X
Vehicle sensors	X	X	X
Payload sensors	[optional]	[optional]	[optional]
Tooling	[optional]	[optional]	[optional]

with the power-hungry component requirements dramatically increasing the vehicle's size and power budget. OCROVs require cameras, lights, and low-power thrusters, all powered through short lengths of small gauge (typically), low-voltage conductors. All deepwater vehicles require high-voltage transfer systems (typically AC for long-distance conduction) with several electrical and mechanical converters at the tether management system (TMS)/vehicle combination. Some common components requiring power (note: electric power drives both the electrical components directly as well as the mechanical pump to drive the hydraulic components) are listed in Table 7.1 by size class. Note: there are a few fully electric WCROVs (i.e., no hydraulic components on the vehicle), but the vast majority is hydraulic; therefore, only hydraulic WCROVs are considered here. Also, the LARS is typically on its own power circuit.

7.1.3 Power system arrangements

The general arrangement of an ROV power delivery system is designed to comply with the full vehicle power required at the initial junction box (J-Box) on the vehicle (i.e., where the power first arrives at the vehicle). Should there be a TMS upstream of the vehicle, those requirements are added to the power budget demand. From that point, power is converted and then sent to separate "bus" circuits to provide the necessary power to various components and conditioned to the appropriate parameters (e.g., AC or DC, single- or multi-phase, sinusoidal or square wave, nominal voltage). Power is split into high-voltage (HV) and low-voltage (LV) circuits for driving the high-demand items with the HV power, while lowering the voltage (and rectifying to DC) for the sensors and electronics. The power can be converted on the surface, at the TMS, or on the vehicle itself before distributing to the various bus circuits, depending upon the configuration of the system. Lights, for instance, may be powered via a DC or AC source depending upon the vehicle power available. Cameras and sensors are typically DC-powered while hydraulic pumps, thrusters, and camera pan and tilt (P&T) units can be either AC or DC (but will certainly be on an HV circuit).

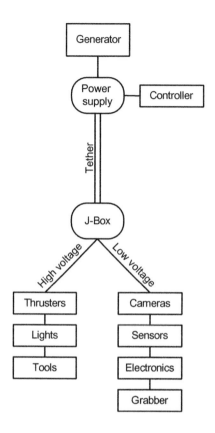

FIGURE 7.4

General power arrangement for an OCROV.

Figure 7.4 depicts a general arrangement for a DC-powered OCROV. Figures 7.5 and 7.6 depict MSROVs running on both AC, as well as DC (respectively), for powering their high-power consumption components. The OCROV unit is straightforward as is the shallow-water MSROV (it is typically very similar to the OCROV in arrangement). On the deepwater MSROV diagrams, the power is generated at the source and then routes through a power distribution unit (PDU). The PDU then splits the various power feeds into their components for conditioning and transfer. For the WCROV (Figure 7.7), the assembly is quite similar to the MSROV but the HV circuit immediately drives a hydraulic pump for powering movable parts (e.g., thrusters, P&T, manipulators, or tooling).

7.1.4 Rotary joints (a/k/a "slip rings")

7.1.4.1 Electrical slip ring

The rotary joint (commonly referred to as a "slip ring") is an electrical (but can also contain fiber-optic junctions in the same unit) junction for converting electrical signal/power from a static

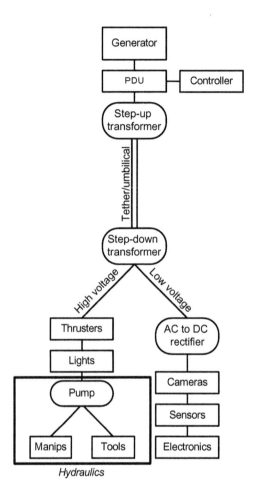

FIGURE 7.5

General arrangement for an AC-powered MSROV.

(non-moving) structure to a rotary structure (e.g., deck power lead to a rotating winch or the frame of a TMS to the tether reel). The frame of the winch or tether reel will hold the slip ring housing with its stationary contacts (Figure 7.8), while the rotary joint, with its complementary contacts, rotates with the drum.

7.1.4.2 Fiber-optic rotary joints

The function of the fiber-optic rotary joint (FORJ) is to provide an uninterrupted passive union between a stationary and a rotary fiber (typically for optical transfer from a stationary frame to a rotary winch). The FORJ is a very precise optical instrument for aligning a very small fiber element with another separate fiber element which is rotating. The basic principal of operation is to refract

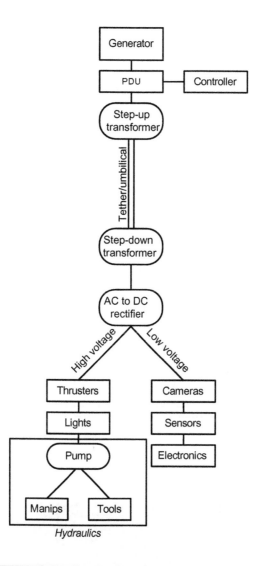

FIGURE 7.6

General arrangement for a DC-powered MSROV.

the beam of a fiber into a graded index lens (GRIN) to pass through some medium (air or clear fluid) to another GRIN lens for reversing the process. The number of fiber optical paths determines the complexity of the joint. The smaller the number of optical paths passing through the FORJ (termed "passes") (typically), the lower the loss of transmission due to increased simplicity as well as shortened optical path through the air or fluid medium. Figure 7.9 depicts the basic operation of a single-pass fiber FORJ while Figure 7.10 shows a multi-pass FORJ (the fiber "cells" can be extended to n number of cells) for multi-mode fiber. For the multi-pass single-mode FORJ, the optical path channeled through fiber is replaced with an optical path through air or fluid.

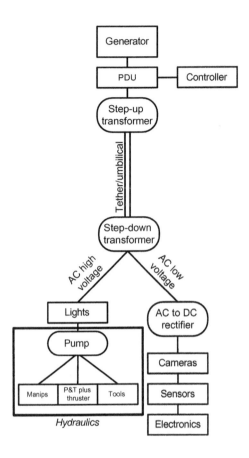

FIGURE 7.7

General arrangement for a WCROV.

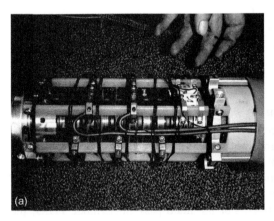

FIGURE 7.8

Electrical slip ring (a) assembled unit and (b) contact between stationary and rotary junction.

(*Courtesy SeaTrepid.*)

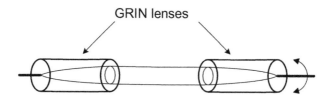

FIGURE 7.9

Diagram of a single-pass FORJ.

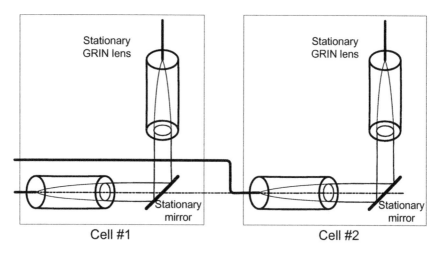

FIGURE 7.10

Diagram of a multi-mode multi-pass FORJ.

7.1.5 The ubiquitous ground fault

It is said that electricity is lazy as it will take the least possible resistance within a circuit. The conductive properties of seawater are such that as the salinity increases so does its conductivity. Any break of the insulation through the power circuit of an ROV will (instead of continuing through the circuit) immediately seek ground state by faulting to ground. It is essential for electrical safety to have grounding protection by way of a ground fault interrupt (GFI) circuit to protect the vehicle from dangerous circuit leakage.

In Figure 7.11, the basic concept of operation of a GFI circuit is depicted. During normal operations, the live (energized) line (L) conducts the same current as the return via the neutral line (N). As the flows are balanced, no magnetic flux is generated into the transformer core (3). Should an imbalance occur between the L and N lines, flux is generated in the core (3) and a current is thus induced into the surrounding coil (2). Should the current exceed the nominal value, the sensor unit (1) will then break the circuit. The tester switch (4) completes the unit. The GFI is also referred to as a ground fault circuit interrupter, residual-current device, or residual-current circuit breaker depending upon local preference.

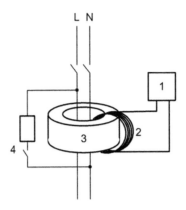

FIGURE 7.11

Basic GFI circuit.

On the smaller OCROVs, a single GFI circuit typically protects the entire system, while on larger more complex systems, a GFI circuit could isolate each bus so as not to take the entire system offline should one circuit fault to ground.

7.1.6 The tether

The tether and the umbilical are essentially the same item. The cable linking the surface to the cage or TMS is termed the "umbilical," while the cable from the TMS to the submersible is termed the "tether." Any combination of electrical junctions is possible in order to achieve power transmission and/or data relay. For instance, AC power may be transmitted from the surface through the umbilical to the cage, where it is changed to DC to power the submersible's thrusters and electronics. Further, video and data may be transmitted from the surface to the cage via fiber optics (to lessen the noise due to AC power transmission) and, then changed to copper for the portion from the cage to the submersible, thus mitigating the AC noise problem. Figure 7.12 is an example of the neutrally buoyant tether for the Outland 1000 observation-class ROV system (courtesy of Outland Technology).

The umbilical/tether can be made up of a number of components:

- Conductors for transmitting power from the surface to the submersible
- Control throughput for telemetry (conducting metal or fiber optic)
- Video/data transmission throughput (conducting metal or fiber optic)
- Strength member allowing for higher tensile strength of the cable structure
- Lighter-than-water filler that helps the cable assembly achieve neutral buoyancy
- Protective outer jacket for tear and abrasion resistance

Most observation-class ROV systems use direct current power for transmission along the tether to power the submersible while MSROVs and WCROVs use AC for transmission and then a combination on the vehicle. The tether length is critical in determining the power available for use at the vehicle. The power available to the vehicle must be sufficient to operate all of the electrical equipment on the

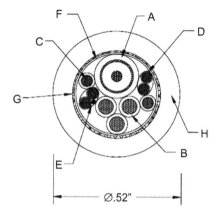

FIGURE 7.12

Tether configuration for OCROV.
A—75-ohm mini coax. Cap. 16.6 pf/ft 22-7 TC, XLPO-Foam, Alum,1 spiral T.C. shield
B—3 × 3 #18 (19/30) T.C. (OR/BL/PUR)
C—2 × 3 #22 (19/34) T.C. (WH)
D—1 × 3 #24 AWG TP, (19/36) T.C.
E—1 × 3 #24 AWG TSP, (19/36) T.C. polypropylene insulation 0.010 wall
F—MYLAR tape
G—KEVLAR weave
H—0.045 Green cellular foam polyurethane

(Courtesy Outland Technologies.)

submersible. The electrical resistance of the conductors within the tether, especially over longer lengths, could reduce the vehicle power sufficiently during high-load conditions to effect operations.

The maximum tether length for a given power requirement is a function of the size of the conductor, the voltage, and the resistance. For example, using a water pipe analogy, there is only a certain amount of water that will flow through a pipeline at a given pressure. The longer the pipe, the higher the internal resistance to movement of the water. As long as the water requirements at the receiving end do not exceed the delivery capacity of the pipe (at a given pressure), the system delivery of water will be adequate. If there were to be a sudden increase in the water requirement (a fire requiring water, everyone watering their lawn simultaneously, etc.), the only way to get adequate water to the delivery end would be to increase the pressure or to decrease the resistance (i.e., shorten the pipe length or increase the diameter) of the pipe. The same holds true in electrical terms between tether length, total power required, voltage, and resistance (Figure 7.13).

Ohm's law deals with the relationship between voltage and current in an ideal conductor. This relationship states that the potential difference (voltage) across an ideal conductor is proportional to the current through it. So, the voltage (V, or universally as E) is equal to the current (I) times the resistance (R). This is stated mathematically as $V = IR$. Further, power (measured in watts) delivered to a circuit is a product of the voltage and the current.

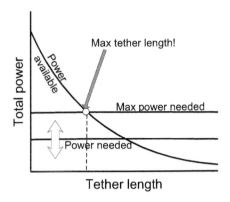

FIGURE 7.13

Graphical depiction of maximum tether length.

Table 7.2 Standard Copper Wire Gauge Resistance over Nominal Lengths	
Wire Gauge (Approximately)	**Ohms/1000 ft**
20	10
18	6
16	4
14	2.5
12	1.5
Source: Example and Table Courtesy of DeepSea Power and Light.	

Thus, based on Ohm's law, the voltage drop over a length of cable can be calculated by using the formula, $V = IR$, where V is the voltage drop, I is the current draw of the vehicle in amps, and R is the total electrical resistance of the power conductor within the tether in ohms. The current draw of a particular component (light, thruster, camera, etc.) can be calculated if the wattage and voltage of the component are known. The current draw is equal to the component wattage divided by the component voltage (or amps = watts/volts).

For example, referring to the table of electrical resistances for various wire gauges (Table 7.2), the voltage required to operate a 24-V/300-W light at 24 V over 250 ft of 16-gauge cable can be calculated as follows: The current draw, I, of a 24-V/300-W lamp operating at 24 V is 300 W/ 24 V = 12.5 A. The resistance of 16-gauge wire is approximately 4 Ω/1000 ft (Table 7.2). Since the total path of the circuit is from the power supply to the light and back to the power supply, the total resistance of the cable is twice the length of the cable times the linear resistance, or for this example, $R = (2 \times 250 \text{ ft}) \times (4 \text{ }\Omega/1000 \text{ ft}) = 2.0 \text{ }\Omega$. Since $V = IR$, the voltage drop, V, is equal to 12.5 A $\times$ 2.0 Ω = 25 V. This means that 25 V is lost due to resistance, so the power supply will need to provide at least 49 V (the 24 V necessary to operate the light plus the additional voltage loss of 25 V) to power this 24-V/300-W light over a 250-ft cable.

Further consideration to the makeup of cables and connectors is discussed in detail in Chapter 8.

7.1.7 Power source

7.1.7.1 Electrical power

The ROV system is made up of a series of compromises. The type of power delivered to the submersible is a trade-off of cost, safety, and needed performance. Direct current (DC) allows for lower cost and weight of tether components; since inductance noise is minimal, it allows for less shielding of conductors in close proximity to the power line. Alternating current (AC) allows longer transmission distances than that available to DC while using smaller conductors.

Most operators of ROV systems specify a power source independent of the vessel of opportunity. The reason for this separation of supply is that the time the vessel is in most need of its power is normally the time when the submersible is most in need of its power. Submersible systems attempting to escape a hazardous bottom condition have been known to lose power at critical moments while the vessel is making power-draining repositioning thrusts on its engines. This can cause entanglement of the vehicle. With a separate power source, submersible maneuvering power is separated from the power needs of the vessel.

With the advent of the lightweight microgenerators for use with small ROVs, the portability of the ROV system is significantly enhanced. Some operators prefer usage of the battery/inverter combination for systems requiring AC power. Also, some smaller systems use only DC as their power source. Either method should have the power source capable of supplying uninterrupted power to the system at its maximum sustained current draw for the length of the anticipated operation.

7.1.7.1.1 AC versus DC considerations

Electrical power transmission techniques are an important factor in ROV system design due to their effect upon component weights, electrical noise propagation, and safety considerations. The DC method of power transmission predominates the observation-class ROV systems due to the lack of need for shielding of components, weight considerations for portability, and the expense of power transmission devices. On larger ROV systems, AC power is used for the umbilical due to its long power transmission distances, which are typically not seen by the smaller systems. AC power in close proximity to video conductors could cause electrical noise to propagate due to electromotive force (emf) conditions. The shielding necessary to lower this emf effect could cause the otherwise neutrally buoyant tether to become negatively buoyant, resulting in vehicle control problems. And the heavy and bulky transformers are a nuisance during travel to a job site or as checked baggage aboard aircraft.

Larger work-class systems normally use AC power transmission from the surface down the umbilical to the TMS (the umbilical normally uses fiber-optic transmission, lowering the emf noise through the video) since the umbilical does not require neutral buoyancy. At the TMS (if provided), the AC power is then rectified to DC to run the DC-powered components through the neutrally buoyant tether that runs between the TMS and the vehicle.

7.1.7.1.2 Data throughput

The wider the data pipeline from the submersible to the surface, the greater the ability for the vehicle to deliver to the operator the necessary job-specific data as well as sensory feedback needed to properly control the vehicle. With the cost of broadband fiber-optic transmission equipment dropping into the range of most small ROV equipment manufacturers' budgets, more applications and sensors should soon become available to the ROV marketplace. The ROV is simply a delivery platform for transporting the sensor and tooling packages to the work location. The only limitation to full sensor feedback to the operator will remain one of lack of funding and imagination. The human–robot interaction (the intuitive interaction protocol between the human operator and the robotic vehicle) is still in its infancy; however, sensors are still outstretching the human's ability to interpret this data fast enough to react to the feedback in a timely fashion. This subject is probably the most exciting field of development for the future of robotics and will be of considerable interest to the next generation of ROV pilots.

7.1.7.1.3 Data transmission and protocol

Most small ROV manufacturers simply provide a spare shielded twisted pair (STP) of conductors for hard-wire communication of sensors from the vehicle to the surface, while on larger vehicles the sensor output is routed through the vehicle's telemetry system or with a separate break-out fiber. The strength of the break-out method is that the sensor vendor does not need engineering support from the ROV manufacturer in order to design these sensor interfaces. The weakness is that unless the sensor manufacturers collude to form a set of transmission standards, each sensor connected to the system "hogs" the data transmission line to the detriment of other sensors needed for the task. A specific example of this problem is the need for concurrent use of an imaging sonar system and an acoustic positioning system. Unless the manufacturers of each sensor package agree upon a transmission protocol to share the single STP data line, only one instrument may use the line at a time. A few manufacturers have adapted industry standard protocols for such transmissions, including TCP/IP, RS-485, and other standard protocols. The most common protocol, RS-232, while useful and seemingly ubiquitous in the computer industry, is distance limited through conductors, thus causing transmission problems over longer lengths of tether. On larger ROV systems, a common (albeit expensive) technique is to place a multiplexer into a separate circuit for gathering sensor data feed into a central point and then transmitting through a separate copper or fiber line. More on this technique in Chapter 17.

7.1.7.1.4 Underwater connectors

The underwater connector is said to be the bane of the ROV business. Salt water is highly conductive, causing any exposed electrical component submerged in salt water to "leak" to ground. The result is the "ubiquitous ground fault" as discussed in Section 7.1.4. The purpose of an underwater connector is to conduct needed electrical currents through the connector while at the same time squeezing the water path and sealing the connection to lower the risk of electrical leakage to ground.

The underwater connector is lined with metal or synthetic rubber that blocks the ingress path of water while allowing a positive electrical connection. Connectors sometimes experience cathodic delamination, causing rubber peeling and flaking from the connector walls.

Even when the contacts are right and the connector has good design features, the connector must be appropriate for the intended use and environment. The connector materials must be able to withstand the environmental conditions without degradation. For example, extended exposure to sunlight (ultraviolet energy) will cause damage to neoprene, and many steels will corrode in seawater. Check that the connector will fully withstand the environment.

The connector must not adversely affect the application. For example, all ferrous materials (steel, etc.) should be avoided in cases where the connectors' magnetic signature might affect the system. In extreme cases, even the nickel used under gold plating could have an effect and should be reviewed.

The physical size of the connector, its weight, ease of use (and appropriateness for the application), durability, submergence (depth) rating, field repairability, etc. should all be assessed. The use of oil-filled cables or connectors should be considered.

Ease of installation and use is especially important, so realistically appraise the technical ability of those personnel who will actually install or use the equipment. If they are inexperienced, a more "user-friendly" connector may be a better choice. And, if possible, train operators in the basics of proper connector use: Use only a little lubricant, avoid over-tightening (this is an especially common and annoying problem), note acceptable cable bending radii, provide grounding wires for steel connectors in aluminum bulkheads, etc.

Splicing and repairing underwater cables and connectors, while quite simple, require some basic precautions to avoid water ingress into the electrical spaces, thus grounding the connection. See Chapter 8 for a detailed discussion of cables and connectors.

7.1.7.2 Electrohydraulic power

While the OCROV has electrical energy exclusively driving moving components, both the MSROV and WCROV typically involve some form of fluid component drive (i.e., hydraulics) for a portion or all of its moving components. On all hydraulic components, the electric circuit drives an electric motor which in turns drives a hydraulic pump at the design specifications of the pump manufacturer. On MSROVs, this typically means that the vehicle manufacturer modifies (or directly uses) a thruster motor to turn the low-volume pump, while the WCROV manufacturer will specially design an HV motor (or design the hydraulic system around a COTS HV motor) to achieve the system's design objective involving a large number of hydraulic components.

As an example, the Schilling design philosophy for driving its WCROV power system involves a single high-capacity electrical motor with double-ended splines to drive the vehicle's two easily replaceable pumps—one high-capacity pump for the main hydraulic system (for thrusters, manipulators, valve packs, and (optionally) P&T unit) and the other for the auxiliary/tooling system (Figure 7.14). The purpose of the dual hydraulic systems is to maintain the cleanliness of the main

Aux pump

Main pump

LP filters

HP filters

FIGURE 7.14

Schilling design for single motor driving a two-pump system for separate circuits.

(Courtesy Schilling Robotics.)

hydraulics as tooling is typically interchangeable and will often induce contaminated oil into an otherwise sanitary hydraulic system. Clogged valves for manipulators or thrusters will produce unexpected movements, possibly compromising safety.

Other manufacturers use variations on this combination with single hydraulic circuits but introducing multiple layers of filtering to reduce the possibility of oil contamination from tooling components.

7.2 Control systems

The control system manages the different functions of the ROV, from the propulsion system to driving valve packs and to the switching on and off of the light(s) and video camera(s). From simple relay control systems in the past to today's digital fiber optics, these systems are equipped with a computer and subsystem control interface. The control system has to manage the input from the operator at the surface and convert it into actions subsea. The data required by the operator on the surface to accurately determine the position in the water is collected by sensors (sonar and acoustic positioning) and transmitted to the operator.

Over the past 10−15 years, computers utilized for these purposes have been designer computers with sophisticated computer programs and control sequences. Today, one can find standard computers in the heart of these systems. There has been a shift back to simpler control systems recently with the commercial advent of the PLC (Programmable Logic Computer). This is used in numerous manufacturing processes since it consists of easily assembled modular building blocks of switches, analog in/outputs, and digital in/outputs.

7.2.1 The control station

Control stations vary from large containers, with their spacious enclosed working area for workclass systems, to simple PC gaming joysticks with personal head-mounted displays for some micro-ROV systems. All have in common a video display and some form of controlling mechanism (normally a joystick, such as that in Figure 7.15). On older analog systems, a simple rheostat controls the variable power to the electric motors, while newer digital controls are necessary for more advanced vehicle movements.

With the rise of robotics as a subdiscipline within electronics, further focus highlighted the need to control robotic systems based upon intuitive interaction through emulation of human sensory inputs. Under older analog systems, a command of "look left/go left" was a complex control command requiring the operation of several rheostats to gain vector thrusting to achieve the desired motion. As digital control systems arose, more complex control matrices could be implemented much more easily through allowing the circuit to proportionally control a thruster based upon the simple position of a joystick control. The advent of the modern industrial joystick coupled with programmable logic circuits has allowed easier control of the vehicle while operating through a much simpler and more intuitive interface. The more sensors available to the "human" that allow intuitive interaction with the "robot," the easier it is for the operator to figuratively operate the vehicle from the vehicle's point of view. This interaction protocol between operator and vehicle has become known as the human−robot interaction and is the subject of intensive current research. Look for major developments within this area of robotics over both the short and long terms.

FIGURE 7.15

Common joystick.

(Courtesy VideoRay.)

7.2.2 Motor control electronics

Since OCROV and MSROV systems use mainly electronic motors for thruster-based locomotion, a study of basic motor control is in order.

A basic control of direction and proportional scaling of electrical motors is necessary to finely control the motion of the submersible. If only "On" or "Off" were the choices of motor control via switches, the operator would quickly lose control of the vehicle due to the inability to make the fine corrections needed for accurate navigation. In the early days of ROVs, the simple analog rheostat was used for motor control. It was quite a difficult task to control a vehicle with the operation of three or four independent rheostat knobs while attempting to fly a straight line. Later came digital control of electric motors and the finer science of robotics took a great leap forward.

The basic electronic circuit that made the control of electronic motors used in robotics and industrial components so incredibly useful is known as the "H-bridge." An understanding of the H-bridge (discussed later in this section) and the digital control of that H-bridge will help significantly with the understanding of robotic locomotion.

Consider the analysis of a simple electric circuit (Figure 7.16). As discussed earlier in this chapter, Ohm's law gives a relationship between voltage (V), current (I), and resistance (R) that is stated as $V = IR$. In Figure 7.16, the current and voltage are known, and thus the resistance can be calculated to be 3.0 Ω.

7.2.2.1 Inductors

An inductor is an energy storage device that can be as simple as a single loop of wire or consist of many turns of wire wound around a core. Energy is stored in the form of a magnetic field in or around the inductor. Whenever current flows through a wire, it creates a magnetic field around the wire. By

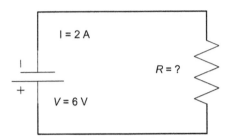

FIGURE 7.16

Sample electric circuit.

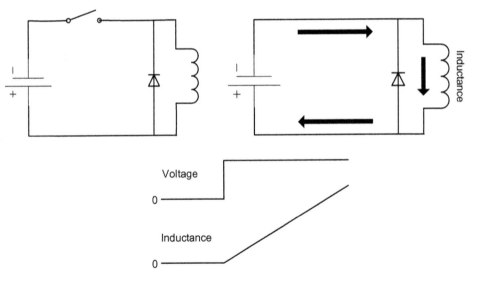

FIGURE 7.17

Inductor—voltage interaction without resistance.

placing multiple turns of wire around a loop, the magnetic field is concentrated into a smaller space, where it can be more useful. When applying a voltage across an inductor, current starts to flow. It does not instantly rise to some level, but rather increases gradually over time (Figure 7.17).

The relationship of voltage to current versus time gives rise to what is known as "inductance." The higher the inductance, the longer it takes for a given voltage to produce a given current—sort of a "shock absorber" for electronics. Whenever there is a moving or changing magnetic field in the presence of an inductor, that change attempts to generate a current in the inductor. An externally applied current produces an increasing magnetic field, which in turn produces a current opposing that applied externally and hence the inability to create an instantaneous current change in an inductor. This property makes inductors useful as filters in power supplies. The basic mathematical expression for inductance (L) (with the SI units in *henries*) is $V = L \times dI/dT$.

The ideal world and the real world depart, since real inductors have resistance. In this world, the current eventually levels out, leaving the strength of the magnetic field plus the level of the stored energy as proportional to the current (Figure 7.18).

So, what happens when the switch is opened? The current dissipates quickly in the arc (Figure 7.19). Diodes are used to suppress arcing, allowing the recirculation currents to dissipate more slowly. The current continues to flow through the inductor, but power is dissipated across the diode through the inductor's internal resistance. This is the basis of robotic locomotion electronic dampening (sort of a shock absorber for your thruster's drive train).

Permanent magnet DC motors can be modeled as an inductor, a voltage source, and a resistor. In this case, the torque of the motor is proportional to the current and the internal voltage source is proportional to the rpm (Figure 7.20) or back emf (when the current is released and the motor turns into a generator as the motor spools down). The stall point of an electric motor is at the point of highest torque as well as the point of highest current and is proportional to the internal resistance of the motor.

That leads to the concern over back emf within thruster control electronics design—once the DC motor is disengaged, the turning motor mass continues to rotate the coils within the armature. This rotating mass converts the electric motor into a generator, rapidly reversing and spiking the voltage in the reverse direction. Unless there is some circuit protection within the driver board circuit, damage to the control electronics will often result (Figure 7.21).

7.2.2.2 The H-bridge

The circuit that controls the electrical motor is known as the "H-bridge" due to its resemblance to the letter "H" (Figure 7.22). Through variation of the switching, as well as inductance filtering (described earlier), the direction and the ramp-up speed are controlled through this circuit in an elegant and simplistic fashion.

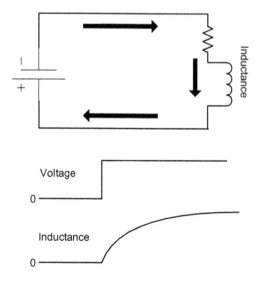

FIGURE 7.18

Inductor–voltage interaction with resistance.

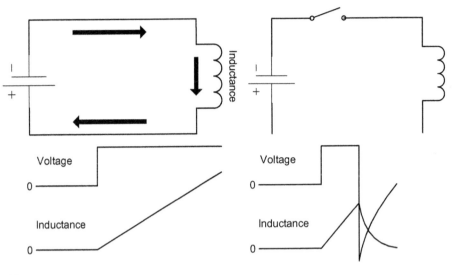

FIGURE 7.19

Inductor–voltage interaction as circuit is opened.

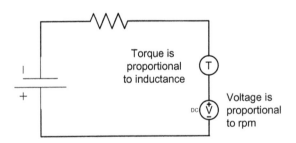

FIGURE 7.20

Diagram depicting inductance–torque interaction.

7.2.2.3 PWM control

Pulse width modulation (PWM) is a modulation technique that generates variable-width pulses to represent the amplitude of an analog input signal. The output switching transistor is on more of the time for a high-amplitude signal and off more of the time for a low-amplitude signal. The digital nature (fully on or off) of the PWM circuit is less costly to fabricate than an analog circuit that does not drift over time.

FIGURE 7.21

Damaged driver board due to back emf.

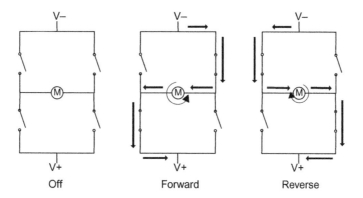

Off   Forward   Reverse

FIGURE 7.22

H-bridge diagram depicting basic operation.

PWM is widely used in ROV applications to control the speed of a DC motor and/or the brightness of a lightbulb. For example, if the line were closed for 1 μs, opened for 1 μs, and continuously repeated, the target would receive an average of 50% of the voltage and run at half speed or the bulb at half brightness. If the line were closed for 1 μs and open for 3 μs, the target would receive an average of 25%.

There are other methods by which analog signals are modulated for motor control, but OCROV and MSROV systems predominate with the PWM mode due to cost and simplicity of design.

Cables and Connectors*

This chapter has been written by three invited co-authors who have addressed underwater connector selection from three distinct perspectives: the end-user, the manufacturer, and sales engineering. Our deepest thanks go to Kevin Hardy, President of Global Ocean Design, Cal Peters, Director of Engineering for Falmat, and Brock Rosenthal, President and founder of Ocean Innovations. They have lectured on the subject of "cables and connectors" at numerous Oceans MTS/IEEE conference tutorials. (These tutorial presentation materials form the basis of this chapter. Additional information on this team is provided in the Acknowledgments section.)

8.1 Introduction

Underwater cables and connectors provide system flexibility, ease of service, and other design advantages for undersea equipment—including ROVs. The primary purpose of underwater cables and connectors is to provide a conductive path, without leakage, in a pressure-resistant or pressure-tolerant package. Cables and connectors allow for simple system configuration (Figure 8.1).

Underwater connectors are used to connect the umbilical to a tether management system (TMS) and then from the TMS to the ROV. In addition, these items interconnect separate components on the ROV to form a functioning integrated system. Connectors enable these components to be disconnected for easy removal for servicing, repairs, or upgrades.

A mated connector pair forms a unique small pressure case. All pressure case design criteria then apply. A bulkhead connector becomes a mechanical part of a pressure hull and thus should be looked at critically.

It is important to the success of any subsea project that both end users and manufacturers speak the same language from the beginning.

In this chapter, readers will:

1. Learn the language of underwater cables and connectors common to the underwater industry;
2. Learn a practical approach to specifying underwater cables and connectors for their specific application; and
3. Raise their understanding of what can be expected from cable and connector manufacturers.

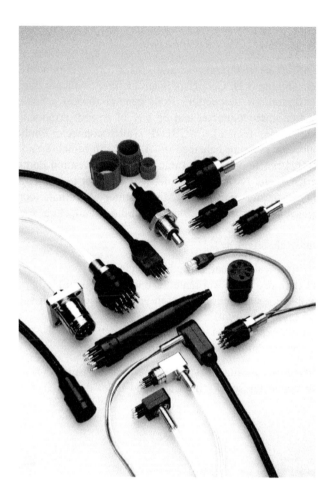

FIGURE 8.1

Cable and connectors provide flexibility in electric power and signal distribution.

(Courtesy MacArtney.)

8.2 Definitions

It is important to speak with the same terminology as your supplier. Jargon can vary somewhat among different industries, such as offshore oil and gas and the military. Generally, common terms are universally understood among applications.

The basics

Connectors are, initially, a mechanical problem. A mated connector pair functions as a small pressure case. All pressure case and seal maintenance rules thus apply. So our definitions will start with the housing.

Plug is the male body.

Receptacle is the female half.

Inside the mated housing are electrical *pins* and *sockets*. Most commonly, plugs have pins, and receptacles have sockets (Figure 8.2). Connector manufacturers typically have naming conventions for their connector configurations descending from general toward specific arrangement. As an example, a four-pin female microconnector can be any of several arrangements such as MCIL4F (microconnector in-line 4-pin female), MCBH4F (microconnector bulkhead 4-pin female), MCPBOF4F (microconnector pressure-balanced oil-filled 4-pin female), and MCDC4F (microconnector dummy connector 4-pin female). The same naming convention continues with low profile (LP), standard circular (SC), etc. As a general rule, it is best not to mix connector manufacturers' products for the same connector unit (e.g., Manufacturer A's male connector plugging into Manufacturer B's receptacle). When in doubt, check with the manufacturer for compatibility between connectors.

Connectors are broadly classified by:

- Connector type (i.e., electrical, optical, or optical/electric)
- Mating environment (i.e., dry mate, wet mate, or underwater mate)
- Voltage (e.g., low voltage, medium voltage, high voltage, or very high voltage)
- Amps (i.e., low/medium/high current)
- Number of contacts (self-explanatory)
- Pressure rating (i.e., low/medium/high/extreme pressure)

Bulkhead (BH) connectors can attach to a pressure housing using O-rings to make the seal. A threaded post bulkhead connector screws into a tapped hole. A spot face (i.e., a smooth, flat or

FIGURE 8.2

Bulkhead connector receptacle (left) and plug (right).

(Courtesy Ocean Innovations.)

conformal curved, accurately located surface) provides the sealing surface for the O-ring face seal. A threaded post bulkhead connector can also be mounted using a simple clearance through-hole and secured with a nut and washer on the interior side. A flange-mount bulkhead connector is attached to a pressure housing using machine screws that secure the connector. The screws pull the flange mount against the pressure housing, compressing the O-ring against a spot face on the pressure housing, and thereby making the watertight seal. Particular attention must be paid to avoid dissimilar materials and anaerobic corrosion.

Threaded post-type connectors
BCR, bulkhead connector receptacle
BCP, bulkhead connector plug

Flange mount-type connectors (Figure 8.3)
FCR, flange connector receptacle
FCP, flange connector plug

Inserts (Figure 8.4)
Fits inside connector shell
CIR, connector insert receptacle (sockets)
CIP, connector insert plug (pins)

In-line connectors or CCP/CCR, whip or pigtail (Figure 8.5)
CCP, cable connector plug
CCR, cable connector receptacle

Dummy connectors and sealing caps
These keep electrical contacts clean when the connector is not mated.

FIGURE 8.3

Flange mount bulkhead connectors.

(Courtesy Ocean Innovations.)

(a)  (b)

FIGURE 8.4

(a) Metal body insert with glass seal of pins. (b) Molded epoxy body insert with pins. Note alignment key molded into base of insert.

(Courtesy SeaCon.)

(a)  (b)

FIGURE 8.5

(a) In-line rectangular body connectors. The rectangular body is also referred to as "low profile." (b) In-line circular body connectors.

(Courtesy Ocean Innovations.)

Dummy plugs (Figure 8.6(a)) can be rated for medium depth or deepwater. Be sure the dummy plug is rated for the chosen application.

Dust caps are for surface use only (Figure 8.6(b)).

Hermaphroditic connectors

A hermaphroditic-style connector (Figure 8.7) has both male pins and female sockets on each connector.

Fiber-optic connectors (Figure 8.8)

There are two approaches used in connecting two optical fibers in a connector: expanded beam optics and physical contact. The expanded beam process uses ball optics to transmit light across the medium. Precise alignment is not required but the losses are higher. The physical contact approach utilizes

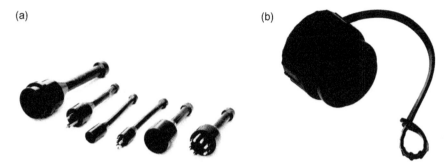

FIGURE 8.6

(a) Circular dummy connectors. (b) A splash-proof dust cap.

(Courtesy Ocean Innovations.)

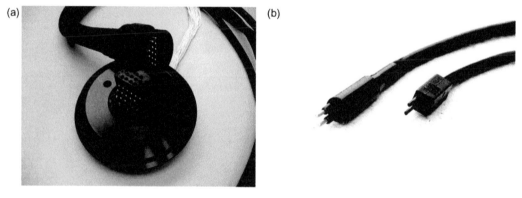

FIGURE 8.7

(a) A hermaphroditic-style bulkhead connector and in-line connector. (b) Hermaphroditic-style in-line connectors with circular and square bodies.

(Courtesy Ocean Innovations.)

polished optical fiber ends inside alignment ferrules. The mating faces must be held in contact to minimize losses, which are lower than the expanded beam approach.

Hybrid connectors

Hybrid connectors can provide contacts for power and data, as well as optical fiber passes.

PBOF

The pressure-balanced, oil-filled (PBOF) connector (Figure 8.9) provides the end user with the flexibility of making and servicing his/her own cables. By running electrical conductors or optical fibers through an oil-filled hose, the connector is easily serviced without the need to destroy a cured connector. In addition, custom underwater cables are not needed and, if required, conductors can be reconfigured or replaced.

FIGURE 8.8

Fiber-optic connectors provide optical fiber passes with the ability for disconnect.

(Courtesy of GISMA.)

FIGURE 8.9

An example of a PBOF connector.

(Courtesy MacArtney.)

Some connectors incorporate a valve for oil filling and bleeding the connector interface. The mating PBOF bulkhead connector is then rated for open face pressure since the connector itself is at ambient pressure (i.e., little or no pressure differential across the connector). Other manufacturers have adapted a PBOF back shell to their standard line of rubber-molded connectors, as the one illustrated in Figure 8.9.

The plastic or rubber hose is slid over a hose barb fitting on the back of the connector and secured with a hose clamp. Then the tube is filled with oil. A second connector is installed on the opposite end in the same manner as the first, excess air is squeezed out, and the second connector is pressed into the oil-filled tube. A hose clamp secures the tube to the back of the second connector.

Penetrators

Penetrators (Figure 8.10) bring wires to the interior of the pressure housing but cannot be disconnected.

Cable glands (Figures 8.11 and 8.12)

- Allows cable to enter an enclosure.
- Splash proof.
- Not recommended for deep submerged use. Check with the manufacturer for depth recommendations.
- Not for dynamic applications where the cable will be flexed at the gland.

Jumper

A jumper (Figure 8.13) is a cable assembly with male to female connectors, much like a common household extension cord.

Adapter cable

An adapter cable (Figure 8.14) transitions from one connector type to another.

(a)   (b)

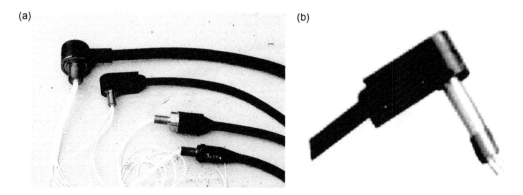

FIGURE 8.10

(a) Right angle and in-line penetrators with short threaded post for installation in threaded endcap. (b) Right angle rubber molded with extended threaded post for use with internal draw-down nut.

(a) (Courtesy Ocean Innovations.) (b) (Courtesy MacArtney.)

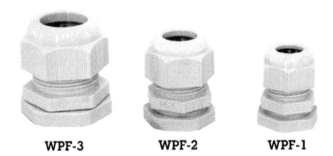

WPF-3   **WPF-2**   **WPF-1**

FIGURE 8.11

Compression seal cable gland with strain relief.

(Courtesy Newmar.)

FIGURE 8.12

Compression cable gland, exploded view.

(Courtesy Conax Technologies.)

FIGURE 8.13

This jumper uses circular profile connectors on each end of an electrical cable.

(Courtesy Ocean Innovations.)

FIGURE 8.14

An adapter cable is a specialized double-ended jumper cable using different in-line connectors. A field-installed cold splice is shown in the lower image.

(Courtesy Ocean Innovations.)

(a)

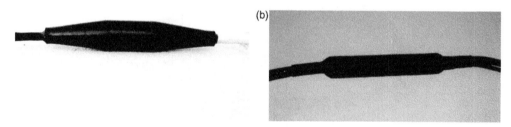

(b)

FIGURE 8.15

(a) A polyurethane-molded splice. (b) A vulcanized rubber-molded splice.

(Courtesy Ocean Innovations.)

Barrel mold

A barrel mold (Figure 8.15), a/k/a "Hot Dog" splice, is a cylindrical-shaped mold joining two or more cables together. There is very limited pull strength in a barrel mold joint.

T-splice, Y-splice

A T-splice and Y-splice are shown in Figures 8.16 and 8.17, respectively.

Breakouts

An example of breakouts is provided in Figure 8.18 where T-splices with short cables are terminated with in-line connectors for attachment to threaded post connectors intended for installation in separate pressure cases.

FIGURE 8.16

A T-splice of three cable segments.

(Courtesy Ocean Innovations.)

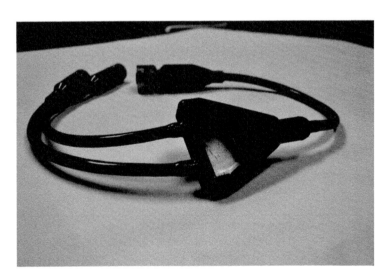

FIGURE 8.17

A Y-splice of three cable segments.

(Courtesy SeaCon.)

Junction box

A junction box, or "J-Box," is a container for electrical connections. These may be enclosed within a 1 atmosphere pressure-proof housing or an oil-filled pressure-compensated design operating at ambient pressure. Junction boxes (Figure 8.19) are used in place of hard-wired splices and

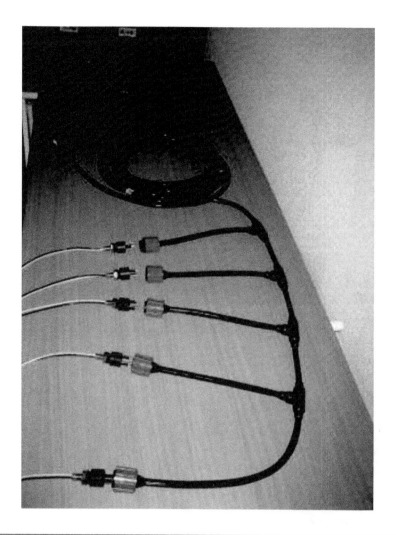

FIGURE 8.18

Main trunk cable with molded branches.

(Courtesy Ocean Innovations.)

Y-molds because they are reconfigurable and easily serviced. They do, however, take up more space than a Y-mold.

Field installable termination assembly

Unlike other termination assemblies, the field installable termination assembly (FITA) (Figure 8.20) can be assembled on site without the need for molding. The system joins two multi-conductor cables in a pressure-compensated oil-filled environment. FITA terminations can be performed by trained operators on board, making it possible to repair broken cables without lengthy down time from sending the materials back for onshore repair.

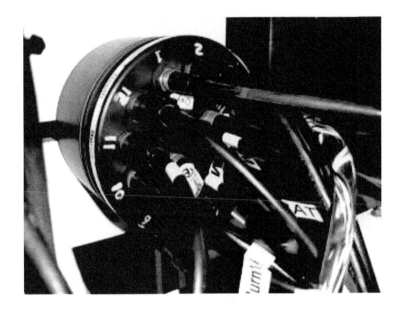

FIGURE 8.19

An example of a junction box.

(Courtesy Ocean Innovations.)

FIGURE 8.20

FITA terminations.

(Courtesy MacArtney.)

FITA terminations are often oil compensated, thus maintaining the same pressure inside the termination as outside. This allows operation at full ocean depth without the need for thick, heavy shells.

Because each conductor is isolated by an individual boot seal (should water ingress the oil-filled termination), it can still continue to operate fully flooded in salt water. Boot seals have been successfully tested to 20,000 psi (1360 bar).

FIGURE 8.21

An epoxy or rubber insert (right) is soldered to wires, then overmolded with vulcanized rubber or polyurethane to produce the connector assembly (left).

(Courtesy Ocean Innovations.)

FIGURE 8.22

A stainless steel guide pin assures correct polarity alignment prior to mating.

(Courtesy Ocean Innovations.)

Over-molding

Over-molding (Figure 8.21) is a secondary molding operation.

Guide pins

Guide pins (Figure 8.22) are nonconducting pins used for alignment.

Retaining devices

Retaining devices include:

- Locking sleeves or lock collars (Figures 8.23 and 8.24)
- Straps (Figures 8.25 and 8.26)

Strain reliefs

Strain reliefs (Figures 8.27 and 8.28) transition the stress from the point where the cable meets the connector along a greater length of cable. These are also known as "bending strain reliefs." Excessive flexing at the junction between the cable and connector is the most common failure mode of underwater connectors.

Cable grips

Cable grips transfer a load from the cable to another object (see Section 8.8.13).

Strength ratings

The following strength ratings apply to cables, terminations, cable grips, shackles, and handling equipment:

- BS- break strength or material yield
- SWL- safe working load
- Safety factor- BS/SWL and varies with application (e.g., manned versus unmanned)

(a)

(b)

FIGURE 8.23

(a) In-line locking sleeve pairs have matching internal and external threads and come in different sizes to fit different connector body sizes. (b) Locking sleeves may be color coded for circuit identification.

(Courtesy Ocean Innovations.)

FIGURE 8.24

A single piece step locking sleeve (left) is installed before the connector is bonded to the cable. A snap ring locking sleeve (center and right) may be installed or replaced at any time. The snap ring may be subject to long-term corrosion.

(Courtesy Ocean Innovations.)

FIGURE 8.25

(a) Loose retaining strap. (b) Loose retaining strap installed.

(Courtesy Ocean Innovations.)

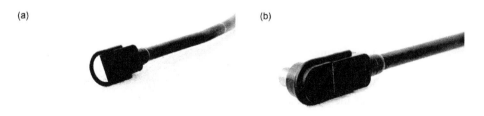

FIGURE 8.26

(a) Molded in-place retaining strap. (b) Molded in-place retaining strap in use.

(Courtesy Ocean Innovations.)

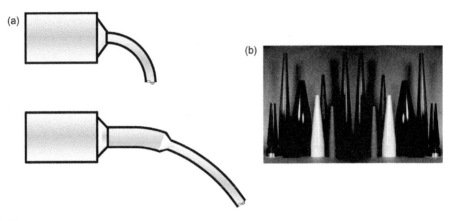

FIGURE 8.27

(a) The molded connector (above left) has little strain relief and subsequently prone to wire breakage behind the connector body. The lower molded connector distributes the load over a greater length, resulting in longer service life. (b) Polyurethane-molded boot provides strain relief for any size cable.

(*a*) (*Courtesy SeaTrepid.*) (*b*) (*Courtesy Ocean Innovations.*)

FIGURE 8.28

Cable gland with strain relief.

(*Courtesy Strantech.*)

Lubrication

Lubrication is needed for O-rings and rubber surfaces. While opinions vary, the authors recommend the use of silicone grease such as Dow Corning DC-4 or DC-111 for the O-ring and 3M Food Grade Silicone Spray Lubricant for the connector body. Food grade silicone spray (Figure 8.29) is important as the propellants are noncorrosive.

Dry versus wet mate connectors

- Dry mate connectors must be mated in a dry environment before they are deployed.
- Underwater mateable/pluggables can be mated or demated underwater but with power off.

FIGURE 8.29

Dow Corning and 3M are well-known brands of silicone lubricants.

(Courtesy Ocean Innovations.)

- Wet mateable/pluggables can be mated or demated in a wet environment, such as on the deck of a ship, before they are submerged.

8.3 Applications and field requirements, writing specifications

Underwater system designers must answer the following questions in order to select the most appropriate connector for their application. The use of these selection criteria becomes the basis for the purchasing specification.

Underwater connector selection criteria:

1. Will the connector transfer power, signal, or both?
2. What is the expected deployment duration or system service life?

3. What is the expected operating service depth and design safety factor?
4. What are the expected environmental conditions? (cold, anaerobic, saline, etc.)
5. What is the available mounting space? Does it allow room for torque wrench installation?
6. Should the connector have seal or contact redundancy?
7. Must the connector mate to an existing system?
8. Must the connector retrofit to an existing installed connector?
9. Must the connector be underwater mateable? With power on?
10. Is the connector field serviceable? If so, what level of technician skill is necessary?
11. Are dissimilar materials present between the connector and housing (which may lead to unacceptable galvanic corrosion)?
12. Is the package designed to be handled in the field without danger of incidental damage to the connector?
13. What is the cable type to be wired and sealed to the connectors (i.e., twisted single pair, parallel bundle, coaxial, electromechanical (EM))? Are there construction fillers, jacket materials, molding, or other considerations?
14. Are there any MIL-SPEC requirements to be met? Are there any other special requirements to be met (i.e., fiber optic, neutrally buoyant, PBOF)?
15. Are locking sleeves or retaining straps an option?
16. Are there cost and delivery constraints?

Use of this checklist during the early design phase of a new underwater system will help the designer or program manager avoid predictable problems with these fundamental underwater system components. Work with the intended supplier as they want the project to be successful too. Retrofitting connectors on the backside of development and deployment can be a costly and complicated process.

8.4 Underwater connector design

A bulkhead connector passes electric signals or power across the pressure barrier. The bulkhead connector becomes an integral part of the pressure housing. Its selection is critical to mission success. A penetrator brings wires through without a demateable connector. A bulkhead connector does the same but provides the option of simple disconnect.

It is always a mechanical problem first; therefore, attention must first be paid to material selection. To avoid galvanic corrosion or delamination, match the housing and connector materials or select a nonmetallic body. Use of isolation washers has helped some designers mix materials, but there is still a potential for cathodic delamination. Material strength, cost, and availability are also important.

A mated in-line or bulkhead connector pair is itself a small pressure case. All pressure case design rules apply, including O-ring seal grooves. The *Parker O-Ring Handbook* (ORD-5700) is the unmatched, unquestioned authority on O-ring seal design. You would be aghast to find how many old connector designs fail to meet "Parker spec." See "References" at the end of this chapter to get a free copy of the *Parker O-Ring Handbook*.

Metal shell bodies use nonmetallic inserts or compression glass to electrically isolate pins from the body. The nonmetallic inserts may rely on an O-ring for sealing against open face pressure. This may require servicing at some point.

Epoxy connectors are a good choice for medium depth. Their greatest weakness is to side loading. A means of protecting the cable (attached to the epoxy connector) from being pulled, especially against side loads, is an important consideration. This may be as simple as black taping it to the ROV frame.

Electrical contacts are selected for power, considering the anticipated voltage and current requirements. The mechanical design should prevent the engagement of the pin and socket before the key–keyway engages. Sharp interior edges, such as on a keyway, should be recessed from any radial O-rings. Contacts passing through epoxy bulkhead connectors must be primed to assure good bonding for open face pressure.

Weaknesses in epoxy connectors include susceptibility to damage from side loads, flammability in the event of high-voltage shorting of pins, and exposed contacts inside the female receptacle. Water retained behind the radial seal of the male connector plug when vertical can drip onto the exposed pins as the plug is retracted.

8.5 COTS underwater connectors

Commercial-off-the-shelf (COTS) underwater connectors come in a wide variety of standard configurations for both electrical and fiber-optic conductors. These "standards" can be classified into the following generic categories:

Rubber-molded connectors

These are molded, typically in Neoprene, in a number of different configurations including straight, right angle, miniature, and others. They can be capable of high pressures and are typically low cost. They can be cleanly and inexpensively molded onto jacketed cables and are easy to use.

One of the oldest rubber-molded connector designs is known by the initials of its original manufacturer, Electro-Oceanic. EO connectors (Figure 8.30) are available from several manufacturers with 2–8 contacts and are rated to 10,000 psi (690 bar) or more. Electrical rating per contact is 115 V/6.5 A or 230 V/15 A.

A vent hole to the exterior in the electrical socket allows EO connectors to be mated or unmated at any depth, though not with power on. The male pins each have two circular contact bands on them. The male pins are flared at the tip and oversized to the inside diameter of the female socket. This insures a wiping action and positive seal by forcing water through the length of the female socket and out the vent.

SubConns (manufactured by SubConn), Wet Cons (manufactured by SeaCon), or Wet Pluggables (manufactured by Teledyne Impulse) are among the most popular types of rubber-molded connectors (Figure 8.31). These are available with 1–25 contacts. The female receptacles have one or more rings molded into them that seal to the pins on the mating plug. Keep in mind, the more contacts, the more force required for mate/demate. Also, as the number of contacts increase, the outer pins are more susceptible to demating (loss of electrical connection) should the connector experience side loading. Variations of these connectors have been used to ocean trench depths.

Epoxy-molded connectors

Rigid epoxy compounds add greater strength and dimensional control to molded connectors. These assemblies perform well at pressure and are moderately low priced. Some still call this style

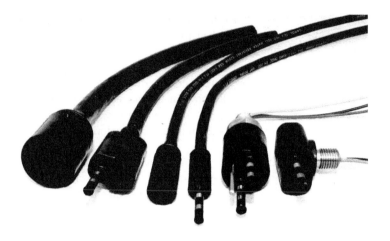

FIGURE 8.30

Examples of EO-style connectors.

(Courtesy Ocean Innovations.)

FIGURE 8.31

Rubber-molded connectors are among the most popular of underwater connectors.

(Courtesy MacArtney.)

"Marsh Marine connectors," referring to the original manufacturer of these first commercial underwater connectors.

Rubber molded to metal bodies

By molding a rubber connector into a metal body, greater strength and stability are added to the product along with positive and stronger keying and locking. The connector is more robust and able to withstand more abusive environments. This configuration is often implemented as a metal-bodied bulkhead mating to a rubber-molded plug connector.

Metal shells with molded inserts

Molded connector inserts can be O-ring sealed into mating metal shells. This configuration offers secure O-ring sealing technology and isolation of the connector elements from the working environment. They are available in a wide variety of sizes and with a large range of inserts at moderate cost.

Metal shells with glass-to-metal seal inserts

Electrical contacts can be glass sealed into metal inserts for more demanding applications of pressure and temperature. In addition, they can withstand high open face pressures. Because elastomeric seals are not used, these connectors do not typically degrade over time, making them popular in military applications where long-term high reliability is crucial. Due to their tolerance to high temperature, glass-to-metal seal connectors are also common in down-hole applications. Due to the sophisticated manufacturing process, glass-to-metal seal connectors are moderately expensive.

Fluid-filled underwater mateable

This style of connector is typically a PBOF assembly incorporating redundant sealing barriers to the environment. This enables the connector to be wet mated by an ROV. Available in electric, optic, hybrid configurations and in high-voltage and/or high-amperage models, they are often used in offshore oil and gas installations as well as the nodes of ocean observatories. Fluid-filled underwater connectors are at the high end of the cost spectrum.

Penetrators

Where systems do not need to be rapidly disconnected, a penetrator may serve as a cable termination and interface to equipment modules. Penetrator construction follows roughly the same configurations as mentioned above for connectors.

8.5.1 Mated pairs

Mated pairs are the assembly created by combining matching halves of a connector combination. In hard shell connectors (both metallic and epoxy), there is an engagement sequence that is followed. An alignment mark assists with initial orientation. Following that, the engagement sequence should follow the pattern: plug in bore, key aligns with keyway, pins and sockets engage, locking sleeve installed, final tightening to set face seal.

Retention

Once mated, the connector pair should be sufficiently restrained to avoid inadvertent disconnect due to stresses on the cable and connector. Retention mechanisms range from simple friction devices to screwed collars and latches.

Locking sleeves are used to prevent the mated pair from accidental disconnection. Locking sleeves are made from Delrin, stainless steel, or glass-filled nylon. Nylon 6/6 alone is hygroscopic and will swell with constant immersion in water, resulting in disengagement of the locking sleeve threads.

8.5.2 Advanced designs

8.5.2.1 Recent trends and future developments

Miniaturization

Smaller versions of many standard connectors and new designs for lower power electronic systems have been recently introduced (Figure 8.32).

High-speed data connectors

Specialized connectors are available for Ethernet communications with data rates of up to 1 GB/s (Figure 8.33).

Smart connectors

Three unique approaches to smart connectors may be seen with the Schilling SeaNet Connector, MBARI PUCK, and Integral Data Converters.

Schilling Robotics SeaNet connector

Concentric circular contacts allow this connector (Figures 8.34 and 8.35) to be rotated to change orientation so the cable exits in the desired direction during final assembly. The design is *not* intended for a rotary application in the manner of a slip ring. Internal indicator lights provide visual feedback of power and data status. The PBOF design makes it usable at any depth.

FIGURE 8.32

Microconnector series take up less space.

(Courtesy MacArtney.)

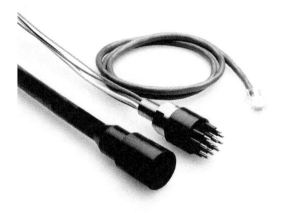

FIGURE 8.33

Example of a high-speed Ethernet data connector.

(Courtesy MacArtney.)

FIGURE 8.34

The Schilling Robotics SeaNet Connector installed on a manipulator.

(Courtesy Schilling Robotics.)

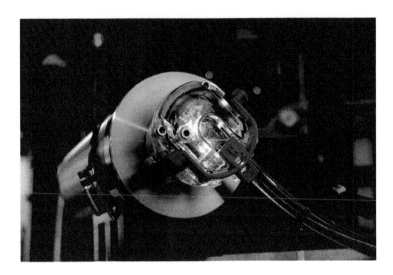

FIGURE 8.35

Schilling Robotics SeaNet Connector indicator lights provide feedback on operation.

(Courtesy Schilling Robotics.)

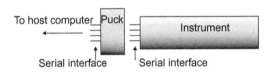

FIGURE 8.36

MBARI PUCK schematic design.

(Courtesy MBARI.)

MBARI PUCK

Monterey Bay Aquarium Research Institute (MBARI) developed the Programmable Underwater Connector with Knowledge (PUCK). The MBARI PUCK (Figure 8.36) has the advantages of being a simple, small storage device for instrument-related information (e.g., unique ID, instrument driver, other metadata). Instrument information is automatically retrieved by the host computer when the PUCK and instrument are plugged into the host.

Integral data converters

Solid-state convertor electronics are pressure tolerant and can be built into a connector or a splice. The convertor transitions electric to optical, serial to Ethernet, or RS-232 to RS-485.

Connectorless data transfer

Data noncontact connections include:

- Inductively Coupled Modem
- BlueTooth (Wi-Fi)
- Optical Infrared (IR)

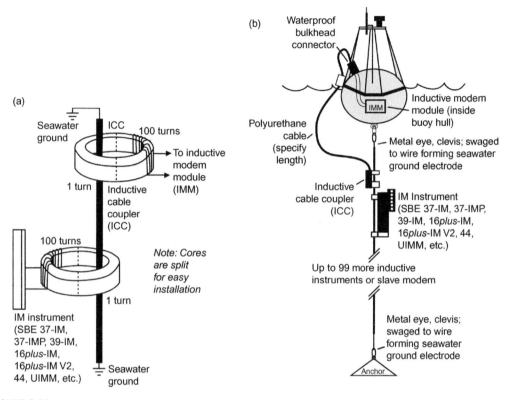

FIGURE 8.37

(a) SeaBird inductively coupled modem detail. (b) SeaBird inductively coupled modem installed.

(Courtesy SeaBird Electronics.)

- Acoustic Modems
- RF

Inductively coupled modem

An example of this design is the SeaBird inductively coupled modem (Figure 8.37). This device uses a plastic jacketed wire rope mooring cable to transmit data.

Wi-Fi

One design team at MBARI selected the 2.4 GHz RF (Wi-Fi) as its subsea communications device. The team modified and tested the Whip, Patch, and Helical antenna designs, selecting the whip as the most effective. The resultant system provided 9.8 Mbps over the few-centimeter gap between spheres underwater (Figure 8.38).

Optical data transmission

Using technology adapted from free space optics (Figure 8.39), underwater optical data transmission can achieve data rates of up to 100 mbps at distances up to 130 ft (40 m). The transmission range is highly dependent on water clarity, including turbidity and biomass.

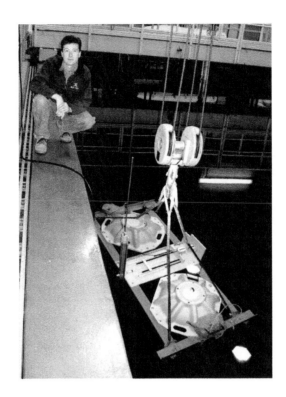

FIGURE 8.38

MBARI tests bluetooth between glass spheres.

(Courtesy MBARI.)

FIGURE 8.39

Ambalux Optical Data Transmission system in a test tank.

(Courtesy Ambalux.)

Infrared

Through an optical coupler, sapphire plate, or borosilicate sphere, designers can utilize IR for data and control signal transfer. One example is the IRTrans Wi-Fi, which provides a USB, an RS-232, or an Ethernet interface. The system includes both an infrared transmitter and receiver.

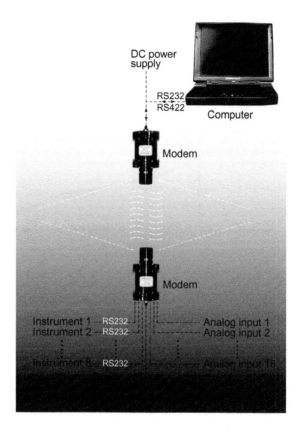

FIGURE 8.40

LinkQuest acoustic modem connection diagram.

(*Courtesy LinkQuest.*)

Acoustic modems

Acoustic modems (Figure 8.40) are commercially available from several vendors, providing reliable and long-range data rates up to 38.4 kb/s. Characterization of the water path is needed for higher data rates. Changing environmental conditions may also affect data rates.

Radio frequency

WFS Technologies has demonstrated RF systems (Figure 8.41) that provide data rates of up to 10 Mb/s at very close ranges (1−2 m), decreasing to 10 bps at longer ranges. Longer ranges are possible in fresh water or through a ground path. An air path is also possible under site-specific conditions. Unlike acoustical and optical communications, transmission through water is not affected by bubbles, thermal stratification, ambient noise, or turbidity.

Connectorless power transfer

An inductive recharge system (Figure 8.42) developed by Florida Atlantic University (FAU) is up to 88% efficient, operating at 4.25 kHz. The unit can transfer up to 1000 W, 0−70 V, 0−30 A.

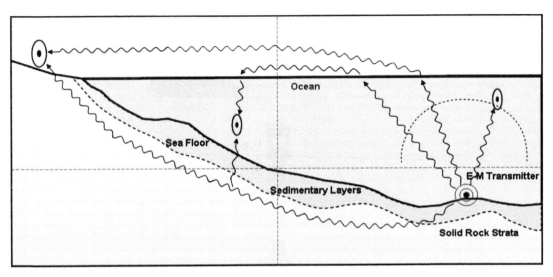

FIGURE 8.41

RF transmission occurs through water, ground, and air.

(Courtesy WFS Technologies.)

8.6 Reliability and quality control

A design starts with engineering and then proceeds to manufacturing. The two departments work together in the product development phase. Quality Control (QC) is an independent department that assures the part is made as it was designed. Quality is not added by the QC department, only validated.

Production Quality Assurance (QA) involves mechanical tests including visual, dimensional, and pressure testing. Electrical tests are done where appropriate, including Hi-pot, conductivity, and Megger (see Section 8.9 below). They may also perform environmental tests such as shock, thermal, suspended material, and corrosion tests.

8.7 Field maintenance

With seals in general, *"Cleanliness is next to Godliness."* The wise technician will put together a field kit containing:

1. Spare O-rings (with replacement use chart)
2. Wood or plastic toothpicks or brass picks, as supplied by Parker Seal, for O-ring removal
3. Lint-free wipes
4. Isopropyl alcohol

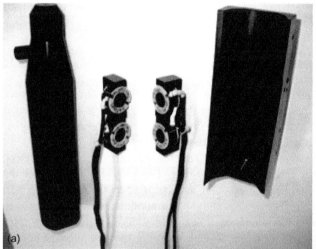
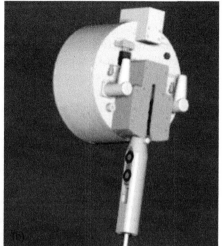
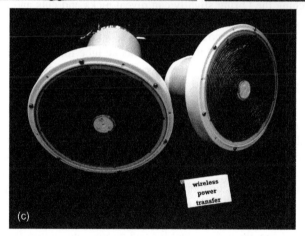

FIGURE 8.42

(a) The FAU "Flying Plug" inductive recharge system. (b) The FAU Inductive Recharge system engaged. (c) WFS Technologies Inductive Power Link.

(a) and (b) (Courtesy FAU-OE.) (c) (Courtesy Ocean Innovations.)

5. Silicone grease

6. Silicone spray, food grade

When applying silicone grease to the O-ring, remember that grease does not seal. Go light. Always carefully inspect O-rings and mating surfaces for defects.

Clean dirt off with soap and water (deionized or distilled). Alternatively use isopropyl (rubbing) alcohol. Use cotton swabs on female contacts. If contacts are tarnished, try using white vinegar. Oxidation may be removed from male contacts by using #800 wet/dry emery paper cut in strips equal to or less than the width of the contact and rubbing lightly. For female contacts use a 0.22 caliber bore brush with nylon bristles.

O-rings: inspection

Carefully inspect O-rings before every use. O-rings on a bulkhead connector body only need to be checked when the connector is removed from the housing.

Remove the O-ring from the recessed groove where it may be inspected. Wooden toothpicks are a safe way to remove an O-ring without the fear of damaging the O-ring groove. Never use dental picks, awls, safety pins, or other steel objects for removing O-rings! These materials with higher hardness values can easily damage the surface of the groove, the O-ring, or both.

One at a time, run each O-ring between your fingers and feel for any dirt or foreign material. You should also feel a thin film of silicone lubricant on the O-ring. This provides lubricity for the O-ring to slide under pressure into its final sealing position and helps keep the O-ring flexible. If you feel dirt or grittiness, clean the O-ring. The surface should be smooth, continuous, and clean. If not, replace it with a new O-ring. *"When in doubt, throw it out."*

Inspect for roundness. Some O-rings could flow and flatten, reducing the compression needed for a good seal. Stretching the O-ring slightly can reveal nicks and cuts and indicate loss of elasticity as evidenced by lack of re-bound.

Carefully inspect the recessed groove as well. There should be no foreign material there. Dirt, hair, sand, salt crystals, or any material that is not silicone lube for the O-ring means you will need to clean both the O-ring and the groove. Remove any excess silicone grease as too much grease can prevent the O-ring from properly seating, resulting in a leak path.

O-rings: cleaning

For O-ring cleaning you will need lint-free wipes, such as Kimwipes, isopropyl alcohol, and Q-Tips. Holding the cloth in your hand, grab the O-ring and gently pull it through the cloth. Pull the entire loop through several times until all the foreign material has been removed and the silicone lubricant is gone. Once again inspect the O-ring under good lighting as described above, since imperfections, such as cuts or nicks, may be seen more easily without the silicone lubricant coating.

Using a Q-Tip or the lint-free cloth, thoroughly clean the recessed groove. Press the Q-Tip deeply into the groove to access corners and remove residual silicone.

After the groove has been cleaned, inspect for and remove any cotton fibers that may have been shed from the Q-Tip.

O-rings: lubrication and installation

Now you are ready to lubricate the O-rings. For this you will need pure silicone lubricant or silicone grease.

Put a small amount of silicone between your thumb and forefinger. Then run the entire loop of the O-ring between your fingers several times. You need to coat the entire surface of the O-ring with a thin film of lubricant. The lubricant film should be thin, uniform, completely cover the o-ring, and have a "wet" feel.

Place or stretch the O-ring back into its recessed groove and be sure it is well seated. Give the O-ring one last look for hair or other dirt that may have fallen on your work.

If you do not need to service an O-ring, it is best to leave it undisturbed. A sealed O-ring will remain so unless disturbed (such as by cleaning).

8.8 **Underwater cable design**

ROVs require a cable to handle both electrical and mechanical functions. The EM cable transfers the mechanical loads, power, and/or communications between the ship and the vehicle. It must be possible to spool the EM cable out and in over and around sheaves and drums, be directed by a cable lay system, and operate in the intended environment. Interconnect cables inside the vehicle transfer power and signal between components.

It is important to characterize the operating environment of the cable with respect to:

- Depth
- Duration
- Temperature
- Salinity
- Suspended particulate matter
- Water chemistry
- Anaerobic condition
- Current profile

Some advanced ROVs carry their own power sources and only require a communication link to the surface vessel through an expendable fiber-optic microcable.

Most ROVs therefore have three general categories of cable to consider: umbilical cable, tether cable, and interconnect cable. Umbilical and tether cables are EM cables, containing the strength, power, and signal components. Interconnect cables seldom carry mechanical load and have only power and signal components in their construction.

The diameter of the cable is the dominant factor in overall vehicle drag. Therefore, minimizing cable diameter is an important part of ROV design and operation.

8.8.1 **Umbilical and tether cables**

The umbilical cable connects the ship to the ROV or TMS, while the tether cable connects the TMS to the ROV. The umbilical cable is generally steel jacketed, while the tether cable uses synthetic fibers to maintain neutral buoyancy. Use of a low-density jacket may help offset the negative weight of copper wire.

Initial cable design considerations include: (i) power requirements, (ii) signal requirements, and (iii) strength and weight requirements.

The power and signal requirements are usually the most important considerations because the cable's primary purpose is to transmit power to the ROV and return signals from it. However, the strength and weight are also important considerations. Therefore, a proper solution to designing an ROV cable requires a concurrent view at these multiple variables.

8.8.2 **Power requirements**

Transferring electrical power through a cable involves four factors:

- Voltage
- Phase

- Amps
- Duration

Transmitting electrical signals through a cable involves four different factors:

- Frequency
- Bandwidth
- Impedance
- Capacitance

When determining the power ($P = IE$) a cable can transmit, it all comes down to amperes. The current capacity of a particular wire gauge does not reference volts. Thus, to increase power down a given size wire, a designer needs to step up the voltage as current is fixed. For each ampere it is necessary to have enough material to conduct the power to the far end. Most conductors have resistance to electrical energy flow. This creates a voltage drop, and it is necessary to keep this value as small as possible to provide power to the source. Therefore, it is necessary to use material with as low a resistance as possible.

The most common material for providing power through a cable is copper. The most common form for electrical cable conductors in ROV cables is electrolytic tough pitch copper with a tin coating. Other coatings are available for special purposes, but most increase the resistance to electrical current. Oxidation of bare copper makes terminations more troublesome.

Another consideration is insulation on the conductors to contain the electrical energy. There are two general insulation families: thermoplastic and thermoset.

Thermoplastic is a material that repeatedly softens or melts when heated and hardens when cooled. Some examples are as follows:

- Polyethylene (PE)
- Polypropylene (PP)
- Polyvinyl chloride (PVC)
- Polyurethane (PUR)
- Nylon
- Fluorocarbons (Tefzel™ and Teflon™)

A thermoset material reacts to heat, changing forever into its final molecular form. Some examples are as follows:

- Cross-linked PE (XLPE)
- Chlorosulfonated PE (Hypalon™)
- Chlorinated rubber (Neoprene™)
- Ethylene propylene rubber (EPR)
- Ethylene propylene diene rubber (EPDM)
- Styrene butadiene rubber (SBR)

ROV cables usually use thermoplastic materials. They process more easily than thermoset materials and thermoplastics cover a broad range. Thermoset materials require special processing equipment. However, because thermoplastics soften or melt with heat, it is important to know both the operating environment and the current requirements. The cable designer needs to look at all these parameters in choosing the proper material.

The operating voltage is another consideration in the cable design. It is important to limit voltage stress on the insulation. If this is too high it can cause the insulation to fail and the electrical energy to exit the conductor before it reaches its objective, which can (to say the least) create a hazardous condition. Further, should the insulation break down (through damage or some other mechanism) and a ground fault develop, the flash grounding could produce an instantaneous high-temperature arcing, thus melting through all parts of the tether. This converts a tethered vehicle to an (expensive) untethered (and unpowered) floating/sinking vessel. This is not comical as it happens all too often. Therefore, it is important for the cable design to address the insulation voltage stress. Also, a separate conductor for an emergency ground is common as a safeguard in case there is a breakdown in the insulation.

8.8.3 Signal requirements

The signal requirements translate to attenuation losses. The signal, whether electrical or optical, attenuates through both the conductor and the insulator. This loss varies with both the signal transmission media and the frequency.

Signal transmission can be either analog or digital and either electrical or optical. The system usually dictates the signal transmission type. It is important for the cable designers to understand the media and as many parameters about the signal transmission as possible so they can select the proper conductor for the signals.

Copper conductors with thermoplastic insulation are also common for electrical signals, similar to power conductors. Signal transmission wires frequently require a shield from electromagnetic interference and radio frequency interference. Also, it is common to group the signal transmission wires separate from the power conductors.

As more fully explained in Chapter 13, there are both balanced and unbalanced electrical transmission schemes, and the system determines this requirement. Typical balanced lines are twisted pairs, and unbalanced lines are coaxial.

You can also transmit signals over optical fibers. Fiber optics come in various types:

— Multi-Mode
    — 50/125
    — 62.5/125
— Single-Mode
    — Dispersion shifted
    — Nondispersion shifted

The system requirements determine the optical fiber type. Some parameters to consider in any type fiber optic are as follows:

— Attenuation
— Bandwidth
— Wavelength

There are two different ways to package the fiber optics in a cable:

— Loose-tube buffer
— Tight-buffer

The optical fibers can have either an individual buffer on each fiber optic, or they can have a common housing for all the optical fibers. The user will need to weigh the trade-off with the different approaches for the specific application because there is no single design that is correct for all applications.

Several standard cable design specification sheets follow throughout this section and are representative of state-of-the-art cables provided to the offshore industry by Falmat. A successful innovator in the ROV cable industry, they have been designing, manufacturing, and testing such cables for over 25 years.

8.8.4 Strength requirements

The strength member provides the mechanical link to the ROV. It usually has to support the cable weight, the ROV, and any additional payload and handle any dynamic loads. Also, the cable size can influence the load on the cable due to drag. Therefore, there are many variables to consider when choosing the cable strength.

Mechanical strength of a cable must consider:

- (Anticipated) working load
- Maximum peak dynamic load
- Minimum bend radius/diameter
- Expected cycle-life performance

The cable design must also consider the handling system (see Chapter 9), including:

- Deployment/retrieval scheme
- Drums, sheaves, and level wind, which may have restrictive bend radii
- Heave/motion compensation

Steel is the most common strength member material for umbilical cables. This material is usually a carbon steel wire with a galvanizing coating on the outside to protect the steel from corrosion. This material's tensile strength, modulus, and abrasion resistance protect the cable from damage in service. Typically, an ROV umbilical will be double helix wrapped (in opposing directions) to balance torque under load, thereby reducing the umbilical's tendency to rotate as it is payed out and taken up.

Synthetic fibers, such as Kevlar™ from DuPont and Spectra™ (UHMWPE or ultra-high-molecular-weight polyethylene) from Honeywell can reduce weight. Synthetic fibers are frequently necessary in tether cables and also in umbilical cables for deepwater systems. Synthetic fiber strength members usually require an overmolded or woven outer jacket, such as Dacron™, for abrasion resistance. A synthetic strength member is generally more expensive than steel, but the weight difference is significant. In many cases, this is the only way to get to the necessary depth. For very deepwater applications, the in-water weight of a steel umbilical will be beyond the steel's tensile strength required to support its own weight—much less the weight of the vehicle−TMS combination.

There are reasons to consider both strength member materials for different applications; these issues should be discussed with your cable manufacturer.

8.8.5 **Construction**

Designers must consider the mission requirements, including:

- Power, signal (copper or fiber), strength, water weight
- Torque balanced
- Cross talk (shielded conductor, twisted pairs)
- Need for neutral buoyancy
- Water blocked
- Flexing, including bend-over-sheave and minimum bend radius

If no COTS alternative is available, a designer has choices of how to proceed with a custom cable, including:

- Modifying a COTS cable
- Adapting a similar design
- Creating a completely custom design

Some alternative ways to modify a COTS cable include:

- Bundle multiple COTS cables together
- Add water blocking to increase the depth rating
- Add an overbraid strength member to support mechanical loads
- Add a flotation jacket to reduce water weight

The common ways to modify a similar design include:

- Minor modifications such as:
  - Improve electrical/mechanical performance
  - Different strength member and/or location
  - Add a flotation jacket to reduce cable weight in water
  - Create a completely custom design
- Designers can optimize cable performance by:
  - Using smaller AWG than standard COTS cable
  - Manufacture special performance components
  - Adjust weight and diameter to suit application
  - Use special materials to increase the operating temperature range
  - Use nonstandard twist rates to maximize flexibility

8.8.6 **Cable design methodology**

Designing a cable requires the engineer to identify the unique system requirements, including:

- Length or depth minimums
- Electrical or optical requirements
- Strength or dimensional limits
- Existing cable handling system restrictions, including bend radius
- Temperature and other environmental conditions

A designer should investigate all potential cable design solutions, including:

- Cable designs for similar applications
- Modify a similar application cable construction
- Modify a COTS cable
- Design a unique EOM cable specifically for the application
- Review the cable design solution to ensure it satisfies the requirements
- Manufacture a prototype and/or production cable design

As noted above, a custom cable design can provide variations in:

- conductor materials and strand numbers,
- insulation materials,
- jacket/sheath materials and construction, and
- the strength member.

8.8.7 Conductors

Conductors are the heart of the electrical cable (Figure 8.43). The most common wires are as follows:

- Copper
  - Tin-plated copper provides the best soldering and availability and is the most common material used in ROV cables.
  - Bare copper provides for lowest DC resistance at the termination, but surface oxidation creates problems with soldering, while availability is poorer.
  - Silver-plated copper is the best choice for higher temperature and signal frequency but comes at a higher cost and longer lead time.
- Stranding (Figure 8.44)
  - Use solid wire for non-flexing applications only.
  - Higher number of strands provides greater flexing.

8.8.8 Insulation

As noted above, there are two families of insulation materials: thermoplastics and thermoset plastics. Materials vary on cost, operating temperature range, insulation value, and bonding ability (Figures 8.45 and 8.46).

8.8.9 Jacket/sheath

The *sheath* is the inner core wrapping that binds the cable assembly together prior to extruding the outer jacket. Materials widely used include:

- Polyolefin (PE and PP)
- Polyurethane
- Thermoplastic elastomer (TPE)

Copper Conductor Data

The conductors used by Falmat Wire meet the applicable requirements of ASTM specifications B-3, B-33, B-172, B-173, B-174, and B-286 and Federal Specification QQ-W-343.

The following data covers the more commonly used conductor constructions in the electrical electronics industry. Special constructions, not shown, are available or can be designed to meet specific requirements. It is suggested that the Falmat Wire product engineering department be contacted before a specification is finalized.

AWG	Stranding	Type Stranding[1]	Diameter[4] in.	mm	Area circ. mils	sq. mm	Weight lbs./M'	kg./km.	D.C. Resistance 20°C[2] Tin Coating[3] ohms/M'	ohms/km	Bare or Silver Coating ohms/M'	ohms/km.
32	7/40	Co or Bu	.0096	.254	100	.051	.21	.31	176	577	-@	-
30	Solid	-	.010	.254	100	.051	.30	.45	113	371	104	340
	7/38	Bu	.012	.305	112	.057	.35	.52	106	348	92.6	303
28	Solid	-	.01264	.321	159	.081	.48	.72	70.8	232	65.3	214
	7/36	Co	.015	.381	175	.089	.55	.82	67.5	221	59.3	194
27	Solid	-	.0142	.361	202	.102	.61	.91	55.6	182	51.4	169
	7/35	Co or Bu	.017	.432	220	.111	.69	1.04	53.8	176	-	-
26	Solid	-	.016	.404	253	.128	.77	1.14	44.5	146	41.0	135
	7/34	Co or Bu	.019	.483	278	.141	.87	1.29	42.5	139	37.3	122
	10/36	Bu	.0193	.490	250	.127	.78	1.15	47.3	155	40.4	133
	19/38	Bu or Co	.021	.533	304	.154	.97	1.44	38.9	128	34.1	112
24	Solid	-	.0201	.511	404	.205	1.22	1.82	27.2	89.2	25.7	84.2
	7/32	Co or Bu	.024	.610	448	.227	1.38	2.05	25.7	84.2	23.1	75.9
	16/36	Bu	.024	.610	400	.201	1.25	1.64	29.5	96.8	27.5	90.2
	19/36	Co or Bu	.025	.635	475	.241	1.48	2.20	24.9	81.7	21.8	71.6
22	Solid	-	.025	.643	643	.324	1.94	2.89	16.7	54.8	16.2	53.2
	7/30	Co or Bu	.030	.762	700	.355	2.19	3.26	16.6	54.4	14.8	48.6
	19/34	Bu or Eq	.0315	.800	754	.382	2.35	3.50	15.5	50.8	13.8	45.1
20	Solid	-	.032	.813	1,020	.519	3.10	4.61	10.5	34.4	10.1	33.2
	7/28	Co or Bu	.038	.965	1,111	.562	3.49	5.19	10.3	33.8	9.33	30.6
	10/30	Bu	.037	.940	1,000	.507	3.14	4.67	11.4	37.4		34.0
	19/32	Co, Bu or Eq	.040	1.02	1,216	.616	3.84	5.71	9.48	31.1	8.53	28.0
	26/34	Bu	.039	.940	1,032	.523	3.28	4.88	11.3	37.1	-	-
19	Solid	-	.0359	.912	1,290	.653	3.90	5.80	-	-	8.05	26.4
18	Solid	-	.0403	1.024	1,620	.823	4.92	7.32	6.77	22.2	6.39	21.0
	7/26	Co or Bu	.048	1.22	1,770	.897	5.55	8.26	6.45	21.2	5.55	19.2
	16/30	Bu	.0475	1.207	1,600	.810	5.01	7.45	7.15	23.4	6.48	21.3
	19/30	Co, Bu or Eq	.050	1.27	1,900	.963	5.95	8.85	6.10	20.0	5.46	17.9
	41/34	Bu	.049	1.244	1,627	.824	5.09	7.08	7.08	23.2	6.60	21.6
16	Solid	-	.0508	1.29	2,580	1.31	7.81	11.6	4.47	14.7	4.16	13.6
	19/29[4]	Bu or Eq	.057	1.45	2,426	1.23	7.52	11.2	4.82	15.8	4.27	14.0
	19/.0117	Bu	.0585	1.50	2,601	1.32	8.02	11.9	4.39	14.4	4.13	13.5
	26/30	Bu	.0606	1.54	2,600	1.32	8.15	12.1	4.39	14.4	3.99	13.1
	65/34	Bu	.060	1.52	2,581	1.31	8.20	11.9	4.47	14.7	4.16	13.6
14	Solid	-	.0641	1.63	4,110	2.08	12.4	18.5	2.68	8.79	2.52	8.28
	7/.0242	Bu	.073	1.85	4,100	2.08	12.7	18.9	-	-	2.61	8.56
	19/27[4]	Co, Eq or Un	.071	1.80	3,831	1.94	12.0	18.0	3.05	10.00	2.71	8.88
	19/.0147	Bu	.074	1.88	4,106	2.08	12.7	18.9	-	-	2.61	8.56
	41/30	Bu	.077	1.96	4,100	2.08	12.9	19.2	2.81	9.22	2.53	8.30
12	Solid	-	.0808	2.05	6,530	3.31	19.8	29.5	1.69	5.54	1.59	5.21
	7/.0305	Bu	.092	2.34	6,512	3.30	20.2	30.1	-	-	1.64	5.38
	19/25[4]	Co, Eq or Un	.0905	2.299	6,088	3.08	19.4	28.9	1.87	6.13	1.70	5.59
	19/.0185	Bu	.0925	2.35	6,503	3.30	20.2	30.1	-	-	1.64	5.25
	65/30	Bu	.094	2.388	6,500	3.29	20.8	31.1	1.82	5.97	1.64	5.25
10	Solid	-	.1019	2.588	10,380	5.26	31.4	46.8	-	-	1.00	3.28
	7/.0385	Co	.116	2.95	10,376	5.25	32.0	47.6	-	-	1.00	3.28
	19/.0234	Bu	.117	2.97	10,404	5.27	32.0	47.6	-	-	.98	3.21
	37/.0169	Co	.112	2.84	9,361	4.74	29.2	43.4	-	-	1.25	4.10
	105/30	Bu	.126	3.20	10,500	5.32	33.8	49.2	1.10	3.61	.99	3.24
8	7/.0486	Co	.146	3.71	16,534	8.38	50.1	74.5	-	-	.65	2.13
	19/.0295	Bu or Eq	.144	3.66	16,535	8.38	50.0	74.4	-	-	.65	2.13
	133/29	Ro 19x7/29	.169	4.293	16,983	8.61	54.0	80.4	.71	2.33	-	-
	168/30	Ro 7x24/30	.174	4.42	16,800	8.51	53.4	79.0	.70	2.30	-	-
6	19/.0374	Bu	.188	4.775	26,576	13.33	81.1	121	-	-	.40	1.30
	133/27	Ro 19x7/27	.213	5.41	26,818	13.60	84.1	125	.43	1.41	-	-
	266/30	Ro 7x38/30	.222	5.64	26,600	13.49	83.2	124	.44	1.44	-	-
4	133/25	Ro 19x7/25	.257	6.53	42,615	21.61	135	201	.29	.95	-	-
	420/30	Ro 7x60/30	.270	6.850	42,000	21.29	140	208	.28	.92	-	-
2	665/30	Ro 19x35/30	.338	8.59	66,500	33.72	213	317	.18	.59	-	-

[1] Bu — Bunched; Co — Concentric; Eq — Equilay; Ro — Rope; Un — Unilay
[2] Typical D.C. Resistance values for uninsulated wires. Multiply by 1.04 for typical values after insulation
[3] Values are for tinned, heavy tinned, prefused, overcoated or topcoated conductors
[4] Does not meet UL conductor stranding requirements.

FALMAT INC. 1873 Diamond St. • San Marcos, CA 92069 • (619) 471-5400 • FAX (619) 471-4970   **M5**

FIGURE 8.43

Conductor size, resistance, and weight.

(Courtesy Falmat.)

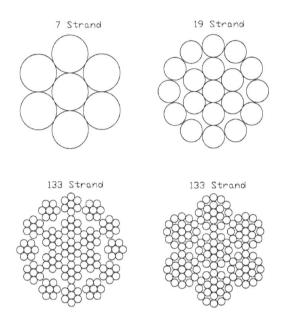

FIGURE 8.44

Strand configurations.

(Courtesy Falmat.)

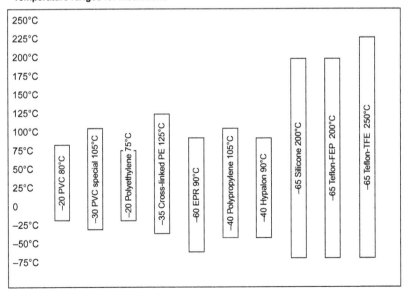

FIGURE 8.45

Operating temperature ranges for various insulations.

(Courtesy Falmat.)

Plastic Insulations

Property Considered	Cellular Polyethylene	High-Density Polyethylene	Low-Density Polyethylene	Nylon	Polypropylene	Polyurethane	PVC	Teflon
Acid Resistance	G to E	G to E	G to E	P to F	E	F	G to E	E
Abrasion Resistance	G	E	F to G	E	F to G	O	F to G	G to E
Alcohol Resistance	E	E	E	P	E	P	G to E	E
Alkali Resistance	G to E	G to E	G to E	E	E	F	G to E	E
Benzol (Aromatic Hydrocarbons) Resistance	P	P	P	G	P to F	P	P to F	E
Degreaser Solvents (Halogenated Hydrocarbons) Resistance	P	P	P	G	P	P	P to F	E
Electrical Properties	E	E	E	F	E	P to F	F to G	E
Flame Resistance	P	P	P	P	P	P	E	O
Gasoline, Kerosene (Aliphatic Hydrocarbons) Resistance	P to F	P to F	P to F	G	P to F	F	G to E	E
Heat Resistance	G to E	E	G	E	E	G	G to E	O
Low Temperature Flexibility	E	E	G to E	G	P	G	P to G	O
Nuclear Radiation Resistance	G	G	G	F to F	F	G	P to F	P to F
Oil Resistance	G to E	G to E	G to E	E	E	E	E	O
Oxidation Resistance	E	E	E	E	E	E	E	O
Ozone Resistance	E	E	E	E	E	E	E	E
Water Resistance	E	E	E	P to F	E	P	E	E
Weather—Sun Resistance	E	E	E	E	E	F to G	G to E	O

P=Poor F=Fair G=Good E=Excellent O=Outstanding
Above ratings are based on average performance of compounds. Any specific property can often be improved by the use of selective compounding.

FIGURE 8.46

Plastic insulation properties.

(Courtesy Falmat.)

The *jacket* is the outer, overall covering of the cable. It is snag and cut resistant and keeps the interior bundle from kinking and hockling. Materials widely used include:

— Polyurethane
— Polyolefin (PE and PP)
— Thermoplastic elastomer (TPE)

8.8.10 Strength member

The most common material used in *umbilical* construction is steel armor wire, including:

— Galvanized improved plow steel (GIPS)
— Galvanized extra-improved plow steel (GEIPS)
— High-tensile alloys for greater corrosion resistance (i.e., Nitronic-50)

The most common materials used for *tether* construction are synthetic yarns, including:

— Aramid fiber (Kevlar® and Twaron®)
— Liquid crystal polymer (Vectran®)
— PBO (Zylon®)

Examples of EM cable composite constructions include:

• Steel wire strength member (Figure 8.47)
• Synthetic strength member (Figure 8.48)

Example steel wire stength member

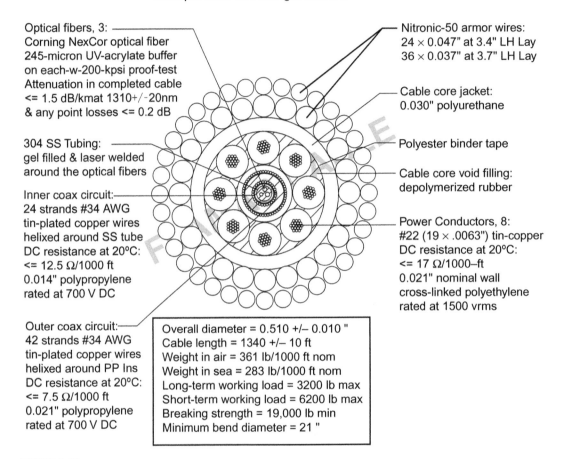

Optical fibers, 3:
Corning NexCor optical fiber
245-micron UV-acrylate buffer
on each-w-200-kpsi proof-test
Attenuation in completed cable
<= 1.5 dB/kmat 1310+/-20nm
& any point losses <= 0.2 dB

304 SS Tubing:
gel filled & laser welded
around the optical fibers

Inner coax circuit:
24 strands #34 AWG
tin-plated copper wires
helixed around SS tube
DC resistance at 20°C:
<= 12.5 Ω/1000 ft
0.014" polypropylene
rated at 700 V DC

Outer coax circuit:
42 strands #34 AWG
tin-plated copper wires
helixed around PP Ins
DC resistance at 20°C:
<= 7.5 Ω/1000 ft
0.021" polypropylene
rated at 700 V DC

Nitronic-50 armor wires:
24 × 0.047" at 3.4" LH Lay
36 × 0.037" at 3.7" LH Lay

Cable core jacket:
0.030" polyurethane

Polyester binder tape

Cable core void filling:
depolymerized rubber

Power Conductors, 8:
#22 (19 × .0063") tin-copper
DC resistance at 20°C:
<= 17 Ω/1000–ft
0.021" nominal wall
cross-linked polyethylene
rated at 1500 vrms

Overall diameter = 0.510 +/– 0.010 "
Cable length = 1340 +/– 10 ft
Weight in air = 361 lb/1000 ft nom
Weight in sea = 283 lb/1000 ft nom
Long-term working load = 3200 lb max
Short-term working load = 6200 lb max
Breaking strength = 19,000 lb min
Minimum bend diameter = 21 "

FIGURE 8.47

Steel wire strength member.

(Courtesy Falmat.)

- Near neutrally buoyant (Figure 8.49)
- Positively buoyant (floats) (Figure 8.50)
- Water-sampling tow-cable

Standard cables include:

- CAT-5, CAT-5e, CAT-6, etc.
- Coax components:
    - 50 or 75 Ω characteristic impedance
    - 30 or 20 pF per foot capacitance
    - MIL-SPEC coaxes per MIL-C-17

Example synthetic strength member

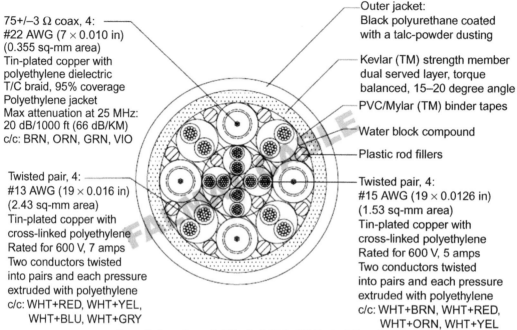

75+/–3 Ω coax, 4:
#22 AWG (7 × 0.010 in)
(0.355 sq-mm area)
Tin-plated copper with
polyethylene dielectric
T/C braid, 95% coverage
Polyethylene jacket
Max attenuation at 25 MHz:
20 dB/1000 ft (66 dB/KM)
c/c: BRN, ORN, GRN, VIO

Outer jacket:
Black polyurethane coated
with a talc-powder dusting

Kevlar (TM) strength member
dual served layer, torque
balanced, 15–20 degree angle
PVC/Mylar (TM) binder tapes

Water block compound

Plastic rod fillers

Twisted pair, 4:
#13 AWG (19 × 0.016 in)
(2.43 sq-mm area)
Tin-plated copper with
cross-linked polyethylene
Rated for 600 V, 7 amps
Two conductors twisted
into pairs and each pressure
extruded with polyethylene
c/c: WHT+RED, WHT+YEL,
   WHT+BLU, WHT+GRY

Twisted pair, 4:
#15 AWG (19 × 0.0126 in)
(1.53 sq-mm area)
Tin-plated copper with
cross-linked polyethylene
Rated for 600 V, 5 amps
Two conductors twisted
into pairs and each pressure
extruded with polyethylene
c/c: WHT+BRN, WHT+RED,
   WHT+ORN, WHT+YEL

Overall diameter = 1.170 +/– 0.06 in (29.7 +/– 1.5 mm)
Weight in Air = 756 lb/1000 ft (1125 KG/KM) nominal
Weight in Sea = 278 lb/1000 ft (414 KG/KM) nominal
Specific gravity = 1.62 +/– 0.1 gm/cc density
Breaking strength = 31,000 lb (14 Tonne) minimum
Peak tension load = 4400 lb (2 Tonne) maximum
Bend diameter = 40 in (1 M) minimum
Depth rating = 1500 ft (450 M) = 650 PSI (4.5 Mpa)
Operating temperature = 32 to 104°F (0 to 40°C)
Storage temperature = –40 to +158°F (–40 to +70°C)

FIGURE 8.48

Synthetic strength member.

(Courtesy Falmat.)

- — Miniature, high-speed, low-loss coaxes
- • RS-232, RS-422, RS-485, etc.

   CAT-5 cable overview (Figure 8.51):

- • Typically contains four pairs
  - — Note: Most applications use only two pairs
  - — However, gigabit Ethernet uses all four pairs

Example near neutrally buoyant

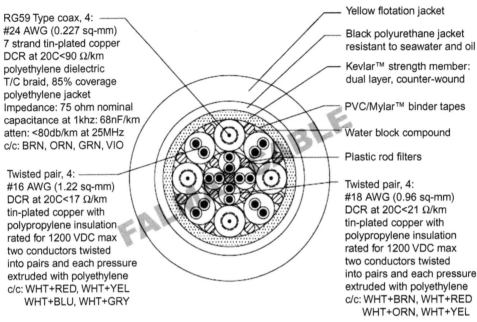

RG59 Type coax, 4:
#24 AWG (0.227 sq-mm)
7 strand tin-plated copper
DCR at 20C<90 Ω/km
polyethylene dielectric
T/C braid, 85% coverage
polyethylene jacket
Impedance: 75 ohm nominal
capacitance at 1khz: 68nF/km
atten: <80db/km at 25MHz
c/c: BRN, ORN, GRN, VIO

Twisted pair, 4:
#16 AWG (1.22 sq-mm)
DCR at 20C<17 Ω/km
tin-plated copper with
polypropylene insulation
rated for 1200 VDC max
two conductors twisted
into pairs and each pressure
extruded with polyethylene
c/c: WHT+RED, WHT+YEL
WHT+BLU, WHT+GRY

Yellow flotation jacket

Black polyurethane jacket
resistant to seawater and oil

Kevlar™ strength member:
dual layer, counter-wound

PVC/Mylar™ binder tapes

Water block compound

Plastic rod filters

Twisted pair, 4:
#18 AWG (0.96 sq-mm)
DCR at 20C<21 Ω/km
tin-plated copper with
polypropylene insulation
rated for 1200 VDC max
two conductors twisted
into pairs and each pressure
extruded with polyethylene
c/c: WHT+BRN, WHT+RED
WHT+ORN, WHT+YEL

Overall diameter = 1.535 to 1.575 in (39 to 40mm) nominal
Weight in air = 930 lb/1000 ft (1384 KG/KM) nominal
Weight in sea = 64 lb/1000 ft (95 KG/KM) nominal
Density = 1.09 to 1.1 gm/cc (1090 to 1100 kg/m^3)
Breaking strength = 31,000 lb (140 kN) minimum
Peak tension load = 6600 lb (3T) maximum
Bend diameter = 40 in. (1000 mm) minimum
Depth rating = 1670 ft (510M) = 725 PSI (5 Mpa)
Operating temperature = 28 to 104°F (−2 to 40°C)
Storage temperature = −40 to 158°F (−40 to 70°C)

FIGURE 8.49

Near neutrally buoyant cable.

(Courtesy Falmat.)

- Pairs are unshielded twisted pairs (UTP)
- Flexing cables need stranded conductors
- Shielded versions are rare in the United States
- Industrial versions are another variety

8.8.11 Spare conductors

Custom cable construction often allows for the choice to either use "filler" strands or spare conductors to maintain a uniform circular cross-section. Fillers are cheaper, but spare wires can be used

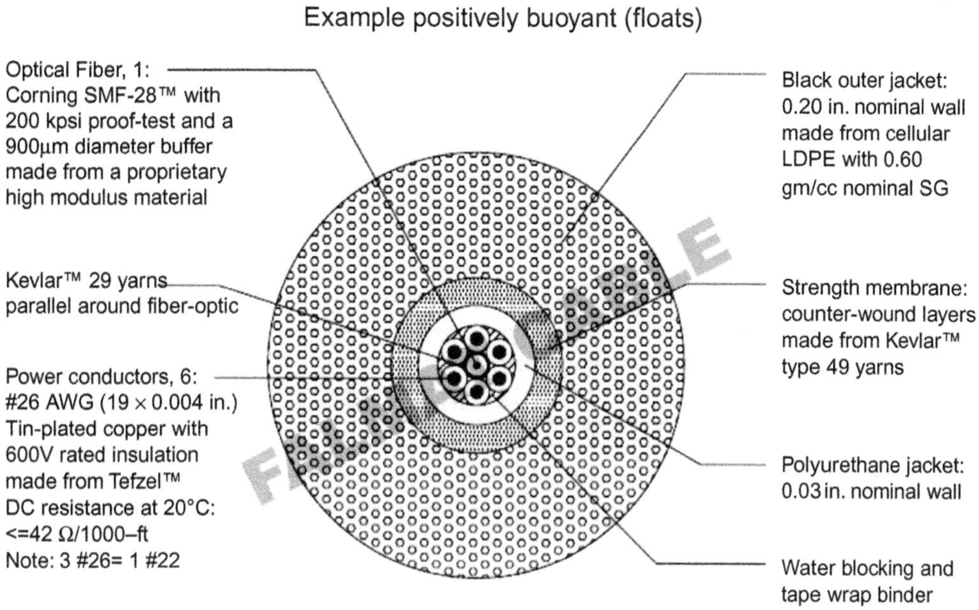

Example positively buoyant (floats)

Optical Fiber, 1:
Corning SMF-28™ with
200 kpsi proof-test and a
900µm diameter buffer
made from a proprietary
high modulus material

Kevlar™ 29 yarns
parallel around fiber-optic

Power conductors, 6:
#26 AWG (19 × 0.004 in.)
Tin-plated copper with
600V rated insulation
made from Tefzel™
DC resistance at 20°C:
<=42 Ω/1000–ft
Note: 3 #26= 1 #22

Black outer jacket:
0.20 in. nominal wall
made from cellular
LDPE with 0.60
gm/cc nominal SG

Strength membrane:
counter-wound layers
made from Kevlar™
type 49 yarns

Polyurethane jacket:
0.03 in. nominal wall

Water blocking and
tape wrap binder

Overall diameter	= 0.65 inches nominal
Density	= 0.73 gm/cc nominal
Hardness	= 87 Shore A nominal
Breaking strength	= 5000 lbs minimum
Breaking strength	= 6000 lbs maximum
Peak tension load	= 1375 lbs maximum
Minimum bend diameter	= 12 inches
Storage hub diameter	= 18 inches
Hydrostatic pressure	= 710 psig

FIGURE 8.50

Positively buoyant cable.

(Courtesy Falmat.)

for emergency field repairs and unforeseen system expansion requirements. However, the weight of these spare conductors must be considered in the neutral buoyancy equation for a vehicle's flying tether.

8.8.12 Interconnect cables

The preceding discussion on Power and Signal applies equally to interconnect cables.

Portable cords are often used underwater for power or signal cable. They are available in either thermoset or thermoplastic constructions. Thermoset rubber material hardens or "sets" under direct application of heat, called "curing," and once it sets, it is not able to be resoftened by heating. Thermoset is a much more durable compound than thermoplastic.

Xtreme-Green "Xtreme-Cat"

Underwater Network Data/Power Cable

Homeland security, Oceanographic, Observation, and other extreme marine enviornments.

P/N	DESC.	Data Comp	Break Strength	Diameter	Wgt/1000'
FMXCAT51205K24	5C- 12AWG	Cat 5E -24awg (7)	2400 lbs	.700"	256 lbs
FMXCAT51207K24	7C- 12AWG	Cat 5E -24awg (7)	2400 lbs	.700"	297 lbs
FMXCAT51606K24	6C- 16AWG	Cat 5E -24awg (7)	2400 lbs	.564"	157 lbs
FMXCAT51610K24	10C-16AWG	Cat 5E -24awg (7)	2400 lbs	.564"	191 lbs
FMXCAT51812K12	12C-18AWG	Cat 5E -24awg (7)	1200 lbs	.500"	143 lbs
FMXCAT51806K12	6C- 18AWG	Cat 5E -24awg (7)	1200 lbs	.500"	110 lbs
FMXCAT52218K12	18C-22AWG	Cat 5E -24awg (7)	1200 lbs	.447"	120 lbs
FMXCAT52218	18C-22AWG	Cat 5E -24awg (7)	0	.433"	117 lbs
FMXCAT52824K8	24C-28AWG	Cat 5E -24awg (7)	800 lbs	.415"	90 lbs
FMXCAT52824	24C-28AWG	Cat 5E -24awg (7)	0	.400"	88 lbs
FMXCAT50000 *	N/A	Cat 5E -24awg (7)	0	.325"	47 lbs
FMXCAT50000K12 *	N/A	Cat 5E -24awg (7)	1200 lbs	.340"	52 lbs

* Cat 5 cable only, with rugged Xtreme-Green Polyurethane, no power conductors, or waterblock.

Data Network/Power Composite Cable
Extreme ruggedness, flexibility with waterblocked construction.
Designed for underwater use with high-speed network data, video, and sensor equipment.
Cables can be used for bottom-laid, vertical, winch systems, and ROV applications.
Cat 5E 4pr stranded conductor. Meets or exceeds TIA 568-B. Suitable for 10Base-T and 100Base-T.
Custom variations of these listed cables can be supplied with added break strengths,
steel armor, additional jackets, or neutral buoyant constructions.

Xtreme-Green Cables are designed for extreme environments with flexibility. These cables feature
Falmat's specially formulated "Xtreme-grade" polyurethane jacket for easier payout, tighter bends, and
better tractor control with extremely low coefficient of friction. The Xtreme-Green reduced diameter cable
offers even smaller bend radius for tighter bends and smaller, portable, handling systems.

FIGURE 8.51

Falmat Xtreme Green CAT-5 cable.

(Courtesy Falmat.)

Thermoset cords include SO cable. From the basic SO cable to the specialized SOOW, these multiconductor cables are readily sourced from a number of manufacturers. The SOOW acronym stands for:

Service—All SO hybrids begin with this word.
Oil resistance—The designation "SO" denotes a cable that is oil resistant.
Other chemical resistance—The cable is impervious to additional chemicals such as acetone and diesel fuels.
Water resistance—A heavy, yet flexible, nonporous casing keeps the cable interior dry. SOW cables are used to power mining applications, bulldozers, conveyors, temporary lighting, submersibles, pump applications, and more.

FIGURE 8.52

Open-End Spelter socket.

(Courtesy Crosby Group.)

The thermoset jackets are readily bonded to in secondary operations to add underwater connector terminations.

Another cable seeing increasing use in moderate depths is the AWM20233. Billed as a "Low Capacitance Communications/Instrumentation Cable," the outer jacket is polyurethane extruded over a copper braid shield. Inside are individual twisted pairs wrapped in aluminum/polyester foil with a stranded, tinned copper drain wire. Nylon rip cord filler provides cross-sectional bulk. End-users report success for yearlong durations at 2.5 miles (4 km).

Additional cable options may be known by your connector manufacturer. Ask their opinion, but use your own best judgment.

8.8.13 EM terminations and breakouts

Mechanical strength terminations are used to transition force from the winch to the umbilical and the umbilical to the TMS or to the tether of an ROV.

The termination type can influence the break strength of the cable. It is important to use similar terminations on each end of the cable. Do not rely on locking sleeves for transferring the load off a cable.

Two common field installable mechanical end-fitting/terminations are as follows:

- Spelter socket
- Kellems grip

8.8.13.1 Socket or spelter socket

A *socket* or *spelter socket* uses molten zinc or epoxy poured in a socket to bond the splayed cable to the fitting (Figures 8.52–8.54). The spelter socket is suitable for steel strength members with

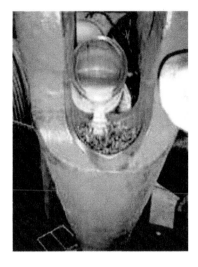

FIGURE 8.53

Spelter socket with wire splayed and technician pouring in epoxy binder.

(Courtesy Ocean Innovations.)

FIGURE 8.54

Pull test of spelter socket termination.

(Courtesy Ocean Innovations.)

high strength-efficiency and repeatability. It is also used for synthetic strength members, but it is about half as efficient and results are highly variable. Still, this is sometimes acceptable because synthetic strength members have higher break strength to working load ratios.

8.8.13.2 Kellems grip
A *Kellems grip* is a woven "Chinese-finger grip": The harder you pull, the tighter it gets. Among the highest strength models is the Hubbell Dua-Pull (Figure 8.55).

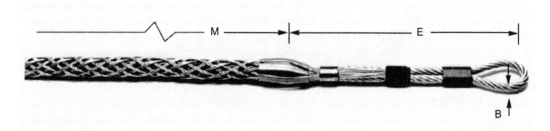

FIGURE 8.55

Hubbell Dua-Pull Kellems grip.

(Courtesy Hubbell.)

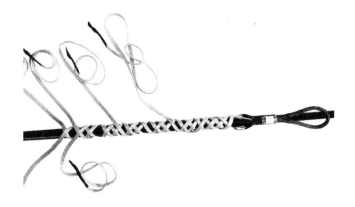

FIGURE 8.56

Yale grip.

(Courtesy Yale.)

8.8.13.3 Lace-up grips

One example is the *Yale grip*, designed to allow easy installation mid-span by braiding four synthetic legs around the cable body (Figure 8.56).

8.8.13.4 Helical termination

A *helical termination* uses gripping wires in a manner similar to the soft woven Yale grip to bind the outside of a matching helix wire rope on a steel armored cable (Figures 8.57 and 8.58).

8.8.13.5 Mechanical termination

A *mechanical termination* is an assembly using conical wedges to trap a splayed cable strength member inside a fitting (Figures 8.59 and 8.60).

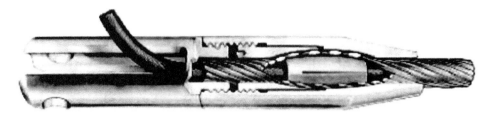

FIGURE 8.57

A "torpedo" wedge helical termination.

(Courtesy Preformed Line Products.)

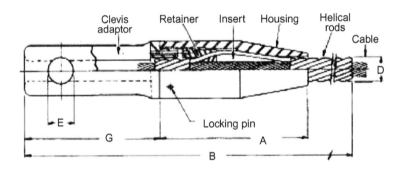

FIGURE 8.58

Helical strength termination.

(Courtesy Preformed Line Products.)

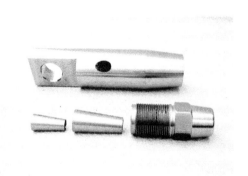

FIGURE 8.59

A conical wedge strain relief.

(Courtesy Ocean Innovations.)

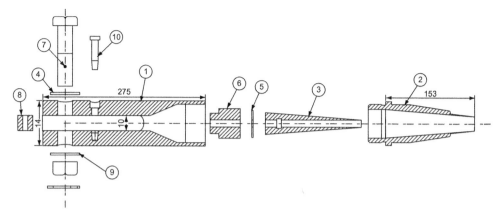

FIGURE 8.60

A conical wedge strain relief cross-section.

(Courtesy Ocean Innovations.)

8.8.14 Bonding

8.8.14.1 Vulcanized rubber splice

The vulcanized rubber splice process requires specialized compression molding equipment that can apply elevated pressure, high temperature, over an extended time. Cable or connector manufacturers, generally not end users, own such machines. The procedure, however, has been refined, tested, and found to produce good adhesion between the uncured thermosetting tapes and a number of polymer cable jackets.

8.8.14.2 Transfer-molded rubber splice

The transfer molding process creates molded joints by transferring an uncured compound onto a mold. Heat and pressure soften the compound, which then flows through the cavity and encapsulates the spliced cable, fusing to the cable jacket. When the part is cooled and removed, the result is a permanent, pressure-proof, reliable splice.

8.8.14.3 Castable polyurethane resin splices

A number of companies manufacturer quality castable polyurethane resin splice materials. One of the most popular, 3M™ Scotchcast insulating polyurethane resins, have been used with great success in submerged applications. Mold forms are available for cylinder and Y-splices, providing a field-ready solution for cable splicing. The electrical and physical properties make them ideal for insulating and protecting electrical connections. Scotchcast 2131 provides excellent adhesion to a number of jacket materials including Neoprene, Hypalon, Nitrile/PVC, PVC, and polyurethane. A caution in assembly is the substrates must be completely clean and dry. In the case of synthetic jacket cables, it is necessary for the resin to be poured immediately after the surfaces are prepared.

Preformed Line Products offers their own version called "RD Encapsulant," a two-component polyurethane compound providing excellent moisture and electrical insulation protection for telecommunications cable splices.

Scorpion Oceanics (UK) produces a similar "SOLRES-01" polyurethane kit that includes the mold forms.

8.8.14.4 Cold splice

For decades, Scripps engineers and technicians have utilized a technique known as "cold splicing" to splice or repair underwater interconnect cables. This technique requires the use of three 3M products: Scotchkote electrical coating, Scotchfil electrical insulation putty, and Scotch 33 black tape. (3M deserves an award from the oceanographic industry for 33 tape alone.) Scripps engineer Frank Snodgrass, a pioneer in free vehicles, passed on this technique to his younger colleagues. Scotch black tapes have near universal loyalty at all US oceanographic institutions for their ability to stick and stretch under every conceivable condition at and under the sea.

The technique for a shielded two-conductor SOW cable follows:

1. Strip back the outer jacket from the end ½ in. for each conductor. The shield counts as one conductor, so strip 1½ in. on each cable end for the two-conductor shielded cable.
2. Clip out the cable bulk filler material, like jute.
3. Unbraid and twist shield wire into a multistrand conductor. Use a knife to scrape the surface clean of any rubber jacket residue on the last ¼ in. to improve solderability.
4. Cut the conductors so the three solder joints will be offset linearly from each other, that is, so they are not located side-by-side. This prevents solder spikes from inadvertently pressing through the insulation under pressure and shorting to a neighboring solder joint. With the first cable end to be spliced, cut the shield ½ in. long. Cut one color conductor (i.e., black) to 1 in., and leave the last color conductor (white) a full 1½ in. long.
5. With the second cable end to be spliced, cut the conductor lengths the opposite, so the shortest in now the longest. Cut the one color conductor (i.e., white) ½ in. long. Cut the other color conductor (black), to 1 inch, and leave the shield a full 1½ inch long.
6. Strip each of the wires back ¼ in. Use shrink tubing on the twisted shield wire as an insulation jacket. Use a heat gun to shrink it down tight on the shield wire.
7. Place approx. ⅝ in. long shrink tubing on wire pairs before soldering.
8. Solder the like color wires together. Perform a continuity check to be sure the conductors go where you expect them to, and not where they should not, by testing each pin to all others.
9. Center the shrink tubing over the solder joint and use a heat gun to shrink it.
10. (Optional) Add a layer of Scotch 33 black tape over shrink tubing for a second layer of insulation if preferred.
11. Using isopropyl alcohol, clean the cable jackets 2 in. to either side of the splice. Similarly, clean the spliced wire jackets and heat shrink. Let dry.
12. Paint the entire cleaned area liberally with Scotchkote. Let it flow into every nook and cranny. Do this over cardboard because it will drip off. Let it dry completely. This is a primer that improves adhesion of the Scotchfil to the cable jacket.
13. Cut a 6 in. length of Scotchfil putty. Remove backing tape and stretch putty to ½ its original thickness. Wrap tape around joint, pulling to create an elastic affect, covering the entire splice

area and 1½ in. over each cable jacket end. Press and massage the Scotchfil to a roughly uniform diameter, slightly larger than the original cable diameter. The overall joint length is then approximately 5½ in. long.

14. Using Scotch 33 black tape, and starting ½ in. beyond the Scotchfil on the cable jacket, begin wrapping the black tape over the joint, pulling so as to create an elastic affect, and overlapping the tape 50% on each turn. When completely covered, cut the tape from the roll with a knife or scissors rather than pulling and breaking the tape. This keeps the bitter end from curling up.

15. Perform a final ohm sift to be sure the connections are still fine.

8.8.15 Cable design summary

A cable for an ROV is a special component because it is the primary link between the vehicle and the operator, providing power, signal, and handling strength. Thus, an ROV cable design must consider all these features.

The vehicle size, weight, and operating depth, as well as the vehicle motors, subsystems, and payload, all combine to determine the cable design, which is usually unique to the vehicle.

These brief descriptions of cable design considerations are just a starting point. Because each ROV has unique requirements, abilities, and limits, it is important to discuss your unique cable requirements with someone who has experience in this area.

8.9 Testing and troubleshooting

8.9.1 Electrical testing, troubleshooting, and predeployment checkout

Cables can get worn, broken, or damaged being moved, in transit, just stored, and in use over time. They should be inspected prior to deployment and possibly after recovery.

Common failure modes include:

Intermittent continuity
Open circuit
Delamination or debonding
Chemical contamination

At a minimum a visual inspection should be performed to identify breaks, cuts, kinking, or fraying.

8.9.2 Ohm sift or continuity test

An *ohm sift*, or *continuity test*, is used to verify the conductivity of each conductor from end-to-end and to verify no conduction (shorts) between the adjacent pins and the housing. This is referred to as "buzzing (or "ringing") out conductors." *DC resistance* precisely measures resistance in ohms of each conductor. Resistance changes with temperature and is typically specified at 20°C (68°F). Measurements at other temperatures must be converted to 20°C (68°F) for comparison.

8.9.3 MegOhm testing or insulation resistance

MegOhm testing or insulation resistance (IR) measures the insulation resistance between wire pairs using a high voltage to verify resistance of the insulation to current flow. This is typically performed at 500 V DC for wire and cable. The values are typically specified in Mega (millions) ohms. This is referred to as "Meggered the cable." IR measurements are influenced by length, temperature, and time.

When interpreting measurements from IR equipment, operators must know:

IR versus length dependent

- IR is inversely proportional to length
  - Shorter lengths have proportionally higher IR
  - Longer lengths have proportionally less IR
- IR is typically specified in MegOhms/1000

  IR versus temperature

- IR is inversely proportional to temperature
  - Measurements at higher temperature have lower IR
  - Measurements at lower temperature have higher IR
  - Rule-of-thumb is that IR halves or doubles each 10°C
- IR is typically specified at 15.6°C (60°F)
- It is important to log the temperature when recording IR measurements

8.9.4 Hi-Pot or voltage withstand test

Hi-Pot or voltage withstand test is a high over-voltage DC test of a wire or cable insulation. It is typically run at 2× operating voltage +1000 V. The higher test voltage is run early in manufacture to verify quality of components. The lower MegOhm test voltage is acceptable at final acceptance. The Hi-Pot is generally derated further after installation.

8.9.5 A time-domain reflectometer

A time-domain reflectometer (TDR) is a device to find the location of a break in a cable by sending a pulsed signal into the conductor and then examining the reflection of that pulse. Parallel wires will have the same time of reflection. A break will reflect the pulsed signal sooner, appearing like a shorter length. That length is the distance to the break. Spare conductors in cables may be available for field repair work.

Other electrical cable tests of potential interest include capacitance, impedance, attenuation, cross talk, and skew.

8.9.6 Mechanical testing and troubleshooting

Mechanical parameters to be measured include the cable's:

- overall diameter and variance
- weight in air and/or seawater

- break strength-with-field termination
- torque and rotation versus applied tension
- bending fatigue cycles simulating anticipated usage conditions including load, sheave diameters, speed, temperature, etc.

Weight in air and seawater

- Weight in air is typically done on a 1-ft sample.
- Weight in seawater is usually calculated from the 1-ft sample weight and diameter.

More elaborate testing is possible, but exercise caution in interpreting results. Additional mechanical testing may include:

- Elongation versus tension (Figure 8.61)
- Torque/rotation versus tension
- Ultimate break strength

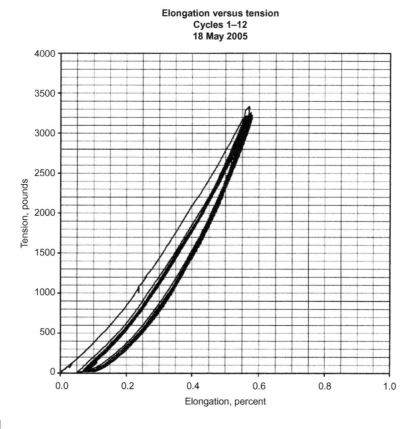

FIGURE 8.61

Elongation versus tension of a towed-array tow-cable.

(Courtesy Falmat.)

8.10 Tips from the field

Words to those who can learn from past suffering.

- Never use a connector if you do not need to. Use the KISS (Keep It Simple, Stupid) principle.
- Consider using connectors that have open face pressure ratings to match your operating depth in the event a connector-to-connector seal leaks.
- Do not mix-and-match connectors from different manufacturers unless there is a well-defined external specification such as a MIL-SPEC to fall back on, especially rigid epoxy or metal shell body connectors. Similar looking parts from different manufacturers cannot be guaranteed to be interchangeable, as one manufacturer cannot rationally guarantee the quality of another manufacturer's product. While there is more tolerance in rubber-molded connectors, as the outer body stretches to form a compression seal, there are many additional dimensions for pin length, pin diameter, pin pattern, and more that can make a critical difference. In a world of pointing fingers, the one who made the "buy" decision will be responsible, not the manufacturers.
- The *Parker O-Ring Handbook* (ORD-5700) is the bible of seal design. Anyone who says differently is playing with the devil. And the devil is in the details.
- Make sure your connector body material is compatible with your housing material. Dissimilar metals exposed to seawater form galvanic cells that will eat away one of the two, and both are important components of the pressure housing. A skirt, cup, or lip seal is one way to avoid creating a galvanic cell by isolating the bulkhead connector threaded post from exposure to seawater.
- Epoxy connectors are brittle and must be well protected from side loads and random impacts.
- Rubber-molded connectors and cable jackets can be seriously degraded if exposed to long-term heat, sunlight, or high ozone levels. Inspect suspect connectors and cables carefully before reuse. There is a reason they call it a "bone yard."
- Do not mate or demate connectors with power "on" as arcing between contacts on demate will likely occur.
- Do not disconnect connectors by pulling on cables as it will stress (and probably break) the internal wire/connector joint. Do not wiggle epoxy or metal shell connectors back and forth when demating. You can get away with a little bit of this with rubber-molded pairs.
- Use a torque wrench to tighten bulkhead connectors to manufacturer specifications, especially epoxy bulkhead connectors. "Tink" is not a happy sound. Some experience is required as thread installation torques can vary with materials, cleanliness of the threads, thread tolerances, and thread lubrication.
- Do not use locking sleeves to mate connector halves.
- Locking sleeves should only be tightened by hand. Remember this point after the vehicle returns to the surface (with the resulting full temperature cycling of the connector/sleeve) only to find the sleeve is practically welded to the connector.
- Avoid sharp bends in cables. Know the "minimum bend radius" of your cables.
- Secure cables with black tape, zip ties, or other means to prevent strumming or movement created by an ROV in motion. Be careful using zip ties as they can bite into the cable jacket or cause sharp bend radii.
- Make sure mating connector halves are clean and properly lubricated.

- Do not apply grease excessively. A thin coat is needed for lubricity only. Grease has no shear strength and does not seal. Too much grease can prevent the O-ring from moving into its optimum position and can even cause hydraulic fracturing of epoxy parts.
- Use only food grade silicone spray. Non-food grade has corrosive volatile propellants that cannot escape a sealed connector pair and can result in "dezincification of brass."
- Keep spares of critical connectors on hand.
- Do not use petroleum-based products, like WD-40, on rubber cables and connectors.
- Any accumulation of sand or mud in the female contact should be removed by flushing with clean fresh water. Failure to do so could result in splaying of the contact and damage to the O-ring seal.
- Ensure that there are no angular loads on the bulkhead connector as this is a sure way to destroy a connector.

8.11 Summary

Underwater cables and connectors provide system flexibility, ease of service, and other design advantages for undersea equipment including ROVs. The primary purpose of underwater cables and connectors is to provide a conductive path without leakage in a pressure-resistant or pressure-tolerant package. They allow simple system reconfiguration.

A mated connector pair forms a unique small pressure case. All pressure case design criteria apply. A bulkhead connector or penetrator becomes a mechanical part of a pressure hull, and it should be looked at critically.

EM cable design is driven by the specific end-user application. Design criteria include power, data, and strength.

Test EM cables both electrically and mechanically simulating the application field conditions. Use field terminations whenever possible. Exercise caution when interpreting results.

Feel free to discuss your requirements with the manufacturer and experience improved operational effectiveness by using the proper cable designed for your application.

Bibliography

1. *Parker O-Ring Handbook (ORD-5700)* is available as a free download at <http://www.parker.com>. Enter "ORD-5700" in the Search box. The Handbook provides details on numerous groove designs, back-up rings, elastomer selection, machining tolerances and finishes, installation procedures, and everything else you might have ever wondered about O-rings. There are generic O-ring groove design charts, and information on how to finesse a design for a particular application. Parker is never wrong. Ever.

2. Bash, J., Ed., *Handbook of Oceanographic Winch, Wire and Cable Technology*, third ed., is available as a free download at <http://www.unols.org/publications/winch_wire_handbook__3rd_ed/index.html>.

3. Busby, R.F., *Remotely Operated Vehicles*, 1979, is available as a free download at <http://voluwww.archive.org/details/remotelyoperate00rfra>.

4. Busby, R.F., *Manned Submersibles*, 1976, is available as a free download at <http://archive.org/details/mannedsubmersibl00busb>. Chapter 7, "Power and Its Distribution," is still a relevant discussion of penetrators, connectors, cables, and junction boxes.

5. Wernli, R.L., *Operational Effectiveness of Unmanned Underwater Systems*, CD-ROM. This book contains details about unmanned systems, their expected performance, including full descriptions and specifications. The CD-ROM contains 700 pages with 390 photographs, charts, and diagrams. Available from the MTS at <https://www.mtsociety.org/shop/view.aspx?product_id = 3415&>.

6. Moore, S.W., Harry Bohm, Vickie Jensen, *Underwater Robotics: Science, Design & Fabrication* is written for advanced high school classes or college and university entry-level courses. The 850-page text is enhanced by hundreds of photos, illustrations, and diagrams of underwater vehicles. In addition, the textbook includes a discussion of subsea vehicle development, resource appendices, an extensive glossary, and a complete index. It is available at <https://www.mtsociety.org/shop/view.aspx?product_id = 3619&pType = 14&>.

7. *Teledyne Impulse*: Additional nomenclature may be found at <http://www.teledyneimpulse.com/download.aspx> and selecting "Glossary-Connector Terminology."

8. *Dummy's Guide to Marine Technology*, a useful description of technologies used beneath the sea, available at <o-vations.com/marinetech/index/html>.

LARS and TMS

CHAPTER CONTENTS

The most sophisticated and capable remotely operated vehicles (ROVs) to date are fully deployed and operational all over the world. However, these vehicles must be launched, recovered, and operated safely and efficiently in order to be functional. The tether is an asset because it provides the capability for the working end of the system to get to the operational location; however, the tether is also a major problem because it can become damaged or entangled, which could result in the loss of the vehicle. Therefore, this section will address two areas: (i) launch and recovery systems (LARS) along with tether management systems (TMS) and (ii) the management of the tether itself. This manual addresses the operation of ROVs of various sizes, and each category of vehicle has its own areas of concern. However, even heavy WCROVs can be crane or A-frame launched (i.e., in free-flying configuration as opposed to TMS-based launch) while even smaller OCROVs can be TMS based. The descriptions below divide the launch and recovery techniques by the presence/

absence of a TMS between the launch platform and the vehicle for managing the transition from the heavy metal umbilical and the soft (and neutrally buoyant) flying tether.

For the smaller observation-class vehicles, the launch and recovery is, in most cases, a much simpler approach. In many cases, the vehicle can be hand launched, use a crane, or be launched in a protective cage with or without a TMS.

The primary objective of the LARS is to move the vehicle or vehicle—TMS combination from the deck through the splash zone and to the working depth safely in a controlled fashion. As the vehicles grow larger, so does the LARS. The equipment can range from a simple overboarding A-frame to a telescoping and/or cursive mechanism with complex motion compensation. Many of the larger LARS are integrated into the ship or platform, which is especially beneficial where heavy sea states are expected.

To provide a complete perspective, the following sections will address launch techniques and concerns from the smallest vehicles up to the largest work-class systems.

9.1 Free-flying vehicle deployment techniques

Deployment methods vary for the free-flying vehicle, but there are a few common methods that have proven successful. The deployment methods can be divided into two main categories: directly deployed and cage deployed. These are described below.

9.1.1 Directly deployed/free-flying

Would it not be great if all ROVs were as small as the micro-ROVs because they could just be lowered over the side by hand (Figure 9.1). However, in many cases for operations with smaller ROVs, the vehicle can be directly deployed from the deck of the boat, if not by hand then through

FIGURE 9.1

Free-fly hand launch for (a) microROV and (b) large OCROV.

(Courtesy VideoRay and SeaTrepid.)

the use of a davit or crane (Figure 9.2). Notice in both Figures 9.1 and 9.2 that there is a soft neutrally buoyant "flying" tether between the surface and the vehicle. Larger vehicles can be directly deployed, but the risk of damage increases as the weight of the vehicle increases. This is due to the vehicle's momentum building through vessel sway while the vehicle is suspended in air between the launch point on the deck and water insertion point in the splash zone (i.e., the vehicle becomes a "wrecker's ball"). Directly deployed vehicles are more vulnerable to any currents prevalent from the surface to the vehicle's operating depth. This is due to the neutrally buoyant flying tether being more subject to current drag than the heavy metal umbilical (as discussed later in this chapter).

9.1.2 Tether management system

Within the ROV industry, there is some confusion as to the exact definition of a TMS. Technically, the TMS is the subsea tether-handling mechanism (only) allowing the soft flying tether to be payed out or taken up from the junction between the clump/depressor weight and the tether. But by common convention, the TMS is typically described as the entire subsea mechanism from the end of the umbilical (umbilical termination to the clump/depressor weight, cage or top hat) to the beginning of the tether. The vehicle handling system (subsea cage or top hat) houses the tether-handling mechanism as well as the vehicle itself and is launched with the vehicle either within the cage or attached to the top hat mechanism. With a clump/depressor weight system (without a TMS), the vehicle is launched first, and then the clump weight is launched once the vehicle's tether is stretched (due to travel distance away from the launch platform).

Simplistically, the TMS can be attached to the so-called clump weight (Figure 9.3) or be part of a cage or top hat deployment system (Figure 9.4). The main function for the TMS is to manage a soft neutrally buoyant tether cable—the link from the TMS to the ROV for electrical power and sensors, including video and telemetry. The tether cable allows the ROV to make excursions at

FIGURE 9.2

Free-fly crane-launched MSROV.

(Courtesy SeaTrepid.)

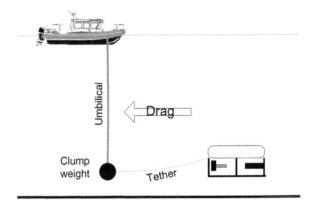

FIGURE 9.3

Clump weight deployed ROV.

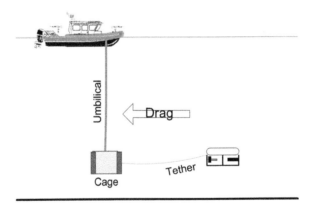

FIGURE 9.4

Cage deployed ROV.

depth for a distance of 500 ft (150 m) or more from the point of the clump weight, cage, or top hat. As discussed above, most operators refer to the TMS as the entire system of cage or "top hat" deployment, tether management, vehicle protection, and junction point for the surface—vehicle link. Technically, the TMS is the tether-handling machinery only.

Clump deployed

The use of a clump weight has become prevalent in the observation-class category. If working on or near the sea bottom, clump weights enable the ROV operator to easily manage the tether "lay" from the insertion point next to the vessel all the way to the clump weight location either mid-water or next to the bottom.

This allows the weight to absorb the cross-section drag of the current, relieving the submersible of the tether drag from the surface to the working depth. The vessel can be maneuvered to a point

FIGURE 9.5

Cage deployed ROV for both (a) OCROV and (b) MSROV.

(Courtesy SeaTrepid.)

directly above the work site, thus locating the center point of an operational circle at the clump weight. In short, the vehicle only needs to drag the tether length between the clump weight and the vehicle for operations on the bottom.

Cage deployed

Cages are used for small and large ROVs to protect the vehicle against abrasions and deployment damage due to the instability of most vessels of opportunity while under way. When operating from a vessel with dynamic positioning (DP), the vessel's thrusters are operating at all times while on DP requiring positive control of the tether from the surface to depth. For a free-flying ROV with neutrally buoyant tether from the surface to the vehicle, the tether naturally drifts with the water flow—possibly straight into the operating vessel's thrusters! Cage deployment systems assist with the positive management of the tether. In Figure 9.5, a cage deployment system without a TMS is displayed. With this type system, the vehicle is overboarded through the splash zone enclosed within the cage with the tether hand tended (or winch tended) through an access guide at the rear of the cage. Once through the splash zone, the vehicle swims out of the cage with the tether controlled/guided through the cage. The benefit of a cage system without a TMS is its simplicity. With a cage enclosed TMS, many operators consider the complexity of that TMS similar to having a second ROV concurrently in the water.

Cages also function as a negatively buoyant anchor to overcome the drag imposed from the cross-section of the cable presented to the current (between the platform and the cage) at shallower depths (Figure 9.4). This allows the weight of the cage to fight the current instead of the vehicle fighting the current. The cage further provides room for a TMS to meter the softer tether in small amounts (as discussed below), thus lowering the risk of tether entanglement. For deployment systems with a TMS, the cage umbilical is normally made of durable material (steel, Kevlar, etc.) with the conductors for the vehicle buried within the core of the umbilical. For deeper diving submersibles, the umbilical encases fiber-optic data links to the cage, requiring digital modem feeds from the cage to the control unit at the surface. For a more detailed discussion of currents and tether management see Section 9.3.

9.2 TMS-based vehicle deployment techniques

The larger ROVs may have more power available at the vehicle, but that just translates to these being able to get into bigger, costlier problems. To ensure safe launch and recovery through the ocean surface, several techniques have been developed. Most of these techniques depend upon the platform itself, whether it is stable (e.g., a large oil rig) or not so stable (e.g., a multipurpose service vessel, MSV). As the vessel's dynamics increase, so does the requirement to add additional equipment to help counter the dynamic forces imposed by the ocean.

At a minimum, a full ROV LARS with a TMS would have the components shown in Figure 9.6 including the HPU, winch, Luffing A-Frame, TMS, vehicle, umbilical, and tether. As the size and depth capability of the vehicle grow, so does the system's footprint, which must be taken into consideration when choosing a system for future use.

The following sections will describe in more detail the various techniques and equipment employed to successfully launch, recover, and operate the TMS-based ROV systems.

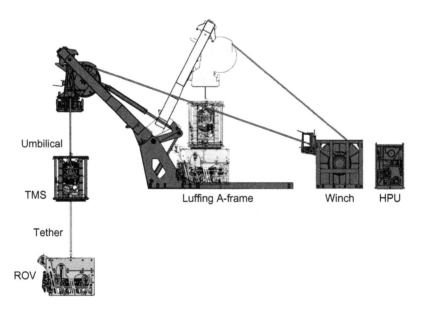

Umbilical

TMS

Tether

ROV

Luffing A-frame

Winch   HPU

FIGURE 9.6

Luffing A-frame LARS.

(Courtesy Schilling Robotics.)

9.2.1 **Winches**

Winches fit into the classification of "simple machines," performing work by transferring one physical force into another. Winches, which have been around for a very long time, were first used to hoist water buckets from water wells.

Winches range from the simple reeling device with use of a hand crank for orderly handling of OCROV tether to the complex heave-compensated winch systems for handling the heavy loads imposed by WCROV systems.

As depicted in Figure 9.7, the winch uses torque ($T_{mechanical}$ or T_{me}—expressed in ft-lb or N-m) as the motive force. A hand, lever, crank, or motor is attached to the cylinder to provide the torque necessary to rotate the shaft attached to the winch drum. The torque applied by the motor is equivalent to a force, F_{me}, applied to a lever of length R. The resistant torque caused by the cable is equivalent to the force F_{cable}, or tension of the cable, acting on the radius of the cylinder (plus the wraps of cable), r, or outer edge of the drum–cable combination. As the cylinder (drum) is rotated, the cable rolls onto or off of the drum. As the cable rolls onto the drum, the size of r increases due to the increased radius of the circle as the cable wraps, thus increasing the resultant torque that the winch motor must overcome.

When the drum is stationary, without using a brake, T_{me} and T_{cable} cancel each other (i.e., the torque caused by the cable is equal and opposite to the torque to drive the drum). Expressed as a formula, $F_{cable} = (R/r)F_{me}$. As an example of static equilibrium, if $R = 3r$, then $F_{cable} = 3F_{me}$, multiplying the torque by a factor of 3. So, as the cable spools onto the drum at a constant *drum* speed

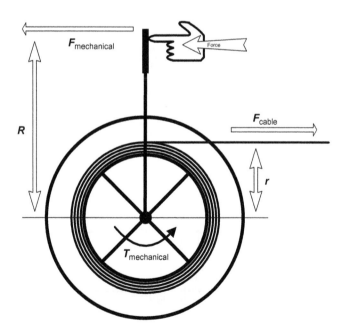

FIGURE 9.7

Simplified winch equation.

of rotation, the mechanical advantage decreases and the *cable* speed increases as the cable wraps onto the larger diameter outer layers. The drive motor's mechanical advantage is much higher with the first wrap (line closest to the drum) than on the last wrap (outermost wrap—assuming multiple cable layers on the drum). Therefore, the maximum torque on the motor for a given cable tension (F_{cable}) will be determined based upon the torque generated by the cable at the outer wrap of the winch. In determining the necessary drive torque of the motor, the force to bend, or wrap, larger diameter cables, fleet angle resistance and safety factors should also be taken into account.

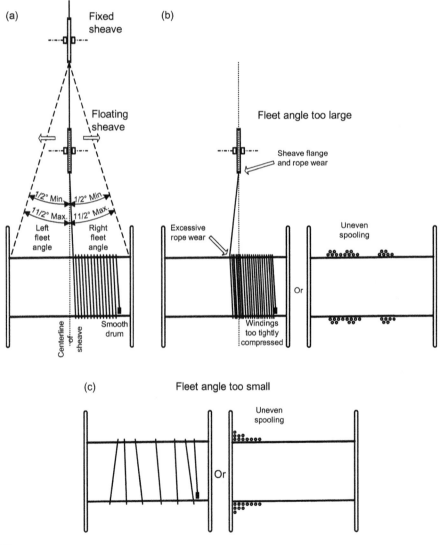

FIGURE 9.8

Fleet angle (a) graphical depiction along with consequences for (b) fleet angle too large and (c) fleet angle too small.

(*Source: Wire Rope Technical Board Wire Rope Users Manual, 3rd. ed.*)

As depicted in Figure 9.8, the *fleet angle* is described as the angle between the projected centerline of the sheave and the cable launch point from the drum. It is important to spool the cable onto the drum in a fashion that assures orderly lay of the cable onto the drum. For wire rope, that angle is normally between 0.5° and 1.5°. An excessive fleet angle bunches the cable onto the drum causing either excessive wear or uneven spooling, while too small a fleet angle causes any number of disorderly modes.

Ingersoll Rand, based on information from the Wire Rope Technical Board, states that to maintain a proper fleet angle (1.5° for a smooth drum, 2° for a grooved drum), the distance from the lead sheave to the centerline of the drum (LDS) is:

LDS = half the drum width (feet) × 38 for smooth drum, 29 for grooved drum

Accordingly, for a smooth drum 40 in. wide, LDS would be 1.66 ft × 38 = 63 ft—obviously, a distance that is not preferred when mobilizing an ROV system. Therefore, to reduce the footprint of the spread and spool the cable onto the drum in an orderly fashion, a mechanical device known as a "level wind" guides the cable onto the drum to maintain a proper fleet angle. Also, in order to distribute the line pull evenly onto the drum (as opposed to focusing the pull at the line anchor point), several wraps of line are left on the drum (typically 10% of the line length, although it would be preferable to have enough line to provide a complete first layer on the drum) so that the line forces distribute over the drum more uniformly.

Winches are "load/lift-path" items requiring certification of the lifting system. Further, winches are not allowed to be in a free-spooling configuration. This requirement forces braking systems to be failsafe with the braking system "failing" into a "braking on" mode should power be lost to the lift motor.

As shown in Figure 9.9, a winch system is composed of several basic components:

1. Winch base
2. Hydraulic motor and brake (or hand crank and manual brake)
3. Brake cylinder and motor support
4. Drum assembly
5. Level wind

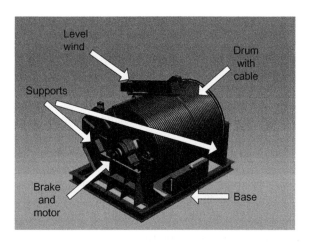

FIGURE 9.9

Winch components.

Typically, the static braking system is hydraulically powered upstream of the winch's hydraulic motor. The brake is only released once sufficient hydraulic pressure is applied to the motor (thus transferring the weight of the load from the brake to the winch).

Winches are specified based upon several parameters including safety margins and winch configuration:

Ratings and capacity

- *Safe working load* (specified at either bare or full drum)—the maximum operating load in order to achieve manufacturers' safety margins
- *Maximum line pull* (specified at either bare or full drum)—typically equal to 1.25 times Safe Working Load and is generally the nominal stalling point of the winch
- *Line speed*—specified in line travel length per unit time—feet per minute (fpm), meters per second (m/s), or meters per minute (mpm) and can be measured at bare drum, mid-drum, or full drum since the line speed changes as a function of distance (i.e., wraps) from the drum
- *Drum capacity*—specified in linear feet (or meters) based upon a stated cable diameter (e.g., 3300 ft (1000 m) of 1.97 in. (50 mm) diameter cable).

Considerable round trip times for the vehicle (predive to postdive) can be experienced, and expensive, depending upon the winch line speed. For instance, if a winch with a line speed of 100 fpm (30 mpm) is used for a 6600 ft (2000 m) operating depth, the round trip time in both directions is 132 min (66 min in each direction). If you add to this a 15-min predive procedure along with 1 min to get through the splash zone (in both directions) and a 15-min postdive procedure, the total nonworking "trip time" of the dive from start to finish is 164 min! If the vehicle arrives at depth only to find that a lightbulb is blown, the round trip to change the bulb is nonproductive time. One oil company is notorious for specifying a nominal line speed in its contracting, computing the theoretical trip time and then comparing it to the actual trip time—*then putting the entire spread on nonrevenue downtime for any variance. Ouch!!!* Beware of the fine print.

Drum dimensions

- Drum core diameter—outer diameter of the drum in feet or meters (*D* in Figure 9.10)
- Drum core width—width from flange to flange in feet or meters (*B* in Figure 9.10)
- Drum flange diameter—flange diameter in feet or meters (*H* in Figure 9.10).

Overall dimensions

- Length—total length of the entire winch system including all ancillary components (base, frame, guard, etc.)
- Width—total width including hydraulic motors, control station, and other winch components
- Height—total height of the winch plus all framing and components. The skid should be added to this to gain a full dimension for the entire LARS plus winch system
- Weight (without cable)—total weight of the winch including all components

Regardless of how the vehicle is deployed, it will require a winch that either handles a heavy-duty tether or armored umbilical. Depending on the depth of operation, the winch can add a large footprint to your spread.

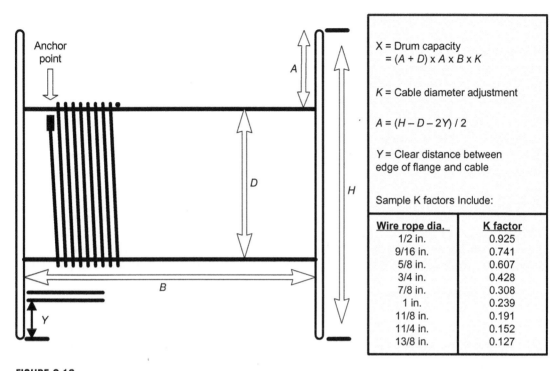

FIGURE 9.10

Drum capacity computations.

9.2.1.1 Standard winch

There are several factors to consider when choosing or designing a winch, such as:

- Safe working load—does the winch meet the requirements of an approved classification society such as Det Norske Veritas (DNV) or the American Bureau of Shipping (ABS)?
- Maximum line pull—the cable weight, vehicle weight, etc. must be considered along with the dynamic forces induced by Mother Nature. Also, in some cases, the vehicle's task may be to lift a heavy object. If so, the in-air weight must be taken into consideration along with the possibility of entrained water.
- Line speed—the faster the recovery, the higher the drag, the greater the forces on the cable and power requirement of the winch.
- Drum capacity—cable diameter, length, etc. will drive the footprint of the winch.
- Hydraulic power—let us not forget the hydraulics. This is another piece of the system that will, in most cases, take up additional space.

All of these will affect the size of the winch along with the following options/considerations:

- Cooling systems—depending on the electrical power being transmitted down the cable, heating of the internal layers of the cable on the drum can become excessive if a cooling system is not available or the number of wraps on the drum is reduced, thus forcing a wider drum width.

- Requirement for an internal slipring (usually the case).
- Grooved drum liner—which can help with the proper wrapping of the cable on the drum.
- Level wind—unless a good fleet angle is available, which is usually not the case on limited space platforms, a level wind will ensure that the cable is being spooled evenly.
- Minimum bend radius—the minimum bend radius of the cable must be taken into consideration when determining the drum core diameter.

An example of a standard winch with a level wind and hydraulic system (to the right) is shown in Figure 9.11. The winch has a maximum line pull of 33,075 lb (15,000 kg) at full drum with a 1.25 safe working load and can hold 9062 ft (2760 m) of cable. The footprint is approximately 16.6 ft (5 m) by 10.3 ft (3.2 m).

9.2.1.2 Traction winch

If higher tension operations are to be encountered with heavy loads, then a traction winch may be required. Dynacon's Model 5521XL (Figure 9.12) traction winch system upgrade to the previous winch brings the maximum line pull up to 49,600 lb (22,500 kg). However, the footprint increases by about 30% and the weight reaches 63,000 lb (28,566 kg), without the cable but with the hydraulic power unit.

9.2.1.3 Heave compensation

If high sea states are expected, then heave compensation options exist that can be stand alone or combined with the winch itself (Figure 9.13). The bottom line is that the umbilical needs to have the slack taken up or additional payed out as the dynamics dictate. If these heave dynamics are not compensated for then the vehicle at the end of the umbilical will not be stable and/or the umbilical can become overstressed and damaged.

Heave compensation systems come in either "Active" or "Passive" versions depending upon their configuration. With active heave compensation (AHC), motion sensors sense vessel movement about the vessel's vertical axis of movement. These sensors measure the amount of heave and

FIGURE 9.11

Dynacon 521XL general purpose ROV winch.

FIGURE 9.12

Dynacon's Model 5521XL traction winch.

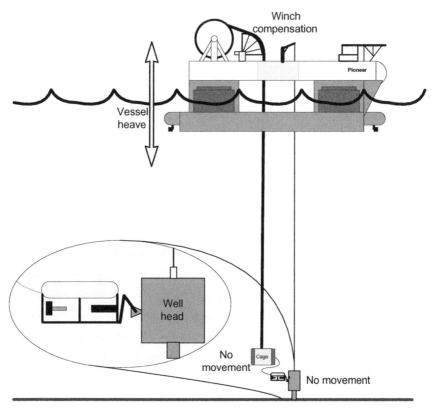

FIGURE 9.13

Heave compensation.

FIGURE 9.14

Dynacon's Model 501 heave compensator.

direct the AHC unit to pay out or take up cable based upon those active readings. With passive heave compensation (PHC), the cable tension is held constant, thus varying the weight to maintain compensation (instead of position). AHC is much more accurate for achieving a constant load position over the bottom as it works on a position reference frame (rather than weight which has no direct reference to motion).

Heave compensation systems can be incorporated directly with overboarding cranes, but most of the larger systems will be using stand-alone or integrated heave compensators with the winches as discussed below.

Stand-alone heave compensator

For use of a stand-alone heave compensator to counter higher sea states and/or ship dynamics, units such as Dynacon's Model 501 heave compensator (Figure 9.14) can be added to the system; however, be sure to factor in the additional footprint, hydraulic power requirements, and acceptable umbilical cable runs.

Integrated winch and heave compensator

If a mechanically cleaner, albeit larger and more complex, winch is desired, there are options for winches that have an integrated active heave compensation system such as the one produced by MacArtney (Figure 9.15).

9.2.2 Tether management systems

In most offshore applications, the ROV umbilical will be handled by a winch system, whether heave compensated or not, that connects to a TMS. The usually neutrally buoyant tether connects the ROV to the TMS, which pays out or takes in the tether as the operator dictates. Control of the

FIGURE 9.15

MacArtney's active heave compensation winch.

tether, as described earlier in this section, is critical to the safety of the ROV and the tether itself. There are typically two types of TMSs: the top hat and caged systems. Both are designed to take the ROV to depth, using the umbilical's strength, and then allowing the ROV to deploy the neutrally buoyant tether as it maneuvers horizontally and vertically to the work site.

And here is where the terminology gets tricky—there are two types of actual subsea tether payout/take-up systems (also known as "TMS"): the slipring system and the baling arm system. With the slipring system, an electrical and/or fiber-optic rotary joint is embedded within the subsea tether drum, allowing for orderly tether payout/take-up which lessens tether wear (but adds a level of complexity due to the rotary joint). (The reader is referred to Moog Components Group, which acquired Focal Technologies, a leading manufacturer of fiber-optic rotary joints and electrical slip rings.) The baling arm system is similar to a spinning fishing reel. A baling arm grabs the line and physically reels the line around the spool (as opposed to simply rolling the drum). With a baling arm system, there is a solid length of wire and/or fiber from the winch to the vehicle with the baling arm gathering the tether onto/off of a spool (thus eliminating the need for a rotary joint at the cage or top hat). The benefit of the baling arm system is the lack of a complex rotary joint with the cost of a typically uneven tether spooling along with added wear and tear on the tether.

Regardless of the design approach, the payout/take-up system can be mounted in either cage deployment (Figure 9.16(a)) or top hat deployment (Figure 9.16(b)).

Top hat TMS

Whether on top of the ROV or integrated within the launch cage, the primary objective is to get the ROV safely to depth, isolate the drag on the umbilical from the ROV, and keep the umbilical clear of obstructions during the operation. Once at depth, the TMS, via the operator, controls the tether deployment. A Schilling HD ROV is shown launched using a top hat TMS in Figure 9.17.

Deployment cage TMS

Unlike the top hat TMS, when a deployment cage, or "garage," is used, the ROV is given additional protection during launch and recovery since it is surrounded by the launch cage itself. Better to damage the cage than the ROV. In addition, the cage can be used as a transport mechanism for additional tooling and/or sensors. In Figure 9.18, SeaTrepid's caged Mohican ROV is launched prior to a platform inspection.

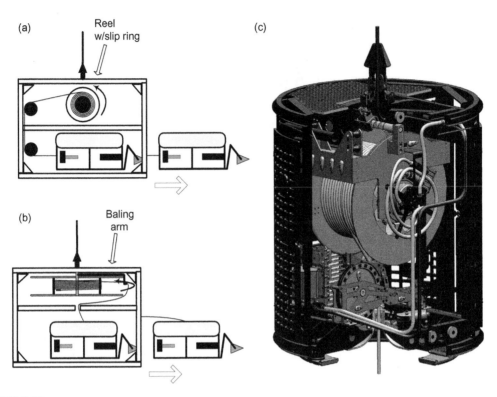

FIGURE 9.16

(a) Reel/slipring TMS versus (b) baling arm TMS in cage deployment. (c) Cutaway of reel/slipring TMS in top hat deployment.

(Courtesy Shilling Robotics.)

9.2.3 Launch and recovery systems

The final part of the launch and recovery equation is the actual overboarding equipment, which is discussed below.

Lock latch/docking head

Once the deployment TMS is hoisted off of the deployment deck and suspended above the deck through to the splash zone, the load becomes a "wrecker's ball" as the suspended weight sways at the end of the umbilical/tether awaiting movement from/to the deck. In high sea states, sway of the load can become quite dangerous.

The answer to this problem is the lock latch (also referred to as a "docking head"). This mechanism (Figure 9.19) grabs the TMS at its attachment point to the umbilical and firmly holds it in place against the A-frame while dampening the motions of the load with hydraulic or pneumatic accumulators about the various axes of sway. More complex docking heads include a "snubber/rotator" to both dampen the load dynamics and rotate the TMS to orient the cage or top hat appropriately from the orientation after retrieval through the splash zone to the proper orientation on the deck.

FIGURE 9.17

Schilling HD ROV with top hat TMS.

(Courtesy Schilling Robotics.)

FIGURE 9.18

A-frame LARS with Sub-Atlantic Mohican (without docking head).

(Courtesy SeaTrepid.)

For recovering free-flying vehicles (without TMS) from the water to the deck, the vehicle must be captured by the lifting device so as to lift the vehicle out of the water. In this instance, a latching device (typically termed a "locking sleeve") is attached to the tether then slid down to lock onto a locking collar attached to the frame of the vehicle (which is still in the water). Once the collar is secured, the vehicle is hauled aboard via the tether winch.

The docking head is composed of three items (two of which are optional):

1. The optional lock latch (Figure 9.20) for holding the suspended load during movement of the LARS arm.

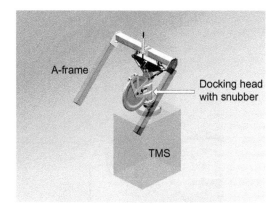

FIGURE 9.19

Lock latch/docking head.

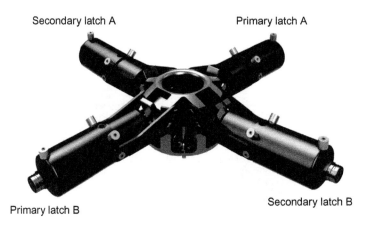

FIGURE 9.20

Locking mechanism for suspended load.

(*Courtesy Schilling Robotics.*)

2. The mandatory load dampener for smoothing the movement of the suspended load (both pneumatic padding at the latch/TMS junction and gas struts for dampening the entire swiveled load).

3. The optional rotator unit for orienting the vehicle–TMS combination onto the deck before lowering.

A-frame

The most typical LARS is the A-frame. The A-frame moves the ROV and its umbilical from the deck to the overboarding position that is clear of the deck as shown previously in Figure 9.6.

Such integrated LARSs are excellent for the smaller ROVs. When the vehicle systems get much larger, then the A-frame (and its impact on the deck) grows (Figure 9.21). The A-frame may be located on the stern or side of a ship or offshore platform.

FIGURE 9.21

Heavy-duty A-frame launch system.

(Courtesy Caley Ocean Systems.)

FIGURE 9.22

SeaBotix CDS.

(Courtesy SeaBotix.)

Fully contained system van

Some firms offer standard transportation vans that contain the vehicle, LARS, TMS, winch, workspace, and control room (Figure 9.22). Such self-contained systems make the deployment of smaller vehicles quick and efficient, often saving deck space and reducing costs. The Containerized Delivery System (CDS) developed by SeaBotix uses a telescoping LARS to deploy their 13,000 ft (4000 m) capable vehicle and its 650 ft (200 m) excursion tether.

FIGURE 9.23

Ship's moonpool ROV launch.

(Todd Walsh© 2009 MBARI.)

FIGURE 9.24

Telescoping LARS.

(Courtesy Dynacon.)

Ship integrated systems

For those vessels that can afford to fully integrate the ROV system there are other LARS options available. One option is the moonpool LARS. Figure 9.23 shows the MBARI marine operations staff launching the ROV *Doc Ricketts* through the "moonpool" on the research vessel *Western*

FIGURE 9.25

Ship door LARS.

(Courtesy Cargotec.)

FIGURE 9.26

Offshore platform LARS.

(Courtesy Forun Energy Technologies.)

Flyer. ROV *Doc Ricketts* can dive to about 13,000 ft (4000 m) below the ocean surface. Other heavy-duty LARS techniques are shown in Figures 9.24 through 9.26. Regardless of the LARS approach, the worst case dynamics and adequate safety factors must be taken into consideration because eventually these conditions will probably be encountered.

9.3 Currents and tether management

Once the vehicle is through the splash zone, tether management becomes critical to the success of the mission. This is especially important for the smaller, lower thrust, observation-class ROVs. Although the more powerful work-class vehicles can encounter many of the same problems, a single problem with tether management and the mission could be a failure and/or the ROV could be lost. The critical considerations of tether management follow.

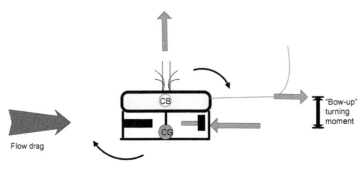

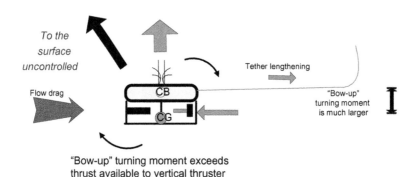

FIGURE 9.27

Vehicle stability considerations.

9.3.1 Tether effects

Tether pull point

Stability testing was performed on a small ROV system at Penn State University's Advanced Research Lab in their water tunnel. The water flow was slowly brought up while observing the vehicle's handling characteristics as well as its computed, versus actual, zero net thrust point.

This particular vehicle had a tether pull point significantly above the line of thrust (Figure 9.27), resulting in a "bow-up" turning moment. As the speed ramped up during the tests, with little tether in the water, the vehicle was still able to maintain control about the vertical plane by counteracting the "bow-up" tendency with vertical thrust down. However, at a constant speed with the tether being lengthened, the tether drag produced an increasingly higher tether turning moment, eventually over-powering the vertical thruster and shooting the submersible to the surface in an uncontrolled fashion.

If the tether is placed in close proximity to the thruster, parasitic drag will occur due to the skin friction and form drag from the thruster discharge flow across the tether. When selecting the tether placement, it is best to design the tether pull point (Figure 9.28) as close to the center point of thrust as possible in order to balance any turning moment due to the tether pull point.

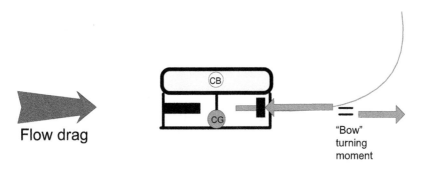

FIGURE 9.28

Minimal bow turning moment with tether on thrust line.

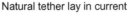

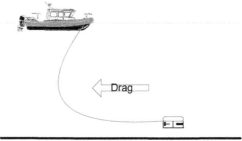

FIGURE 9.29

Natural tether lay behind the vehicle as the speed ramps up.

Tether pull/lay

Some vehicle manufacturers place the tether pull point atop the vehicle. The benefit to this placement is the tether will not lie as easily in the debris located on the bottom, allowing a cleaner tether channel from the vehicle to the surface. This is beneficial if the vehicle is operated in minimal currents with little or no horizontal offset. If either a horizontal offset or a current (or both) is encountered, the vehicle may experience difficulty through partial (if not total) loss of longitudinal and/or lateral stability.

Hydrodynamics of vehicle and tether

The most typical arrangement for an observation-class system involves a clean tether (i.e., without clump weight) following the vehicle to the work site. The tether naturally settles behind the vehicle and slopes in the current as it feeds toward the surface. As the vehicle speed ramps up, the flow drag on the tether correspondingly melds the tether into its wake, forming a "sail" of sorts behind the vehicle (Figure 9.29). A small reduction in the drag due to reduced angle of incidence to the oncoming water flow is more than offset with the additional form and flow drag of the excessive tether in the water. There is an old technique used by surface-supplied commercial divers to counter the excessive umbilical in the water—grab hold of something on the bottom while the tender takes up the slack. The same technique can be applied to ROVs by placing the vehicle on a stationary item on the bottom and then having the tender pull the excessive tether back on deck.

9.3.2 Currents

As stated previously, the objective of operating an ROV is to deliver a camera, instrumentation package, or work system and tools to a place where it can be of use. Under most situations, the item of interest will be submerged in a fixed location. Currents can cause difficulty in flying the submersible to the work site.

The resistance to delivery of the submersible to the work site is directly related to the total drag upon the submersible and tether. As stated mathematically in Chapter 3, hydrodynamic drag upon the wetted surfaces of the submersible system (vehicle plus tether) is affected by the following:

- Density of the seawater
- Characteristic area for both the submersible and the tether
- Velocity of the submersible system through the water
- Nondimensional drag coefficient of the system (essentially the hydrodynamic shape of the system through the water).

Increase any of these items and the submersible will have more difficulty fighting the drag.

Drag caused by current can be mitigated by the following:

- Maintain as little tether in the water as possible to accomplish the task.
- When pulling the tether to the work site, attempt to pull the tether into the current to present the least cross-section drag possible.
- If possible, lay the tether across some obstruction (while assuring an easy egress) between the work site and the pilot's operating station, so that the submersible is not required to continually fight the drag while station keeping.
- Approach a structure from the lee side to allow operation in the turbulent area downstream of the structure of interest.
- Operate at slack tide, or consider delaying the operation until slack tide, if currents prove to be above the submersible's power capability. Or invest in a more powerful ROV system.
- Make liberal use of clump weights while working on or near the bottom so that the weight takes the cross-section drag on the tether instead of the submersible's thrusters.
- When operating on the bottom, make use of the boundary layer near the bottom to stay in the calmer waters below the boundary layer.

9.3.3 Teamwork and proper tether management

Proper coordination between the tether handler and the submersible pilot will do more to prevent tether management problems, as well as entanglements, than any technological solution.

An ROV pilot's success as an operator of robotic equipment will directly depend upon his ability to figuratively place his head inside the submersible. As part of that "feel," the ROV pilot is in a position to gain a "feel" for the amount of tether needed as well as the degree of pull being experienced by the tether at any moment. As experience is gathered, a situational "lay of the land" orientation should be gained, allowing a mental picture of the location/lay of the tether. From this feel, the ROV pilot can direct the tether handler for the best tether management.

In general, the submersible should pull the tether to the work site (as opposed to the tether pushing the vehicle). The tether handler should allow just enough slack in the water so that the submersible is not wasting thruster power pulling the tether into the water. Any excess tether in the water, other than that absolutely necessary to accomplish the job, invites tether entanglements and wastes thruster power due to the drag of the excess tether.

Direct ingress/egress routes to the work site are generally preferable to multiple turn navigation due to tether friction and the higher likelihood of snags. The intended task should be planned in advance so that the best route can be chosen. The extra preparation time to choose the best route could save time later unfouling the tether.

In every ROV pilot/operator's tool kit should be additional floats and weights to compensate for the varying density of water. If an ingress route to the work site takes the submersible over a wall and then under an overhang, a weight at the proper location, followed by a float near the top, will lay the tether properly and avoid unnecessary tether hangs.

A neutrally buoyant tether is best for practically all ROV operating situations—but neutrally buoyant in what water type? Usually, ROV manufacturers specify neutral buoyancy of the tether based upon fresh water at average temperature (i.e., zero salinity and 60°F/15°C water). As the temperature of the water moves down and the salinity of the water moves up, the tether becomes increasingly buoyant. This factor must be taken into consideration and compensation made in the field.

While operating ROV equipment, a negatively buoyant tether may be useful in situations such as an under-hull ship hull inspection (to keep the tether clear of hull obstructions). But a negatively buoyant tether operated near the bottom will drag across any item protruding from the bottom much like a dragline.

Clear communications need to be established between the tether handler and the ROV pilot to avoid misunderstanding under critical situations. Standard terminology includes:

- "Launch vehicle"
- [for predive inspection] (Pilot) "Lights" (Tether Handler) "Check," etc.
- "Pull tether × feet (meters)" for pulling in tether
- "Slack tether × feet (meters)" for slacking tether
- "Pull tether × pounds (kilograms)"
- "All Stop"
- "Hold"
- "Pull until tension"
- "Pull then slack in [frequency] succession"
- "Recover vehicle"

9.3.4 Tether snags

Tether snags range from friction drag across an obstacle to total entanglement. The steps to clear tether entanglements are fairly universal on all ROV platforms and include the following:

1. When the submersible stops moving, the tether is snagged somewhere behind the vehicle.

2. Thrust forward slowly to bring the tether to "taut" condition to isolate the tether between the location of the snag and the vehicle.
3. Reverse slightly to bring slack in the tether line behind the vehicle.
4. Make a 180° turn in place to locate the tether between the vehicle and the snag.
5. Follow the tether to the location of the snag.
6. Work the tether snag out visually in coordination with the tether handler.

Once the tether snag is located, the general procedure for clearing the snag will present itself. There are four general ways of clearing a snag:

1. The preferable choice is to move the submersible to the snag and clear the foul using positive control.
2. If the above is not possible, move the operations platform (i.e., move the tether from the operator's side).
3. It may be possible to move the snag or the item upon which the tether is snagged.
4. The last choice is to cut the tether either physically or via a connection point.

If it becomes necessary to leave the operations area before a snagged tether or stuck vehicle can be freed, the vehicle can be powered down, left in the water, and retrieved at a later time. Steps to accomplish this task are as follows:

1. Power down the vehicle.
2. Pay out all of the tether.
3. Unplug the top-end or intermediate tether connection point and wrap in plastic or some other water-resistant wrapping.
4. Note the location with GPS.
5. Attach a buoy to the tether for easy retrieval upon return.
6. Advise operations that there is a tether in the water, which may pose a hazard to navigation, so that advisories can be made.

9.3.5 Tether guides and ROV traps

There are objects that, due to their shape and size, are more likely to snag a loose tether. Some can be used as tether guides, but some are considered "tether traps."

The typical tether trap has an edge with any converging angle less than 90°. This type of arrangement will allow the tether to slide naturally into the groove made by the angle and, once the tether is tensioned, form a friction lock on the tether (Figure 9.30). At that point, it is likely the submersible is lost.

Tether   Tether   Tether
guide   guide   trap

FIGURE 9.30

Tether guides and traps.

Edges with angles of 90° or more can form a "tether guide." These types of structures are handy during operations since they allow the operator to place the tether into the groove for known placement and allow for low-friction sliding of the cable along the groove.

Any mission that involves operations in or around structures will require a choice of tether lay and guiding.

9.3.6 Clump weights and usage

Hold a piece of yarn in front of your mouth and blow. The yarn is blown by the air from your mouth due to the cross-section drag of the string. Take that same piece of yarn and tie a weight to the end and repeat the exercise. The yarn does not blow in the wind. The cross-section drag is now absorbed by the weight instead of the yarn itself. This is a close analogy to the need for a clump weight while operating any free-flying ROV. The clump weight serves several purposes in ROV operations:

- It allows for orderly lay of the tether from the water insertion point to the work site on the bottom or mid-water.
- It delivers the tether to a known location that can be measured above the level of the bottom so that only the tether from the clump weight to the submersible requires positive management.
- The submersible is only required to drag the amount of tether from the clump weight to the submersible, thereby freeing the submersible for direct work tasks (Figure 9.31).

When conducting a long bottom search or inspection task, it may be more feasible to use a clump weight to bring the tether to the bottom and then move the work platform, rather than anchor and try to swim the submersible along a transect. Consider the appropriate use of a clump weight to more easily complete the mission.

9.3.7 Rules for deployment/tether management

A few rules for tether management are as follows:

- *Never* pull on the tether to clear a snag.

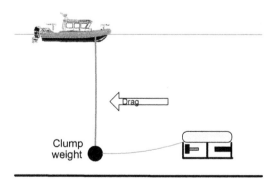

FIGURE 9.31

Use of a clump weight reduces tether drag to only the length between the clump weight and submersible.

- Always follow your tether back out from the work site with the submersible to the insertion point while pulling the tether slowly back with as little free tether behind the vehicle as possible.
- A clean and organized workplace is a safe and productive workplace.
- Tether turns are introduced when the submersible is turned from its original heading. Minimizing the number of tether turns while operating will enhance maneuvering. Remove all turns before submersible recovery.
- Be observant of obstacles located near the submersible that have the potential to snag the vehicle or the tether.
- If the vehicle is at the end of its tether, there may not be enough slack to allow an easy turn-around to follow the tether out. In this case, reverse direction to generate slack, and turn the vehicle around to manage the tether.
- Avoid weaving in and around fixed objects, such as baffles, pipes, pilings, rocks, and anchors. When operating in possibly fouled areas, it is advisable to remain on the surface until the vehicle is approximately above the work site and then dive.
- Attention must be paid to objects standing vertically or horizontally in the water column, such as anchor chains, pilings, lines, pipes, and cables. The vehicle's tether can easily become entangled. In currents, approach such items from the downstream side.
- If operating from an anchored ship, avoid working on the upstream side of the anchor.
- When operating in an area containing obstructions or obstacles that could snag or foul the tether, the operator should endeavor to remember the route taken to get to any one position. Not only will this information be helpful on the return trip, it will be valuable in the event the tether does become snagged!
- If the tether does become entangled, *do not* pull the tether to free it. The best course of action will present itself after careful and cautious consideration of the alternatives. During a recent operation, while performing a penetration of a wrecked aircraft at 425 ft (130 m) of seawater, the submersible became entangled in the wreck a total of 16 times during an 8-hour period. Tether snags happen as a normal course of operations. Plan for them and take positive precautions to manage the tether lay throughout the work site.

Video

CHAPTER CONTENTS

As video is generally considered the main sensor of the ROV, this chapter will delve fairly deeply into the theory and mechanics of video generation and troubleshooting. This section is targeted for a general overview of the various aspects of camera and video technology so as to understand the specifications of the camera used on subsea vehicles. For further explanations of the various technologies involved, please refer to the references at the end of this manual.

10.1 **History**

Motion pictures are, in their most simplistic form, still pictures arranged sequentially and displayed in succession in order for our brains to create a sense of movement. The first "movie" was a stack of pictures shuffled rapidly in sequence so as to create for the viewer a sense of fluid motion. If the pictures are shuffled quickly enough, the brain loses the sense that the picture was changing incrementally and the motion perceptually appears even. The rate at which individual pictures were displayed was (and still is) measured in frames per second (quantified in the metric *hertz* or cycles per second).

In the nineteenth and early twentieth centuries, studies were made on human subjects to determine the threshold of perception for sensing frame change (termed "intermittent light stimuli") between pictures such that the perceived flicker perceptually disappears. The objective for these studies was to determine the frame rate at which the brain loses the sense of frame change and perceives the changing picture as even motion. The concept itself is termed "flicker fusion threshold," and in several landmark studies, the average human perception threshold was measured.

Flicker is the sinusoidal variance of light intensity over time measured in both frequency and intensity. The flicker frequency threshold at which humans can perceive motion averages in the 16 Hz range. In practice, movies today are generally displayed at 24 frames per second (fps), and video is shown at between 25 and 30 fps (depending upon the standard employed). However, human light intensity flicker perception is above the frequency of human perception of motion, thus requiring video to run in the 50−60 fps range in order to avoid viewer objection.

The major breakthrough from simple film motion pictures to television came about when electronic capture sensors (through the photoelectric effect) and circuits were discovered that could sense light and transfer it into radio frequency (RF) signals for transmission. The system was complete when the cathode ray tube (CRT) was invented to convert the RF signals back to a display compatible with human perception.

Early television transmission systems were electromechanical devices projecting light upon photoelectric plates exhibiting low image quality. The first truly functional television system with electronic scanning of both the pickup and the display devices is credited to Philo Farnsworth in the United States in 1928. The 1936 Berlin Olympics were broadcast in monochrome (black and white) via television bringing the first large-scale viewing of television to a world audience.

During the development of this new communication medium, it became obvious that the manufacturers of commercial and consumer television equipment required standards so that all devices could be made to a common communications protocol. In the United States, the National Television Systems Committee (NTSC) was formed within the US Federal Communications Commission (FCC) to issue standards for transmission of television signals in the United States. The original monochrome analog standards were issued in 1941 and were later updated in 1953 to add the analog color transmission standards. The NTSC standard set the frame rate, screen aspect ratio, and line count so that the electronics industry could design equipment compatible with the standard.

In Europe and other parts of the world, other standards (including PAL or "Phased Alternation Line" and SECAM or "Sequentiel Couleur Avec Memoire") evolved for various reasons including improved resolution over NTSC, synchronization of frame rate to standard electrical (alternating current) grid frequency, resolution preferences, and (periodically) nationalistic fervor.

Today, analog standards have given way to new digital television standards worldwide, but the original NTSC and PAL frame rate, line resolution, and aspect remain (defined below). The old NTSC and PAL formats are now folded into different standards which have been expanded to include the new high-definition digital television standards. The current non-governmental organization developing television industry consensus standards in the United States is the Advanced Television Systems Committee (ATSC) located in Washington, DC. Standards issued by the ATSC are typically adapted by most governmental regulatory organizations (e.g., in the United States, the FCC). As the frame rate and the line count are embedded into other standards, the NTSC and PAL designations currently only refer to the method of embedding color coding into the transmission.

Various other international standards organizations are involved with consensus standards for the video and motion picture industries including:

Advanced Television Systems Committee (ATSC—www.atsc.org)
Association of Radio Industries and Businesses (ARIB—www.arib.or.jp)
Cable Television Laboratories (CableLabs—www.cablelabs.com)
Consumer Electronics Association (CEA—www.ce.org)
Digital Video Broadcasting (DVB—www.dvb.org)
European Broadcasting Union (EBU—www.ebu.ch)
European Telecommunications Standards Institute (ETSI—www.etsi.org)
International Electrotechnical Commission (IEC—www.iec.ch)
Institute of Electrical and Electronics Engineers (IEEE—www.ieee.org)
International Organization for Standardization (ISO—www.iso.org)
International Telecommunication Union (ITU—www.itu.int)
Society of Cable Telecommunications Engineers (SCTE—www.scte.org)
Society of Motion Picture and Television Engineers (SMPTE—www.smpte.org)
Video Electronics Standards Association (VESA—www.vesa.org)

10.2 How it works

10.2.1 The camera

The camera functions as a transducer converting light into electrical pulses through sensing of light frequency and intensity with use of the photoelectric effect. Screen position and intensity of light levels entering a camera's lens are measured and converted to electrical signals for output as video signals. With use of a photoelectric converter (converting light into electrons), the electrons are converted sequentially into voltages (per pixel) by the electronic circuit while a synchronizing signal is added to synchronize position information with the voltages for rendering a picture (Figure 10.1). The camera itself is made up of a lens channeling light to a series of light-sensing elements mounted to a sensor plate connected to a conductor (Figure 10.2). The signal is then transmitted to a video capture device (for later viewing) or to a display for real-time viewing.

Traditional electron beam "vacuum tube cameras" have given way to solid-state charge-coupled devices (CCD) and complementary metal−oxide−semiconductor (CMOS) as the optical image sensor (device converting optical images to electrical signals). Both CCD and CMOS sensors are

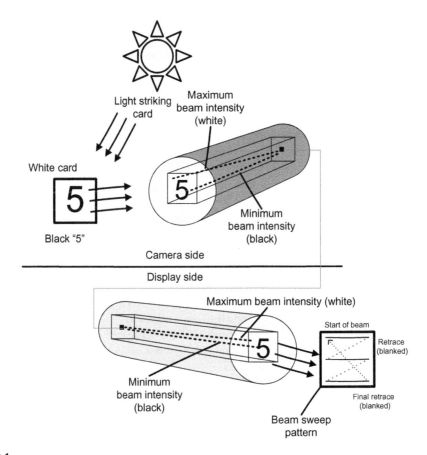

FIGURE 10.1

Camera captures image and transmits it to display.

analog devices converting light signals one pixel at a time; circuits then convert the voltage to digital information (Figures 10.3 and 10.4).

The smallest element of the light sensor is the pixel. In the low-light subsea environment, there may not be sufficient light energy ("illuminance" or total luminance flux per unit area with the SI unit of "Lux" as the metric) falling upon the individual pixels to produce sufficient voltage under normal conditions; therefore, some type of light/video multiplier is needed in order to produce a useful image in low lighting conditions. This is the realm of the "high-sensitivity camera" (HSC).

HSCs enhance the light-gathering capabilities of the sensor elements and are generally grouped into charge integration and electron multiplication types:

Charge integration—essentially, this type of camera enhances sensitivity by extending the exposure time (charge integration time), allowing for signal strength gathering/boosting of the available light energy. An example of this type of camera is a so-called cooled CCD.

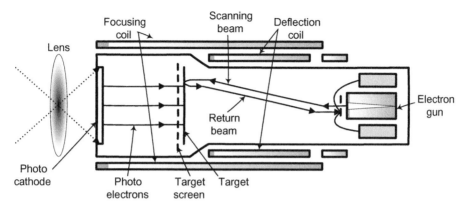

FIGURE 10.2

Schematic of basic camera system.

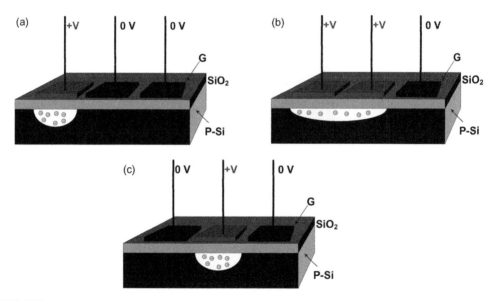

FIGURE 10.3

Schematic of CCD element demonstrating light transfer (a) from first element then (b) transitioning between elements to (c) second element.

Electron multiplication—this method is any of several means of multiplying the received electron signal strength. Examples of this type of camera/method are ICCD (intensified CCD), SIT (silicon intensified target), EB-CCD (electron bombardment CCD), and EM-CCD (electron multiplier CCD).

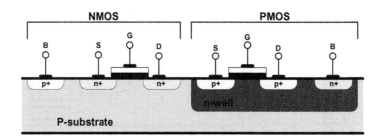

FIGURE 10.4

Schematic of CMOS element.

FIGURE 10.5

Small pinhole camera.

(Courtesy VideoRay.)

The traditional vacuum tube video camera has given way to the solid-state CCD. Likewise, the SIT camera has been the traditional low-light camera in the subsea industry until the late 1980s. The latest high-sensitivity CCDs of the electron multiplication type now have the sensitivity of the early SIT cameras—and the CCD is much more cost effective—therefore, SIT cameras have gone out of vogue in the subsea industry.

In subsea camera systems, the typical camera uses a standard industry capture device enclosed in a pressure housing wired with wet-mate connectors to provide the total deepwater camera system. Examples are provided in Figures 10.5 and 10.6.

10.2.2 Lens optics

The lens is a light-gathering device that collects and then refracts incoming light to project a certain quality/quantity of that light onto a light-gathering device for processing. There are several types of lenses depending upon their shape and light-gathering characteristics (Figure 10.7).

FIGURE 10.6

The Kongsberg OE14-370 Standard Camera System mounted into pressure housing.

(Courtesy Kongsberg Maritime.)

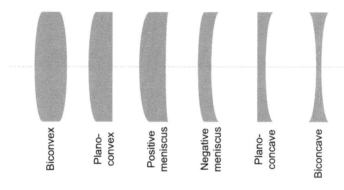

FIGURE 10.7

Lens types by shape.

Lenses are broken into several components including field of view (FOV), focal length, and f-number (also termed "f-stop," "focal ratio," or "relative aperture"). As illustrated in Figure 10.8, the FOV metric is expressed in angular degrees in either width-by-height ($W \times H$) or (more typically) corner-to-corner (or "diagonally").

As illustrated in Figure 10.9, focal length (f) is the measure of the lens's light-gathering/magnification ability in that the more acute the angular deflection of the light rays bent in the lens the higher the light magnification (S_1 is the distance to subject and S_2 is the distance to the light capture mechanism—S_2 and f are required to be synched in order for the object to be in focus). The shorter the focal length (f), the higher the concentration of the light upon a discrete point (where the light-gathering mechanism is placed—e.g., CCD or CMOS). Hence, the measure of focal length is the measure of the optical power of the lens. Longer focal lengths are for higher magnification for telephoto applications. The longer focal lengths allow for higher magnification of the object with a corresponding loss of angular FOV (i.e., the magnification versus angular FOV are inversely proportional as the object is brought closer to the center of view—Figure 10.8). Also see Figure 10.10.

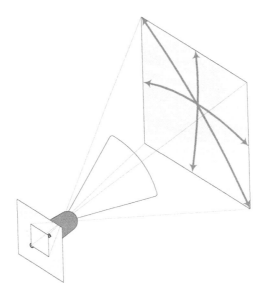

FIGURE 10.8

Camera FOV.

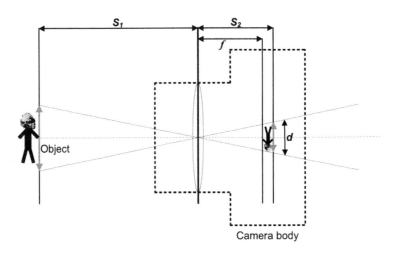

FIGURE 10.9

Camera optics.

The f-number/f-stop is the ratio of the lens focal length to the corresponding lens aperture (or opening through which the light passes). Of course, the amount of light captured at the focal point (where the CCD or CMOS sensor resides) is directly proportional to the size of the aperture—the bigger the hole, the more light that comes through. However, all of the values (FOV, focal length,

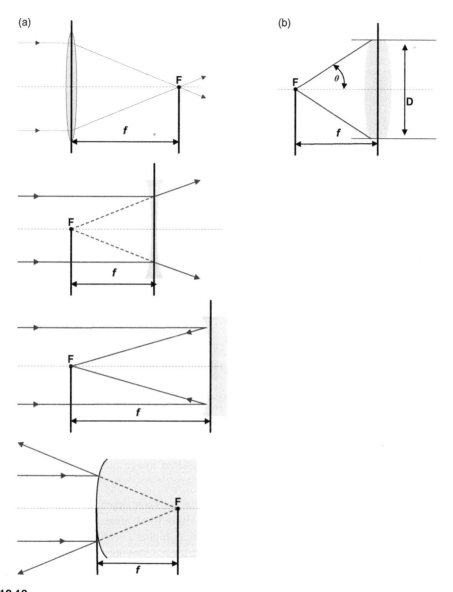

FIGURE 10.10

(a) Focal lengths by various lens types (both convergent and divergent) as well as (b) depiction of the aperture (D) versus the focal length (f) converging upon focal point (F).

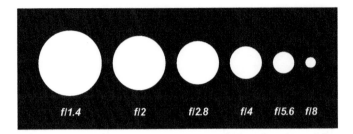

FIGURE 10.11

Examples of f-stop settings from *f*/1.4 (low) to *f*/8 (high).

FIGURE 10.12

Example of two pictures of the same scene—one with low f-stop (a) and the other with high f-stop (b).

(Photos by Pat McCallan.)

and f-number) will vary in proportion to achieve the desired image quality. Some rules of thumb are as follows:

- A small, or high, aperture setting (e.g., *f*/8) is great for a greater depth of field (i.e., nominal distances from the camera lens whereby all objects within those areas are in focus), but bad for both low lighting situations and/or fast shutter speeds. A higher f-stop setting (smaller aperture) is used for daylight shots with a slower shutter speed for nonmoving objects while night photography requires a lower f-stop (larger aperture—e.g., *f*/2) with a higher sensitivity camera (HSC from above) (Figures 10.11 and 10.12).
- A high f-stop (smaller aperture) with a longer focal length (telephoto magnification) requires either a very slow frame rate or some type of HSC in order to gather enough light for a meaningful picture rendering.
- Use a low f-stop and HSC for low ambient lighting subsea conditions.

In Figure 10.12(a), the picture was shot with a low f-stop, causing a narrow depth of field (i.e., distance whereby the object is in focus) while Figure 10.12(b) is shot with a higher f-stop, allowing a great depth of field.

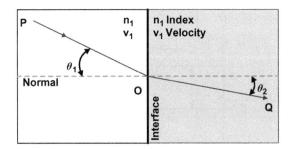

FIGURE 10.13

Snell's Law describes the water–lens–air interface within the camera housing.

Of particular importance to subsea cameras housed within pressure housings is the shape of the lens versus the FOV of the camera. Just as light bends (refraction) upon entering water (as described by Snell's Law, Figure 10.13), so does light bend (refract) upon transferring from water to glass (lens) to air (inside of camera housing). For wider fields of view, a curved pressure housing lens will be required to avoid outer edge distortion of the image while narrower fields of view allow for a flat plate lens to produce acceptable image quality. Further, a camera in air with fixed focus on infinity will not be in focus underwater to infinity with the same focus setting.

10.2.3 The signal

Once the light is captured and converted to electrical energy, the signal must be put into usable information. The output from the camera is an electrical sinusoidal signal with both amplitude and frequency (Figure 10.14). The typical composite video signal is a 1-V peak-to-peak sinusoidal signal transmitted into 75 Ω cable (standard coax) at a signal-to-noise (SNR) ratio of sufficient strength to push the signal through the conductor (the higher the SNR, the better the picture).

The video signal is then divided into luminance (light intensity/quantity) and chrominance (light color/quality). As depicted in Figure 10.1, maximum luminance is with reference to pure white (saturated brightness) while lowest luminance is a reference to black (no brightness). Chrominance will be further discussed below. Further, the positioning information for each of the element (pixel) measurements is laced into the quantity/quality measurement and the information synched so that a picture may be built. As shown in Figure 10.15, a single line of composite video is depicted. The horizontal blanking pedestal syncs the beginning of the line paint to the start of the viewable screen area while the color burst sets the reference color level. The remaining viewable area of the line is laced with luma/chroma information as the line is painted. At the end of the line the front porch closes down the line, and the synch information blanks and retraces the image projector to the next line for the next line generation.

The variables within the composite video signal for picture generation are made up of the various components: "color, video, blanking, and sync" (CVBS). Each of these is required to render a color video picture. The luminance portion (termed the "luma" or "Y" component) of the video signal contains the brightness information while the chrominance (the "chroma" or "C" component) portion contains color, hue (tint), and saturation information (Figures 10.16 and 10.17). On black

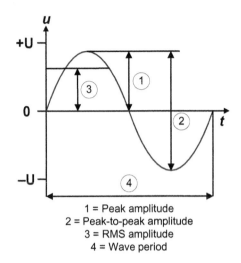

1 = Peak amplitude
2 = Peak-to-peak amplitude
3 = RMS amplitude
4 = Wave period

FIGURE 10.14

A sinusoidal curve of a standard composite video signal.

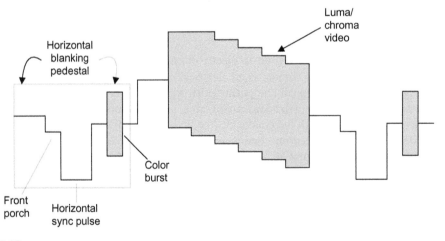

FIGURE 10.15

Analysis of one line of composite video.

and white monitors, only luma information is projected. When troubleshooting color monitors, signs of signal strength degradation are often evident by a color picture losing the chroma signal and reverting to black and white with less image clarity.

Color: The chrominance aspect (C) is a combination of the red, green, and blue (RGB) components of the basic color spectrum. Every color in the rainbow is made up of a

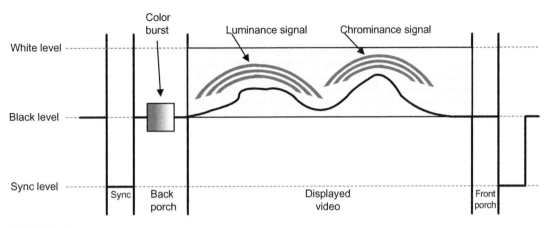

FIGURE 10.16

Video signal schematic of one horizontal screen line.

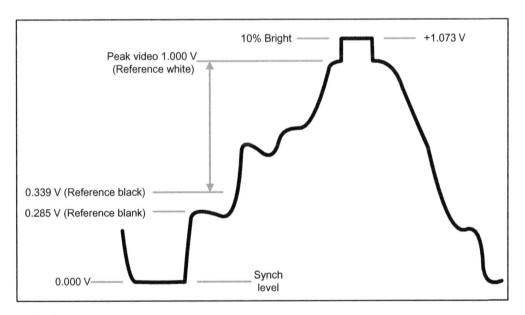

FIGURE 10.17

Closer examination of composite video signal components.

combination of the RGB components (white comprises all three RGB colors combined) and by properly mixing these (adding or subtracting) we may make up any color (Figure 10.18). The C factor is assigned separately from the video/luminance (Y) factor in either an analog or a digital format. In digital format, each of the RGB components is assigned a numerical value between

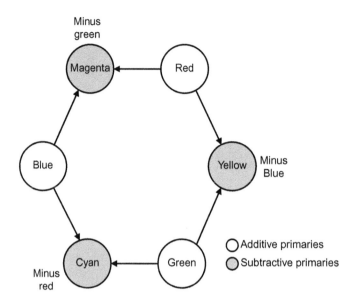

FIGURE 10.18

The color matrix.

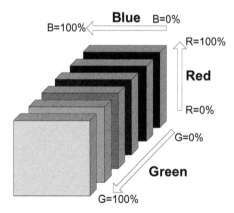

FIGURE 10.19

The color cube.

0 (no mixing or 0%) and 255 (full mixing/saturation or 100%), allowing for any color to be exactly rendered with digital precision (Figure 10.19).

Video: The luminance (Y) factor describes the measure of individual pixel brightness in a purely monochrome fashion. The input to or output from the individual pixel corresponds directly with the amplitude measurement of the light signal measured on the individual sensing unit in the camera.

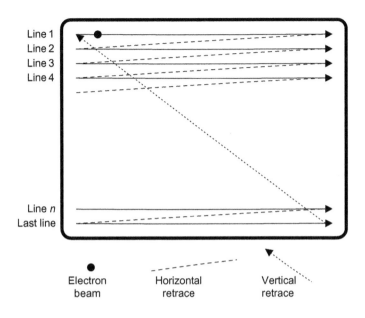

FIGURE 10.20

Video screen render/raster of picture.

Blanking: Lines of video are painted horizontally beginning from the top left corner of the video screen and continuing from left to right further continuing from top to bottom (Figure 10.20). Once a row has been rendered, the electron beam (for CRT displays) must reposition to the next line for continued rendering. The blanking interval is the time delay whereby the beam is turned off for the interval between the end of one line to the beginning of the next (termed "horizontal retrace/blanking") or the end of the frame to the beginning of the next frame (termed "vertical retrace/blanking"). Components of the horizontal blanking interval are the front porch (blanking while still moving past the end of one line), the synch pulse (blanking while moving rapid left for repositioning to the next line), and the back porch (blanking while resuming moving right before the beginning of the next line). Color burst happens on the back porch followed by unblanking at the beginning of the line to begin rendering. Blanking is specifically used to avoid rendering of the retrace line across the screen. In television broadcasts, the vertical blanking interval is used to embed and transmit data such as closed captioning, time codes, and copyright information.

Synch: The synchronization signal is embedded within the video signal to synchronize the timing and alignment of the screen paint. The synch signal is comprised of both horizontal and vertical synchronization.

The video picture is built by the sequential updating of a series of individual pixels arranged horizontally and vertically on a picture screen. Beginning on the first line of the picture (in the top left corner of the screen), the electron beam illuminates each pixel from left to right and then from top to bottom. There is a "blanking" line between each paint line whereby the beam is blanked (or disabled) while repositioning between the right end of one line to the beginning of the next line, much like a typewriter needs to reposition from the end of one line to the beginning of the next before writing.

10.2.4 The display

Displays come in many shapes and sizes. The first functional television displays were the CRTs (Figure 10.21). A CRT consists of a vacuum tube containing an electron gun projected upon a fluorescent screen. Within the tube is an electromagnet for steering the beam so as to project a full display based upon the target painting of the electron beam upon the screen.

CRTs have given way to more advanced and efficient digital display screens such as liquid crystal displays (LCD), plasma screens, and organic light-emitting diode (OLED) displays. However, the basics of display elements are unchanged. With LCD, plasma, and other modern fixed-pixel screens (versus the floating pixel of the CRT), either specific direct addressing or multiplex addressing is used to provide information to each pixel on the display. With specific addressing, each pixel (*x/y*) address is given specific detail for picture generation while multiplexing addresses the rows and columns, further breaking down the information by line and row. The information can then be transmitted (versus individual pixel addressing), allowing for larger screens with higher pixel counts with the same bandwidth of a smaller screen with specific addressing.

As an example, some of the benefits of LCD over CRT are as follows:

- Smaller footprint for LCD (approximately 15% of CRT footprint)
- Lighter weight (typically 20% of CRT weight)
- Lower power consumption (typically 25% of CRT consumption)
- Flat screen incurring no geometrical errors
- Sharper pictures due to digital precision and uniform colors
- No electromagnetic emission!
- Much larger screens are possible
- Longer useful life than CRT

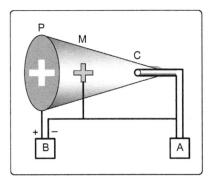

FIGURE 10.21

Basic schematic of a CRT.

Examples of disadvantages of LCDs over CRTs are as follows:

- Higher price of LCD (although the pricing falls with technology development)
- Low contrast and luminance (typically 1:100)
- Low luminance (typically 200 candela per square meter (cd/m^2))

Pixel: The basic element of a screen is the pixel. For the color monitor, the pixel is a combination of the three color elements (RGB) to make up a single light point. The values of Y (luminance or brightness) and C (chrominance or color) are assigned during capture by the camera, transmitted, and rendered at the pixel level on the capture (display) device.

Frame rate: The rendering of one complete picture on a display is termed a "frame." The rate at which a frame is updated is measured in frames per second and is set by the video standard adapted for the specific camera/display combination (e.g., NTSC, PAL, SECAM, etc.). Most frame rates are a part of the video standard adapted within each country and generally correspond to the frequency of the electrical grid within the country (e.g., European countries specify 220 VAC/ 50 Hz grid power and PAL specifies a 50 Hz interlaced (25 Hz progressive) frame rate to avoid inherent video noise).

Aspect ratio: The width to height ratio of the video monitor is termed its "aspect ratio" (expressed in width to height, Figure 10.22). The original NTSC television standards adapted a 4:3 aspect ratio or four units wide by three units high. In high-definition television, a 16:9 ratio has been adapted for higher resolution with wider viewing area to conform to higher cinema quality standards. The aspect ratio is also described by its ratio to a one denominator base assigned to the height (e.g., a 4:3 ratio equates to a 1.33:1). The 4:3 ratio is predominantly used in television broadcasting while the wider aspects prevail in the movie industry. The movie industry may at times provide a modified aspect ratio for viewing of wider aspects on a typical 4:3 viewing screen by cropping the outer edges of the viewing area.

Aspect ratios are further divided into the *display aspect ratio* (aspect ratio of the image as displayed) and the *pixel aspect ratio* (aspect ratio of the pixel element) (Figure 10.23). Typically, if a picture is horizontally distorted, the pixel aspect ratio is set to other than a 1:1 (square) shape.

Resolution or definition: The display resolution (termed "definition") is described by the number of individual pixel elements that are rendered within the display screen. The resolution is defined as screen pixel width count by height count and is generally depicted with the

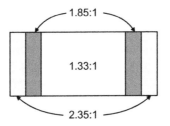

FIGURE 10.22

Screen aspect ratios.

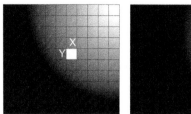

FIGURE 10.23

Pixel aspect ratios of 1:1 and 2:1.

denominator/height count (e.g., 1920 × 1080 for 1080 interlaced "i" or progressive "p" HD video). Further, video definition is described in three levels:

1. Standard Definition (576i (PAL/SECAM) or 480i (NTSC) minimum)
2. Enhanced Definition (576p or 480p minimum)
3. High Definition (720p or 1080i minimum)

The NTSC standard, as originally adapted in 1941, described a 480i resolution at a 60i (60 interlaced is 30 progressive) fps interlaced frame rate. When color was added in 1953, an additional timing factor of 1000/1001 was added to avoid interference between the chroma carrier signal and the sound carrier, reducing the frame rate to 59.94i (60 × 1000/1001) or 29.97p. With the demise of analog television, the need to separate the carrier and sound signals is moot as these are embedded digitally (i.e., mooting the need for the 1000/1001 spacing). Expect further evolution of video standards to reflect the new digital format.

Screen resolution is fixed by the vertical frame line count. Aspect is then adjusted to comply with the desired viewing area. As an example, a 480 line resolution with a 4:3 aspect ratio yields a 640 × 480 viewable pixel count while a 1080 resolution with a 16:9 aspect ratio is 1080 × 1920 (width to height).

Interlaced versus progressive: Interlacing of video is a method of doubling the perceived video frame rate while maintaining the same data bandwidth. As stated above, human perception of individual frames disappears at approximately 16 fps (minimum flicker rate). In early television broadcasts, a method of splitting the picture into two separate interlaced frames allowed the perceived frame rate to go from 30 fps to 60 fps (NTSC) by transmitting one picture frame in two separate half sections. As an example, a 480i picture is 525 (480 viewable) horizontal lines of pixels. If you split these pixels into even and odd number rows, you can achieve the same picture by transmitting the odd number rows (row 1, 3, 5, 7, etc.) totaling 262.5 rows per half-frame followed next by the even number rows (row 2, 4, 6, 8, etc.) (Figure 10.24). This allows the flicker rate to rise to a full 60 fps (59.94, actually, for NTSC), for smoother video quality, while maintaining the same bandwidth of 30 fps. This allows for enhanced motion perception providing a much smoother video while practically eliminating perceived flicker. A screen capture of one interlaced video frame shows a poor image quality of the individual frame (actually, half-frame) while the video picture (aggregate frames displayed in series) quality is high, demonstrating a capture of only one-half of the frame for the frame grab. Progressive video is simply a full painting of the entire frame upon each rendering (i.e., a frame grab of a progressive video is picture quality). Therefore, 30p is essentially equal to 60i, but the picture is of a much higher quality as the image is not distorted by the

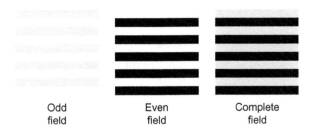

| Odd | Even | Complete |
| field | field | field |

FIGURE 10.24

Interlaced video rendering.

bias of half-frame rendering. Video formats are often expressed as a combination of resolution and frame rate (e.g., 1080p50 is 1920 × 1080 progressive at 50 fps).

Typically, computer monitors and film projectors display progressive images while television and video displays are interlaced. Therefore, interlaced images (video and television) must be converted to progressive for display on computer monitors while progressive images (computer and film) must be converted to interlaced in order to play on a television screen. Also, frame grabs from progressive video are image quality while grabs of interlaced video are typically jagged due to horizontal sync.

10.2.5 Composite (baseband) video

A composite video signal combines the various video components (luminous, chrominance/hue/saturation, and audio) into a single signal for easy transmission. The embedding techniques vary between the various standards (NTSC, PAL, SECAM), but all have the information within the signal for picture rendering (Table 10.1). The composite feed is also termed "baseband" as it is the base level of video signal before modulating for transmission over broadcast services.

In order to accommodate higher resolution video formats (enhanced or high-definition video—EDTV or HDTV), the standard composite video connector is insufficient to accommodate the increased bandwidth demand. The best method for connecting video minimizes the number of video corrections between capture and display. Specifically, the original camera captures analog data. From that capture, the minimum number of format conversions to the ultimate display maintains its video quality. Further, the full video capture/transmission/display system is only as good as its *lowest quality component*.

The single-wire composite video connector transmits the lowest of quality signals. For the higher order analog and digital video signals (EDTV and HDTV), the individual RGB color components are separated and transmitted separately over individual lines. In the early days of television, monochrome (gray-scale) video data was all that was needed. However, as color broadcasting made its way into development, the separate analog RGB components needed to be transmitted but took up three times the bandwidth. A technique was developed to transmit the Y, R-Y (red difference), and G-Y (green difference) information using one signal over the same bandwidth as the original gray-scale (hence the birth of the "composite video signal"). All variations of chrominance relate back to the RGB mathematical formula. The question then becomes how best to transmit the mix.

Table 10.1 SD, ED, and HD Formats' Summary of Major Attributes

Standard Definition Format	NTSC	PAL
Vertical frame lines	525 lines (480 visible)	625 lines (576 visible)
Frame rate	59.94i	50i
Interlaced/progressive	Interlaced	Interlaced
Enhanced Definition Format	**NTSC**	**PAL**
Vertical frame lines	525 lines (480 visible)	625 lines (576 visible)
Frame rate	29.97p	25p
Interlaced/progressive	Progressive	Progressive
High-Definition Format	**NTSC**	**PAL**
Vertical frame lines	720p or 1080i	720p or 1080i
Frame rate	29.97p or 59.94i	25p or 50i
Interlaced/progressive	Either	Either

FIGURE 10.25

Component video feed for digital/analog YCbCr/YPbPr connectors.

(Courtesy Jake Christ.)

S-Video came about for connecting selected consumer video equipment. This method transmits video over two analog signals, the gray-scale (Y) and the R-Y/B-Y color data signal.

Digital color video is transmitted in two flavors, RGB and YCbCr (although YCbCr is a mathematical formulation of the RGB color matrix). The digital RGB is simply the converted analog RGB while the YCbCr (digital) and YPbPr (analog) are variations on the RGB standard (Figure 10.25). When connecting video equipment, a determination of the signal must be made before defining the connection method.

In higher resolution transmissions (above standard definition), the colors are separated into their components for separate transmission over different lines. Care must be taken to determine the compatibility of the video signal with the capture device.

Methods of connecting video capture devices to ultimate monitors vary in quality but are typically ranked in descending order for connecting subsea/terrestrial cameras, DVD (digital video disk) players, digital cable/satellite/terrestrial set-top boxes, or other video feeding media.

1. HDMI (digital YCbCr)
2. HDMI (digital RGB)

FIGURE 10.26

Composite video connectors for (a) consumer (RCA) and (b) commercial (BNC).

3. Analog YPbPr
4. Analog RGB
5. Analog S-Video
6. Analog Composite

Figure 10.26 displays standard composite video connectors for both consumer (RCA) and commercial (BNC) formats.

10.2.6 The transmission (RF modulation)

Once the baseband signal is generated, the various portions of the paired audio/video signal are separated and then modulated for transmission on carrier waves to the receiver for rendering (Figure 10.27). The luminance, chrominance, and audio carrier are transmitted at separate frequencies and modulated at the receiving end into television channels for blending into the final display picture.

The typical consumer television set allows for capture of either composite or modulated video signals. The modulated analog (or, more recently, digital) signal is generally transmitted via coax cable while the composite signal is transmitted via regular copper wire. DVD and various analog video players transmit to a television display via either composite or modulated. The composite feed is generally a straight feed/capture out of the player into the display (with a standard RCA or BNC connector) while the modulated feed will require assigning a television channel (typically channel 3 or 4—manually selected on both the output and the input devices) for display/capture transmitted over a coax cable with F-type connectors (Figure 10.28(a) and (b)). BNC and RCA connectors are typically for baseband coax cabling with signal transmission while modulated signal transmission uses F-type connectors.

10.3 Digital video

Digital video evolved in the 1980s as a simple digitalization of analog video for video-editing purposes. Eventually, it took on a life of its own as standards evolved to streamline this new medium. While analog video comprises a series of video frames, digital video portrays a series of bitmap

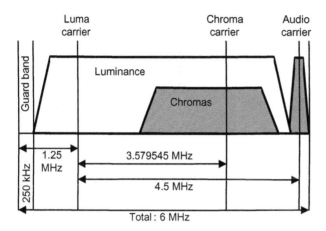

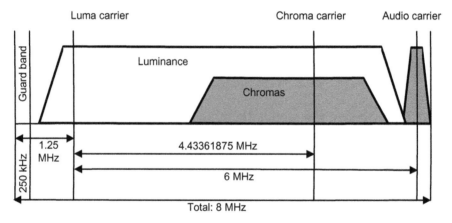

FIGURE 10.27

NTSC (top) and PAL (bottom) composite video signals.

images onto the screen in succession at a nominal rate. These images are also called frames (with the frame rate still reflected in fps), but a digital picture has a much deeper ability for embedding information and manipulating the image. The electron stream of the CRT projected pixels of light onto a fluorescent screen (i.e., the pixel was projected wherever the electron beam landed on the screen and, hence, could move around the screen), but the new digital pixels were fixed upon the screen in individual elements/locations.

Frames of an orthogonal raster bitmapped image (i.e., a raster of pixels), as with analog video frames, are measured as width versus height ($W \times H$). The basic image is also made up of pixels, the arrangement of which determines the frame size/resolution. However, in digital video, pixels only have one characteristic and that is color. The color is represented by a fixed number of bits (e.g., 24-bit color). The color bit count is termed "color depth" (CD) with the depth of color varied by the number of bits (e.g., 8-bit color would have eight color balance intervals between white and

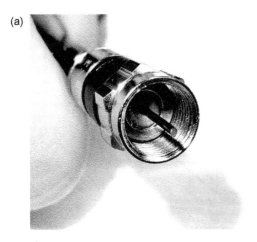
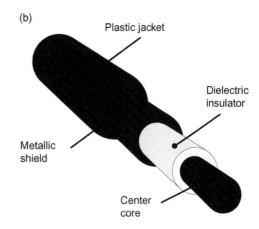

FIGURE 10.28

(a) Coax F-type connector and (b) coax cable schematic.

(Courtesy Jake Christ.)

Table 10.2 The 100% RGB Color Values

	Range	White	Yellow	Cyan	Green	Magenta	Red	Blue	Black
R	0–255	255	255	0	0	255	255	0	0
G	0–255	255	255	255	255	0	0	0	0
B	0–255	255	0	255	0	255	0	255	0

black versus 24-bit having 24 intervals). Also, the portion of the signal containing actual video information is termed "active video" (the remaining portion of the signal contains blanking information and other embedded data).

The data pipeline for video transmission, as well as end capture, must be sized for the amount of data transmitted (i.e., the higher the definition, the higher the data transmitted). An example, the data volume of video stream calculation is as follows:

A 1 hour, PAL formatted, uncompressed video has 3600 seconds of duration. With a typical frame size of 640×480 and an 8-bit CD at 25 fps, the computations are as follows:

Pixels/frame = 640*480 = 307,200
Bits/frame = 307,200*8 = 2,457,600 = 2.46 Mb
Bit rate (BR) = 2.46*25 = 61.44 Mb/s
Data set size = 61 Mb/s*3600 s = 219,600 Mb = 27,450 MB = 27.5 GB/h

The typical analog color cube can be further expressed in digital format by varying the colors within each of the RGB color components for mixing based upon a digital value between 0 (lowest saturation) and 255 (highest saturation) (Table 10.2 and Figure 10.29).

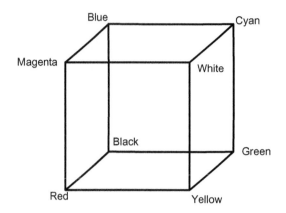

FIGURE 10.29

RGB color cube.

The color sampling formats between the various RGB components (or YCbCr) are expressed into their ratios (e.g., 4:4:4 for sampling of four parts Y by four parts Cb by four parts Cr, 4:2:2, 4:2:0, etc.). Mathematical conversion of the various bit levels, sampling rates, and color saturation are beyond the scope of this text.

Also, the digital interface standards have moved very rapidly since the advent of digital video, making any detailed mentioning of those standards (in all likelihood) obsolete by the time of publication of this text; therefore, this section has concentrated on digital video basics, transmission, capture, and compression. However, a partial listing of the current video interface standards at the date of publication of this text (from various standards organizations with varying references to standard identification) is as follows:

* Serial Digital Interface (SDI)
* FireWire
* High-Definition Multimedia Interface (HDMI)
* Digital Visual Interface (DVI)
* Unified Display Interface
* DisplayPort
* Universal Serial Bus (USB)
* Digital component video
* Digital Video Broadcasting (DVB)
* Asynchronous Serial Communication Interface (ASI)

10.4 Video capture

The end deliverable for a typical ROV assignment is the audio-annotated video along with a video log of the activities on the video (for easy referencing). The video may be captured in any of several formats based upon the client-desired storage medium.

The standard analog tapes of old have given way to various digital storage mediums including digital videotapes (e.g., miniDV, DVCPRO, DVCAM, etc.), DVD (defined as a digital video disc, digital versatile disc, or just a DVD), and DVR (digital video recorder). The original optical disc for recording consumer video was the video CD (VCD) introduced in 1993 using the MPEG-1 (Motion Picture Experts Group) digital compression format. The follow on to the VCD was the DVD's introduction in 1996 integrating the MPEG-2 compression format.

Today, the most widely used format for exchanging digital video is the DVR, which records digital video to a hard drive, flash drive, or some other digital media capture. DVRs record and play in the current industry standard compression formats including MPEG-4, MPEG-2.mpg, MPEG-2.TS, VOB (Video Object), H.264, and ISO images video (with AC3 and MP3 used for audio track formatting). However, the typical DVD is encoded and plays in MPEG-2 format, which is generally a digitalization of analog data with standard compression.

Videos are recorded on capture devices in various capacities/speeds based upon the desired image playback quality requirement and degree of digital capture space. The three consumer-grade recording speeds are standard play (SP), high-quality (HQ), and long play (LP) (the actual acronyms will vary with manufacturer). The various recording speeds allow for differing picture resolutions based upon the image quality preferred. The standard DVD has a 4.7 Gb capacity with the nominal video length per DVD of 2 hours; therefore, the SP will record at 2.35 Gb/h. LP records at a considerably lower data rate (<1 Gb/h), allowing for $10+$ hours of video on the same DVD; however, the quality of the video is of a much lower resolution than it is in the SP mode. The same happens with an HQ mode only in the opposite direction whereby the DVD is full after only 1 hour. A summary is as follows:

LP: "Long play" mode whereby a longer video length per unit storage capacity (e.g., 10 hours per DVD) with substantially lower video resolution (e.g., 0.6 Gb/h of data capture in this mode)

SP: "Standard play" mode with typical length and quality to nominal quality (e.g., 2.35 Gb/h in this mode)

HQ: "High quality" mode whereby the image resolution is enhanced through speeding of the data rate for higher image resolution (e.g., 4.7 Gb/h in this mode).

10.5 Video compression

This section is also termed "how to get 10 pounds of stuff into a 5 pound bag" (i.e., any data that is nonrandom can be compressed).

As stated above, the introduction of the VCD corresponded to the promulgation of the MPEG-1 standard in the early 1990s. The stated purpose was to play back the audio—video combination on a standard compact disc (used extensively in the music industry at that time) with a bit rate of 1.416 Mbps (1.15 Mbps of which would be video). The mathematical compression algorithm for the process of digitizing analog video into digital format is termed a "codec" (for compression/decompression). In the early days of codec development, chaos ensued with many organizations touting the "best codec" in attempts to win general acceptance by industry. Chaos gave way to industry standards with the formation of the Motion Picture Experts Group (MPEG). The early MPEG-1 standard allowed for proprietary compression/decompression algorithms, thus fragmenting

Table 10.3 Various Video Compression Standards over Time

Year	Standard	Issuer	Description of Main Feature
1984	H.120	ITU-T	First video encoding standard
1990	H.261	ITU-T	Video conferencing
1993	MPEG-1	ISO	Video CD
1995	H.262/MPEG-2	ITU-T/ISO	DVD, DVB, Blue-Ray, SVCD
1996	H.263	ITU-T	Expanded video conferencing with cellular
1999	MPEG-4 Part 2	ISO	Video over Internet protocol
2003	H.264/MPEG-4 AVC	ITU-T/ISO	HD DVD over digital video broadcasting
2008	VC-2	ISO	HDTV broadcast

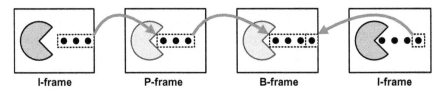

I-frame   P-frame   B-frame   I-frame

FIGURE 10.30

The basic concept of video compression with I, P, and B frames.

the manufacturers' common display interfaces, breaking down the wide acceptance of digital consumer video in its early stages. MPEG-2 was introduced to extend the MPEG-1 to a deeper pool of applications and produce commonality in the display algorithms. The MPEG-4 and H.264 standards further extended the MPEG-2 protocol by allowing deeper data compression (higher quality per unit time at a constant bit rate) and support for video interactivity. To summarize, MPEG-1 provided low-quality video in a CD-based environment, MPEG-2 (still in use today with typical consumer DVDs) tightened the deficiencies of MPEG-1 with medium compression. Then the MPEG-4 followed, allowing deep compression algorithms with interactions such as easy overlaying of graphical and textual information (such as displaying the latest score of other ballgames while watching your favorite game or synthetically overlaying the next down marker into the "field layer" while a player on the field in the "player layer" steps on the synthetic yard marker within the frame).

The MPEG standard is an ISO-based standard while the H.26x standard is an ITU standard. As with most separate standards that gain a wider audience, the separate standards tend to merge over time. The current MPEG-4 and H.264 standards are essentially the same. Various video codec standards over time are shown in Table 10.3.

The basics of video compression are simply to repeat portions of a previous frame into the next frame by comparing two sequential frames and then predicting the frame into the space in between (Figure 10.30).

The concept for this video referencing scheme is to repeat common objects within a frame and predict the occurrence of another object within the next frame. The frame types are termed

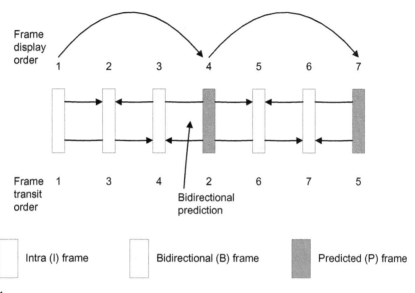

FIGURE 10.31

Sequential ordering of the I, P, and B frames.

"I-frames" (for "intra frame"), "B-frames" (for "bidirectional frame"), and "P-frames" (for "predicted frame"). The sequencing of these frames is quite ingenious as displayed in Figure 10.31.

The highest quality picture is obtained in video compression schemes with slight movement of one object within the frame while the remainder of the frame stays consistent. An example of a high-quality compressed video would be a news announcer who moves with only head gestures with a steady and static background. You will notice from a rapidly moving compressed video signal (for instance, an inexperienced camera operator moving the camera in an unsteady fashion) that substantial smearing of the image occurs. The reason for this phenomena is the P, B, and I frames are not able to generate steady duplication and prediction as the repeating portions of the frame decrease. This will have a substantial implication for lower bandwidth links such as those present while performing high-definition video over long lengths of copper (such as a submersible's tether) or with transmission of video over Internet protocol.

10.6 Video over Internet protocol

The modern customer of subsea services is quite impatient to receive the product of the subsea inspection or intervention tasks—oftentimes requiring immediate video. Thus, the people in the field (who often do not understand the physics behind the subsea object or structure being observed) are tasked with feeding the video quickly back to the customer's office (to the people who designed or are responsible for that structure). With the advent of high-bandwidth wireless and satellite networks, video can be cost-effectively transmitted real time with little time delay or loss in resolution.

Typically, a composite video feed is brought to the surface, converted to MPEG-4 (or equivalent) digital video, and transmitted to a multicasting station for rebroadcast. Each node on the network (e.g., the MPEG-2 converter box) is assigned an IP address whereby either the video is assigned to a port on the multicasting station or an individual may log into the converter at the field (i.e., ROV surface console) location.

Remotely (typically from the client's office ashore) logging into a remote video server aboard the vessel of opportunity (e.g., the converter located within the ROV shack) and then viewing the video individually is certainly acceptable. This is termed "pulling" the video from the converter. The challenge with this method is the video pipeline is limited to the one user as the video requires a certain bandwidth in order to maintain a certain nominal image quality (e.g., 540 kb/s). If a second user logs in remotely to the same port on the converter, the bandwidth is cut by half, which destroys the video feed for both individuals. A far superior method for multiple viewing of video streams is to assign the video feed to a port on a multicasting service whereby the signal is "pushed" to a port on the server of an Internet Service Provider (or within the ROV service company) with the signal then multicast to the population of viewers logging into the multicast session for their viewing pleasure.

The challenge is then to change the outbound video signal by adjusting the frame rate, frame size, and video resolution to conform within the bandwidth constraints of the video connection.

Two methods of video transmittal are used depending upon the tolerance for latency time between video event occurrence and display at the final viewer's location—Transmission Control Protocol (TCP) and User Datagram Protocol (UDP). TCP allows for data packet proof of receipt circle-back for confirmation of packet receipt. The benefit of this method of transmission is that the picture quality is of the highest quality due to resubmission of packets should they be lost in transmission. The problem with this method is that with a dodgy connection, the packet loss is high and the bandwidth is taken up by lost, then retransmitted, packets causing an inordinate amount of latency time between the transmission of the video and the final receipt. The UDP protocol, on the other hand, transmits its packets at full rate with no confirmation of packet receipt by the end receiver. So, TCP is characterized with high video quality accompanied by a certain (perhaps substantial) amount of latency time between picture capture and end display while UDP allows for full data rate video transmission of varying (perhaps very poor) image quality.

10.7 Video documentation

The difference between a film produced by a Hollywood production company and an amateur videographer is less about the equipment and more about the technique. What separates the amateur from the pro is the attention to detail that allows a complete portrayal of the subject matter through moving images. It is not about the image, it is about the message.

It has been said in many different industries: "If you didn't get it on video, it didn't happen." The meaning of this saying indicates that video and still-camera recordings are an integral part of a professional documentation package, forming part of the deliverable at the end of a project. Part of any documentation package is a set of prolific notes documenting the entire operation from start to finish so that as few items as possible are left to guesswork or memory. Field notes, as well as audio annotation, complete the video content and report, allowing a stand-alone document that can "tell the whole story."

As an anecdote to professional technique, consider the freefall photographer. At a parachuting school in the southern central United States, a service offered to customers included freefall video and stills during each tandem skydive. With each new videographer, it took time to learn the new skills of video documentation over and over again. It was a case of having to teach skilled skydivers how to be competent cinematographers—the first few camera runs had the cameraperson with the head-mounted video camera jerking his/her head in all directions with occasional filming of the customer. Not only was the video unacceptable to the customer, in most instances it made the customer airsick just viewing the film.

The same problem applies to underwater photography. Video documentation with an ROV system is a case of an ROV technician being required to learn the skills of a cinematographer. It is important to understand the status of the underwater location under investigation. Thus it is critical that an image documentation of the area/item be made that properly orients the viewer while maintaining his or her interest. All of the principles of land cinematography apply to filming underwater.

Experience gained during prior projects with some of the legends in the underwater photography business is invaluable. These legends included still photographer David Doubilet, the incredible cave cinematographer Wes Skiles, and the man whose accomplishments form the textbook of modern underwater photography, Emory Kristof, who became famous working for *National Geographic*. A few of their rules of thumb are listed below:

- Get as much footage as possible—it can always be edited.
- Keep the camera/submersible stable and still. When taking a mosaic of a subject area, stay on the starting shot for a few seconds, slowly pan from left to right (the direction in which a book is read), and come to rest on the ending shot. Hold the ending shot still for a few more seconds and then move on.
- Go from macro (to get a situational reference) to micro. If going into a structure, get as full a view as possible. Then go to specific items.
- Some ROVs have a camera zoom function. Try not to use the zoom function too frequently while filming, since it makes for poor subject content.
- Vary (and get as many of) the camera angles of the items being filmed as possible. It gives the capability to paste together a full video mosaic of the subject.
- Video is about gathering information in the form of moving images. Get as many close-ups as possible of the item being filmed, along with varying distance/frame content, to enhance the information.
- The job of the cinematographer is to gather image quality and content. It may be better for the operator to do the navigating and allow an observer/supervisor to direct the operation to assure full content, much like the director on a movie set directs the camera operator. In practically all instances, it is best to have a separate note taker to assure proper documentation of the project. Attention to detail is essential.
- Upon encountering an item of interest, leave the item in frame and count to 5 before slowly panning to the next item.

- Make all movements slow, controlled, and deliberate, understanding that the submersible is both an eye and a camera platform.
- Attempt to be consistent with the filming style. When panning, try to always pan from left to right (or right to left). When approaching an item for inspection, attempt to look all around the item for status and structure before going in for close-ups.
- Since the specialty of an ROV system is its on-station loiter capabilities, take the time to fully document the subject before moving on to another location. Do as many "takes" as necessary to get the required shot.
- Go into a subject area with an eye for the final edit and get the footage needed for that edit in mind.
- An audio overlay will assure the video is annotated and allow for the tape to be a stand-alone document. Make sure to have an on/off switch, because an open mike can make for embarrassing playback.

10.8 Documentation and disposition

The end product of any underwater operation is the documentation of the project and the final disposition of the object of interest. Some ROV systems allow digital capture of images directly onto magnetic media; others simply allow screen capture of a frame of video to use in formal documentation.

Any final report to the customer should include a condensed version of all the notes taken during the operation, in a readable format, along with an edited video presentation of the operation.

There are many excellent inexpensive digital video-editing programs available. These enable coarse editing while in the field to allow for a deliverable report immediately at the end of the job. Also available in postprocessing is a much more detailed condensation of the operation. In order to organize a formal, professional presentation to the customer, design a template for the final report that is easily adaptable to different operations and open customer base.

The format of "Who, What, When, Where, Why, and How" allows the work to be in a journalistic format, with complete reporting of the event at an adequate level of documentation. The final report is all about transmitting the information from the underwater project in a clear, concise, accurate, and readable format. The final report becomes a calling card, showcasing the professional capabilities of the team and possibly a ticket to the next interesting assignment.

10.9 Underwater optics and visibility

The physics of underwater lighting/optics will affect the image properties produced by ROV equipment used in any underwater mission. Most people's experience (and frame of reference) with optics centers on light in air; therefore, this section will address the changes that occur when light enters water.

Light refraction and dispersion occurs when a light beam is passed through water (Figure 10.32). When viewing an object in water through a plane window, unless the lens is corrected, refraction causes focus error, FOV error, and distortion as follows.

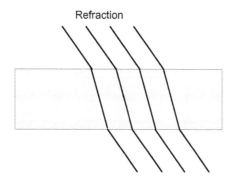

FIGURE 10.32

Light refraction in water.

10.9.1 Focus error

Rays diverging from a point at a distance from the window appear, after they have passed through the window, to come from a shorter distance.

10.9.2 FOV errors

A ray coming to the window at any angle of incidence other than zero will leave it at a wider angle. Thus, an optical system such as a camera, which has a given angular FOV in air, will have a small angular field in water. In addition, an object, which would produce an image of a certain size if it were in air, produces a larger image if it is in water. The lens appears to have a longer focal length, approximately 4:3, than that in air. This is why subjects appear magnified when using a diving mask underwater.

10.9.3 Distortion

Rays from points forming a rectangular grid in water will not seem to come from a rectangular grid after they have passed into air. There will be distortion in a lens that is not corrected for underwater use. Several types of distortion are possible due to the air−water interface distorting the image unevenly across the entire FOV. Nature took this into account by designing the human eye to render light through the lens of our camera (the eyeball) evenly with the capture device (the retina) located on a curved surface. Unfortunately, both the camera lens (at the air−water interface as well as on the actual camera) and the capture device (the CCD or CMOS sensor) are typically both flat (and sometimes distorted due to water pressure), introducing a host of image aberrations. As depicted in Figure 10.33, various possible distortions are likely based upon the lens/capture configuration. Careful consideration should be made while selecting both the housing lens and the FOV for proper matching of lens/camera optical characteristics.

About 4% of the natural light striking the surface at a normal angle of incidence is reflected away. The rest is quickly attenuated by a combination of scattering and absorption (discussed below).

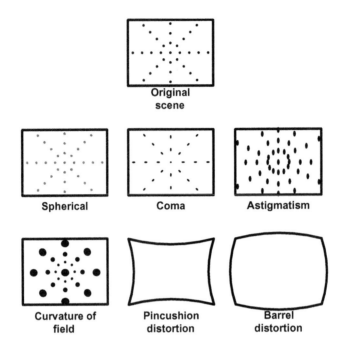

FIGURE 10.33

Possible lens aberration effects due to lens and capture device geometry.

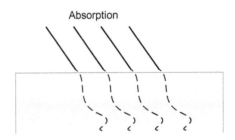

FIGURE 10.34

Water absorption of visible light.

10.9.4 Absorption

Visible light occupies the electromagnetic spectrum between approximately 400 nm (violet) and 700 nm (red). The absorption rate of water varies depending upon the wavelength (Figure 10.34). The ends of the spectrum (the red and the violet ends) are absorbed first, with the maximum penetration/lowest absorption rate in the blue/green spectrum.

Maximum penetration is gained when no particulate matter is suspended in the water (which would cause scattering), such as in the warm tropical waters of the Pacific. As a result, the red

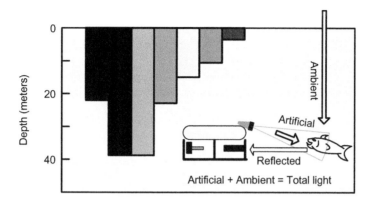

FIGURE 10.35

Light penetration and total illumination of target.

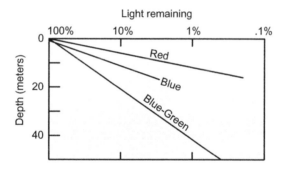

FIGURE 10.36

Light absorption by wavelength.

wavelengths are absorbed within the uppermost 60 ft (18 m) (Figure 10.35), the yellow will disappear within 330 ft (100 m), but green light can still be recognized with the human eye down to more than 800 ft (244 m) below the surface. So, even at shallow depths, objects become monochromatic when viewed through the camera of an ROV system. The only way to bring out the color of the object of interest is to reduce the water column from the light source to the object. If that light source is aboard the ROV system, the light will need to be near the object (taking backscatter into consideration) in order to illuminate it with full color reflection. An example of the absorption spectrum for pure water is shown in Figure 10.36.

10.9.5 ROV visual lighting and scattering

Observers agree that the absorption and scattering in clear ocean water are essentially the same as in clear distilled water, that some dissolved matter increases the absorption, and that suspended matter increases scattering. Both absorption and scattering present difficulties when optical obser-vations are made over appreciable distances in water.

Scattering is the more troublesome, as it not only removes useful light from the beam, but also adds background illumination (Figure 10.37). Compensation for the loss of light by absorption can be made by the use of stronger lights, but in some circumstances, additional lights can be degrading to a system because of the increase in backscatter. These circumstances are analogous to driving in

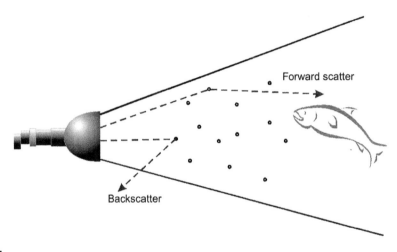

FIGURE 10.37

Illustration of the scatter phenomenon.

(Graphic by Deepsea Power and Light.)

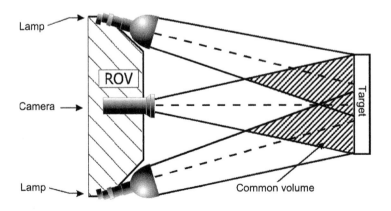

FIGURE 10.38

Separation of the light source from the water column before the camera mitigates backscatter.

(Graphic by Deepsea Power and Light.)

fog; the use of high-beam headlights, in most cases, causes worse viewing conditions than low-beam headlights.

Just as driving an automobile in a fog causes reduced visibility, the lighting aboard an ROV system blinds the camera through backscattering of light hitting the particulate matter suspended in the water column. In highly turbid water conditions (such as in most harbors around the world), reduction of the lighting intensity may be necessary in order to gain any level of visibility. Another consideration to aid the viewing of items underwater is the separation of the light source from the camera so that the water column before the camera is not illuminated, thus eliminating the source of backscattered lighting (Figure 10.38).

As an anecdote to lighting, Hollywood director James Cameron accomplished optimal elimination of the water column's effect during such projects as the filming of the shipwrecks *Titanic* and *Bismarck*. This was accomplished by placing the light source aboard a completely separate manned submersible (the Russian *Mir-1* submersible) from the camera platform (the Russian *Mir-2* submersible). With small ROV platforms, the separation of lighting source from the camera point may not be possible. This may require reduced onboard lighting while taking advantage of the ambient light as back-illumination of your object of interest.

The ability to have a second vehicle like the *Mir* at hand is unlikely. Therefore, in very turbid water, using two or three lower-powered lights positioned efficiently on the ROV, instead of one higher-powered light, may help the situation.

Vehicle Sensors and Lighting

11

In this chapter, we cover what are best described as "housekeeping" items integral to the vehicle. These vehicle subsystems are sensors needed for basic navigation of the vehicle along with health monitoring of the subsystems for a fully operational vehicle system. As part of the camera package (i.e., the visual sensor), lighting will also be examined as the purpose of lighting is to maximize the effectiveness of the optical sensor package.

11.1 Vehicle sensors

Sensor packages aboard an ROV system are broadly divided into *survey sensors* and *vehicle sensors*. The survey sensor and the vehicle sensor may be the exact same item (e.g., pressure transducer depth gauge, an integrated fiber optic gyro), but the two are separated by responsibility. Vehicle sensors are the responsibility of the ROV team while the survey sensors are for the survey team. The ROV team typically gives the survey team a fiber and a power plug (and a data feed on the surface) for their survey bottle along with flotation for the extra payload. The vehicle sensors are normally integrated through the vehicle's telemetry system and displays on the vehicle's pilot console graphical user interface (GUI).

Typically, survey sensors are of a higher quality than vehicle sensors due to surveys' inherent need for better accuracy. Basic vehicle navigation usually does not require surveying's higher tolerance measurements for simple orientation and positioning. Please refer to Chapter 17 for a deeper explanation of this division. For this discussion, the vehicle sensors will be divided into vehicle navigation sensors and vehicle health monitoring sensors.

11.1.1 Vehicle navigation sensors

Vehicle navigation sensors are normally integrated into the vehicle's telemetry system and are ported to the pilot's control console to actively manage the position, orientation, and physical status of the vehicle. Such sensors are typically positional in nature (e.g., compass, depth gauge, and tether in/out). The smaller OCROV systems feature only basic items while the larger MSROV and WCROV integrate increasingly more complex sensors for a full situational status of the vehicle. Table 11.1 details the various levels of "housekeeping" sensors by ROV classification.

For a deeper explanation of sensor theory as well as survey sensors, review Chapters 12 (Sensor Theory) and 17 (Navigational Sensors). Some of the more common vehicle navigation sensors are examined below:

11.1.1.1 Flux gate compass

This simple device is used to measure the ambient magnetic field vector and intensity surrounding the sensor. It is hoped that the only magnetic field surrounding the sensor is that of the Earth's field (so as to measure orientation with regard to magnetic North), but that is not always the case.

The basic sensor of a flux gate compass is a simple coil surrounding the core of some high-permeability magnetic material for measuring the ambient magnetic field. This sensor then becomes either 2D or 3D by orienting a series of coils along two or three (or more) axes. As these magnetometer arrays are rigidly attached to the chassis of the vehicle, they are termed "strapped down" magnetometers. In a perfect world (i.e., one where the host vehicle had no self-induced magnetic field and did not operate near magnetic anomalies such as steel offshore platform legs or ship hulls), the magnetometer would only measure the Earth's magnetic field as there would be no local interference. But that is seldom the case.

In order to calibrate a strapped down magnetometer, the total field must be measured (Earth's magnetic field along with all of the local magnetic noise from the vehicle). Once the field is measured along a single axis (from all sources of magnetic flux), the vehicle must then be rotated along all relevant axes in order to subtract out the variations due to local (i.e., vehicle induced) interference from the variations due to orientation with the Earth's magnetic field.

Earth's magnetic field + vehicle's magnetic field = total measured field

It is assumed that the vehicle's magnetic noise is a fixed sum (although it may vary somewhat due to magnetic noise from electric motors, vehicle frame, and/or power transformers). Once the total magnetic field is measured on each orientation (North/East/South/West), the total variation is

Table 11.1 Housekeeping Sensors by ROV Classification

Sensor	OCROV	MSROV	WCROV
Single channel video	X	X	X
Compass	X	X	X
Depth gauge	X	X	X
Tether turn counter	X	X	X
Rate gyro	X	X	X
Lighting level adjustment	X	X	X
Ground fault interrupt	X	X	X
Multiple channel video		X	X
Camera pan and tilt		X	X
Camera zoom/focus		X	X
Motor current draw		X	X
Tether in/out		X	X
Water ingress alarm		X	X
Ground fault interrupt by circuit		X	X
Obstacle avoidance sonar		X	X
Altimeter		X	X
Low oil level warning		X	X
Valve pack control sensors		X	X
Oil temperature sensor			X
Compensator oil level			X
Lighting adjustment by light			X
System diagnostics			X
System oil pressure by circuit (mains/aux)			X
Inclinometer for vehicle pitch/roll/yaw/trim			X
Integrated INS/DVL for dynamic positioning			X
Multiaxis slaved gyro			X

derived, thus registering the measurement at the cardinal headings and orientations. This is the theory of operation of a simple flux gate compass. The most popular of the flux gate compasses used today in ROV applications is the PNI *TCM2* series of sensors. For further details, visit their web site for technical specifications.

11.1.1.2 Tether turn counter

Tether turn counters are not technically considered sensors. The turn counter is a register within the vehicle's CPU that accepts digital heading readings from the compass and counts the number of times the vehicle turns based upon those readings. It is best to bring the tether turn count back (or near) to zero before recovering the vehicle to the surface.

11.1.1.3 Tether in/out

On TMS-based ROVs, a linear counter is added to the tether management system for measuring the amount of tether payed out from the reel. This measurement is typically displayed on the pilot console to advise the pilot of the tether status.

11.1.1.4 Pressure-sensitive depth gauge

As further discussed in Chapter 12, there are several technologies available to measure ambient pressure. On OCROV systems, inexpensive pressure gauges measure some nominal amount of local pressure (in psi or bar) and then scale the results into water depth. Care must be exercised to accurately calibrate the depth sensor based upon the water density (e.g., fresh water or salt water and cold or warm). The calibration technique typically has the technician setting the sensor pressure to surface reference and then to some nominal reference pressure (e.g., 100 psi or 7 bar). The sensor then scales the output to feet or meters based upon these scaling and density parameters.

11.1.1.5 Rate gyro

As discussed more in-depth in Chapter 17, the rate gyro is primarily used for heading hold routines based upon auto stabilization functions.

11.1.1.6 Obstacle avoidance sonar

As explained in Chapter 15, vehicles are usually equipped with a mechanically scanning single-beam sonar for sensing of obstacles and locating major anomalies surrounding the vehicle. The output is typically displayed on the pilot's console.

11.1.1.7 Altimeter

Along with the pressure-sensitive depth gauge is the vehicle's altimeter, which is used to measure the vehicle's height above the bottom. This sensor comes in a variety of sensitivities. It works much like a boat's fathometer and typically transmits a vertical pencil-beam acoustic signal to bounce off the seafloor and measure the time of flight to determine the distance measurement between the transducer and the reflective surface (e.g., the bottom). Bottom type (e.g., sand, mud, and clay) as well as topography (e.g., flat bottom, canyons, and sea mounts) all play into the measured distance of the echo return, thus affecting the signal strength of the acoustic return (and, hence, the distance resolved). These factors must be considered when interpreting the readings from the altimeter.

11.1.1.8 Inclinometer

Also discussed in Chapter 17 is inclinometer theory. These sensors are used for sensing vehicle orientation for pitch/trim functions. The vehicle orientation will directly affect the interpretation of sensor output including camera angle (e.g., zero camera angle is referenced to the vehicle, but the vehicle could be off of the horizontal plane), sonar interpretation (e.g., your sonar could paint a huge obstacle (i.e., a clear flat bottom) if the vehicle is oriented nose down), and altimeter output. It could also affect vehicle performance (e.g., forward thrusting with the vehicle-oriented nose-down could drive the vehicle into the bottom). The sensor's output is to the pilot console.

11.1.2 **Vehicle health monitoring sensors**

Vehicle health monitoring sensors resolve various measures within the vehicle in order to detect any parameter that strays from its nominal value. The human analog for this would be the measuring of blood pressure, body temperature, heart rate, blood chemistry, and the like (versus navigational parameters such as running speed/distance, speech rate, lifting weight/amount). On an ROV, such parameters are used for a series of diagnostics and alarms to warn of impending faults or damage to system components. Some common vehicle health monitoring sensors are as follows:

Water ingress alarm: Seawater is a conductor. To measure water ingress into an air-filled space, a simple open circuit with a small potential is placed at the low point of a pressure bottle. Should a small amount of water begin to enter the bottle, the water will eventually rise to a level that contacts both conductors, thus allowing a current to pass between them. This sets off the water ingress alarm on the pilot's console, warning of a potentially hazardous state. A word of caution—put the water ingress sensor at the very bottom of the bottle. If it is at the top, it will fry the electronics in the bottle due to board submersion before the alarm ever sounds.

Oil temperature: Oil temperature outside of a nominal range could signal breakdown of the lubricity of the fluid (especially an over-temperature status). The vehicle or lubricant manufacturer will have further information on the normal operating temperature. Alarms may then be set once temperatures fall outside of some nominal range.

Oil pressure: Oil pressure is also a critical parameter for monitoring the health of the system. Any combination of factors could cause a loss or spike in oil pressure and elicit any number of hazardous conditions. Hydraulic circuit pressure should be carefully monitored throughout the mission. On electric vehicles, oil provides lubrication for the thrusters and motive power for the manipulators and tooling; however, it is not critical to the vehicle's operation. On the hydraulic WCROV, oil pressure is life or death to the vehicle's function.

Compensator oil level: On more complex vehicles (especially WCROV systems), compensator oil levels are measured and monitored with output to the pilot's console. Typically, a minimum level is set on the system alarms so that an early warning is given to the operator that the reservoir is low before the compensator's oil is depleted. Compensator systems are discussed further in Chapter 20.

System diagnostics: Along with the vehicle's health monitoring sensors is a logic circuit within the CPU. This software routine constantly monitors system parameters and (theoretically) diagnoses any potential problems before they become hazardous conditions. System diagnostics also assist in resolving any system faults identified through sensor readings. These systems range from the simple to the complex depending upon the vehicle operated and the difficulty of the environment encountered.

11.2 **Vehicle lighting**

In this section, some theory is necessary in order to properly understand lighting specifications issued by manufacturers of lighting products. We will get into a discussion of lighting parameters before delving into the practical applications of underwater lighting technology.

11.2.1 Lighting theory

This section is a condensed version of a technical paper by Cyril Poissonnet of Remote Ocean Systems, which is available for download from the ROS web site. There is much misconception about lighting. Most household lights are measured in wattage—which is a measure of electrical power consumed (*not* light power generated). Thus, wattage is a proper metric sizing the power source but has no direct correlation to the actual amount of light produced. Closer examination of lighting parameters must be considered to achieve the mission's lighting requirements.

Lighting power is measured in SI units as the candela (as originally defined, a *candela* is the light intensity given off by one candle, although that definition was later refined). Visible light is further divided into the various wavelengths as depicted in Figure 11.1. Light is thus discriminated into both quantity and quality.

11.2.1.1 Quantity

The unit candela is an omnidirectional unit that does not take into account directionality and is independent of distance. As with sound, light is subject to a concept known as "spreading loss." As light spreads away from its source, the power is continually spread over a wider area, thus depleting its power per unit area. Thus the term "luminous intensity," or "lumen," arose as a more definitive metric since it measures total flux intensity from a light source. "Lux" is then defined as lumens per unit area. By standard convention, lux is referenced at a range of 1 m (defined as lumen/m^2).

A lighting product is specified by its lumen output and its color temperature, along with a directionality factor for the beam angle. A floodlight illuminates an entire scene while a spotlight focuses on an individual item within the scene. With a constant lumen output from the lighting source, the actual light incident to the surface of the target (its lux) is much higher for the spotlight than the floodlight. The beam is shaped by the reflector, thus focusing and intensifying the lumens onto a discrete surface (Figure 11.2). And with spreading loss, the beam loses intensity as it propagates from the light source (Figure 11.3).

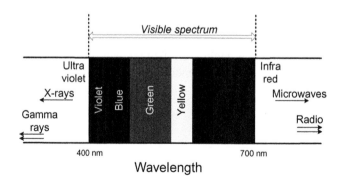

FIGURE 11.1

The visible light spectrum.

11.2.1.2 Quality

From the visible spectrum described in Figure 11.1, one can see that visible light is made up of many differing colors depending upon its wavelength. In a symphony of sound, the total orchestral score is experienced by the mixture of all sounds reaching the ear. It is the same for color

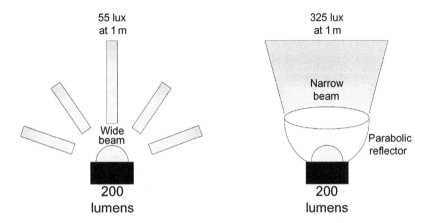

FIGURE 11.2

Lumens to lux.

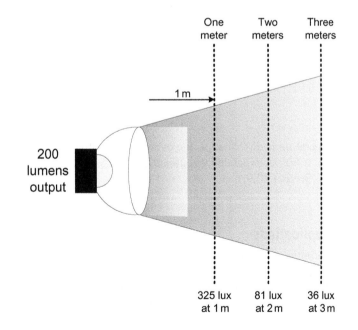

FIGURE 11.3

Spreading losses per unit distance.

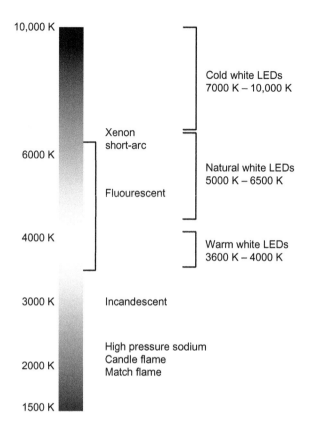

FIGURE 11.4

Color temperature chart.

(From Poissonnet, 2008.)

temperature. Color temperature is described as the overall light output, expressed in a Kelvin scale, as depicted in Figure 11.4. For an expanded discussion of underwater optics and light source placement considerations, see Section 10.9.

11.2.2 Practical applications

This explanation of lighting comes from Ronan Gray, an expert on the subject. The need for underwater lighting becomes apparent below a few feet from the surface. Ambient visible light is quickly attenuated by a combination of scattering and absorption, thus requiring artificial lighting to view items underwater with any degree of clarity. We see things in color because objects reflect wavelengths of light that represent the colors of the visible spectrum. Artificial lighting is therefore necessary near the illuminated object to view it in true color with intensity. Underwater lamps provide this capability.

Table 11.2 Light Source Characteristics

Source	Lumens/Watt	Life (hours)	Color	Size	Ballast
Incandescent	15–25	50–2500	Reddish	M–L	No
Tungsten-halogen	18–33	25–4000	Reddish	S–M	No
Fluorescent	40–90	10,000	Varies	L	Yes
Green fluorescent	125	10,000	Green	L	Yes
Mercury	20–58	20,000	Bluish	M	Yes
Metal halide	70–125	10,000	Varies	M	Yes
High-pressure sodium	65–140	24,000	Pink	M	Yes/I
Xenon arc	20–40	400–2000	Daylight	V	Yes/I
HMI/CID	70–100	200–2000	Daylight	S	Yes/I
Low-pressure sodium	100–185	18,000	Yellow	L	Yes
Xenon flash	30–60	NA	Daylight	M	NA

V, *very small*; S, *small*; M, *medium*; L, *large*; I, *ignitor required*; NA, *not applicable*.
Source: *Courtesy of Deep Sea Power & Light.*

Lamps convert electrical energy into light. The main types or classes of artificial lamps/light sources used in underwater lighting are incandescent, fluorescent, high-intensity gas discharge, and light-emitting diode (LED)—each with its strengths and weaknesses. All types of light are meant to augment the natural light present in the environment. Table 11.2 shows the major types of artificial lighting systems, as well as their respective characteristics.

- *Incandescent*: The incandescent lamp was the first artificial lightbulb invented. Electricity is passed through a thin metal element, heating it to a high enough temperature to glow (thus producing light). It is inefficient as a lighting source with approximately 90% of the energy wasted as heat. Halogen bulbs are an improved incandescent. Light energy output is about 15% of energy input, instead of 10%, allowing them to produce about 50% more light from the same amount of electrical power. However, the halogen bulb capsule is under high pressure instead of a vacuum or low-pressure noble gas (as with regular incandescent lamps) and, although much smaller, its hotter filament temperature causes the bulbs to have a very hot surface. This means that such glass bulbs can explode if broken or if operated with residue (such as fingerprints) on them. The risk of burns or fire is also greater than with other bulbs, leading to their prohibition in some underwater applications. Halogen capsules can be put inside regular bulbs or dichroic reflectors, either for aesthetics or for safety. Good halogen bulbs produce a sunshine-like white light, while regular incandescent bulbs produce a light between sunlight and candlelight.
- *Fluorescent*: A fluorescent lamp is a type of lamp that uses electricity to excite mercury vapor in argon or neon gas, producing short-wave ultraviolet light. This light then causes a phosphor coating on the light tube to fluoresce, producing visible light. Fluorescent bulbs are about 40% efficient, meaning that for the same amount of light they use one-fourth the power and produce one-sixth the heat of a regular incandescent. Fluorescents typically do not have the luminescent output capacity per unit volume of other types of lighting, making them (in many underwater applications) a poor choice for underwater artificial light sources.

- *High-intensity discharge*: High-intensity discharge (HID) lamps include the following types of electrical lights: mercury vapor, metal halide, high-pressure sodium, and, less common, xenon short-arc lamps. The light-producing element of these lamp types is a well-stabilized arc discharge contained within a refractory envelope (arc tube) with wall loading (power intensity per unit area of the arc tube) in excess of 3 W/cm^2 (19.4 W/in^2). Compared to fluorescent and incandescent lamps, HID lamps produce a large quantity of light in a small package, making them well suited for mounting on underwater vehicles. The most common HID lights used in underwater work are of the metal halide type.
- *LED*: An LED is a semiconductor device that emits incoherent narrow-spectrum light when electrically biased in the forward direction. This effect is a form of electroluminescence. The color of the emitted light depends on the chemical composition of the semiconducting material used and can be near-ultraviolet, visible, or infrared. LED technology is useful for underwater lighting because of its low power consumption, low heat generation, instantaneous on/off control, continuity of color throughout the life of the diode, extremely long life, and relatively low cost of manufacture. LED lighting is a rapidly evolving technology and is being widely adapted by ROV manufacturers and users.

Observation-class ROV systems use the smaller lighting systems, including halogen and metal halide HID lighting (although LED systems are now standard equipment for most OEM OCROVs). In the MSROV and WCROV world, LED lights have now become standard equipment.

The efficiency metric for lamps is efficacy, which is defined as light output in lumens divided by energy input in watts, with units of lumens per watt (LPW). Lamp efficacy refers to the lamp's rated light output per nominal lamp watts. System efficacy refers to the lamp's rated light output per system watts, which include the ballast losses (if applicable). Efficacy may be expressed as "initial efficacy," using rated initial lumens at the beginning of lamp life. Alternatively, efficacy may be expressed as "mean efficacy," using rated mean lumens over the lamp's lifetime; mean lumens are usually given at 40% of the lamp's rated life and indicate the degree of lumen depreciation as the lamp ages.

An efficient reflector will not only maximize the light output that falls on the target but will also direct heat forward and away from the lamp. The shape of the reflector will be the main determinant in how the light output is directed. Most are parabolic, but ellipsoidal reflectors are often used in underwater applications to focus light through a small opening in a pressure housing. The surface condition of a reflector will determine how the light output will be dispersed and diffused. The majority of reflectors are made of pure, highly polished aluminum that will reflect light back at roughly the same angle to the normal at which it was incident. By adding dimples or peens to the surface, the reflected light is dispersed or spread out. When a plain white surface is used, the reflected light is diffused in all directions.

Payload Sensors

Sensor Theory

CHAPTER CONTENTS

Since the turn of the new millennium, huge leaps forward in sensor technology have been achieved across the realms of science and industry. From the early days of the simple mercury thermometer to the later evolution of the Bourdon tube to the modern macro and nanoscale sensors of the new century, they have been used to sense physical phenomenon and convert related measurements into information discernible by humans or other machines for further action.

Sensors are everywhere in our lives. In our homes, temperature sensor circuits direct feedback to elements that control our heating−cooling systems as well as cooking appliances, optical and pressure sensors are used in our home security systems, acoustic sensors for our phones, voltage/current sensors for electrical safety, and a host of others. In our cars, wheel turn counters (with use of various technologies) clock our vehicle's speed (and the police use optical and radio frequency sensors to clock our speed as well ...), optical sensors turn on/off headlights, pressure/temperature/fluid level/tachometer/etc. sense engine operating parameters. Boats, power plants, airplanes, space vehicles—and especially underwater vehicles—make extensive use of sensor technology for operation of the machinery we use every day in the modern world.

Sensor technology is a broad and diverse body of knowledge that is rapidly advancing in both scope and depth. An overview of sensor technology elements is provided below. For a more in-depth coverage of the subject, a recommended text on basic sensor technology is the *Sensor Technology Handbook* by Jon S. Wilson (published in 2005—also by Elsevier).

The purpose of this chapter is to acquaint the reader with the basics of sensor theory in order to better understand the specifications of sensor packages deployed on ROV systems.

12.1 Theory

12.1.1 History

"If real is what you can feel, smell, taste and see, then 'real' is simply electrical signals interpreted by your brain" (Morpheus, quoted from the 1999 movie, *The Matrix*).

Sensor technology is pervasive in the biological world. But similar sensors are tuned to different sensitivity levels as well as differing physical phenomenon in order to achieve the measurement objective. As an example, typical human auditory sensitivity is at a frequency level from 20 Hz to 20,000 Hz while dogs can hear up to 60,000 Hz—and bats up to 150,000 Hz. This is a case of the same sensor being tuned to differing sensitivity and amplitude/source levels for various applications. The reason is the sensors are used for different purposes—humans for voice recognition/speech dissemination, dogs for identifying prey, and bats for echolocation.

The first mechanical sensors originated with the ancient Greeks. Archimedes designed a specific gravity sensor, a crude odometer, and various other early measurement devices. As civilization passed into the Renaissance, devices for measurement were devised for heat, pressure, and various other parameters in order to support the budding field of scientific investigation.

While the sensor was reactive to changing physical phenomenon a actual metric for measurement was ill-defined. The early scientist was burdened to come up with a common and accepted scale by which physical characteristics would be measured (e.g., pressure in pascal/bar/atmosphere/torr/psi, temperature in Kelvin/Celsius/Fahrenheit/Rankine/Delisle/Newton/Réaumur/Rømer). In all instances, the scale required a linear relationship (or logarithmic that could be plotted in some linear function) so that the instrument output could scale in a predictable fashion. Regardless of the units used, the end product of sensor functionality is always the same—to measure physical phenomena and then convert those measurements into information translatable by humans or other machines for further action.

12.1.2 **Function of sensors**

For the purposes of this text, the term "sensor" will deal with those mechanisms that sense physical parameters and then turn their measurements into electrical signals (as opposed to simple passive sight systems such as the mercury thermometer or the Bourdon tube, shown in Figure 12.1). A host of sensor technologies can be employed to reach the same conclusion regarding a certain physical

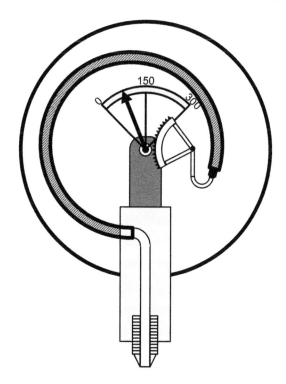

FIGURE 12.1

Typical bourdon tube pressure gauge.

phenomenon (e.g., temperature is the simple measurement of molecular excitation but can be measured via thermistor, fluid pressure, thermocoupler, or resistance temperature detector (RTD)). For example, RTDs are sensors used to measure temperature by correlating the resistance of the RTD element with the temperature. Following this example, various technologies for sensing temperature are sensitive/accurate only within certain ranges. Also, the sensor output has a host of issues involved before a logical and practical use of that output can be employed. An example of accurate but unusable sensor output would be an ambient water temperature sensor mounted onto a high-pressure hydraulic pump aboard an ROV—the sensor would accurately measure the water temperature surrounding the pump, but it would indicate a temperature (in all likelihood) higher than the ambient temperature due to' the heat generated by the pump. Or if an obstacle avoidance sonar transducer head was embedded in the ROV's flotation block, the sonar transducer would accurately sense the acoustic reflection from the flotation foam but not the desired echo from the local obstructions. How you incorporate a selected sensor is as critical as the design of the sensor itself.

The perfect sensor has the following characteristics:

1. The sensor senses only the desired physical phenomenon.
2. The sensor is insensitive to other environmental or physical factors.
3. The sensor does not influence the item being measured.
4. The sensor's electrical output signal is linearly proportional (or can be made as such—e.g., plotted linearly on a logarithmic scale).

Have you ever had your eyes "deceive" you? Accurate interpretation of sensor output, as well as proper sensor placement, is required in order to achieve accurate measurement for either human or machine decision making.

12.1.3 Sensor output

A sensor typically functions as a transducer converting physical phenomena to electrical current for later conversion into machine or human-readable information. With regard to sensors, the term "Garbage In/Garbage Out" is certainly applicable as sensors are prone to a host of calibration errors, sensor performance issues, and user interpretation problems.

Sensor performance varies by device. Sensor performance characteristics are measured with the following qualifiers:

- *Transfer function*: This defines the relationship between the physical input to the device (i.e., direct/indirect sensing of physical phenomena) and the electrical output from the device. An input/output graph is normally generated for the sensor so that the user of the information may interpret the output of the sensor (e.g., temperature sensor is 4–20 mA DC or 10–50 mA DC) and either display the output (for a visual display) or input the voltage (for machine control) into an intelligent device for commands or other usage.
- *Sensitivity*: This characteristic (sometimes termed "gain") is the scaling of the sensor's input to its output. This parameter is further described as the ratio of a change in output magnitude to a corresponding change in steady-state input (which caused the output). This, of course, is expressed in ratio form. A temperature sensor has a high sensitivity if a small change in

temperature (sensor input) results in a large change in voltage (sensor output). Some sample output units are volts/Celsius, millivolts/dB sound change, etc.

- *Dynamic (or span) range*: This metric defines the range of inputs from physical phenomena whereby the sensor output falls within acceptable accuracy levels when the sensor input is *within limits, i.e., within its dynamic range*. Should the sensor input exceed these ranges—either over or under—it may have inaccurate measurements because it is *outside of the dynamic range*. The "range" of a sensor is the region between the limits within which a quantity is measured. The upper range value is the highest quantity measurable by the subject sensor while the lower range is its lowest. The "span" of a sensor is the algebraic difference between the sensor's upper and lower range values. An example of this concept is an optical sensor adjusted to the wavelength of visible light (e.g., a human eye) that observes wavelengths outside of its dynamic range (e.g., infrared or ultraviolet (UV) light). In this example, the sensor is receiving input from both within and outside of its dynamic (or span) range and is interpreting the visible light while discarding (or inaccurately measuring) the light waves outside of its dynamic range. The difference between range and span is demonstrated by the following example—if a gauge measures the pressure in a closed tank, the range of this gauge could be 100−180 psi (7−12 bar) while the span is 80 psi (5.5 bar).
- *Accuracy*: A comparison is made between the actual output of a sensor to the output with an ideal sensor with the difference (or expected error) defined as its accuracy. Typically, this is expressed as a fraction of the full-scale output (FSO)/reading. For instance, the manufacturer of a pressure sensor may guarantee the accuracy of the sensor to 3% of FSO, meaning that it will output a value within 3% of its true value. Sensors will never be "all things to all people" in that typically the wider a range for a sensor the less resolution (defined below) and accuracy that will be reflected. As an example, two pressure-sensitive depth gauges of similar quality will have a resolution/accuracy inversely proportional to their range (i.e., a 0−330 ft (0−100 m) range sensor will have twice the resolution/accuracy of a 0−660 ft (0−200 m) range sensor with all other parameters being equal). The accuracy is typically stated in percentage (which, of course, would not change as a percentage of output between like sensors). The resolution and the accuracy of a sensor are generally related, but they are not directly linked, i.e., a high-quality instrument may have a low resolution but be highly accurate (e.g., a deep-rated digital quartz depth gauge as used on a WCROV), while a low-quality instrument may have a high resolution with low accuracy (e. g., a shallow-rated pressure-sensitive depth gauge as used on an OCROV).
- *Hysteresis*: The "hysteresis" of a sensor is its dependence upon the output (for any given change in input) value's history of prior excursion of the input and the direction of the current excursion (Figure 12.2). This means simply that the output values obtained while increasing the input will be different from the output values obtained while decreasing the input. Hysteresis is based on the inherent physical characteristics of the materials used to construct the instrument. Hysteresis can be mechanical or magnetic. As a sensor is cycled up and down its dynamic range, the track of the sensor's output may not exactly correlate over time or scale with actual conditions. In other words, hysteresis is the measure of the lead/lag time of the sensor's input to the sensor's output and/or the differing accuracy of the sensor as the dynamic range is cycled from top to bottom versus bottom to top. An example of the lead/lag hysteresis would be a pressure sensor on an ROV still in the TMS that is being lowered vertically to the job site at a rate of 30 m/min (meters per minute) versus 90 m/min. With a 0.05% hysteresis, if the winch is

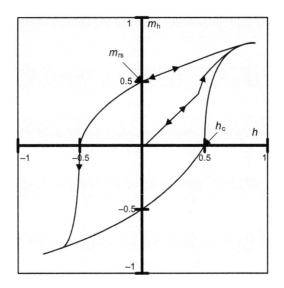

FIGURE 12.2

Typical hysteresis plot of a sensor output.

stopped when the pressure/depth sensor output reads 1000 m, the vehicle would actually be at 1005 m at the 30 m/min rate—and at 1015 (or more) with a 90 m/min rate. For the same pressure sensor, the sensor range hysteresis may be 1% at 100 bar and 5% at 1000 bar. The root of the word "hysteresis" comes from the ancient Greek word meaning "shortcoming" or "deficiency." Hysteresis can be "rate dependent" or "rate independent" depending upon the sensor used or the number of cycles the sensor has experienced. A rate-dependent sensor experiences lead/lag errors inherent in the sensing technology while rate-independent sensors may be due to either sensor age/cycle degradation or affinity/aversion to a nominal sensor value. The term "hysteresis" is also applied to a material's characteristics of retentivity and remanence whereby a sensor is not completely relieved of its input stress once a zero input is again reached. An example of this is a magnetic sensor receiving an input intensity that is varied in both the positive and negative directions and then back to zero. The output of the sensor is not completely canceled out when zero input is again reached since some amount of magnetic flux is retained by the core material. Hysteresis is a rather complex instrument issue due to its technology dependency (i.e., there is mechanical hysteresis, magnetic hysteresis, electrical, contact angle) and should be examined while choosing and integrating a sensor for a particular application.

- *Linearity (or nonlinearity)*: This characteristic defines adherence to (or deviation from) the linear transfer function over the dynamic range of the sensor. This characteristic is easy to define, but a bit more difficult to consistently measure as the scaling between sensors must be comparable. A linearity graph (Figure 12.3) is plotted on the chart showing the sensor output for an accelerometer over the various points throughout the dynamic range of the sensor. The actual output is plotted (using regression methods) against an ideal output with the measure of deviation as the metric.

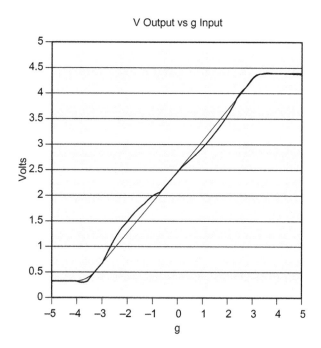

FIGURE 12.3

Plot of actual versus theoretical sensor output for an accelerometer.

- *Noise*: All sensors produce output noise as a by-product of the signal generation process. It is a product of the mechanism that has to be dealt with. The noise could be anything—local magnetic variations inducing noise in the vehicle's electronics, noise in the electrical generation system, thruster or electrical motor inductance, etc. Noise is most problematic with instrumentation that must have high gain to develop a coherent signal. An example of such an instrument would be the thermocouple, which must scale from millivolts to volts—a theoretical gain of 1000. If any noise is present it could significantly distort the output of the instrument. The higher quality sensors, of course, produce the lowest amount of internally generated noise and/or the highest *signal to noise ratio*.
- *Resolution*: The minimum detectable signal variation (or threshold) defines the sensor resolution. The floor of detection ability of the sensor varies with the technology used as well as the quality of the sensor. As an example, the thermocouple in a turbine jet engine measures a wide dynamic range of temperatures (e.g., 0–1000°C), but the minimum resolution of the sensor may only be as low as 50°C (low resolution) while a lab instrument measuring the freezing point of water may have a low dynamic range (<10°C) but a high resolution of 0.00001°C. Electrical noise can be a limiting factor to a sensor's resolution. The resolution of the sensor must also apply to the bandwidth of the sensor, which is not always the case. A high-resolution specification that is outside the bandwidth where the measurement will be made may not be useful.

- *Bandwidth*: Sensor bandwidth, sometimes referred to as "frequency response," is the sensor's ability to respond to instantaneous changes in the physical conditions being measured, i.e., how the sensor responds at different frequencies. The spread between the higher and the lower frequency limits of the frequency components (for significant amplitudes in the spectrum over which the gain remains reasonably constant) is called the "bandwidth" of the signal and is expressed in hertz.

The typical electronic sensor starts up with the following logical steps:

1. Power is applied to the sensor.
2. The firmware within the system reads the instructions to start up the sensor.
3. As part of the startup sequence, a sensor measurement is taken in analog format.
4. The reading is put into a digital format (normally in ASCII format in either a proprietary protocol or an industry standard format—e.g., National Marine Electronics Association standard NMEA 0183).
5. The output is placed into serial format (e.g., 8-N-1 or eight (8) data bits, no (N) parity bit, and one (1) stop bit) at a specified rate (a default value of 4800 b/p/s for a typical GPS signal and transmitted down a standard communications line—e.g., RS-232 cabling).

As an example of a common industry sensor data transfer protocol, the NMEA 0183 standard specifies a series of output "sentences" used by common marine sensors to output sensor data by one sensor (the "talker") to a single or multiple receiver(s) (the "listener(s)") in a format acceptable to sensor manufacturers all the way to end users. The standard data output is in ASCII characters in serial communication protocol. The sentence begins with the "$" character and then defines the sentence type via the next few ASCII characters with data fields separated (delimited) by comma separators (","). NMEA sentences are structured with the following rules:

a. Each sentence starts with the "$" character (position 1).
b. Positions 2 and 3 identify the transmitting device (see below for sample list).
c. Positions 4, 5, and 6 identify the type of message.
d. All fields are comma delimited.
e. Where data is unavailable, the field contains *NUL* bytes between the comma delimiters (i.e.,",,").
f. <CR> <LF> ends the sentence.
g. The last character following the last data field is an "*" if a checksum is present and is followed by a two-character checksum.

So, consider a GPS outputting a GGA sentence:

$GPGGA, [and then the comma-delimited data ...]

A short sample list of output device ("talker") designations (at positions 2 and 3) is:

- GP—Global Positioning System (GPS)
- HC—Heading—Magnetic Compass
- HE—Heading—North Seeking Gyro
- HN—Heading—Non North Seeking Gyro
- SD—Sounder, Depth

- SS—Sounder, Scanning
- TI—Turn Rate Indicator

A short sample list of sentence definitions (from NMEA) for a simple output device (such as a compass) and a bit more complicated device (such as a GPS):

COMPASS

HDG—Heading—Deviation & Variation

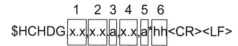

$HCHDG x.x x.x a x.x a*hh<CR><LF>

Field Number
1. Magnetic Sensor heading in degrees
2. Magnetic Deviation, degrees
3. Magnetic Deviation direction, E = Easterly, W = Westerly
4. Magnetic Variation degrees
5. Magnetic Variation direction, E = Easterly, W = Westerly
6. Checksum

GPS

GGA—Global Positioning System Fix Data—Time, Position, and fix related data for a GPS receiver.

$GPGGA,hhmmss.ss,1111.11,a,yyyyy.yy,a,x,xx,x.x,x.x,M,x.x,M,x.x,xxxx*hh<CR><LF>

Field Number
1. Universal Time Coordinated (UTC)
2. Latitude
3. N or S (North or South)
4. Longitude
5. E or W (East or West)
6. GPS Quality Indicator
    - 0 = Fix not available
    - 1 = GPS fix
    - 2 = Differential GPS fix (values above 2 are 2.3 features)
    - 3 = PPS fix
    - 4 = Real Time Kinematic
    - 5 = Float RTK
    - 6 = Estimated (dead reckoning)
    - 7 = Manual input mode
    - 8 = Simulation mode
7. Number of satellites in view, 00-12
8. Horizontal Dilution of precision (meters)

FIGURE 12.4

Typical compass card (PNI TCM 2.5 compass module).

(Courtesy PNI.)

9. Antenna Altitude above/below mean-sea-level (geoid) (in meters)
10. Units of antenna altitude, meters
11. Geoidal separation, the difference between the WGS-84 Earth ellipsoid and mean-sea-level (geoid), " − " means mean-sea-level below ellipsoid
12. Units of geoidal separation, meters
13. Age of differential GPS data, time in seconds since last SC104 type 1 or 9 update, null field when DGPS is not used
14. Differential reference station ID, 0000-1023
15. Checksum

An example of an NMEA 0183 sentence for a simple flux gate compass on a typical ROV (e.g., a PNI Corporation TCM 2.5 Magnetic Compass Card, Figure 12.4) would be:

$HCHDM, < compass>,M*checksum < CR > < LF > [for magnetic heading]

For example:

$HCHDM,182.3,M*checksum < CR > < LF > [for magnetic heading = 182.3°]

12.1.4 Types of sensors

Sensors do not, as part of the mechanism, typically produce voltages. Instead, the sensor reacts to the physical phenomenon in some way and then an electrical circuit measures the electrical parameters of the sensor. The electrical parameters measured are (any or all of) the quality/quantity of resistance, capacitance, and/or inductance.

12.1.4.1 Resistance measurement circuits

Ohm's law defines the relationship between electrical current, voltage, and resistance. A resistance measurement circuit measures voltage output from a measured resistance difference between system voltage (as measured across a system resistor—typically the "load" or item either doing the

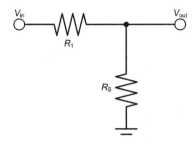

$$V_S = (R_S/(R_1 + R_S))V_{IN}$$
$$\text{if } R_1 >> R_S \text{ then } V_S = (R_S/R_1)V_{IN}$$

FIGURE 12.5

Basic voltage divider.

work or a sensor—R_1 in Figure 12.5) and a reference resistor (R_S). The measurement point is downstream of the measurement/load resistor. The resistance at the load resistor is much higher than at the reference resistor, amplifying the output voltage difference. The voltage downstream of R_S is then a highly amplified measurement of the resistance across R_1, allowing the resistance to be easily measured with a galvanometer.

In practical circuits, a Wheatstone Bridge Circuit (Figure 12.6) is used for more accurate and balanced measurement of the resistance values over a wider range of resistance measurements. This circuit is ubiquitous in sensor circuitry, and the reader will benefit from a deeper knowledge of this circuit. For a better understanding of the function of this circuit, please refer to a basic circuits text.

In the circuit depicted in Figure 12.6, $R_1 = R_2$. R_u is the sensor and the value of R_v is usually matched to the 0% value of R_u for circuit calibration.

A typical resistance circuit would be used on an ROV's pressure-sensitive depth gauge. The output resistance on the sensor circuit, e.g., the "squeeze" on the piezoelectric circuit, will vary the output resistance quantity on the reference resistor and thus vary the output voltage to be measured. Another example of an element used as a variable resistance sensor is the strain gauge.

12.1.4.2 Capacitance measurement circuits

Capacitance measurement circuits are also used quite frequently as the output measurement parameter for a given sensor. Resistive circuits are very effective in a DC environment; however, in an oscillating circuit, excessive heat may be built up, destroying both the signal (through noise generation) and the efficiency of the circuit. However, most circuits where this is an issue will use a temperature compensation circuit to counteract this phenomenon. A capacitive measurement circuit measures the impedance through an oscillating circuit. Since a capacitor is an effective break in a circuit within a DC environment, charge builds up on the plates of the capacitor, causing a charge imbalance—and then the system stabilizes. In an AC environment, as the circuit oscillates, the charge ramps up and down over the capacitor. This causes the impedance measurement characteristics across the circuit to behave in a similar fashion to a resistive circuit that can then be outputted and sensed (with the impedance quantity as the measured output to the data

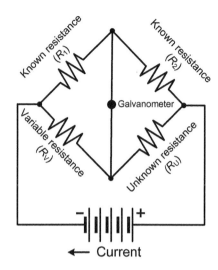

FIGURE 12.6

A Wheatstone bridge circuit.

capture/acquisition device). Also, capacitive measurement circuits are more adaptive to digital circuits due to their reactance to clock speeds of rapidly oscillating circuits.

12.1.4.3 Inductance measurement circuits

Inductance circuits and resistance circuits are very similar in measurement techniques. These circuits typically measure voltage induced into reference coils located within the sensor. These types of sensors have limited usage due to their expense and bulk. An example of an inductance-type sensor used in an ROV environment is a cable tracker for sensing cable power frequency through the induced current through sensor coils from the cable's AC power conductors (Figure 12.7). These types of inductance sensors are very effective at discriminating power frequency and are used in subsea power cable and pipeline tracking (when a reference "tone" (i.e., nominal or "reference" current passed through the cable/pipeline at some discrete frequency) placed on either the cable/pipeline or nominal grid power is applied at 50/60 Hz grid frequency).

The types of sensors measuring each category of physical parameters vary greatly. Each type of sensor has its own set of characteristics. Each must be examined for its performance characteristics given the sensing range and sensor application. As an example, a list of differing sensor types for temperature measurement includes:

- Thermocouple
- Thermistor
- RTD
- Pyrometer
- Langmuir probe (for plasma electron temperature)
- IR

FIGURE 12.7

The TSS 350 inductive coil passive cable tracking system.

(Courtesy Teledyne TSS.)

- Gas thermometer
- Traditional mercury glass thermometer

For measuring pressure, magnetic flux, resistance, radiation, essentially any physical phenomenon, the measurement technologies used will vary depending upon the application.

A *sensor* is normally defined as an instrument that measures a local phenomenon and converts its measurement into an electrical signal, while a *transducer* is defined as a device that converts one form of energy to another. In practice, however, the two terms are normally considered synonymous.

Further, sensors may be defined by their source of excitation (*active* sensors receive an outside source of excitation while *passive* sensors produce their own output signals). Two types of sensors (one active and one passive) used in the ROV environment are the TSS 440 pipe tracker (active) and the Innovatum Smartrak pipe tracker (passive), shown in Figures 12.8 and 12.9, respectively. These sensor differences are more fully explained in Chapter 18. A sample listing of sensors by type and output is provided in Table 12.1.

Sensors themselves have inherent limitations—particularly with each measurement technology:

- Lead resistance—the lead wires and circuitry have resistance of their own. This resistance may induce errors within the system sufficient to yield sensor measurement errors. However, the use of three and four wire RTDs eliminates this issue, since there is a compensation lead.
- Sensor position—RTDs must be seated properly in their thermal wells or the sensor will not indicate properly (or will have long lag times).
- Output impedance—the measurement network has a resident resistance, itself placing a lower limit on the resistance value the system is able to sense.
- Stray capacitance—all wires have a finite capacitance value with respect to ground. Excess capacitance due to circuit components has the ability to introduce errors in the output of the sensor.

It is best to know your sensor's parameters and circuit characteristics before selecting your sensor for use.

FIGURE 12.8

The TSS 440 active coil array pipe tracking system.

(Courtesy Teledyne TSS.)

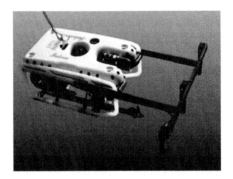

FIGURE 12.9

Innovatum Smartrak passive gradiometer array pipe tracking system.

(Courtesy Saab SeaEye.)

12.1.5 Data acquisition

The process of data acquisition entails the acquiring of sensor measurements and then processing the output into a human or machine readable format. Sensors acquire measurements in analog format and through the process of signal conditioning and analog to digital conversion transforms the electrical waveforms to digital outputs for final use.

Table 12.1 A Sample Listing of Sensors by Type and Output

Property	Sensor	Output	Active/Passive
Light	Photodiode	Current	Passive
Position	LVDT	AC voltage	Active
Acceleration	Accelerometer	Capacitance	Active
Force/pressure:	Piezoelectric	Voltage	Passive
	Strain gauge	Resistance	Active
Temperature:	Silicon	Voltage/current	Active
	RTD	Resistance	Active
	Thermistor	Resistance	Active
	Thermocouple	Voltage	Passive

Typical data acquisition systems include:

1. A sensor converting physical phenomena into electrical signals.
2. Signal conditioning circuits for converting raw electrical signals to smooth waveforms for further conversion.
3. Analog to digital conversion to convert the sensor output into its final form for transmission (analog to digital conversion may occur either before or after signal conditioning).

The process by which signals are conditioned into coherent output for transmission (i.e., further down the line toward the capture device) is another aspect of the sensor system worthy of consideration. As the output from most sensors involves a small voltage variation (or resistance changes/current measurements), further processing must occur in the form of signal conditioning. An entire class of circuits, termed "signal conditioning circuits," evolved to address these needs.

Sensor output is typically nonlinear (with respect to sensor stimulus) and requires some form of conditioning in order to scale the signal to a coherent output for correct measurement. This conditioning may be performed in the resident analog output, but more recent advances have the signal conditioned in digital format by first converting the signal to digital via an ADC (analog to digital converter) and then conditioning.

Electronic filters are used to isolate the output signal into its measurement value and noise (then filtering out the noise). Some sample filter circuits are as follows:

- Low-pass filter (Figure 12.10)—this uses a typical voltage divider circuit to filter out high frequencies and pass lower frequencies. The best way to understand this circuit is to analyze the charge time on the capacitor. Charging and/or discharging the capacitor through the resistor takes some nominal amount of time. At the lower frequencies, the capacitor has more than ample time to charge up to an equal voltage to the input voltage. However, at high frequencies, the capacitor only has time to charge a small amount before the input alternates (i.e., switches direction due to the AC current). The capacitor, therefore, destroys the signal of all frequencies higher than the nominal amount set by the capacitor by not allowing the higher frequencies to fully charge the capacitor in a timely fashion.

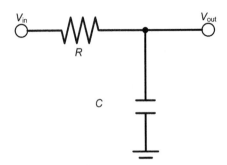

FIGURE 12.10

Typical low-pass filter circuit.

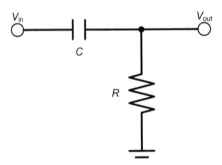

FIGURE 12.11

Typical high-pass filter circuit.

- High-pass filter (Figure 12.11)—this circuit is the mirror of the low pass with use of the divider circuit to pass high frequencies and filter out low frequencies. A high-pass filter can also be used in conjunction with a low-pass filter to make a band-pass filter. With this circuit, the capacitor functions as a low impedance device for an AC circuit, charging and discharging as the current alternates. The resistor drags the lower frequency signals down (attenuates) due to the imbalance between the sides of the capacitor (further impeding the AC flow through the capacitor). Applications for this device include blocking DC from the circuit or with RF tuning devices.
- Band-pass filter—a band-pass filter is a circuit that filters or "passes" frequencies within a certain range and attenuates frequencies outside that range, therefore blocking "passage." The band-pass circuit uses a combination of high and lower pass filters in conjunction to allow a preset oscillation frequency to pass. Figure 12.12 demonstrates the function of both the high-pass portion of the circuit along with the low-pass portion to filter out all but the narrow (pass band) frequencies. The circuit diagram further depicts the component arrangements.

Amplifier circuits take the output from the filter's circuits to amplify the conditioned signal for further transmission down the line.

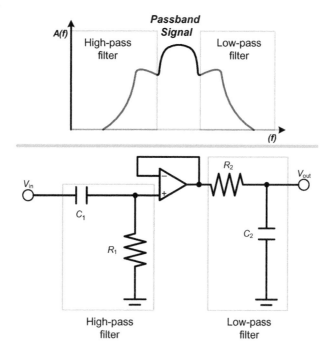

FIGURE 12.12

Typical band-pass filter circuit.

12.1.6 Systems

A sensor system comprises all aspects of the sensor from the initial sensing unit to the ultimate capture and disposition of the information (Figure 12.13). In Chapter 13, the transmission protocols will be discussed in more depth. For this section, a simple description is in order.

For sensor transmission, the data can either be channeled through the vehicle's telemetry system as a communication channel or it may be completely separated into its own transmission conduit. A typical industry standard ROV telemetry system will have as standard equipment a number of serial communication channels. As an example, a common mid-sized fiber-based ROV system (in this case, a Sub-Atlantic Mohican) has the following data channels provided through the telemetry system:

- $3 \times$ RS-232
- $1 \times$ RS-485
- $1 \times$ Single Mode Fiber

A typical ROV umbilical–tether system combination will provide one or more fibers along with spares encased within a metal or plastic tube embedded in the umbilical/tether. As an alternative to use of the vehicle's telemetry system (due, perhaps, to the sensor data requirement exceeding the capacity of the vehicle's telemetry channel), one of the spare fibers can be broken out from the fiber bundle for routing to the sensor package. The benefit is a clean and clear line for the sensor. The cost

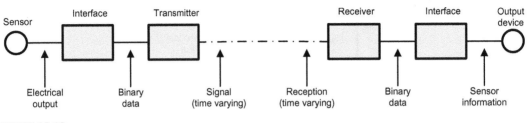

FIGURE 12.13

Simplified model of a sensor system.

is the requirement for a fiber-optic multiplexer on both the vehicle and the surface for transmit and receive (TX/RX) functions on the spare fiber as well as the requirement for another pass on the fiber-optic rotary joint (FORJ or "slip ring").

12.1.7 Installation

The final consideration for sensor system integration is the placement of the sensor on the vehicle necessary to achieve the desired outcome. As previously discussed, the sensor may be fully functional and still output erroneous and/or nonusable information. The sensor must be placed in a location that allows for sensing the environment. It must be free from the parasitic influences of the vehicle and be allowed to assess the local conditions with the full dynamic range of the sensor's capabilities.

12.2 Sensor categories

It would be difficult to include all possible ROV-deployed sensors within this section; however, the basic categories of subsea sensors typically used in the subsea environment will be examined. Sensors break down into categories based upon the type of physical phenomena being sensed.

12.2.1 Acceleration/shock/vibration

This sensor measures acceleration and outputs the measurement into various formats depending upon the sensor's application. The output of this category of sensor is typically used for vehicle control, position, orientation, or other motion-based parameters.

The most common sensor technology in this category is the piezoelectric accelerometer due to the sensor's wide linear amplitude, range, and rugged durability (Figure 12.14). The piezoelectric properties of the sensor allow for a proportional electrical output signal due to the stress applied to the materials. Using this general sensing method upon which all accelerometers are based, acceleration acts upon a known mass that is restrained by a spring or suspended on a cantilever beam and converts a physical force into an electrical signal. Before the acceleration can be converted into an electrical quantity, it must first be converted into either a force or displacement.

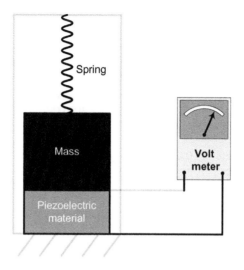

FIGURE 12.14

Configuration of a typical accelerometer used in seismic sensing.

The active element is the piezoelectric crystal which allows for a varying electrical signal passing through the crystal as the force applied to the crystal increases or decreases. The voltage is measured as a highly sensitive value allowing for close measurement of the stress applied to the sensor. The sensor can be arranged in various orientations based upon the plane of motion to be sensed.

The piezoelectric sensor can be adapted to several applications including a depth sensor (measuring stress on the crystal due to ambient pressure), inertial navigation system sensor (measuring acceleration through all planes of motion), vibration sensing (sensor mounted to the motor housing for sensing mechanical degradation), or other uses.

An example of an accelerometer would be an airbag sensor on a car or a "G-meter" measuring the stresses on an airplane (and the pilot for the old-fashioned manned vehicles) during aerobatic maneuvers. An accelerometer is used on an ROV for sensing momentum change such as through contact or collision with a subsea structure by various vehicle components (e.g., collision with the TMS, subsea structure, or vessel hull).

Typical accelerometer types and characteristics (Table 12.2) are as follows:

- IEPE (internal electric piezoelectric) piezoelectric accelerometers
- Charge piezoelectric accelerometers
- Piezoresistive accelerometers
- Capacitive accelerometers
- Servo accelerometers

12.2.2 Biosensors

Biosensors are used extensively in the medical industry but have a particular application in the subsea industry for scientific applications. In recent years, subsea biosensor applications have

Table 12.2 Typical Accelerometer Characteristics

Accelerometer Type	Frequency Range	Sensitivity	Measurement Range, gravity	Dynamic Range, dB	Size/Weight, grams
IEPE piezoelectric	0.5 Hz–50 kHz	0.05 mV/g–10 V/g	0.000001–100,000	~120	0.2–200+
Charge piezoelectric	0.5 Hz–50 kHz		0.000001–100,000	~110	0.14–200+
Piezoresistive	0 Hz–10 kHz	0.0001–10 mV/g	0.001–100,000	~80	1–100
Capacitive	0 Hz–1 kHz	10 mV/g–1 V/g	0.00005–1000	~90	10–100
Servo	0–100 Hz	1–10 V/g	<0.000001–10	~120	>50

Source: From Wilson, 2005.

expanded to use in various nonscientific fields including aquaculture, military/homeland security applications, environmental monitoring, and industrial process controls.

A biosensor is the combination of a bioreceptor plus a transducer in order to combine the components into a single sensor unit, allowing one to measure a target analyte without using a reagent. The specific application for the sensor is the detection of various products in order to infer the condition of the biological system being sensed. An example of a product to sense in order to infer the condition of a biological system would be the dissolved oxygen of an aquaculture ecosystem used to determine the fish or bacterial capacity of the water volume.

12.2.3 Chemical sensors

Chemical sensors isolate and detect specific chemical elements/compounds along with their concentration within the environment (in this case, water) in order to extrapolate the sample concentrations to the larger environment.

Various methods and technologies are used for sensing different chemical components depending upon the desired application. A peculiar property of matter involves the rotational and vibrational modes of molecules. In its simplest form, molecules are typically sensitive to energy in the optical range of the radio magnetic spectrum; therefore, molecules (once excited) tend to either absorb or emit photons of light based upon several factors. This absorption/emission is typically in the infrared range; therefore, infrared spectroscopy is extensively used to identify certain molecules. As an example, consider the fluorescence emission process for oil in water.

Oil in water can be stimulated (excited) with UV (or near UV) light in order to have the sample emit fluorescent light (Figure 12.15). UV light is directed to the sample at a wavelength (λ_{EX}). The molecules in the sample absorb the UV light and then rise from their ground energy level (E_0) into a new higher energy state (E_2). The excited molecules lose their energy previously absorbed (termed "relaxation"), reducing the molecule to a lower energy state (E_1). The molecule returns to ground state through emission of a photon of light (through fluorescence or $E_1 - E_0$). Light energy is inversely proportional to light wavelength; therefore, the emitted light wavelength (λ_{EM}) is longer than the input wavelength (λ_{EX}). The sample water can either be sensed *in situ* or be pumped

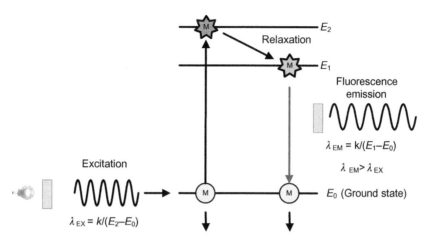

FIGURE 12.15

Infrared spectroscopy depicted.

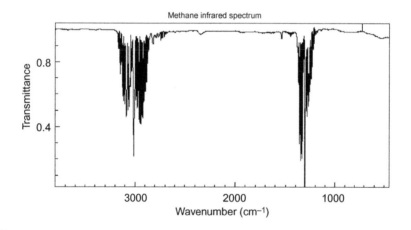

FIGURE 12.16

Methane infrared spectrum.

(From NIST webbook.)

through the instrument to sample a higher quantity of water and then filtered to remove larger sample contaminants.

In light of the April 2010 disaster aboard the transocean *Deepwater Horizon*, new regulations on hydrocarbon emissions have been enacted requiring further discussion of hydrocarbon-in-seawater detection techniques. Typical hydrocarbon sensors emit light at infrared frequencies with the sensor measuring the emission spectra. Hydrocarbons luminate at a discrete frequency based upon the composition of the compound. A sample spectral resonance graph is shown in Figure 12.16 for methane as well as for benzene (Figure 12.17). The optical hydrocarbon fluorometer tuned for discrete compounds

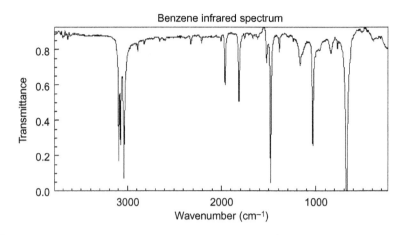

FIGURE 12.17

Benzene infrared spectrum.

(From NIST webbook.)

simply emits light at one frequency and senses fluorescence at the second resonant frequency to positively identify the compound.

Other methods of sensing chemical compounds in water are through electrochemical detection techniques such as particle diffusion, gas chromatography as well as many other technologies. Chemical testing has a much broader capability for sampling in air, but water (being the universal solvent) allows trace elements to be suspended in solution for sensing at low chemical concentrations. Consult a basic text on chemical sensing for further information on this topic (such as Wilson, 2005).

12.2.4 Capacitive/inductive sensors

Capacitive/inductive sensors measure conductive potential between two conductors that are separated by nonconductive materials. The advantages of this noncontact technique include higher dynamic response to moving targets, higher sensor resolution, more rugged sensor design, and minimal hysteresis.

Both capacitive and inductive sensors consist of a probe as well as a driver. The probe generates the sensing field while the driver drives the probe's electronics. These types of sensors are used extensively in applications whereby the sensor is needed to measure proximity to certain conductive materials (e.g., paint coating thickness) or for measurement of metal integrity (e.g., eddy current measurement).

Capacitance probes are reactive to three parameters:

1. The relative surface sizes of the probe/target combination
2. The size of the gap between the probe and target
3. The makeup of the material within the gap

Examples of capacitive/inductance sensors used in the ROV industry are touchscreen monitors for human—machine interface (capacitive), label readers (capacitive), position indicators (capacitive or inductive), and proximity switches (capacitive or inductive). Stray capacitance accounts for most of the errors in capacitance measurement. Shielding the capacitance signal and grounding the shield can mitigate the effects of stray capacitance (which, if not guarded, can fluctuate between 10 pF and 10 nF). Further, locating the sensing electronics close to the sensor electrodes will minimize the effects of stray capacitance from local noise.

Capacitive and inductive sensors work best in a vacuum or in air but have limited use in a fluid due to the relatively high attenuation of electromagnetism through water.

12.2.5 Electromagnetic sensors

Electromagnetic sensors make use of perturbances in the electromagnetic field of sensor hardware in order to measure certain changes in the field. Danish physicist Hans Christian Ørsted first discovered the relationship between magnetism and electric current when he noticed that a current was induced into a coil when the magnetic field surrounding the coil changed. This opened a whole gamut of applications for this technology. The direct result of this discovery was the follow-on invention of the electric motor and a host of sensor technologies measuring the voltage/current induced by the motion through an electrical field.

The add-on principal enabling sensor development is best captured in Lenz's Law, which posits that the combined effect of all of the electromagnetic interactions on a field is to resist the modification to the field.

Any number of usages for electromagnetic sensors can be found. Specifically, in the ROV industry you will find sensors with use of electromagnetic field measurement of:

- inductive proximity sensors (e.g., camera position indicator on a tilt motor such as the linear variable differential transformer (LVDT))
- low-frequency motion (e.g., a geophone for oil and gas seismic measurements)
- magnetism (e.g., magnetometer for sensing ferrous metals or the simple flux gate compass for measuring vehicle orientation with reference to the earth's magnetic field)
- tachometer for measuring thruster motor speed

The electromagnetic sensor field is quite in-depth and rapidly evolving. The reader would benefit from a study of this field from a basic text on the subject.

12.2.6 Flow/level

The flow sensor is the basic measure of fluid volume passage. As an add-on to the measurement of flow, level indication can be deduced from the measured flow. All fluids—whether air, water, metals, or some combination—flow. The technologies used to measure this flow vary with the characteristic being measured. There are different methods peculiar to use in gas as well as in liquid.

Flow rate is measured by the simple velocity measurement of fluid through a conductor (such as a pipe, aqueduct, or other fixed-volume structure) and then multiplying the fluid speed versus time through the fixed-volume conductor to provide a volume quantity.

Thermal anemometers measure the transfer/removal of heat from a heated sensor (also referred to as "hot wire") to the flowing fluid to compute flow based upon the heat transfer rate. These sensors are typically used in air flow measurements.

Differential pressure measurement (DPM) sensors are the most common type of liquid flow sensors currently used. The DPM sensor isolates the pressure drop through the meter. The pressure drop across the meter is proportional to the square of the flow rate (i.e., the square root of the pressure differential). Most of these types of measurement devices have multiple sensor stations for sensing differentials over a range of locations in order to mitigate the effects of local measurement variations for the single pick-up (i.e., measurement) point. DPM sensors can either measure velocity pressure through tube insertion into flow (e.g., Pitot tube) or pressure drop across a restriction (e.g., Venturi tube).

Examples of these types of sensors are as follows:

1. The Pitot tube (measuring kinetic versus static pressures)—this device offers the least pressure drop from the incoming fluid stream. The most common use of this type of sensor is for airspeed on aircraft, but it is somewhat inaccurate compared to other devices. The instrument's basic function measures both dynamic fluid flow pressure (from the oncoming fluid stream), then compares this to static fluid pressure (from the ambient/static fluid reservoir), and outputs the pressure differential. The sensing tube is subject to obstruction (as with the 2009 Air France 447 accident involving Pitot tube blockage due to ice), which blocks the sensing unit. Aircraft manufacturers typically use this type of sensor because of its simplicity, durability, and cost. Subsea vehicle engineers use this sensor for travel speed through the water column on high-speed vehicles and pipeline engineers for fluid flow speed through a pipe.
2. The Venturi tube (measuring pressure drop through a restriction)
3. The concentric orifice
4. The flow nozzle

Vortex-shedding sensors make use of the Von Karman principal by measuring the vortices shed downstream of an object due to the eddying effect of turbulent flow. Vortex-shedding frequency is measured as it is proportional to the velocity of the flowing fluid, thus deducing the fluid volume by measurement of the velocity through a fixed volume.

Positive displacement flow sensors separate the fluid into measured volumes and then send the fluid on after precise volumetric measurement. This type of sensor is the most energy inefficient (as it requires mechanical sample separation), but it is the most accurate since it precisely measures the volume during each sample.

Turbine-based flow sensors force the fluid flow through a fixed volume that turns a turbine or propeller, which in turn spins at a rate proportional to the fluid flow rate. The turn rate is then measured to derive the fluid flow.

Mass flowmeter sensors make use of the Coriolis effect to directly measure the mass of the fluid flowing through the meter. The fluid volume is vibrated at a known frequency as it passes through the sensor. The vibrating fluid induces a twisting moment into the volume of water that is then measured as torque in the piping. The torque is directly proportional to the mass flowing through the pipe and is a direct measurement of the fluid's mass.

Ultrasonic flow sensors are either Doppler or time-of-travel/flight meters. With Doppler sensors, the meter measures the frequency shift caused by the fluid's flow through the meter across an acoustic sender—receiver pair. The sender transducer transmits at a known frequency, and then the receiver measures the shift in that frequency based upon the flow through the meter. This method works well in homogeneous pure liquids but is highly attenuated by dissolved or suspended gas within the liquid due to multipath reflection of the gas—liquid bubble interface.

There are other flow sensor technologies available and more are coming onto the market as the march of technology steadily progresses.

Level sensors are quite prevalent in the subsea industry. These make use of any number of methods to sense fluid levels in various mediums. Some examples of level sensor applications are as follows:

1. Tank level indicators for measurement of potable water, fuel, or chemical levels in storage tanks
2. Hydrocarbon level measurement in sunken ships for environmental remediation (e.g., the diesel level in the tanks of the *Deepwater Horizon* was measured before removal after the rig sank in April 2010)
3. Fluid transfer control in various applications.

Sensors for this application fall loosely into the following categories:

a. Hydrostatic—this instrument measures differential specific gravity of a liquid in order to sense the level at various points within the sample fluid, thus deducing the level by accurately measuring the depth at the sensor (located at a known position).
b. Ultrasonic—frequency shifting under the same principles as ultrasonic flow above.
c. RF capacitance—the measurement of capacitance between the sensor and the tank/vessel wall as the tank fluid level changes.
d. Magnetostrictive—in magnetostrictive materials, an external magnetic field can be induced that reflects electromagnetic waves in a waveguide, which are then metered for level measurement.
e. Through-air radar—divided further into "frequency-modulated continuous wave" radar and pulsed wave time of flight, these methods measure the air space above the liquid in a known volume and reverse compute the liquid volume.
f. Guided wave radar—pulses of energy are transmitted down a waveguide (e.g., a rod or cable) and the dielectric constant changes at the air—liquid interface, reflecting part of the signal back to the transmitter and thus computing the distance to the air—liquid interface.
g. Air bubbler—a flow of air is pumped down a submerged tube to its exit point. The pressure of the air injected into the tube at the surface is proportional to the exit point density and the pressure at depth, thus very accurately measuring the depth of the exit point due to the hydrostatic pressure on the water column evacuated from the tube. This sensor is used in the gas diving industry (termed a "pneumofathometer" or, more commonly, the abbreviated "pneumo") for very accurate depth readings.
h. Radiological neutron backscatter—a neutron emission source transmits neutrons through a containment vessel and measures the opposite side for backscatter from either liquid or gas, thus deducing the interface point for level measurement due to differing backscatter characteristics between different compounds (e.g., diesel fuel and seawater).
i. Heated junction thermocoupler—the principle behind thermocouplers (Figure 12.18) is that dissimilar metals conduct varying degrees of current dependent upon the temperature

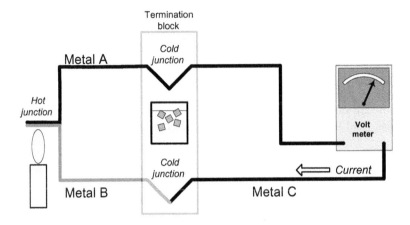

FIGURE 12.18

Simple diagram of a thermocoupler.

differential between the metals. Heated junction thermocouplers, however, amplify this phenomenon by placing a thermocoupler next to a heat source and another farther away. The thermocoupler next to the heat source will measure a much higher reading while in gas (or steam), versus while in fluid, due to gas conducting heat much more efficiently than fluid. Thus, the two thermocouplers are compared and sense the level (or degree thereof) by comparing the degree of in-fluid temperature/conductance on the primary versus the reference secondary.

12.2.7 Force/load/weight

For the purposes of this section, the term "force" measures the interaction between bodies, "load" measures force exertion, and "weight" measures gravitational acceleration forces. Sensors in this category most commonly measure these parameters based upon either piezoelectric quartz crystal or strain gauge sensing elements. While piezoelectric sensors are best for a dynamically changing load situation, the strain gauge is best suited for a static load environment.

12.2.7.1 Quartz sensors

Quartz crystals have the unique characteristic of generating an electrostatic charge when force is applied (or removed). This is quite useful in a host of force/load/strain measurement functions such as depth gauges where vehicle depth is constantly changing. The quartz sensor uses the piezoelectric effect to measure the pressure strain upon the transducer. This technology is typically used for the measurement of highly dynamic pressures (such as those encountered in deepwater applications).

12.2.7.2 Strain gauge sensors

With strain gauge sensors, the transducer forms a variable resistor based upon the load applied to the sensor (typically a foil-type transducer). The load applied is directly proportional to the deformation of the transducer and (considering that a voltage is applied to the transducer element) the resistance measurement directly correlates to the load applied. The sensor element can be any of a series of sensor types including a "bending beam," column-type load cell, or shear-web. All have their strengths and weaknesses depending upon the particular application.

12.2.8 Humidity

Humidity is the measure of water vapor in air or other gases and is measured as a ratio of the temperature/dew point spread or relative humidity (ratio of moisture content to the saturation point considering a constant pressure/temperature combination—abbreviated "RH").

As with other sensors described above, RH sensors can make use of several technologies including capacitive RH sensors, thermoset polymer-based capacity RH sensors, and resistive humidity sensors.

Humidity sensors are sometimes used in subsea vehicles as moisture sensors inside air-filled electronics bottles to sense for water intrusion.

12.2.9 Optical and radiation

The sensing of optical and near-infrared radiation is useful for many applications using light detection sensors. Light comes in discrete particles called "photons" that have an energy and wavelength—the determination of which yields some peculiar characteristics useful in the subsea industry.

Light detectors can be characterized as either quantum detectors (converting radiation directly into electrical signals in semiconducting materials) or thermal detectors (absorbing thermal energy for indirect measurement of radiated energy).

Optical sensors do not see widespread use in the subsea industry other than through light-sensing devices tied to artificial lighting.

12.2.10 Other sensors

Other sensors that find limited use in the subsea industry include radiation sensors such as the so-called Geiger Counter. The Geiger–Müller tube (commonly referred to as "GM tube") detects alpha, beta, or gamma rays through ionization produced in a low-pressure gas. These instruments detect the presence of radiation (quantity but not magnitude/energy) by sensing the ionization of the gas. The gas becomes momentarily conductive when ionized through the "Townsend avalanche effect," thus conducting a current through the gas—the event of conduction is then counted per unit time (or cumulative) by electronic circuits. This is useful with primary and secondary cooling tank inspections within nuclear power plants (which are often inspected via ROV). Gamma radiation detectors of various types have been developed for ROV use.

There is a wide array of additional sensors available for use in the subsea industry. The requirements for use of these sensors are simply to provide power for the sensor's electronics and a communications channel for transmission of the data to the surface for dissemination and discrimination.

12.3 Common ROV sensors

Although sensor technology is a large and growing field, the majority of ROV-deployed sensors falls easily into just a few types based upon the job function being performed.

12.3.1 Sensor by job type

"It's not about the vehicle, it's about the sensors and tooling" (Phil Montgomery, NAVSEA ca. 2001).

The ROV is simply the "ride to the job site" for most sensors. The choice of sensor will certainly evolve around the client's final deliverable product from the operation. The decision to purchase or lease a particular sensor will be a combination of economics (return on investment) and logistics (opportunity cost for not having the sensor or backup available in the field should you need it). The following are just a few applications requiring sensors. The ultimate sensor selection will rest with the end user of the data gathered during the field operation.

12.3.2 Aquaculture

Aquaculture operations are directed toward the growth of aquatic species for animal or human use (either for direct consumption or use of the products from harvest). The objective is to make the environment as friendly to breeding as economically possible. That means enhancing the "stock-friendly" factors and mitigating the "stock-unfriendly" factors. Examples of stock-friendly environmental factors include high dissolved oxygen, optimal breeding temperatures for the species cultivated, proper pH balance, low parasite and disease content, and other things that will enhance cultivation. Mitigation of stock-unfriendly factors would include net protection from stock natural predators, disease inhibitors, feeding cycle recognition, and water chemistry analysis.

A sample sensor load-out of an aquaculture vehicle would include:

- Dissolved O_2 meter
- Alkalinity and pH sensor
- CTD for temperature and salinity measurement by depth
- ORP (oxidation reduction potential)
- Rhodamine
- Turbidity
- Chlorophyll

12.3.3 Construction

The main sensor requirements for subsea construction projects involve dissemination of the physical characteristics of the subsea environment at the construction site along with conditions for completion of the work scope. Such job tasks involve setting subsea structures, measuring distance offset between items (termed "metrology") for physical connections, placing anchors for proper mooring of large floating structures, and a host of other tasks. Metrology is defined by the International Bureau of Weights and Measures (BIPM) as "the science of measurement, embracing both experimental and theoretical determinations at any level of uncertainty in any field of science and technology."

A typical sensor load-out of a vehicle tasked for a subsea construction project includes:

- Single-beam obstacle avoidance sonar
- 3D multibeam sonar for *in-situ* characterization of subsea structures
- HD camera for documenting the job
- CTD for sound velocity profiling (for acoustic positioning)
- USBL or LBL subsea acoustic positioning
- High-tolerance fiber-optic gyro for orientation of subsea structures
- ISO and/or API standard tooling interface with torque and turn counter sensors

12.3.4 Salvage

Since salvage operations are typically "smash and grab" operations involving the removal of items on the seafloor, few sensors are typically needed other than navigational and survey-type sensors.

A typical sensor load-out of a vehicle tasked for a salvage project includes:

- Fuel level sensor for removing any hydrocarbons present
- Single-beam obstacle avoidance sonar
- 3D multibeam sonar for *in-situ* characterization of subsea structures
- Standard definition camera for documenting the job (or HD if the job is a documentary)
- CTD for sound velocity profiling (for acoustic positioning)
- USBL or LBL subsea acoustic positioning

12.3.5 Science

For science assignments, the typical investigator (sometimes called the "beaker") will be quite demanding of the sensor load-out as the gathering of information is of utmost importance.

A typical sensor load-out of a vehicle tasked for a science project includes:

- Dissolved O_2 meter
- Alkalinity and pH sensor
- CTD for temperature and salinity measurement by depth
- ORP (oxidation reduction potential)
- Rhodamine
- Turbidity
- Chlorophyll
- Hydrocarbon and chemical sensor tuned for chemical spectroscopy
- Water and bottom sampling devices
- Laser scaling and measuring devices
- Single-beam obstacle avoidance sonar
- Sub-bottom profiler
- 3D multibeam sonar for *in-situ* characterization of subsea structures

- Standard definition camera for documenting the job (or HD if the budget and principal investigator allows)
- CTD for sound velocity profiling (for acoustic positioning)
- USBL or LBL subsea acoustic positioning

12.3.6 Structural inspection

The inspection, repair, and maintenance (IRM—also termed IMR (depending upon the European or American accepted acronym)) is a critical job for the subsea industry as more and more offshore structures for oil and gas, as well as renewable energy generation, proliferate. Typical structures are placed with a life expectancy of at least 20 years. During that time, storms, oxidation, vessel traffic, and typical wear and tear require periodic inspection and maintenance.

A typical sensor load-out of a vehicle tasked for an IRM (IMR) project includes:

- Cathodic potential probe for measuring passive anode health
- HD and/or standard definition camera for documenting the inspection
- CTD for sound velocity profiling (for acoustic positioning)
- USBL or LBL subsea acoustic positioning (for pipeline or other subsea operations with no exposed surface structure)
- Pipe tracker (for buried pipelines)
- Single or multibeam sonar for location of structure (and avoidance of hazards)
- NDT sensors for structural metal degradation testing

12.4 The future

As stated in the preamble to this chapter, sensor technology is where the real subsea technological developments are taking place. The future for sensor development in the medium term involves sensor fusion for automation of subsea functions (allowing for decoupling from the operator) as well as gains in sensor sensitivity for more accurate measurement of physical phenomena.

Speculation within the oil and gas industry (specifically, senior members of the Marine Technology Society) projects that the deepwater subsea field will be populated with autonomous vehicles docked to power stations that maintain an oil and gas production system to full ocean depth with minimal human intervention. As we move into the Arctic environment, such technology will become even more important. More discussion of this subject is available in Chapter 23.

Seventy-one percent of the earth's surface is covered by water. There are vast lakes of liquid water underneath the ice caps of Greenland and Antarctica. There is speculation that there is more water on Jupiter's moon Europa than there is on the earth. Water is the environment in which ROVs live. The future of our industry is limited only to our own imagination.

Communications

In this chapter, the subject of data communications as it relates to electronic communication with sensors delivered by a remotely operated vehicle (ROV) will be addressed. This is a very important topic for underwater vehicle control and utility. The entire purpose of the underwater vehicle is to deliver a sensor or tooling package to a work location to perform work of some type or to sense the environment. In order to place the vehicle there, it must be controlled through the use of instructions transmitted to the vehicle, which are based upon data received from the vehicle's sensors. The constant flow of data and instructions from/to the vehicle and its sensors allows the operator to complete the task safely and efficiently. In two-way communications between humans, it is sometimes stated that "it does not matter how you say it as ong as it is understood." But the easy flow of interpersonal communications does not measure up to the exacting requirements of intelligent machine communications.

13.1 Overview

13.1.1 What is communication?

Imagine yourself in a room where everyone is talking simultaneously in a different language. Is communication taking place (probably not)? Will communication take place if only one person spoke at a time but still in a nonhomogeneous language? How about one person speaking in a common language, only with a different dialect (to better understand this concept, imagine a Bostonian traveling to the bayous of southern Louisiana and ordering lunch!)? Try listening to a speaker who is lecturing in a very slow or rapid manner. College professors are notorious for speaking in low tones while facing the blackboard and writing unintelligible signs using undefined terms—and the professor is shocked when students fail the subsequent test because of a basic failure to communicate.

Communications between spouses, organizations, governments, civilizations—and machines—require a common set of protocols in order for messages to be transmitted, received, and understood. For communication to take place, ideas must be conveyed using commonly accepted symbols (speech, body language, data, or other acceptable forms) through some common medium at a rate and with a delivery format that allows these symbols to be both transmitted and received for information dissemination. This is applicable to people, animals, and intelligent machines.

13.1.2 Evolution of data communication

Speech evolved over millennia, eventually allowing for direct human communication over short distances. Technology further allowed humans to communicate over vast distances through the use of differing mediums. Native Americans made use of the smoke signal, tribes in Africa used drums, sailors used flags as well as signal lamps, and church bells signaled the faithful to service (or slow

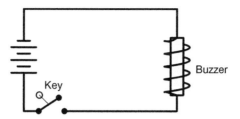

FIGURE 13.1

Example of a basic telegraph circuit.

tolling of the bells for a death announcement). The Greeks used signal fires to announce the fall of Troy (i.e., signal fire = "Win"—no signal fire = "No Win").

Modern electrical signaling techniques came about with the advent of the telegraph. This basic signaling mechanism was designed over a number of years. But it was not until Samuel Morse came up with a commonly accepted code for information interchange that the telegraph was finally able to embed intelligible messages within the transmitted signal. The telegraph quickly gained wide acceptance, and electrical data communication was born!

The basic telegraph allowed a key to be activated, closing a DC electrical circuit and causing a current to be induced in a coil (Figure 13.1). This caused an iron rod to strike a plate and make a sound each time the key was depressed. With Morse's series of dots and dashes (long versus short signal timing), alphanumeric characters could be tapped out and transmitted over long lengths of conductor spanning huge distances.

Later, voice was impressed over wires, thus evolving the entire telecommunications industry. The transmission of data was further refined with the advent of the teletype (from which many of the standards still employed today were derived) allowing for direct alphanumeric transmission between common terminals.

In 1969, the US Department of Defense's Advanced Research Projects Administration (predecessor to today's DARPA) introduced the first packet-switching network, named "ARPANET," designed to connect the Department of Defense's computer network with several major universities. In the 1980s, ARPANET reached its structural capacity and the US National Science Foundation (NSF) took over management of the budding national data communication infrastructure with its own network infrastructure known as "NSFNET." Soon, the network freed itself from the NSF toward an open source industry steering committee, allowing for faster development and forming the Internet we know today.

13.1.3 Data networking and ROVs

Data communications and networking, with regard to the subsea vehicle, followed the model of land-based communications. The same principles of electronic communication over conductors evolved with the use of the vehicle tether for communication lines and power transmission. The original ROVs made use of separate conductors for each component on the ROV. In its simplest form, the early ROVs used a separate conductor wire controlled via rheostat for each motor and controlled each motor separately. The camera power and signal used another discrete conductor. This made for a dreadfully difficult control regime. There was no *per se* "data communication" with the vehicle—and the resulting tether size was huge! Later, as circuits further evolved, analog control of the vehicle and communication with the sensors led to digital circuits and control.

ROV-mounted sensors communicate with the surface through four separate modes (Figure 13.2):

1. Communication through the vehicle's tether and telemetry system (a)
2. Communication through the vehicle's tether but bypassing the telemetry system (b), for example, with a separate fiber breakout within the junction box routed to a survey pod (or directly to the sensor)
3. Communication outside of yet attached to the vehicle (c), for example, physically attaching (via plastic or fiber wraps) a separate communication line to the tether
4. Communication separate and apart from the vehicle (d), that is, with a separate down line from the vehicle and tether

Examples of each of these modes follow:

1. *Mode 1*: Most commercial ROV systems have accommodation for multiple serial ports within the telemetry system. A sonar mounted to one of these ports allows for communication through the telemetry system to an output on the vehicle's control panel.
2. *Mode 2*: Most tethers of commercial ROV systems have extra conductors or fibers located within the tether specifically for spare capacity for sensor communication or for backup should a primary telemetry conductor fail. For instance, a typical fiber-based ROV system has four to eight fibers housed within a steel tube (or some other fiber stiffener arrangement for limiting microbending) located within the tether. One fiber is typically used for the telemetry system leaving three (or more) extra unterminated fibers available for use with sensors. These are easily broken out from the fiber bundle and routed from the vehicle's junction box directly to the sensor through a fiber multiplexer/modem. But if the ROV will use a LARS or TMS (Launch and Recovery System or Tether Management System), another pass on the fiber-optic rotary joint (FORJ) will be required.
3. *Mode 3*: A sensor is typically mounted to the ROV's frame. But for convenience, the sensor's power and communications cable is mated to the outside of the tether, thus avoiding the time and trouble of integrating (or the sensor's power and/or bandwidth needs exceed the ROV system's capabilities) the sensor through the ROV's tether. An example of this is a high-throughput 3D sonar (which exceeds the vehicle's telemetry and power capabilities). The sensor is mounted to the vehicle's frame, but the sonar's cable is mated to the outside of the vehicle's tether with tape or binder.
4. *Mode 4*: Some sensors are simply mounted to the ROV with no other attachment to, or communications with, the vehicle. An example of this is a vehicle-mounted, battery-operated, acoustic transponder. The transponder is self-powered and the sensor's transducer communicates directly with the hydrophone mounted to the vessel. The vehicle provides no power to the sensor and no sensor communication medium is linked to the vehicle.

13.1.4 Transmission versus communication

The data transmission versus communication question has a very close analogy with pipelines. The pipeline can carry water, gas, oil, gasoline, diesel, or a host of other gases or liquids. The actual process of transmission of the liquid is a completely different function from the fluid itself.

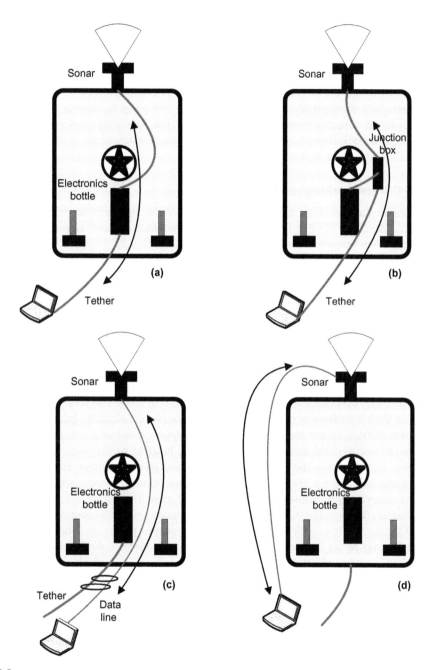

FIGURE 13.2

Modes of connecting ROV sensors to the surface.

The pipeline is designed with the core gas or liquid in mind (e.g., small pipe for low-volume gas, large pipe for high-volume liquid), but even within the various sample liquids vast differences of materials are likely (e.g., oil may be light sweet crude or sour heavy sludge, water may be fresh or salty, clean or dirty, hot or cold, etc.).

The *transmission function* of data transfer involves the physical lines, radio waves, data communication channels, and wiring that transmit the data from one point to another. The data *communication function* is the actual data content transmitted. In the following sections, we will consider the data communication function separate from the physical transmission and then pull it together into a set of standards for use in the ROV World.

13.1.5 The OSI networking model

The data communication function runs the full gamut from the simple light switch (turning on and off the light in a room) to the full intergalactic data communication web (for you *Star Trek* fans). In order to make some sense of this vast range from the simple to the complex, and to help manage the data communications question, the International Organization for Standardization (ISO) developed its famous "Open Systems Interconnection" (OSI) model. As the reader will notice, the model layers the various aspects of hardware to software in levels of complexity. They begin with the physical layer (in its basic form, the "transmission" layer) through the data-link layer (the beginning of "communication"), all the way through to the applications layer. It is in the applications layer where the protocols of communication smooth the flow of data through disparate locations and channels allowing the machines to talk the same language in an effective fashion.

While this model's applicability to ROVs is somewhat limited to the lower layers in the smaller OCROV sizes, as vehicle complexity evolves in the later iterations (e.g., the larger modern WCROVs display complex computer networking of various sensors, subsystems, and components) the full range of data layering evolves into a complex data communications platform.

In a small OCROV, a simple heading or depth hold application has a routine whereby the compass "talks" to the horizontal thrusters to hold a specific heading or the vertical thruster "talks" to the depth sensor to hold a specific depth. These are lower level functions. But as the complexity evolves, a Doppler velocity log (DVL) may "talk" to a positioning system whose acoustic beacon is "talking" to other beacons while further data "feeds" flow from a gyro (slaved to a magnetic compass), altimeter, depth gauge, CTD (conductivity/temperature/depth) sound velocity sensor, etc. The processor for the thrusters may be controlled from a central processor aboard the vehicle or the processing power may be on the surface with all subsea components linked as networked nodes.

In the modern complex MSROV and WCROV, sensors "talk" to vehicle controllers in order to autonomously track pipelines, map grid lines, maintain station, and operate tooling as the control function of the vehicle communicates with vehicle sensors to determine path decisions so as to instruct movers/controllers in a lavish dance of data communications. The OSI model is ISO's attempt to organize this function into an easily discernible hierarchy.

The OSI model, instead of solving the overall data communications problem, seeks to break the complex question of subsystem communications into levels and components for two reasons:

1. to link disparate elements of a networked system and
2. to mitigate the ripple effect of modification to hardware and software components by compartmentalizing and layering these various components

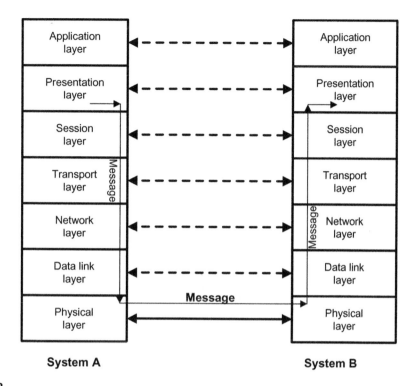

FIGURE 13.3

The OSI reference model.

This model is not *per se* a "standard" but a model by which other standards organizations may concentrate within the various layers and compartments.

ISO uses the term "peers" to define communications among related layers of various systems. Figure 13.3 shows that communications between peers (in this case, Systems A and B) are only made directly through the physical layer and are indirect at all other layers. Communications, instead, travel from the original "talker" down the various layers to the physical layer. The signal is then transmitted between devices and back up through the hierarchy of the second "receiver" device of the corresponding peer.

Communications between two devices can be either as a "connection service" or a "connection-less service." In a connection service, the message is being relayed from one peer (in Figure 13.3, System A) of the system down the various layers within that system to the physical layer, on to the physical layer of the corresponding system (in this case System B), and then up the layers to the corresponding peer in a direct communication session. As shown in the figure, communications are represented by dashed lines through all other layers except the physical layer due to this indirect communication between peers. An example of this method is a Telnet connection whereby a "session" is opened so that data terminal equipment (DTE—in this case a personal computer emulating a standard VT-100 terminal) "talks" directly to a host computer. On an ROV, the vehicle's sonar, communicating directly (via an RS-422 combination physical/data-link layer) with a surface computer (for display) through a separate twisted pair conductor within the tether (direct continuity of the conductor can be checked between the sonar head and the serial connector at the surface) would actually be communication within a single system. In a connectionless service, all aspects are the same except that the data being transmitted is wrapped into a data "packet" and addressed to a node within the network. The data is then received through the physical layer, transmitted to the peer and then received, reassembled, and disseminated (much like a letter that is written, put into an envelope addressed to the recipient, and mailed through the postal service). A simplified example of this is an e-mail routed through the Internet. Network protocols are further explained at Section 13.3.4.

13.2 Transmission

Data transmission is the physical/electrical transfer of a signal, message, or other form of intelligence from one location to another. Just as fluids flow through pipelines, data flows through a series of conductors, ports, connections, and links. In this section, the issues involved with the physical conduction of data flow will be addressed.

13.2.1 Basic data transmission model

Data transmission is the basic exchange of information between two "agents" linked together through a series of data pipelines (see Figure 12.13 for a graphical depiction of this concept). This exchange can be between a machine and a human sitting at a terminal or it can be between a machine (program) exchanging data with another machine (program). Reliability, integrity, and intelligibility of the message are achieved through the processing of this information from transmission to receipt as an integral part of this process. The source must produce a time varying signal of binary data couched in a format suitable for the medium being used. The signal received at the far end of the cycle must be similar to the signal transmitted at the source; therefore, the essence of this model has a time varying signal generated that emulates the initial source input (say, a sonar reflection off a gold-laden Spanish galleon protruding off the deep seafloor [to make this interesting]). The sonar reflection is transferred to time varying electrical signals through the sonar's transducer, which in turn is modulated to a carrier signal that is converted (through the interface) to binary data and sent down the line. On the far end, the process is reversed to deliver the information to the capture device for final processing. The end sonar rendering of a beautiful square-rigged sailing ship (filled with valuable loot) is generated on the computer display as the dollar signs light up in the sonar operator's eyes (with dreams of fast cars, lavish houses, and expensive vacations dancing in his/her head). But I digress. . . .

13.2.2 Electrical signal transmission

The two basic methods of data transmission relate to the data sequencing. This sequencing comes in two broad categories, parallel and serial data flows.

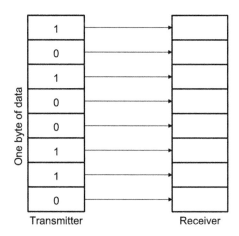

FIGURE 13.4

Parallel data transmission.

Parallel data transmission (Figure 13.4) involves the transfer of discrete bytes of data in separate yet parallel channels to be reassembled at the receiving device. As the timing and sequencing of the data must be closely timed and rigorously in step, practical parallel transmission requires the transfer distance between the agents to be over a very short distance (e.g., with a parallel printer cable). For the ROV application (with its long lengths of cable), parallel data transfer is typically impractical.

Serial data transmission, however, is the preferred method of data transfer for the ROV and will be extensively examined in this chapter.

In serial transmission, data bits are transferred one after another in a serial fashion usually with the least significant bit (LSB) transmitted first (bit significance follows the numeric analog with the "one" position as least significant—for example, with 1234.56, the 4 is the least significant digit). While parallel transmission requires eight (or more) separate paths for data to flow, serial transmission only requires a single path or circuit connecting the devices. As the data is transmitted and received sequentially, the timing of the receipt is far less critical than in parallel transmission and is much more suitable to transmission over long distances (such as through ROV tethers and/or umbilical).

As reflected in Figure 13.5, bits are transmitted as electrical signals in either a neutral or polar waveform depending upon the changing charge state. The discrete states of 0 and 1 are represented by two voltage levels and are categorized in three types: unipolar, polar, and bipolar. As depicted in Figure 13.6 below, unipolar waveform is a simple on/off whereby the binary bit is transmitted as a simple positive voltage. The polar waveform is either a + voltage or a − voltage and the bipolar signal always returns to neutral upon bit representation.

The rate at which these bits of data are transmitted is termed the "bit rate" (expressed in bps, or bits per second), with the bit duration being 1/bps (i.e., the inverse of bit rate—expressed in seconds). The received bits are subject to a host of issues experienced during their travels, such as the timing of both transmissions, noise experienced during the transmission, clock accuracy, and decoding of the received signal. This may necessarily introduce errors into the transmission process at a

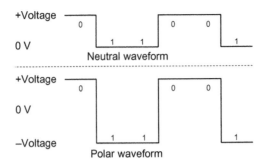

FIGURE 13.5

Binary data with use of voltage conversion.

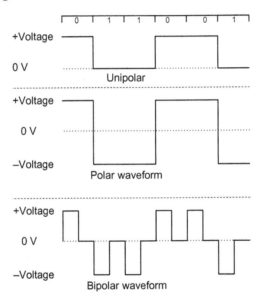

FIGURE 13.6

Unipolar, polar, and bipolar waveforms.

nagging regularity termed the "bit error rate" (BER—bit errors per unit time). A graphical illustration of BER is depicted in Figure 13.7.

Data is transmitted via two modes: asynchronous transmission and synchronous transmission.

- *Asynchronous transmission mode* (ATM): In serial data transfer, ATM (Figure 13.8) has data flowing back and forth between terminals at nonscheduled times. The flow of data is controlled in the sequence through the use of start and stop tags embedded within each signal to overcome the challenges involved with synchronizing the messaging services. This mode is used extensively in the ROV industry for sensor platforms as it allows for robust communications through dedicated channels.

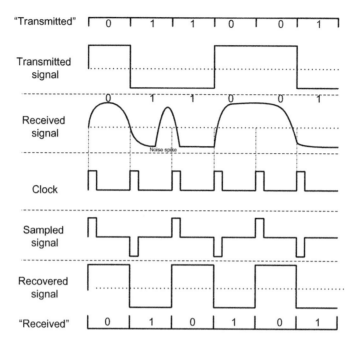

FIGURE 13.7

Bit error propagation.

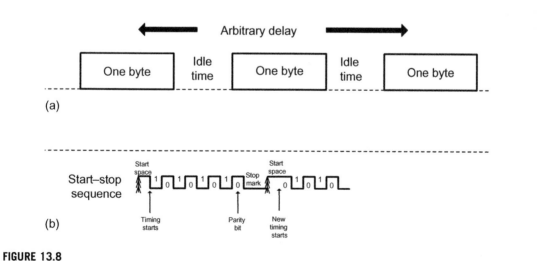

FIGURE 13.8

Asynchronous transmission: (a) depicts timing, while (b) displays tagging sequences.

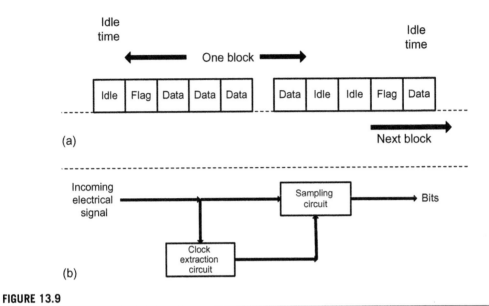

FIGURE 13.9

Synchronous transmission: (a) depicts data blocks and (b) displays bit recovery.

- *Synchronous transmission mode* (STM): While similar to ATM, STM (Figure 13.9) is carried out under the auspices of a timing source. All bit transactions are carried out with clock synchronization regardless of the bytes they belong to. There are not start–stop bits and the data is transmitted in a continuous stream. While this method is quite effective for large mainframe computers with scheduled data transfers, ATM remains the method of choice in the subsea industry.

Figure 13.10 provides a human voice analogy of synchronous versus asynchronous transmission. Data is transmitted in discrete information packets (called "frames"). Within each frame (Figure 13.11) are the various tags, flags, and addresses used for messaging and delivery along with the actual message wrapped in the frame.

Data is transmitted to its destination through a data transmission channel. These data paths transport the electrical signal from the transmitter to the receiver. The channel is characterized by its bandwidth and signal-to-noise ratio (SNR) parameters. Further, baseband transmission is centered on this data channel and is modulated via a signal coding scheme (discussed above) and is typically a DC signal modulated by one of these schemes.

Anything that is nonrandom can be compressed through some type of data compression method. The purpose of data compression is to allow for reducing the number of actual bits transmitted while maintaining the message integrity, making the compression function a part of the transmission.

As a last consideration, communications may be further broken down into point-to-point transmissions or point-to-multipoint transmissions. These configurations will determine if a two-way communication link can be established or a more wide area "broadcast" arrangement is made.

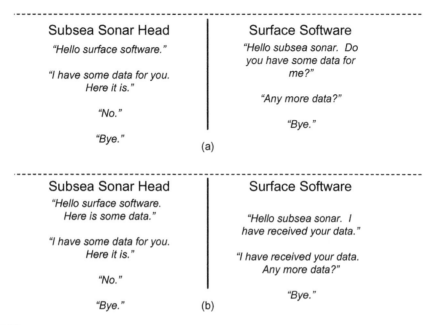

FIGURE 13.10

Synchronous versus asynchronous transmission: (a) depicts synchronous transfer and (b) displays asynchronous transfer.

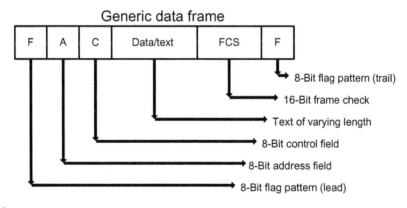

FIGURE 13.11

A generic data frame. Flag pattern = unique field. Data/text is the information.

13.2.3 Line characteristics

Next, we consider the actual transmission data pipeline in order to determine the strengths and weaknesses of each transmission method.

Transporting electrical signals over distances requires some type of conductive medium through which to send these signals. There are four types of transmission media:

1. Wire pair
2. Coaxial cable
3. Fiber-optic cable
4. Radio

Each of these media has its usages, strengths, and weaknesses. As radio frequency (RF) transmission has limited use in the subsea environment, our discussion will be limited to the metallic and optical fiber transmission media.

13.2.4 Metallic transmission media

The classic metallic transmission line has arrangements consisting of two parallel metallic conductors placed either side-by-side or coaxially. Each has its strengths and weaknesses. The actual conducting media can be any of a number of conductive substances (gold and platinum are much better conductors than copper, but the cost in all but a few cases is prohibitive for normal use). However, the most widely used conductive metal is copper. We will define a conductor here as any substance that conducts electricity, while a capacitor is two conductors separated by an insulator.

The conducting line characteristics can be described using what are termed "primary" and "secondary" parameters. Primary parameters (Figure 13.12) further allow us to describe the secondary parameters and are detailed as follows:

1. Series resistance per unit length (R) of the two conductors
2. Inductance per unit length (L)
3. Capacitance per unit length (C)
4. Leakage conductance per unit length (G) which primarily accounts for the dielectric losses (insulation losses are generally considered to be negligible)

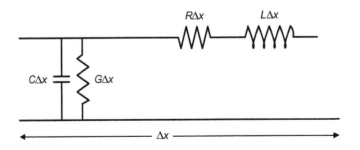

FIGURE 13.12

Primary parameters.

Secondary parameters describe two main characteristics: characteristic impedance and propagation constant. *Characteristic impedance* is the input impedance of an infinitely long transmission line and is given by the formula:

$$\mathbf{Z_0} = \sqrt{(\mathbf{R} + j\omega\mathbf{L})/(\mathbf{G} + j\omega\mathbf{C})}$$

where $\mathbf{R}$ is resistance of conductors in series per unit length, $\mathbf{L}$ is inductance per unit length, $\mathbf{G}$ is dielectrical conductance per unit length, $\mathbf{C}$ is capacitance per unit length, j is an imaginary unit, and ω is angular frequency. Or simply,

$$\mathbf{Z_0} = \sqrt{\mathbf{L}/\mathbf{C}}$$

for a lossless line since $\mathbf{R}$ and $\mathbf{G}$ go to zero.

If a transmission line is terminated in its characteristic impedance, its input impedance is $\mathbf{Z_0}$ (note: some texts reference the characteristic impedance function with the notation $\mathbf{Z_c}$) irrespective of its length. As an example, the characteristic impedance of a coaxial cable is typically 75Ω— hence the standard convention "75Ω Coax."

The *propagation constant* (γ) determines the attenuation and the phase change of the time varying sinusoidal wave traveling along the transmission line given by:

$$\gamma = \alpha + j\beta = \sqrt{(\mathbf{R} + j\omega\mathbf{L})(\mathbf{G} + j\omega\mathbf{C})}$$

where α is attenuation of the signal having frequency ω over a unit distance, and β is the phase change in the signal when it travels a unit distance. Both α and β are called the "attenuation constant" and the "phase constant" of the transmission line, respectively.

Theoretically, a linear transmission line should not distort an electrical signal (Figure 13.13). However, it can cause the received signal to differ by having been multiplied by some constant and be delayed for a period pending reception. Therefore:

1. The line will introduce attenuation distortion if its attenuation constant α varies with frequency (in a linear fashion) and
2. phase distortion if the phase constant β is nonlinear with the frequency (i.e., all frequency components are delayed by the same amount).

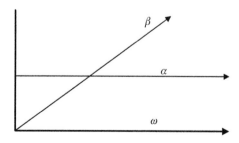

FIGURE 13.13

Attenuation and phase characteristics of distortion-less transmission.

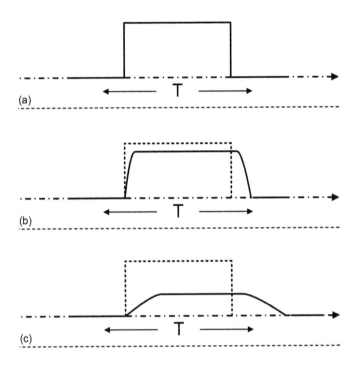

FIGURE 13.14

Pulse distortion over time. (a) Input pulse, (b) output pulse in short line, and (c) output pulse in long line.

Other problems crop up within the time domain as binary signals degrade as they travel down a line. For instance, a binary input pulse waveform is subject to both attenuation and phase shift as it travels down a conductor (Figure 13.14). This phenomenon can significantly degrade signal code without signal conditioning.

Crosstalk: When data transmission conductors are placed in close proximity to one another, they interfere with one another resulting in crosstalk due to three types of couplings (Figure 13.15) between the lines:

1. *Galvanic coupling* ((a) in Figure 13.15): common resistance between two lines (especially when the lines have a common return line)
2. *Capacitive coupling* (b): due to capacitance between the two lines
3. *Inductive coupling* (c): due to mutual inductance between the lines

This phenomenon is a substantial problem in the ROV application due to the varying geometrical arrangement of the tether (which holds the conductors) and the necessary proximity of the wires (in this case, both are encased within the tether jacket). The crosstalk problem is further exacerbated should the data communications line be placed adjacent to an AC power line (e.g., on larger AC-powered ROVs, the only vehicle telemetry option available is through a fiber since copper would induce significant crosstalk due to high-frequency AC-induced noise). As depicted in Figure 13.16, the crosstalk problem progresses either to the near end (termed "near end crosstalk,"

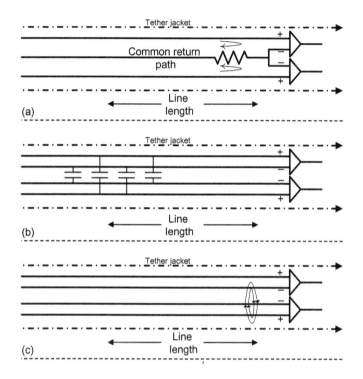

FIGURE 13.15

Crosstalk. (a) Galvanic, (b) capacitive, and (c) inductive.

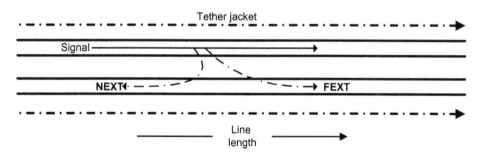

FIGURE 13.16

Near end crosstalk versus far end crosstalk.

or NEXT) or to the far end (termed "far end crosstalk," or FEXT). NEXT has little correlation to the tether/line length; however, the FEXT is directly proportional to the length of the line and forms a substantial problem in ROVs without proper wire shielding.

Twisted pair: A "balanced pair" wire set is a two-wire transmission line with two identical conductors having the same capacitance and leakage conductance to ground. The pair is electrically balanced specifically to reduce the crosstalk phenomenon from galvanic, capacitive, and inductive coupling. Further, wire pairs in close proximity induce current in adjacent lines, causing ghosting (or echoing) in other lines. This was a real problem in the early days of the telephone industry with other conversations echoing through to a telephone conversation and rendering the line useless. To curb this inductance impairment, twists in the amount of 6−36 turns per meter are induced into the line to inhibit the inductance field (hence the name "twisted pair," or TP). Also, shielding was introduced in the form of either foil or metallic braided mesh weaved around the TP to further mitigate inductance noise as well as the problem of crosstalk. What has come to be known as shielded twisted pair (STP) and unshielded twisted pair (UTP) are standard conductors in the tethers of all complex ROVs.

UTPs predominate in the telecommunications industry where conductors can be routed to avoid high inductance noise areas. Electronics Industries Alliance/Telecommunications Industry Association (EIA/TIA) standards have evolved for transmission of data over UTP cables for use in telecommunications and have categorized UTP cables in six categories (the first two categories are for voice transmission and low speed data while the remainder is suitable for data transmission). This standard is widely used in ROV cabling and has broad applicability for defining data throughput capabilities of twisted pair wires—the most common of which is the Category 5e (simply named "Cat 5e") and Category 6 ("Cat 6"). Cat 5e and Cat 6 cabling have a nominal maximum operating length of 100 m. However, in spite of the nominal certified cabling length, some OCROV manufacturers still make use of Cat 5e and Cat 6 cabling for telemetry and sensor data transmission over longer (>300 m) lengths of tether.

The useful bandwidth (Hertz) of a twisted pair varies with the type/quality of wire as well as its length. Typical UTP wiring used in the telephony application can support a 2 MHz bandwidth over about 1.5 km of length. Cat 5e twisted pair displays a 67 dB loss at 100 MHz over a length of 300 m (the typical maximum tether length of a small OCROV).

Coaxial cable: The other wire pair used in data communication is the coaxial cable, named for its arrangement with a single centralized conductor held in place by insulating material and surrounded by a cylindrical tube. This arrangement is advantageous when high frequencies are involved. The nominal (characteristic) impedance of the coax is 75Ω with specialized cabling available with 50Ω impedance. From the mid-1950s to the mid-1980s, coax cabling was widely used for long-distance telephony. These days, coax is mostly used for the last 30 m for cable television from the termination point at the curb to the television set. Frequency response for coax is exponential, making its loss increase drastically as frequency is increased—thus equalization is required in order to level out the frequency response. With all of these various issues with copper-based data transmission, it is a wonder that fiber is not exclusively used for all data transmission applications (especially in ROV applications) as it is superior in practically all parameters. However, coax is still used in some ROV tether applications for telemetry.

13.2.5 Optical fiber

Guiding light by refraction, the principle upon which all optical fiber transmission is based, has its roots in nineteenth century physics. Optical refraction was first introduced by Daniel Colladon and Jacques Babinet in France during the early 1840s through their demonstration of their "light fountain" or "light pipe" (Figure 13.17).

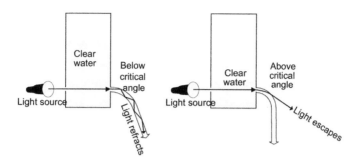

FIGURE 13.17

Basic function of Colladon and Babinet's "light fountain."

The waveguide, or "light pipe," principle can easily be demonstrated with a common bucket, some clear liquid, and a flashlight (a laser flashlight works best). Fill the bucket with 10–15 L of water, place a puncture hole at 0.5–1.0 cm in the bucket just above the bottom of the bucket (on its side), and shine the laser horizontally through the hole (if the bucket is opaque, mount a transparent window opposite the hole so that the light can shine down the outflow stream) as the water flows out of the bucket. The refractive index (the scientific notation for refractive index through a medium (e.g., air, water, and glass) is n) of water is 1.33, making the critical angle (θ_C) for the water–air interface as (arcsin of 1/1.33 in degrees) 49°. With the bucket full, the water pressure keeps the stream coming out of the bucket at an angle below 49° from the horizontal. However, as the water level decreases due to decreased water pressure, the angle of the water streaming out of the bucket will slowly increase until the "critical angle" of 49° when the light can no longer follow the water's path. At that point, the light will escape the water stream (Figure 13.18(a)). The same principle works for optical glass fiber (Figure 13.18(b)), only the critical angle changes to 41°.

The refractive index of transparent materials varies based upon the wavelength/frequency of the light passing through the medium. A prime example of this is white light passing through a prism (Figure 13.19) whereby the light is dispersed into its various component colors.

Some representative refractive indexes (n) and critical angles (θ_c) through various materials at a wavelength of 589 nm are provided in Table 13.1.

To compute the critical angle between two transparent media (e.g., air/water, water/glass, glass/glass (of different optical properties), etc.) simply compute the arcsine between the two refractive indexes and convert into degrees.

Optical fiber data transmission was developed by physicists as a superfast method of data transmission over long distances. The bandwidth of fiber-optic cabling can be measured in the terahertz (THz) range! Fiber has such huge frequency response that the entire RF spectrum can be conducted over one strand approximately the diameter of a human hair. The data rates achieved are such that one serial bit stream can be transmitted at 10 Gbps (gigabits per second) or through wave division multiplexing (WDM) at greater than 100 Gbps. Further, a single fiber can transmit more than a whole bundle of copper.

With optical fiber, light waves are conducted through extremely high-purity glass or plastic (due to its internal refractive properties), allowing the light to be channeled to its destination.

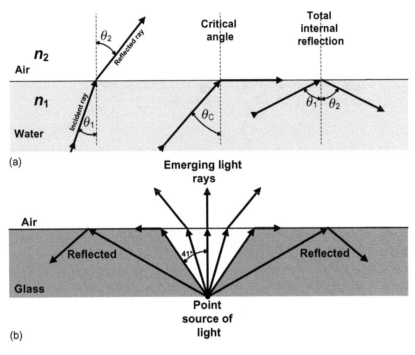

(a)

(b)

FIGURE 13.18

Depiction of (a) water–air interface and (b) glass–air interface.

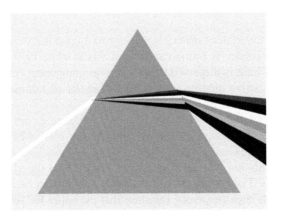

FIGURE 13.19

Light dispersion through glass splits into component colors.

Table 13.1 Refractive Indexes of Various Translucent Materials

Medium	Refractive Index (n)	Critical Angle (θ_c) with Air
Vacuum	1.000 (per definition)	Undefined
Air	1.0003	90°
Water	1.33	48.77°
Ethanol	1.36	47.35°
Benzene	1.501	41.79°
Fused silica	1.458	43.32°
Crown glass	1.52	41.15°
Polycarbonate	1.5849	39.13°
Diamond	2.42	24.41°

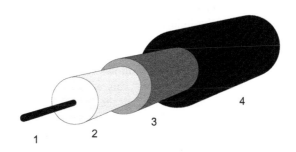

FIGURE 13.20

Components of a typical single mode optical fiber (1, core at 9 μm; 2, cladding at 125 μm; 3, buffer at 250 μm; 4, jacket at 400 μm).

The glass or plastic core is surrounded by a cladding of optical material (Figure 13.20) with differing refractivity index than the core, allowing the light to be "trapped" (termed "total internal reflectance") within the core of the fiber. The outer buffer protects the core and cladding from abrasion and moisture and is the coating that is stripped off during cable termination. The cladding is of a different refractivity index so as to either bend the light back toward the core (thus "steering" the light down the core's "channel") with single or graded index fiber or to totally reflect it within the core (for multimode fiber) to keep it encapsulated within the core.

The term "mode" with regard to optical fiber describes the path taken by a photon of light as it transits down the fiber. In multimode fiber, the photon will take any number of paths as it is bounced around within the core toward the receiver. This multimode technique is much cheaper than the other techniques as the fiber need only be coated with an imprecise reflective coating for "bouncing" the reflections down the fiber. The core and cladding sizes are typically 50 and 125 μm, respectively (Figure 13.21). With graded index fiber, the cladding is layered optical material with gradually changing characteristics to bend the light back down the channel in a pattern

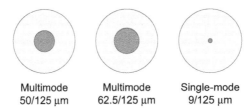

Multimode
50/125 μm

Multimode
62.5/125 μm

Single-mode
9/125 μm

FIGURE 13.21

Typical sizes of core and cladding of fiber-optic lines.

that approximates a parabolic shape. This method is more precise than multimode, but allows for several different paths the light may follow as it travels down the fiber. Multimode and graded index fibers are used infrequently in the subsea environment. Single mode fiber is by far the most precise as the light follows only one channel (a single mode) down the line, allowing for precise transmission and reception timing. The fiber core is in the 9 μm range; with a line of this small size, the light has little chance to change paths as the channel is too narrow. Further, the cladding is of a layered consistency to assure the light is bent back toward the fiber axis as it is channeled down this narrow path.

A light pulsed down the fiber of a multimode or graded index fiber is allowed to take any number of paths down the channel (Figure 13.22). This allows for the pulse at the transmission end to be received at differing times at the receiving end. This causes a pulse distortion termed "modal dispersion," and it increases in severity as the length of the fiber increases. This is why most multimode fiber systems are limited to short runs. Modal dispersion is limited with graded index fiber and virtually eliminated with single mode fiber. For the subsea application, the single mode fiber is the preferred technique for transmitting data over long lengths of umbilical and tether.

Also, as the frequency of light varies between the chosen modes, a further dispersion is encountered. This dispersion, termed "chromatic dispersion," is dependent upon the frequency of light employed and, due to increased errors, limits the bit rate. With chromatic dispersion, the light is propagated at differing speeds down the fiber based upon the wavelength emitted by the source. The arrival time of the light at the other end of the fiber varies (due to the differing propagation speeds) causing a higher BER due to degradation of the wave pattern.

Fiber-optic light transmission can be loss or dispersion limited based upon the loss of optical energy or data error propagation through the cable. The link can withstand losses to the point where the receiver can no longer detect the light in a coherent enough fashion that intelligence can be discerned. These losses are a function of:

1. Type of fiber
2. Transmitted wavelength of the signal (note: in fiber optics, wavelength is used as the standard measurement metric for frequency as opposed to "frequency")
3. Bit rate/BER
4. Power output of the light transmitter
5. Sensitivity of the optical receiver

Loss limitation is due to impediments to light energy transmission from impurities in the conducting medium, connector or splice losses, and/or power of the output transmitter device versus the

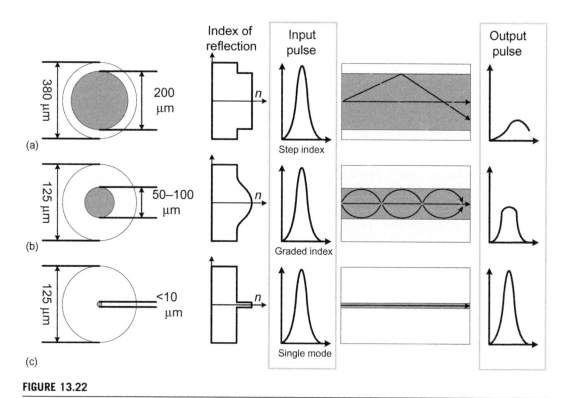

FIGURE 13.22

Light paths (termed "modes") of optical fiber for (a) multimode, (b) graded index, and (c) single mode.

sensitivity of the receiver. The two general types of losses (termed "attenuation") are due to the combination of two factors—scattering and absorption. Scattering is the largest of the attenuation factors and occurs when the photons of light collide with the individual atoms in the fiber. The photons of light can be deflected slightly, thus bouncing off the cladding and continuing down the line, or they can be bounced back to the transmitter. Scattering is a function of the wavelength and the longer the wavelength, the lower the scattering loss. Absorption loss, however, is due to the absorption of light into the molecules within the glass and the light energy is converted into heat energy (which cannot be sensed by the optical receiver). The main components in the fiber that absorb the light energy are the dopants and residual OH+ impurities used during the fiber manufacturing process to manipulate the refractivity index of the fiber; therefore, the absorption (Figure 13.23) is reactive within a narrow range of wavelengths (1000, 1400, and >1600 nm). As a result of this absorption phenomenon, standard wavelengths used in fiber-optic data communications equipment are 850 or 1300 nm (for multimode (MM)) and 1310 or 1550 nm (for single mode (SM)).

"Dispersion limitation" involves the accumulation of errors (excessive BER) due to signal corruption as the length of the fiber conductor within the vehicle's umbilical/tether increases, thus degrading the optical wave pattern. The higher the transmitted optical power, the more likely it will be received on the far end. But the receiver has a certain minimum optical power for sensing the coherent light signals; therefore, BER will decrease as the received optical power is increased until

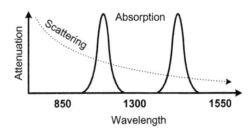

FIGURE 13.23

Combined attenuation within the typical optical fiber.

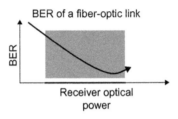

FIGURE 13.24

Bit error rate (BER) as a function of receiver optical power.

the optical power begins to overpower the receiver (Figure 13.24). For our purposes, BER decreases proportionally with the increase in received optical power.

Typical fiber types and specifications are provided in Table 13.2.

13.2.5.1 Enough with the theory! What's important?

Fiber-optical data communication systems used in ROVs (Figure 13.25) consist of the following components:

1. The optical transmitter (LED or laser)
2. The cable (MM-GI, MM-SI, SM, POF)
3. Connectors (temporarily join strands of fiber)
4. Splices (permanently join strands of fiber)
5. Hardware (for mounting of connectors, splices, and routing cables)
6. Test equipment (OTDR, test light, etc. used to test continuity and performance)

The *transmitter* (Figure 13.26) comes in two varieties—light emitting diodes (LED) and semi-conductor lasers (laser). The LED transmitter produces incoherent light with a fairly wide spectral width (30 − 60 nm) and is of a much lower performance than the laser and is generally used only with MM fiber systems. As the SM fiber technique will typically be employed in the subsea industry, the focus will be on the laser transmitter with the use of SM fiber. In the laser diode, an electric current is injected (as opposed to optically pumped for the LED) into a p−n junction, thus emitting light at a very narrow-band and discrete frequency/wavelength in a coherent fashion.

Table 13.2 Specification Summary of Fiber Types

Multimode Graded Index (MM-GI)

Core/Cladding	Attenuation @850/1300 nm	Bandwidth	Application
50/125	3/1 dB/km	500/500 MHz/km	Gb Ethernet LAN
62.5/125	3/1 dB/km	160/500 MHz/km	Typical LAN fiber
100/140	3/1 dB/km	150/500 MHz/km	Obsolete

Multimode Step-Index (MM-SI)

Core/Cladding	Attenuation @850 nm	Bandwidth	Application
200/240	4–6 dB/km	50 MHz/km	Slow LAN

Single Mode (SM)

Core/Cladding	Attenuation @1310/1550 nm	Bandwidth	Application
9/125	0.4/0.25 dB/km	∼100 THz!	Long high-speed cables

Plastic Optical Fiber (POF)

Core/Cladding	Attenuation @650 nm	Bandwidth	Application
1 mm	∼1 dB/m	∼5 MHz/km	Very short links

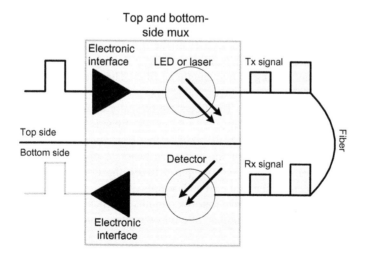

FIGURE 13.25

Components of a fiber-optic system.

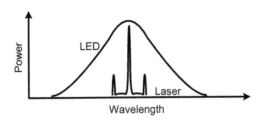

FIGURE 13.26

Comparison of LED to laser frequency response.

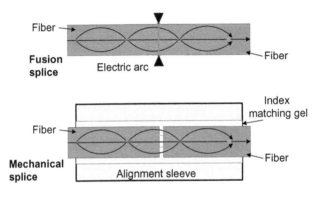

FIGURE 13.27

Comparison of fusion versus mechanical splicing.

Once the light is transmitted in the fiber, the losses are subject to attenuation issues as outlined above. Between the transmitter and the receiver are temporary and permanent connectors and splices, respectively, that are introduced as part of the installation process. Each of these connections introduces optical power losses for various reasons.

The *splice* is a permanent joining of two lengths of fiber. The two types of splices (Figure 13.27) are fusion and mechanical. The fusion splice "welds" the two fibers together with the use of an electrical arc to fuse the two parts together. The fibers are typically aligned mechanically with an automatic splicer since the alignment tolerance of a 9 μm SM fiber is very tight indeed. This type of splice has several advantages including high strength, low loss, and low back reflection through the splice. Mechanical splices make use of mounting hardware to align the fibers; the final link is through index matching gel, thus bridging the gap between the fibers through the gel. The fusion splice is superior in all aspects, other than cost and complexity.

Connectors (Figure 13.28) temporarily join two lengths of fiber in a variety of connector types. Connectors are chosen based upon a number of environmental, functional, and fault-tolerance factors including:

1. Expected optical power loss
2. Operating environment (temperature, humidity, mechanical load, etc.)

FIGURE 13.28

Connector examples: (a) ST and (b) SC connectors.

3. Reliability
4. Fiber internal reflective properties
5. Ease of termination
6. Cost

Both the splice and the connector are subject to a range of transmission issues inducing optical losses across the junction between the two fibers. These issues are practically all mechanical in nature whereby there is a mismatch of optimal fiber end-to-end mating. The mating should be as close as possible to a perfect end-to-end matching with clean and clear end-caps, but in practice a combination of contaminants, less-than-parallel end points, and fiber size mismatch cause a host of losses, as shown in Figure 13.29.

Cable splicing and connector mating are quite simple and reliable, once proper training and practice are received. It is highly recommended to take a basic hands-on fiber termination course on the nuances of fiber terminations. The connection process, while simple, is quite unforgiving for lack of attention to detail.

13.2.5.2 Pay attention—this part is important!

The total power loss through the conductive medium (fiber/connector/splice/etc.) versus the optical power input is its "optical power budget" (Figure 13.30). As a rule of thumb in the ROV industry (based upon the industry standard Focal Technologies 900-series fiber-optic multiplexer's power budget), a signal loss of less than 20 dB between the optical transmitter and receiver will be sufficient for data transmission—that is, above (greater than) 20 dB loss, a typical fiber-based ROV system will begin experiencing telemetry and sensor losses due to exceeding the optical power budget.

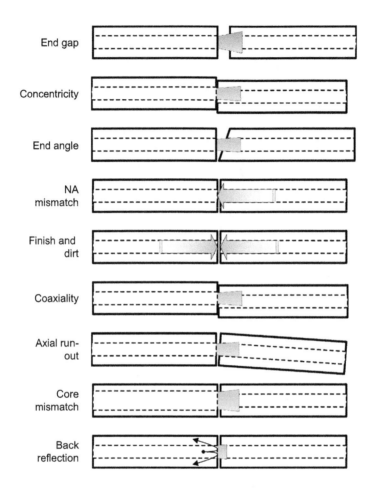

FIGURE 13.29

Splice and connector issues causing losses.

13.2.5.3 Troubleshooting fiber-optic systems

The fiber tool kit of the ROV technician should include the following testing equipment items needed for troubleshooting and verifying the optical power through the cable: a source and power meter, reference test cable (as a baseline), fiber tracer/visual fault locator, cleaning/polishing materials, connector inspection microscope, and an optical time-domain reflectometer (OTDR, Figure 13.31). Line continuity is quickly checked with the visual fault locator. A total line power loss is ascertained with the power meter. Once it is determined that the loss is above the acceptable level, the OTDR is able to isolate the location along the line of the fault. This is quite a simple process, but it will take some training and hands-on experience to quickly and effectively troubleshoot fiber issues. With modern tools and training, fiber optics' usage in ROV systems has become the staple of the industry. The data pipeline is huge, the transmission system is robust, and

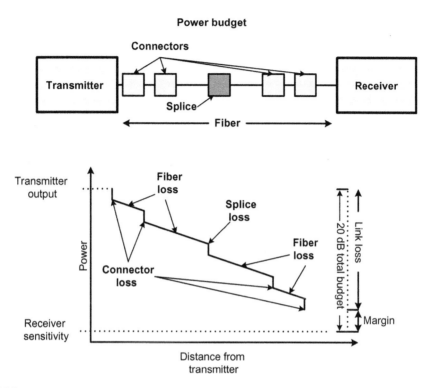

FIGURE 13.30

Optical power budget and various losses between transmitter and receiver.

(with rapid development of tools and techniques for troubleshooting) field techniques allow for reliable telemetry with subsea components.

13.2.6 Radio

As stated above, RF communications with subsea vehicles have a very limited usage due to the high attenuation of RF waves through water. However, new technologies are evolving for data transmission over short ranges through water. Look for further technological movement of subsea telemetry based upon RF as the deepwater minerals extraction industry develops automation techniques.

13.2.7 The ubiquitous decibel

The decibel (dB) is a unit of relative power. The base unit is the "bel" with the decibel as 1/10 of the unit of bel (named in honor of Alexander Graham Bell). The dB is a common unit of measurement and is ubiquitous in the telecommunications, electronics, and broadcasting industries.

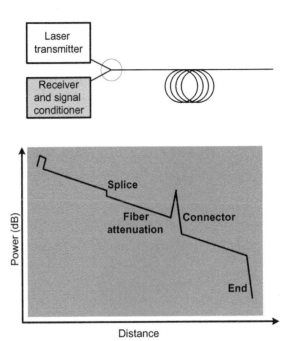

FIGURE 13.31

OTDR diagram.

In simple terms, the dB is the ratio between two power levels expressed in logarithmic terms with relation to some reference level. For example, if given two known power levels, P2 and P1, the relative value of P_2 with respect to P_1 in dB is given by:

$$dB = 10 \log_{10}(P_2/P_1)$$

So, if P_2 is 10 times P_1, the dB gain calculated will be 10 dB (or P_2 is 10 dB more than P_1). Thus, we can express power levels in both gain (through an amplifier) and loss (through an attenuator).

Let us throw some numbers into this equation with an example—a 10 mW signal is transmitted through a UTP cable and is received at a 5 mW power level. What is the attenuation of the conductor?

dB = 10 $\log_{10}$ (P_2/P_1)
P_2 = 5 mW, P_1 = 10 mW
Attenuation = 10 $\log_{10}$ (5/10) = −3 dB

To put this into perspective:

1. Human audio response (Table 13.3)
2. Fiber optics attenuation over a 1 km fiber system (Table 13.4)

Table 13.3 Human Audio Response

Sound Type	Sound Power Level
Hearing threshold (reference)	0 dB
Whisper	+30 dB
Air-conditioner	+50 – 70 dB
Two-person conversation	+50 – 70 dB
Rush-hour traffic	+60 – 85 dB
Jet aircraft	+120 – 140 dB
Led Zeppelin concert	>+130 dB

Table 13.4 Fiber-Optic System Component Attenuation

Source Type	Power Level
Transmitter (reference)	0 dB
Fiber attenuation	−0.5 dB/km @ 1310 or 1550 nm
6 Connectors	−0.75 dB per connector or −4.5 dB
2 Splices	−0.3 dB per splice or −0.6 dB
Total loss from transmitter to receiver	−5.6 dB

Table 13.5 Reference dB Power Levels for Optical Circuits

dBm	Watts	dBm	Watts
+10	10 mW (milliwatts)	− 10	100 μW
+3	2 mW	− 20	10 μW
0	1 mW	− 30	1 μW
−3	500 μW (microwatts)	− 40	100 nW (nanowatts)
−6	250 μW	− 50	10 nW

This is quite a handy measurement unit. In telephony and optical power, the unit of measure is dBm (decibels referenced to 1 mW), in radio systems the dBW (decibels referenced to 1 W) is used, in sonar systems the acoustic dBSPL (or sound power level) unit is used (typically measured at the 1 m distance from the transducer), and in television electronics systems dBmV (decibels referenced to 1 mV) is used. Some reference optical power levels are listed (Table 13.5) along with an optical power conversion chart (Figure 13.32).

Some rules of thumb to memorize with regard to the dB unit of measure are:

1. An increase factor of 10 in power is a 10 dB change.
2. A change in power level by a factor of 2 is approximately a 3 dB change.
3. Halve the power of a circuit or transducer results in a −3 dB change.
4. Any transmission media will function as an attenuator decreasing the power level.

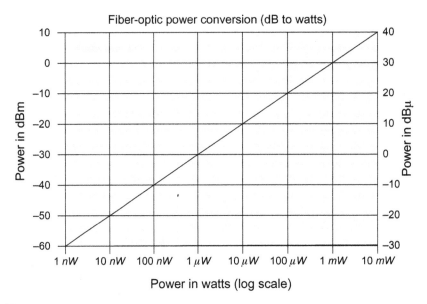

FIGURE 13.32

Fiber-optic power conversion chart.

5. Any amplifier should increase the power level (positive dB gain).
6. Reference of 10 dB = 10X
7. Reference of 3 dB = 2X
8. Reference of 0 dB = 1X
9. Reference of −10 dB = 0.1X

13.2.8 Baseband transmission

Baseband transmission is described as the raw electrical signal transmission typically surrounding some zero value. A signal is considered "baseband" when it includes frequencies at or near 0 Hz up to the highest frequency in the signal. For instance, a baseband digital transmission signal emanates a series of 0s and 1s from a PC to some other device. For an analog device, such as a telephone, human voice impresses an alternating current over a conductor of a phone line to transmit sound to the receiver. Baseband signals can be distance limited; therefore, some type of modulation scheme is necessary in order to condition the signal for long-distance transmission. The physical device used for modulating and demodulating a carrier signal is typically colocated on to the same device called a modem (or *mo*dulator/*dem*odulator).

13.2.9 Modulation

IEEE defines *modulation* as "a process whereby certain characteristics of a wave, often called the carrier, are varied or selected in accordance with a modulating function." Modulation is the process

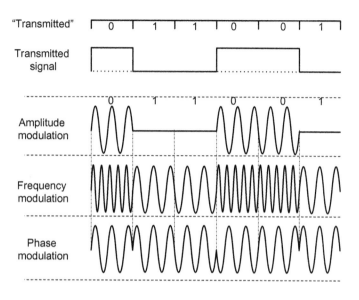

FIGURE 13.33

Comparison of the three methods of modulation.

of varying a high-frequency waveform (or "carrier signal") with a modulated signal containing the information to be sent (as opposed to the baseband signal centering its frequency on some zero value). On the transmitting side, a radio carrier is generated with the carrier described by a central frequency. The central frequency carries no useful information until such time as a time varying signal is impressed upon the central carrier frequency. *Modulation* is the process of putting that useful information upon the carrier frequency while *demodulation* reverses the process to bring the signal over the carrier back to baseband (or some other discernible format) for dissemination to the end user. Either a sinusoidal or a square pattern signal can be used for transmission.

The three methods of modulation (depicted graphically in Figure 13.33) are:

1. *Amplitude modulation* (AM) carrier is varied in amplitude according to the information in the baseband signal.
2. *Frequency modulation* (FM) carrier is varied in frequency with the baseband signal.
3. *Phase modulation* (PM) carrier is varied in phase in accordance with the information in the baseband signal.

13.2.10 Multiplexing

Multiplexing is a communication technique used so that multiple communications channels can fit onto a common line. In the early days of trans-Atlantic telegraphing, only a single message could be tapped out at a time—that was an extremely expensive message! Later, the messages were modulated to different carrier frequencies and multiplexing was born.

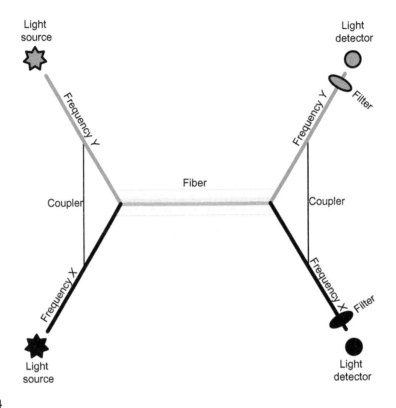

FIGURE 13.34

Example of FDM with a fiber-optic WDM simultaneous transmission.

As the demand for more and more bandwidth in our information-hungry industry expands, there are just two options to meet the higher requirements: install more cabling (thus substantially increasing cost) or pack more bandwidth onto individual lines through multiplexing. Multiplexing is quite common in the ROV industry (especially in fiber-optics communications). The two general methods of multiplexing are:

1. multiplexing in the frequency domain and
2. multiplexing in the time domain

Frequency division multiplexing (FDM) makes use of various nonoverlapping frequency slots in order to carry information simultaneously over a single line. As fiber optics uses the moniker "wavelength" (as opposed to "frequency" in all other medium), the acronym WDM is used. Both are the same concept of varying the frequency for capturing the communications channels since the wavelength of a signal is defined as the inverse of its frequency. Figure 13.34 depicts a WDM diagram breaking out two communication channels, coupling them for transmission and then decoupling them at the receiving end. For an example of a typical industry standard course wave division multiplexer (CWDM) board, the Focal 907-CWDM spaces its channels 20 nm apart allowing for further data pipelines over the same fiber. The 8-channel version of the board has communications at 1471, 1491, 1511, 1531, 1551, 1571, 1591, and 1611 nm.

Time division multiplexing (TDM) makes use of assigned time slots to allow information to take its turn down the communications line. This technique, however, has not gained much traction in the subsea industry.

13.2.11 Binary signals

The essence of the binary signal is the concept of discreteness. Whereas the analog signal implies continuity of signal (as in the form of a sine wave of a constantly varying nature), the digital signal is quantized as either on/off, yes/no, up/down, 1/0, or presence/absence of voltage (to name a few). The basic unit of information, termed a "bit", is encapsulated within this binary information stream and will be covered more extensively in Section 13.3.1.

13.2.12 Directionality

There are three possible directionality terms of data exchange:

1. Transfer in both directions simultaneously (termed "full duplex")
2. Transfer in both directions with only one direction transmitting at a time (termed "half duplex")
3. Transfer in one direction only (termed "simplex")

The synchronous versus asynchronous discussion is often confused with the directionality question; they are mistakenly viewed as synonymous.

13.3 Communication

As mentioned above, the communication function differs from the transmission function in that communication is the actual information being transferred from transmitter to receiver. In this section, the elements and symbols of the information being communicated will be examined, thus leading into the final section on typical communications standards used within the subsea industry.

13.3.1 Of bits and bytes

The *binary digit* (shortened to "bit") is the most elemental unit of information and in data communications is represented as either a 1 or a 0. In and of itself, the single bit does not carry much information, but it allows for very distinct states of being. For instance, a light is either on or off, a motor running or not and (in Schrödinger's example, for us physics fans) the cat is either alive or dead. The bits come alive as the information is expanded into further combinations of bits. While one bit has only two possible states (1 or 0), two bits have four possible states (00, 01, 10, and 11) and three bits have eight possible states (000, 001, 010, 100, 110, 011, 101, and 111). Therefore, the possible states of bits can be expressed as 2^n number of states. The magic number for data transmission in the subsea world for serial communications is 2^7 number of possibilities (2^7 is 128 possible states or characters) with one bit as a checksum. This leads us into the worldwide standard

				b7	0	0	0	0	1	1	1	1
Bits				b6	0	0	1	1	0	0	1	1
				b5	0	1	0	1	0	1	0	1
								Column				
b4	b3	b2	b1		0	1	2	3	4	5	6	7
0	0	0	0	0	NUL	DLE	SP	0	@	P	`	p
0	0	0	1	1	SOH	DC1	!	1	A	Q	a	q
0	0	1	0	2	STX	DC2	"	2	B	R	b	r
0	0	1	1	3	ETX	DC3	#	3	C	S	c	s
0	1	0	0	4	EOT	DC4	$	4	D	T	d	t
0	1	0	1	5	ENQ	NAK	%	5	E	U	e	u
0	1	1	0	6	ACK	SYN	&	6	F	V	f	v
0	1	1	1	7	BEL	ETB	'	7	G	W	g	w
1	0	0	0	8	BS	CAN	(	8	H	X	h	x
1	0	0	1	9	HT	EM	)	9	I	Y	i	y
1	0	1	0	10	LF	SUB	*	:	J	Z	j	z
1	0	1	1	11	VT	ESC	+	;	K	[	k	{
1	1	0	0	12	FF	FC	,	<	L	\	l		
1	1	0	1	13	CR	GS	-	=	M	]	m	}
1	1	1	0	14	SO	RS	.	>	N	^	n	~
1	1	1	1	15	SI	US	/	?	O	_	o	DEL

FIGURE 13.35

ASCII character set.

of character sets called the American Standard Code for Information Interchange (or ASCII). The basic unit of serial data is termed the "byte" and is normally eight bits long (seven bits with a checksum for error control). This has evolved into the *standard data byte* of "8-N-1" for eight (8) data bits, no (N) parity bit, and one (1) stop bit. This abbreviation is typically given along with the line speed (e.g., "9600/8-N-1" for 9600 baud (baud = symbols per second)). The standard serial byte will be examined further below in the section on standard serial protocols.

13.3.2 Data representation

The ASCII character set was first issued for use with teletype machines in the early 1960s and has evolved as the worldwide standard of symbols for use in telecommunications. The character set displays 2^7 possible characters that represent both the characters themselves as well as machine states for use in machine control (Figure 13.35). As an example of characters, 1000001 = "A" while 1000011 = "a." For a machine control example, 0000000 = "NUL" while 1101100 = "ESC."

With this table and a little ingenuity, one can refer back to Section 12.1.3 and derive the exact digital stream to send over a line for a GPS receiver. Looking at the first seven characters in a typical NMEA 0183 transmission for a GPS receiver ("$GPGGA,") in sequence, the result is:

0010010 1110001 0000101 1110001 1110001 1000001 0011010

During troubleshooting of an input device (e.g., a GPS, heading indicator, or other sensor outputting a digital stream) on a Windows® platform, the ASCII stream can be easily viewed with the HyperTerminal tool by setting the output to the proper data speed and protocol (again, typically 9600, 8-N-1) and reception with the same speed/protocol combination.

13.3.3 Error control

As shown in Section 13.2.2, signaling errors can crop up at unexpected locations. A totally error-free communications channel is impossible as it defies the second law of thermodynamics. All things tend toward the disorganized, making some measure of errors unavoidable. In the early days of telegraph, high error rate lines required the operator to transmit critical words twice, thus halving the transmission speed in order to reduce error rates. This doubled the effective cost of data transmission per unit communicated.

As a result of these inherent errors, modern standard data communications protocols have evolved with use of the "parity check" (also referred to as the "vertical redundancy check" or "VRC"). Error detection identifies bit errors that are received while error correction corrects these bit errors received at the far end of the line. Standard data streams transmitting the ASCII character set involve a 7-bit character set with the last bit being the parity bit for a total of an 8-bit byte. This sacrifices 1/8 of the raw data bandwidth to error control as redundancy. For a further explanation on error correction techniques, please refer to a basic text on the subject.

For a nice low-noise line with high transmission clarity, a low error rate can be achieved. But for a high-noise environment, the SNR will degrade to such a point that the error rate will slow the transmission due to error correction hogging up the computer processor power. Examples of this phenomenon are (i) a UTP line running along an AC power source in a subsea tether or (more typically) (ii) a data line running along the side of an AC power cord (or thermal noise, generator circuit noise, or any of a host of noise spikes) in the control room.

13.3.4 Protocols

IEEE defines a protocol as "a set of rules that govern functional units to achieve communication." Certain rules, procedures, and interfaces are established within the data communications framework in order to get the most out of the network. These rules establish the common language of the network and are termed "protocol." These protocols are based upon the network topology/architecture, transmission media, switching, and network hardware/software selected.

Typical basic networking protocol functions (for both connection and connectionless sessions) are:

- *Segmentation and reassembly* (SAR): the breaking up of messages or files into blocks, packets, or frames into quantized (i.e., nominally measured) packets. Some standards refer to this as "packets" of information, others as "cells" (as used in asynchronous transfer mode—ATM) of information. On the receiving end, the packets or cells are reassembled in the reverse process of segmentation for putting the messages back together for dissemination.

- *Encapsulation*: This function is the process of adding headers to describe the information packet/cell.
- *Connection control*: The three basic steps of a connection controlled in this sequence are (i) connection establishment, (ii) data transfer, and (iii) connection termination. In more sophisticated protocols, various aspects of error control, connection interrupt/recovery, and other session parameters can be controlled.
- *Ordered delivery*: The packets/cells are sent (much like mailed letters sent from the same address to the same subsequent address through the mail) in sequence, but the packets may not arrive in the same order as transmitted. This function reassembles the data into ordered and sequential segments for processing through use of a numbering plan with a simple numbering sequence.
- *Flow control*: This function manages the data flow from source to destination assuring the data does not overflow the buffers and memories along the way while maintaining the network components at full capacity.
- *Error control*: The networking error control function is the "lost and found" service of the computer world. It allows for the recovery of lost data packets identified during the reconstruction process to assure a complete data set on the receiving end. This function is broken down into four possible subfunctions, (i) numbering of packets, (ii) incomplete octets (number of bits—e.g., less than an 8-bit set), (iii) error detection/correction, and (iv) receipt acknowledgment back to the data source.

13.4 Standard protocols

Standards organizations are crucial in data communications platforms. These organizations allow disparate manufacturers of sensors and processing equipment to design to the same languages for interoperability. This benefits the consumer (specifically, the operators of ROV equipment) by allowing all of the technicians to train to the same standards and equip to the common interface. As an example, serial communications devices can exchange information on a common connector speaking in the same language based upon just a few protocols, such as RS-232, RS-422/485, or Ethernet.

The most common serial protocols in the subsea industry parallel the standards in the computer industry. The most common of these used with ROVs are covered below.

13.4.1 TIA/EIA standards

The Telecommunications Industry Association (TIA) is a Washington, DC (USA)-based standards organization sanctioned by the American National Standards Institute (ANSI) to develop voluntary, consensus-based industry standards for a wide variety of telecommunication products. These products include radio equipment, cellular phones/towers, data terminals, satellite equipment, telephone terminal equipment, VoIP devices, mobile device communications, multimedia multicast, and machine-to-machine communications (among others). The Electronics Industries Alliance (EIA) was a trade organization that is typically mentioned in parallel with the TIA, but it ceased operations in February 2011. The common communications standards mentioned below are from consensus standards promulgated by the TIA/EIA over the past few decades. These are "Recommended Standards" that are cataloged as "RS" followed by the standard number (e.g., RS-232) that were jointly adapted by the two organizations. These standards are widely used today for serial data communications in the subsea industry.

13.4.2 RS-232

The EIA232 standard (commonly referred to as "RS-232") is actually a series of standards for serial binary single-ended data and control signals for linking data terminal equipment (DTE) and data circuit-terminating equipment (DCE). See Figure 13.36 for a graphical depiction of the link. The standard specifies signal voltage/timing/function, a protocol for information exchange and wiring/pin-out for mechanical connectors.

If all devices exactly followed the RS-232 standard, all cables would be identical with no possibility that an incorrectly wired cable could be used. Unfortunately, the standard varies depending upon a number of factors.

Signal functions in the EIA232 (RS-232) standard can be subdivided into six categories:

1. Signal ground and shield
2. Primary communications channel (including flow control signals)
3. Secondary communications channel (for control of a remote modem, requests for retransmission due to errors, and control over the setup of the primary channel)
4. Modem status and control signals (modem status for establishing voice or data communications channel)

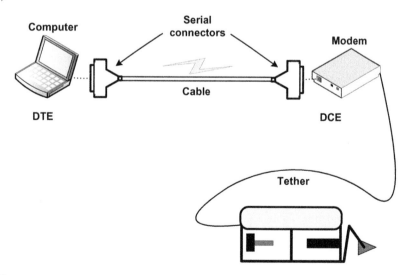

FIGURE 13.36

Typical RS-232 component link.

5. Transmitter and receiver timing signals (providing timing information for the transmitter and receiver—which may operate at different baud rates)
6. Channel test signals

The connectors for the RS-232 standard are traditionally of the 9-pin or 25-pin variety, but any number of connector types may be used as long as the function protocols are closely followed. The functions within the connector require the cabling technician to examine the functions for proper wiring of the communications channel in order for the transmit (Tx) and receive (Rx), as well as other typical functions, to be matched. Refer to the manual of each component in order to assure wiring continuity.

The EIA232 standard is applicable to data rates of up to 20,000 bps (the typical upper limit is 19.2 kb, or "kilo baud"). Typical baud rate values are 300, 1200, 2400, 9600, and 19,200 baud. Also, a maximum length for an RS-232 link is not specified in the standard, but typical maximum lengths are 8 m. Some ROV manufacturers stretch their RS-232 lengths up to 75–100 m with signal loss. As such, the RS-232 protocol is length-limited and is typically only used in short runs at the surface or through the vehicle's telemetry system when communications are to/from a fiber-optic link (i.e., short cable run from the sensor or surface computer to the fiber-optic multiplexer). The RS-232 standard, however, has been essentially replaced by the universal serial bus (USB) standard for connecting to DCE.

13.4.3 RS-422/485

The RS-422/485 protocol is part of a set of standards using "differential signaling." Differential signaling is a technique whereby complimentary information is sent electrically over two wire pairs (termed a "differential pair"). The signals are compared at the receiving end. External interference tends to affect both wire pairs simultaneously; therefore, the information is embedded in the difference between the wires (thus canceling out the surrounding noise). This technique is used in the RS-422/485 standards along with other standards including USB, Ethernet over twisted pair, serial digital interface (SDI), high-definition multimedia interface (HDMI), and Firewire.

The RS-422 standard defines the signal level for extending the range of serial devices to up to 1500 m. Often times, an RS-422 device is used in conjunction with an RS-232 device in order to extend the effective range of the signal. The RS-422 standard allows for unidirectional/nonreversible, terminated or nonterminated transmission lines, point to point, or multidrop. In other words, the RS-422 standard allows for multiple receivers while the RS-485 allows for multiple drivers/ "talkers" for use in balanced digital multipoint systems.

The RS-422 standard is used extensively with sensors over copper for the copper-based OCROV systems as a simple extender while the RS-485 protocol is used for two-way communications with various sensors (such as single-beam sonar systems). A comparison of the RS-422 and RS-485 protocols is provided in Table 13.6.

13.4.4 Ethernet

The Ethernet is an ANSI/IEEE (as adopted by ISO/IEC (International Electrotechnical Commission) in standard 8802-3) standard and is covered in IEEE 802.3 (as amended). It is the most widely used

Table 13.6 Comparison of the RS-422 Versus RS-485 Standards

Parameter	RS-422	RS-485
Physical media	Twisted pair	Twisted pair
Network topology	Point-to-point/multidrop	P2P/multidrop/multipoint
Max devices	10 (1 driver/10 receivers)	32–256 devices
Max distance	1500 m	1200 m
Mode of operation	Differential	Differential
Max baud rate	100 kbs to 10 Mbps	100 kbs to 10 Mbps
Voltage levels	± 6 V	−7 V to +12 V
Available signals	Tx + , Tx − , Rx + , Rx − (full duplex)	Tx+ /Rx+ , Tx− /Rx− (half duplex)
		Tx+ , Tx− , Rx+ , Rx− (full duplex)
Connector types	Any	Any

local area network (LAN) protocol worldwide due to its simplicity, reliability, and relatively low cost.

As mentioned in Section 13.2.2, the essence of Ethernet transmission in through sending of frames of information through a "medium access control" (MAC) protocol for directing data traffic. The theory behind the Ethernet relay system is within the OSI model physical layer whereby the individual units of data are switched from dealing with bytes to directing traffic on ever-increasing sizes (and numbers) of frames. The original Ethernet started with a data rate of 10 Mbps and has now grown to over 100 Gbps (and growing!).

Ethernet was originally designed for transmission over coaxial cables but quickly adapted to the more cost-effective and malleable medium of twisted pair. Ethernet can now be transmitted over copper or fiber with fiber optics, gaining the most ground in data rates due to its huge capacity.

The version of Ethernet is designated based upon the cable-carrying capacity and medium used. For instance, a 10Base-T1 line denotes 10 Mbps transmitting a baseband signal over two twisted pair. Speeds of 10 Mbps are considered standard Ethernet while 100 Mbps is "fast Ethernet" and >1000 Mbps is considered "gigabit Ethernet." Some other transmission standards are listed in Table 13.7.

Ethernet transmission over varying lengths of cable are distance limited without some type of signal amplifier. Check the standard being used in order to determine if your cable tether length is suitable for Ethernet transmission without degradation of the signal.

13.4.5 Universal serial bus

The USB standard has become ubiquitous for connecting and powering small devices of a broad range (from cameras to smart phones to gaming consoles—it has even replaced many power chargers for low-powered devices) over very short distances (<5 m). The latest iteration of USB is 3.0 (or later). With its release in November 2008, the standard has a theoretical transmission speed of up to 5 Gbps. The typical usage of the USB in subsea operations is for connecting serial devices to computing equipment.

Table 13.7 Sample Ethernet Transmission Standards

10Base-T	10 Mbps baseband over two twisted pair
10Base-F	10 Mbps baseband over fiber-optic cable
100Base-TX	100 Mbps baseband over twisted pair
1000Base-T or 1GBase-T	1000 Mbps baseband over Cat 5e or higher cable

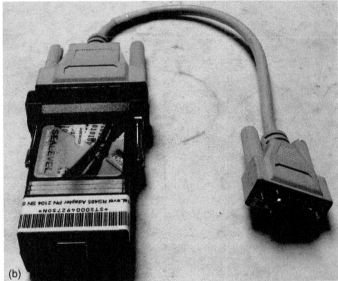

FIGURE 13.37

Example of protocol converters (a) fiber to Ethernet and (b) USB to DB-9.

13.4.6 Protocol converters

Many times, the computer running the sensor software does not have a compatible adapting device as most modern computers are sold with USB ports only. In order to adapt to some industry standard sensors, some type of protocol converter must be adapted. For instance, the typical sonar system used with an ROV will transmit any of several protocols (RS-232, RS-422, RS-485, or Ethernet). In order to capture these into the surface computer, a protocol converter is required to convert between the serial standards (Figure 13.37).

The wiring on these converters must be closely examined as the various configurations of the peripheral equipment will determine the device's wiring.

Underwater Acoustics

CHAPTER CONTENTS

Mathematics is the language of science. All of the properties of underwater acoustics can be expressed in mathematical terms. Later in this chapter, as well as in Chapter 15 on sonar, two of the practical applications of underwater acoustics (acoustic positioning and imaging/profiling sonar) will be explored. In the initial sections of this chapter, the theoretical basis of underwater acoustics is presented to segue into the acoustic positioning and sonar applications.

This section is fairly technical and is math based for those who would like to explore the theoretical aspects of underwater acoustics to gain a better insight into positioning and sonar. For those less interested in the technical background and more concerned with the practical applications of these technologies, proceed to Chapter 15 for sonar as well as Chapter 16 for acoustic positioning.

This chapter on underwater acoustics is based on a publication by Kongsberg Maritime entitled *Introduction to Underwater Acoustics*. A special thanks goes to Arndt Olsen with Kongsberg Maritime for obtaining permissions to include this material.

14.1 Introduction

Acoustic sound transmission represents the basic techniques for underwater navigation, telemetry, echo sounder, and sonar systems. Common for these systems are the use of underwater pressure wave signals that propagate with a speed of approximately 1500 m/s through the water (Figure 14.1).

When the pressure wave hits the sea bottom, or another object, a reflected signal is transmitted back and is detected by the system. The reflected signal contains information about the nature of the reflected object.

For a navigation and telemetry system, the communication is based upon an active exchange of acoustic signals between two or more intelligent units.

Transmission of underwater signals is influenced by a number of physical limitations, which together limit the range, accuracy, and reliability of a navigation or telemetry system.

The factors described in this section include:

- Transmitted power
- Transmission loss
- Transducer configuration
- Directivity and bandwidth of receiver
- Environmental noise
- Requirements of positive signal-to-noise ratio for reliable signal detection
- Ray bending and reflected signals.

The signal-to-noise ratio obtained can be calculated by the sonar equation.

14.2 Sound propagation

14.2.1 Pressure

A basic unit in underwater acoustics is pressure, measured in μPa (micropascal) or μbar. The Pa (Pascal) is now the international standard. It belongs to the MKS system, where $1\ \mu Pa = 10^6\ N/m^2$. The μbar belongs to the CGS system.

$$1\ \mu bar = 10^5\ \mu Pa$$

$$0\ dB\ re\ 1\ \mu bar = 100\ dB\ re\ 1\ \mu Pa$$

The μbar is a very small unit so negative decibels will rarely occur, if ever. To convert from μbar to μPa, simply add 100 dB.

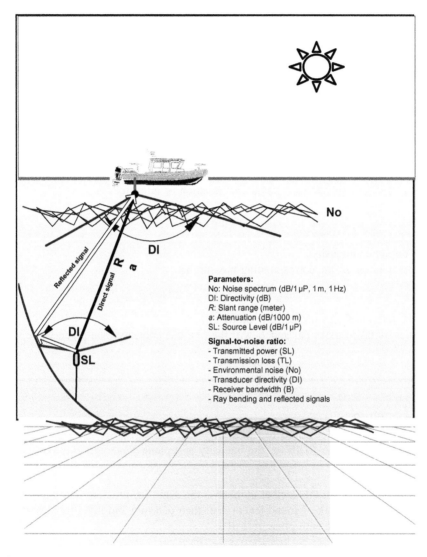

FIGURE 14.1

Parameters of the sonar equation.

14.2.2 **Intensity**

The sound intensity is defined as the energy passing through a unit area per second. The intensity is related to pressure by:

$$I = p^2 / \rho c$$

where

I = intensity,
p = pressure,
ρ = water density, and
c = speed of sound in water.

14.2.3 Decibel

The decibel is widely used in acoustic calculations. It provides a convenient way of handling large numbers and large changes in variables. It also permits quantities to be multiplied together simply by adding their decibel equivalents. The decibel notation of intensity I is:

$$10 \log I/I_o$$

where I_o is a reference intensity.

The decibel notation of the corresponding pressure is:

$$10 \log(p^2/\rho c)/(\rho_o^2/\rho c) = 20 \log p^2/p_o$$

where p_o is the reference pressure corresponding to I_o.

Normally, p_o is taken to 1 µPa, and I_o will then be the intensity of a plane wave with pressure 1 µPa.

EXAMPLE

A pressure $p = 100$ µPa
  In decibels: $20 \log 100/1 = 40$ dB re 1 µPa
  The intensity will also be 40 dB re "the intensity of a plane wave with pressure 1 µPa."

As shown in the example, the decibel number is the same for pressure and intensity. It is therefore common practice to speak of sound level rather than pressure and intensity. The reference level is in both cases a plane wave with pressure 1 µPa.

14.2.4 Transmission loss

14.2.4.1 Geometrical spreading

When sound is radiated from a source and propagated in the water, it will be spread in different directions. The wave front covers a larger and larger area. For this reason the sound intensity decreases (with increasing distance from the source). When the distance from the source has become much larger than the source dimensions, the source can be regarded as a point source, and the wave front takes the form as a part of an expanding sphere. The area increases with the square of the distance (Figure 14.2) from the source, making the sound intensity decrease with the square of the distance.

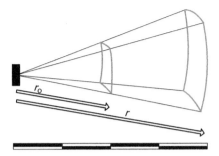

FIGURE 14.2

Transmission loss due to geometrical spreading.

Let I and I_o be the sound intensities in the distances r and r_o. Then:

$$I_o/I = (r/r_o)^2$$

Expressed in decibels, the geometrical spread loss is:

$$TL_1 = 10 \log I_o/I = 20 \log r/r_o$$

Usually a reference point is taken 1 m in front of the source. Setting $r_o = 1$ m we get:

$$TL_1 = 20 \log r$$

where r is measured in meters.

14.2.4.2 Absorption loss

When the sound propagates through the water, part of the energy is absorbed by the water and converted to heat. For each meter, a certain fraction of the energy is lost:

$$dI = -A \cdot I dr$$

where A is a loss factor. This formula is a differential equation with the solution:

$$I(r) = [I(r_o)/(e - Ar_o)] \cdot e^{-Ar}$$

$I(r_o)$ is the intensity at the distance r_o:

$$TL_2 = 10 \log I(r_o)/I(r) = \alpha(r - r_o)$$

where $\alpha = 10 A \log (e)$.

Expressed in decibels, the absorption loss is proportional to the distance traveled. For each meter traveled, a certain number of decibels is lost.

If r_o is the reference distance 1 m, and if the range r is much larger than 1 m, the absorption loss will approximately be:

$$TL_2 = \alpha r$$

where α is named the absorption coefficient. Figure 14.3 shows absorption loss coefficient as a function of frequency. The value of α depends strongly on the frequency. It also depends on salinity, temperature, and pressure.

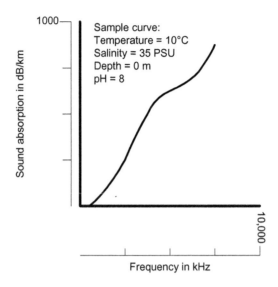

FIGURE 14.3

Absorption loss.

14.2.4.3 One-way transmission loss

The total transmission loss, which the sound suffers when it travels from the transducer to the target (Figure 14.4), is the sum of the spreading loss and the absorption loss:

$$TL = 20 \log r + \alpha r$$

where r is measured in meters and α is measured in dB/meter.

14.3 Transducers

14.3.1 Construction

A modern transducer is based on piezoelectric ceramic properties, which change physical shape when an electrical current is introduced (Figure 14.5). The change in shape, or vibration, causes a pressure wave, and when the transducer receives a pressure wave, the material transforms the wave into an electrical current. Thus, the transducer may act as both sound source and receiver.

14.3.2 Efficiency

When the transducer converts electrical energy to sound energy or vice versa, part of the energy is lost in friction and dielectric loss. Typical transducer efficiency is as follows:

- 50% for a ceramic transducer
- 25% for a nickel transducer

The efficiency is defined as the ratio of power out to power in.

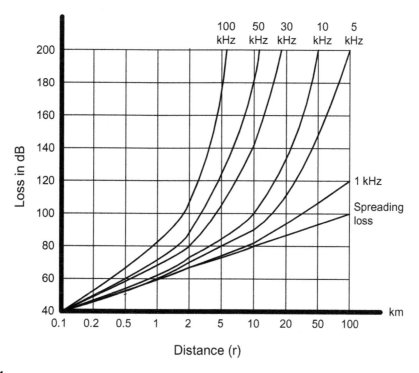

FIGURE 14.4

Propagation loss versus range (one-way transmission loss $TL = 20 \log r = \alpha r$).

14.3.3 **Transducer bandwidth**

Normally a transducer is resonant. This means that they offer maximum sensitivity at the frequency they are designed for. Outside this frequency the sensitivity drops. Typically, the Q-value is between 5 and 10, where

$$Q = \text{center frequency/bandwidth (between 3 dB points)}$$

14.3.4 **Beam pattern**

The beam pattern shows the transducer sensitivity in different directions. It has a main lobe, normally perpendicular to the transducer face. The direction in which the sensitivity is maximum is called the *beam axis*. It also has unwanted side lobes and unwanted back radiation.

An important parameter is the beam width, defined as the angle between the two 3 dB points. As a rough rule of thumb, the beam width is connected with the size of the transducer by:

$$\beta = \lambda/L$$

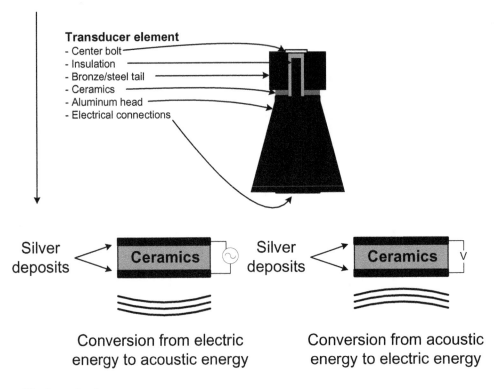

Principle

Transmit:
Electrical signal [converts to] mechanical ceramic vibrations [converts to] acoustic signal

Receive:
Acoustic signal [converts to] mechanical ceramic vibrations [converts to] electric signal

FIGURE 14.5

Cross-section of a typical ceramic transducer.

where

β = beam width in radians,
λ = wavelength,
L = linear dimension of the active transducer area (side for a rectangular area, diameter for a circular).

This rule is not valid for very small transducers, i.e., $L < \lambda$.

The theoretical beam pattern of a continuous line transducer is a sin x/x function, namely:

$$b(\theta) = (\sin(\pi L/\lambda \cdot \sin \theta)/\pi L/\lambda \cdot \sin \theta)^2$$

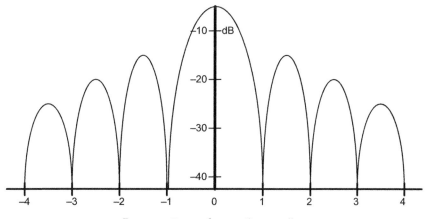

Beam pattern of a continuous line

FIGURE 14.6

Beam pattern of a continuous line.

where

θ = angle from the beam axis,
L = line length,
b = transducer power response.

Figure 14.6 shows this pattern in a Cartesian plot. Note that the side lobes are gradually decreasing. The first side lobe is 13 dB below the point of maximum response.

A transducer having a rectangular active area vibrating uniformly as a piston will have this beam pattern in the two planes parallel to the sides.

In many transducers, the side lobes are reduced by a technique called "tapering."

14.3.5 Directivity index

With reference to transmission, the directivity index (DI) of a transducer can be defined by:

$$DI = 10 \log(I_o/I_m)$$

where

I_o = the radiated intensity,
I_m = the mean intensity for all directions (including back radiation).

Both intensities are measured in the same distance from the transducer.

The mean intensity I is equal to the intensity we would get from an omnidirectional source, if this was given the same power and had the same efficiency as the transducer. We could therefore also define the DI as the ratio between the radiated intensity at the beam axis and the intensity an omnidirectional source would have given at the same point.

Table 14.1 DI Examples

Omnidirectional source	DI = 0 dB
Transducer with equal radiation everywhere in one half-plane and zero back radiation	DI = 3 dB
Typical echo sounder transducer	DI = 25 dB
Wide beam transducer	DI = 4 dB
Medium beam transducer	DI = 9 dB
Narrow beam transducer	DI = 15 dB
USBL Surface Station	DI = 25 dB

A transducer, which has a rectangular active area vibrating uniformly as a piston, will have this beam pattern in the two planes parallel to the sides. As shown in Table 14.1, the narrower the beam, the higher the DI.

The DI for a transducer with beam pattern $b(\theta, \Phi)$ and the mean intensity is found by integration over all directions, with solid angle element $d\Omega$, and division by the total solid angle 4π:

$$I_{\mathrm{m}} = (1/4\pi) \int_{4\pi} I_0 \cdot b(\theta, \Phi) d\Omega$$

According to the definition of DI:

$$DI = 10 \log(4/\int_{4\pi} b(\theta, \Phi) d\Omega)$$

Calculation of DI after this formula is, however, no easy job, not even for the simplest transducer.

If the transducer side or diameter is larger than λ, the DI is approximately:

$$DI = 10 \log (4\pi A/\lambda^2)$$

where A is the active transducer area, $A = L^2$.

When the beam width is known, another approximate formula can be used. For a rectangular transducer, the beam pattern in the two planes parallel to the sides are sin x/x function as mentioned previously.

The response is 3 dB down at:

$$(L/\lambda)\sin \theta_{3 \text{ dB}} = 0.443$$

Inserting this in the formula above gives:

$$DI = 10 \log(2.47/[\sin(\beta_1/2) \cdot \sin(\beta_2/2)])$$

14.3.6 Transmitting response

The transmitting power response (S) of a transducer is the pressure produced at the beam axis 1 m from the transducer by a unit electrical input. The electrical input unit may be volt, ampere, or watt. A typical value for the transmitting response for a ceramic transducer is:

$$S = 193 \text{ dB re } 1 \text{ } \mu\text{Pa}/W$$

14.3.7 Source level

The source level (SL) of sonar or an echo sounder is the sound pressure in the transmitted pulse at the beam axis 1 m from the transducer. If the transmitting response (S) is known, then the source level is:

$$SL = S + 10 \log P$$

where

P = transmitter power,
S = transmitting power response.

A widely used formula is:

$$SL = 170.9 + 10 \log P + E + DI$$

where

SL = source level in dB re 1 µPa,
P = transmitter power in watts,
$E = 10 \log \eta$,
η = transducer efficiency,
DI = directivity index.

The constant 170.9 incorporates conversion from watts to pascals and can be derived as follows:

$$SL = 10 \log(1/4\pi) + 10 \log P + E + DI$$

The factor $1/4\pi$ represents the source level from an omnidirectional source, supplied with 1 W electrical power and with 100% efficiency. When 1 W sound power is distributed over a sphere with radius 1 m and surface 4π m^2 the sound intensity will be:

$$(1/4\pi) \; W/m^2$$

The connection between pressure and intensity is:

$$p = \sqrt{I\rho c}$$

where

I = intensity,
ρ = density of water,
p = pressure,
c = sound velocity (in water).

Seawater at temperature 10°C with salinity 35 (PPT) at the sea surface has the following values:

$\rho = 1027$ kg/m^3
$c = 1490$ m/s

A sound density of 1/4 W/m^2 will in this environment correspond to a sound pressure of:

$$p = 349 \times 10^6 \; \mu Pa = 170.9 \; dB \; re \; 1 \; \mu Pa$$

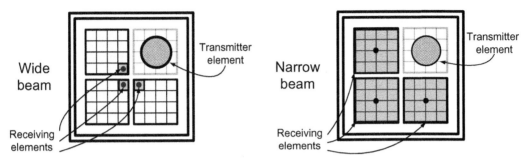

FIGURE 14.7

Multielement transducers.

14.3.8 Medium beam/narrow beam transducer

The transducer used in USBL (ultrashort baseline—Europeans use the term SSBL or "super short baseline") mode (acoustic positioning) normally consists of three different groups of elements. This is to be able to calculate a three-dimensional bearing to the transponders.

- The beam width of a transducer can be changed during operation. This is achieved by combining more transducer elements in series/parallel.
- A more narrow beam gives higher directivity (higher gain and higher noise suppression from outside the beam) but will give a smaller signal "footprint."
- For the *narrow beam transducer* in Figure 14.7, the 3 dB point in wide beam mode is 160°, while in narrow beam mode at the 3 dB point it is 30°.
- When using the transducers in SBL (short baseline) or LBL (long baseline) mode, no angle measurements are done and only the R-group (reference) is used. Dedicated SBL/LBL transducers containing only one element or one group can be used.

14.4 Acoustic noise

14.4.1 Environmental

Noise from thrusters and propellers from surface vessels is the dominating environmental noise source. This noise is approximately 40 dB above normal sea noise. Common for all noise sources is that the noise level drops approximately 10 dB per decade with increasing frequency.

14.4.2 Noise level calculations

The noise level at the system detector is calculated by the following equation:

$$N = (N_0 - 10 \log(B) - \text{DI})$$

where

B = detector bandwidth,
DI = directivity of transducer.

14.4.3 **Thruster noise**

The noise from the thruster is changing depending on the thruster. On pitch-controlled thrusters (fixed rpm), the noise level is actually higher when running idle (0% pitch) than running with load. In addition, the impact of the thruster noise is determined by the direction of the (azimuth) thruster.

Running a thruster on low rpm and high pitch normally generates less noise than a thruster on high rpm and low pitch. In general, thrusters with variable rpm/fixed pitch generate less noise than thrusters with fixed rpm/variable pitch.

14.4.4 **Sound paths**

The velocity of sound is an increasing function of water temperature, pressure, and salinity. Variations of these parameters produce velocity changes, which in turn cause a sound wave to refract or change its direction of propagation. If the velocity gradient increases, the ray curvature is concave upward (Figure 14.8). If the velocity gradient is negative, the ray curvature is concave downward.

The refraction of the sound paths represents the major limitations of a reliable underwater navigation and telemetry system. The multipath conditions can vary significantly depending upon ocean depth, type of bottom, and transducer—transducer configuration and their respective beam patterns. The multipath transmissions result in a time and frequency smearing of the received signal as illustrated.

There are several ways of attacking this problem. The obvious solution is to eliminate the multiple arrivals by combining careful signal detection design with the use of a directional transducer beam. A directional receiving beam discriminates against energy outside of the arrival direction and a directional transmit beam projects the energy, so that a minimum number of propagation paths are excited.

14.4.5 **Sound velocity**

From Bowditch (2002), the speed of sound in seawater is a function of its density, compressibility, and, to a minor extent, the ratio of specific heat at constant pressure to that at constant volume (Figure 14.9). As these properties depend upon the temperature, salinity, and pressure (depth) of seawater, it is customary to relate the speed of sound directly to the water temperature, salinity, and pressure. An increase in any of these three properties causes an increase in the sound speed. The converse is also true.

The speed of sound changes by $3-5$ m/s/°C temperature change, by about 1.3 m/s/PPT (PSU) salinity change, and by about 1.7 m/s/100 m depth change. A simplified formula adapted from Wilson's (1960) equation for the computation of the sound speed in seawater is:

$$U = 1449 + 4.6T - 0.055T^2 + 0.0003T^3 + 1.39(S - 35) + 0.017D$$

where U is the speed (m/s), T is the temperature (°C), S is the salinity (PSU), and D is depth (m).

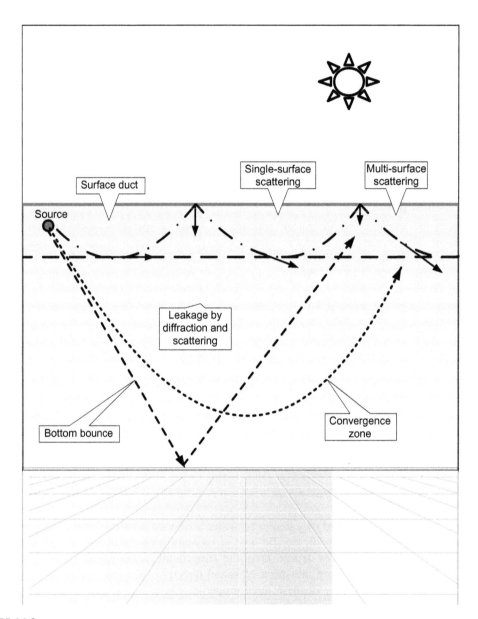

FIGURE 14.8

Ray diagrams for deepwater propagation.

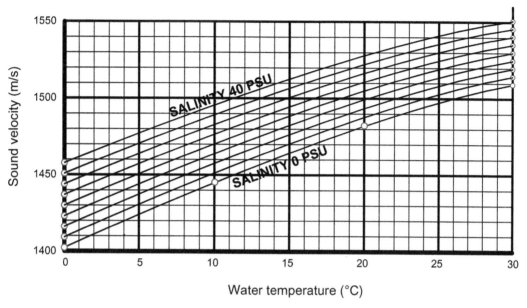

FIGURE 14.9

Example of velocity change with changing temperature and salinity.

14.4.6 Reflections

Reflections can be caused when the signal bounces off a subsea structure, seabed, riser, ship's hull, or surface (Figure 14.10). Normally the reflection is not perfect, meaning the reflected pulse has less energy than the direct pulse and should not cause problems. Sometimes the pulse is so strong it might cause problems for the pulse detection in the receiver.

When two pulses (sine waves) are added, the result can be stronger or weaker than a single pulse. Adding two signals that are 180° out of phase and of equal strength will create no signal at all (Figure 14.11).

If the signal path of the reflection is 0.5λ (or multiples of this) longer, the above situation might occur.

Example (refer to Figure 14.11):

$$f = 30 \text{ kHz} \approx \lambda = 0.05 \text{ m}$$
$$d = 100 \text{ m}$$

180° phase shift (1/2λ):

$$d_x = \sqrt{((100.025/2)^2 - (100/2)^2)} = 1.12 \text{ m}$$

180° phase shift (2 + 1/2λ):

$$d_x = \sqrt{((100.125/2)^2 - (100/2)^2)} = 2.50 \text{ m}$$

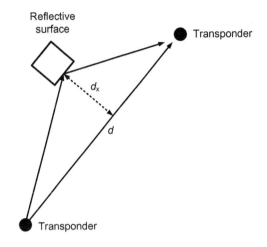

FIGURE 14.10

Two LBL transponders measuring baseline.

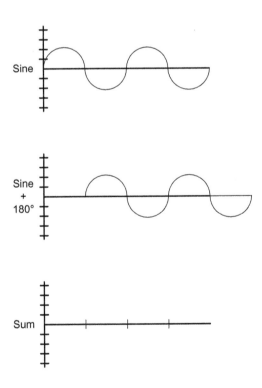

FIGURE 14.11

Sine wave interference.

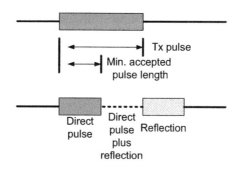

FIGURE 14.12

Received pulse.

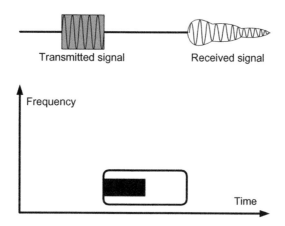

FIGURE 14.13

Multipath transmission: time and frequency smearing of acoustic signals.

This shows that a surface with 100% reflection will create no resulting signal at the receiver given the distances. The same problem can be caused by ray bending or reflections from risers or a ship's hull.

Even reflections that are slightly delayed might cause problems. The receiver has certain criteria in order to accept the signal as a pulse. One of them is the pulse length. As Figures 14.12 and 14.13 illustrate, a worst-case condition will be that the direct pulse length is too short to be acknowledged as a pulse before the reflection cancels it. Then the remainder of the reflection is just long enough to be accepted as a pulse.

Sonar

15

CHAPTER CONTENTS

In this section, the subject of sonar is presented with emphasis on "imaging sonar" systems as deployed aboard many remotely operated vehicles (ROVs). The field of sonar comprises a broad and very in-depth body of knowledge lightly dealt with in this chapter. For further reading on this subject, refer to the bibliography.

This chapter relies heavily upon contributions by experts in this field. The explanation of compressed high intensity radar pulse (CHIRP) technology comes courtesy of Maurice Fraser with Tritech International Limited. Special thanks go to Richard Marsh for arranging for its inclusion. The explanation of acoustic lens technology comes from a technical paper entitled "Object Identification with Acoustic Lenses" courtesy of Edward Belcher of Sound Metrics. And special thanks go to Willie Wilhelmsen, Jeff Patterson, and Mitch Henselwood of Imagenex Technology Corporation for their extensive contributions to this chapter over a 2-year period.

15.1 Sonar basics

15.1.1 Why sound?

Sound propagates readily through water well beyond the range of light penetration (even at light's highest penetration wavelengths). Sound propagates best through liquid and solids, less well through gases, and not at all in a vacuum. As light reflections differentiate between objects by varying levels of reflection (light intensity) as well as changing wavelengths (light color), so does sonar characterize targets through reflected sound frequency and intensity. Through proper data interpretation, target information may be discerned to identify the object through active or passive sound energy.

Some typical applications for sonar technology include:

- Echo sounding for bathymetry
- Side-scan sonar for bathymetry and item location
- Underwater vehicle-mounted imaging sonar for target identification
- Geophysical research
- Underwater communications
- Underwater telemetry
- Military listening devises (passive sonar) for submarines and shipping identification
- Position fixing with acoustic positioning systems
- Fish finding
- Acoustic seabed classification
- Underwater vehicle tracking over bottom
- Measuring waves and currents

15.1.2 Definition of sonar

Sound transmission through water has been researched since the early 1800s. The technology rapidly matured, beginning in the 1930s. With the explosion of technology in the fields of beam-forming transducers and digital signal processing, today's sonar system encompasses a wide range of acoustical instruments lumped under the general heading "Sonar."

The term "sonar" is derived from "*so*und *na*vigation and *r*anging." The purpose of this technology is to determine the range and reverberation characteristics of objects based upon underwater sound propagation.

15.1.3 Elements required for sonar equipment

For sonar equipment to function, three key elements are necessary:

1. *Source*: A sound source is needed to produce the pulse energy for reflection (in active sonar systems) and/or reception (for passive sonar systems).
2. *Medium*: In the vacuum of space, sound does not travel. Some type of medium is required to transmit the sound wave energy between the source and receiver.
3. *Receiver*: Some type of receiver is needed to transform the mechanical energy (sound waves) into electrical energy (electrical signal) in order to process the sound into signals for information processing.

The source vibrates (whether it is a sonar transducer, machinery noise from an engine aboard a vessel, or the mating call of whales in the ocean) causing a series of compressions and rarefactions, thus propagating the sound through the transmission medium.

15.1.4 Frequency and signal attenuation

Beginning in 1960 and culminating with the Heard Island feasibility tests in 1991, sound propagation tests were conducted off the coast of Australia with listening stations placed around the world's oceans. What was discovered was the extreme distance low-frequency sound in seawater would travel. As the frequency increased, however, sound attenuation increased dramatically. Table 15.1 presents a general working range of various sound frequencies in seawater.

Lower frequency sound propagates through higher density materials better than higher frequency sound waves. Very low-frequency sound is used in applications such as seismic surveys of sub-bottom rock formations in search of hydrocarbons, since the sound waves will penetrate more dense strata before reflecting off of hard substrate. Low frequency is used to penetrate through mud by sub-bottom profilers. In the very high-frequency ranges, sound reflection produces only a

Table 15.1 Sonar Two-Way Working Ranges

Frequency	Wavelength	Distance
100 Hz	15 m	1000 km or more
1 kHz	1.5 m	100 km or more
10 kHz	15 cm	10 km
25 kHz	6 cm	3 km
50 kHz	3 cm	1 km
100 kHz	1.5 cm	600 m
500 kHz	3 mm	150 m
1 mHz	1.5 mm	50 m

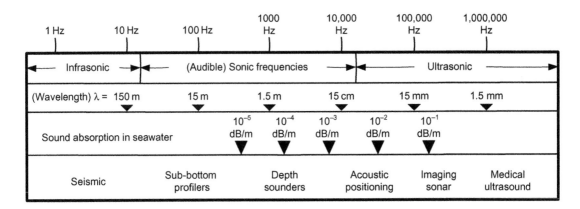

FIGURE 15.1

Acoustic spectrum.

surface paint of the target but gives higher target surface detail. An analysis of the sound frequency spectrum is provided in Figure 15.1.

15.1.5 Active versus passive sonar

Range and bearing from a sound source can be derived either actively or passively. The transducer/ receiver receives sound from some source and then (through a series of computations) arrives at a range and bearing solution. *Passive* sonar essentially uses listening *only* to derive these computations. *Active* sonar (in a method similar to radar) uses a transmitter/receiver arrangement to send out an acoustic signal and then listens for a reflection (echo) of that signal back to the receiver over time to derive a range and signal strength plot.

Beam forming of sonar signals provides for various acoustic properties of the reflected signal (for active sonar), thus allowing analysis of the backscatter for target characterization. For imaging sonar, a so-called fan beam is used to depict small details, thus building a clear pictorial image of the target being insonified (Figure 15.2). For depth sounding sonar, a broad conical beam gives an indication of the closest distance to the bottom over a broader area. The typical usage of a conical beam is for depth sounding. For ROV-mounted sonar applications, a fan beam is used for the imaging sonar while a conical beam is used for the "altimeter" to determine the submersible's height above the bottom.

15.1.6 Transducers

This section will delve further into the transducer and its use in beam forming for gathering various backscatter characteristics of the target. By definition, a "transducer" is any of a number of devices that convert some other form of energy to or from electrical energy. Examples of devices that fall under the category of "transducer" are photoelectric cells, common stereo speakers, microphones, electric thermometers, any type of electronic pressure gauge, and an underwater piezoelectric transducer. This chapter will concentrate on the piezoelectric transducer and its usage with sonar

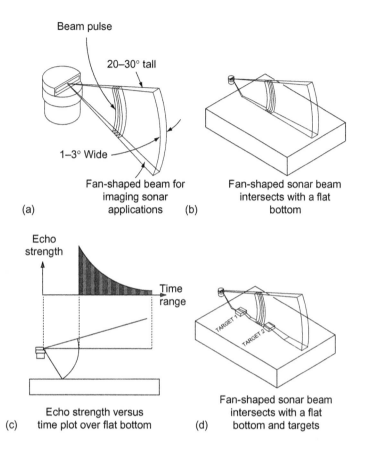

FIGURE 15.2

Single beam sonar (a) graphically defined, (b) pulsed over bottom through water column, (c) echo strength plotted over time, and (d) depicted with sample targets.

systems. This analysis begins with the familiar land-based transducer, the home stereo speaker, and then progresses into the unfamiliar underwater transducer.

In a stereo speaker, any number of technological twists is used to convert electrical energy to acoustic energy. The most basic means involves the varying of a voltage within a coil wrapped around an electromagnet moving a paper cone. The cone vibration produces acoustic energy in the form of sound waves propagating in an omnidirectional fashion as depicted in Figure 15.3. As shown in the figure, the sound moves through the medium (in this case air) through a series of compressions and rarefactions as the wave travels at sonic velocity away from the sound source. As the cone vibrates, it produces sound in a bidirectional pattern. In order to change the sound pattern from an omnidirectional wave pattern to a more focused sound projection, enclosures and sound projectors are used to focus the sound beam to a desired direction. If a home stereo speaker is disassembled, one would find three basic speakers (the woofer, midrange, and tweeter), each with frequency response based upon its size and nominal resonance frequency.

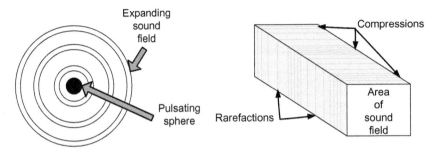

FIGURE 15.3

Sound pulse traveling in wave fronts through a medium.

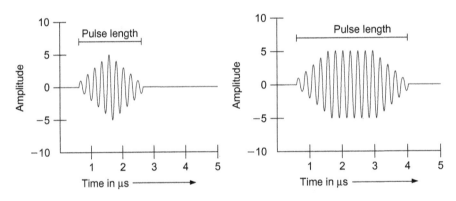

FIGURE 15.4

Pulse length as a function of time.

The most prevalent underwater transducer is the polycrystalline, "piezoelectric" material PZT (lead zirconium titanate), which was discovered in the 1940s to be an effective transducer. As discussed in Chapter 14, piezoelectric underwater transducers emit acoustic pulses through vibration when an electric signal is sent through the silver deposits at the poles of the ceramic core. Piezoelectric underwater transducers are resonant; therefore, a transducer is sized to match a certain nominal narrowband frequency in which it is most sensitive. The sonar system is then designed around this nominal frequency.

On a typical active transducer, the transducer will vibrate at the frequency of the electrical signal being applied, not necessarily the resonant frequency. If that frequency is near the resonant frequency, the transducer is freer to respond, giving a larger amplitude signal than if the frequency is far from the resonant frequency. The length of time the transducer is activated determines the pulse length generated (Figures 15.4 and 15.5). A short pulse length allows better discrimination between targets, but may not allow enough acoustic energy to reflect off targets as distance increases.

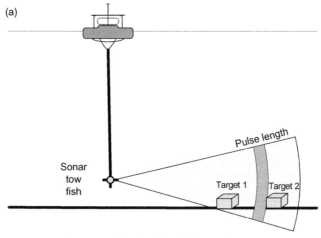

(a)

Sonar
tow
fish

Pulse length

Target 1   Target 2

Short pulse length = Two distinct reflections

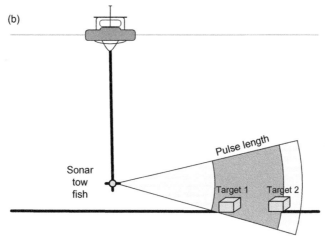

(b)

Sonar
tow
fish

Pulse length

Target 1   Target 2

Long pulse length = No distinction between targets

FIGURE 15.5

Pulse length for discrimination between targets with (a) short pulse length and (b) longer pulse length.

The directivity index for a transducer defines the ratio between an omnidirectional point source, that is, a circular (two-dimensional, 2D) or spherical (three-dimensional, 3D) sound source with no directivity, to the source level intensity on the beam axis of a directional sound pulse. The reason this factor is important is that the beam form is defined by its directivity index as well as its vector (Figure 15.6).

As the directivity index of a sound beam becomes more focused, some side effects become pronounced, thus requiring consideration. The main beam of a directional sound source is called the "main lobe" and is the primary pulse directed toward the target. As depicted in Figures 15.7 and 15.8,

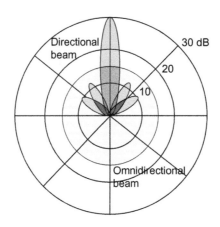

FIGURE 15.6

Directivity index for a circular as well as a directional sound beam.

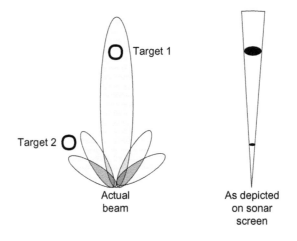

FIGURE 15.7

Target in the side lobe of this beam will produce an echo at the receiver.

a by-product of this main lobe, side lobes, develop that are also subject to backscatter off secondary targets. An analogy to this concept is the simple push of the hand on the surface of a swimming pool, directing a wave pulse in a specific direction. As the hand is pushed in one direction, it can be noticed that the main pulse pushes in the direction of the wave but side waves emanate on either side of the main wave.

15.1.7 Active sonar

The majority of civilian sonar systems falls under the classification of active sonar. Active sonar systems make use of beam forming, frequency shifting, and backscatter analysis to characterize

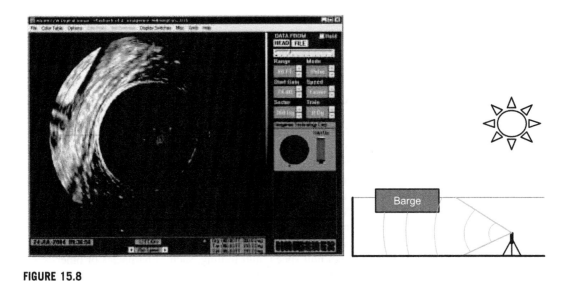

FIGURE 15.8

Mechanically scanning sonar viewing barge with scalloped lines from side lobe backscatter.

targets insonified. There are a vast number and category of active sonar products on the market to fit various applications (with the technology advancing daily). Any number of these technologies is available for mounting on ROV equipment for delivery to the work site.

15.1.8 Terminology

The following provides an explanation of the basic terms used to describe sonar techniques:

- *Angle of incidence*: The angle at which a sound reflects upon a surface (in degrees). Example: If a flashlight is shone on a mirror, the light will reflect directly back to the flashlight if there is a zero angle of incidence, reflected 90° with a 45° angle of incidence, etc. The same principle applies to a sound reflector.
- *Beam forming*: The principle of forming the sound propagation wave to provide the desired data from the wave front.
- *Color*: The different colors used to represent the varying echo return strengths.
- *Echo*: The reflected sound wave.
- *Echo return*: The time required for the echo to return to the source of the sound.
- *Pulse width*: The width of a sound propagation wave (in feet, meters, or microseconds) generated from some source. This factor has an influence on the quality of the image generated by the sonar receiver.
- *Target*: The object requiring characterization.
- *Target strength (TS)*: A measure of the reflectivity of the target to an active sonar signal.
- *Reflectivity*: The ability of a material to reflect sound energy.

- *Transmission (sound) loss (TL)*: Total of all sound losses incurred between the sound source and the ultimate receiver. Transmission losses come in two main types: spreading loss and attenuation loss.
- *Spreading loss*: Sound energy loss due to geometrical spreading of the wave over an increasingly large area as the sound propagates. Spreading losses are considered on either a 2D cylindrical plane (horizontal radiation only, or thermal layer, or large ranges compared to depth) or a 3D spherical profile (omnidirectional point source).
- *Attenuation loss*: Generally considered to be the lumped together sum of losses produced through absorption, spreading, scattering, reflection, and refraction.
- *Absorption loss*: The process whereby acoustic energy is absorbed by a material, thus producing heat. Absorption loss increases with higher frequency.
- *Scattering and reverberation*: Sound energy bouncing or reflecting off items or surfaces.
- *Noise level (NL)*: Total noise from all sources potentially interfering with the sound source reception. Therefore, Noise level (NL) = Self-noise (SN) + Ambient noise (AN).
- *Self-noise (SN)*: Noise generated from the sonar reception platform potentially interfering with reception of the sound source. Examples of this are machinery noise (pumps, reduction gears, power plant, etc.) and flow noise (high hull speed, hull fouling, such as barnacles or other animal life attached to hull), and propeller cavitations.
- *Ambient noise (AN)*: Background noise in the medium potentially causing interference with signal reception. Ambient noise can be either hydrodynamic (caused by the movement of water, such as tides, current, storms, wind, rain, etc.), seismic (i.e., movement of the earth—earthquakes), biological (i.e., produced by marine life), or by ongoing ocean traffic (i.e., noise caused by shipping).
- *Source level (SL)*: Sound energy level of sound source upon transmission.
- *Directivity index (DI)*: The ratio of the logarithmic relationship between the intensity of the acoustic beam versus the intensity of an omnidirectional source.
- *Signal-to-noise ratio (SNR)*: The ratio to the received echo from the target to the noise produced by everything else.
- *Reverberation level (RL)*: The slowly decaying portion of the backscattered sound from the sound source.
- *Detection threshold (DT)*: The minimum level of received signal intensity required for an experienced operator or automated receiver system to detect a target signal 50% of the time.
- *Figure of merit (FOM)*: The maximum allowable one-way transmission loss in passive sonar and the maximum two-way transmission loss in active for a detection probability of 50%. These are termed "active" (AFOM) and "passive" (PFOM). This concept is especially pertinent to the anti-submarine warfare community.
- *Shadows*: The so-called shadowgraph effect is an area of no sonar reflectivity due to blockage. The best analogy for this would be to shine a light onto an object a few feet away. The object will block the light directly behind it and cast a shadow. A shadow is depicted on a sonar display by an area behind a blocking mechanism where there is no sonar reflection.
- *Sonar*: The principle used to measure the distance between a source and a reflector (target) based on the echo return time.
- *Sound losses (absorption, spreading, scattering, attenuation, reflection, and refraction)*: Sound energy loss through various factors influencing the reception and display of reflected sound waves.

15.1.9 **Sonar equations**

Sonar equations look at signal losses compared to signal sources to determine the likelihood of detection of the sound source. For the passive sonar, only one-way sound transmission and noise levels are considered in deriving the minimum detection threshold, whereas on active sonar systems the sound must travel first from the sound source (the transducer of the active sonar system) through the water column to the target and then (after the backscatter) back to the receiver (Figure 15.9).

The passive sonar equation:

$$SL - TL - NL + DI \geq DT$$

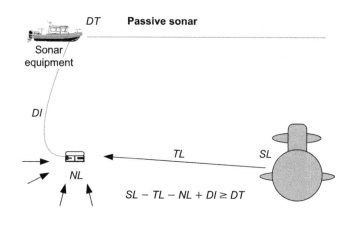

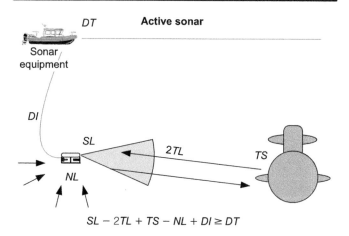

FIGURE 15.9

Sonar equations for (a) passive and (b) active sonar.

The active sonar equations:
Ambient noise limited:

$$SL - 2TL + TS - NL + DI \geq DT$$

Reverberation noise limited (reverb > ambient noise):

$$RL > NL + DI$$
$$SL - 2TL + TS - RL \geq DT$$

The figure of merit equations:

$$PFOM = SL - NL + DI - DT$$
$$AFOM = SL + TS - NL + DI - DT$$

15.1.10 Reflectivity and gain setting

Different materials reflect sound with different efficiency (Table 15.2). Mud will reflect sound very poorly while water will not reflect sound at all. The closer the substance's consistency is to water, the lower the reflectivity index. Therefore, a very good feel for the target's makeup can be gained simply by the target's level of reflectivity.

The gain setting on the sonar system will allow the operator to pick up detail within the reflection. If the gain setting is set high while surveying a sandy bottom, the screen will display no contrast between targets, since everything will show a high reflectivity value. Likewise, if the gain is set too low with a mud bottom, no detail will display, since practically all of the reflections from the bottom will be below the display setting and will be rejected.

In Figure 15.10, differing compositions on this combination of sand and mud bottom allow for discrimination between bottom makeup as well as targets standing proud of the bottom with differing makeup. The bottom of this figure displays a practice course as well as two targets easily discernible from the mud bottom. This figure depicts differing bottom composition through varying echo strength with a fixed gain setting.

In Figure 15.11, an example of a swimming pool is used to illustrate the effect of varying angles of incidence on sonar reflectivity. The tripod-mounted sonar was placed at the center of a varying shaped pool (Figure 15.11c). The walls of the pool have a very high reflectivity at the zero incidence point with much lower backscatter on the higher angles of incidence.

As a further example of incidence versus reflectivity, consider a sonar source placed in an underwater room (Figure 15.12). The sonar will insonify the room with the highest target strength

Table 15.2 Sample Reflectivity Indexes	
Substance	**Reflectivity**
Mud	Low
Sand	Medium
Rock	High
Air/air-filled	Very high

coming from the zero angle of incidence along with another "bright spot" located in the corners. The corners will not be depicted as square due to the sound multipath as the beam approaches the corner point. Instead of a square corner, the sonar display will depict a rounded corner due to the sound reflection as the beam sweeps the corner.

15.1.11 Sound backscatter

Backscatter is the reflected sound from any object being insonified—the analysis of which is the subject of active sonar systems. Backscatter is analyzed for any number of parameters to solve particular operational needs. Some examples of backscatter analysis follow:

- Multiple bounce depth sounder backscatter is analyzed with acoustic seabed classification systems to determine the texture and makeup of the sea bottom (sand, mud, rock, oyster bed, kelp, etc.) for environmental monitoring as well as vessel navigation.
- Doppler shift backscatter is used for vehicle speed over ground as well as current/wave profiling.
- Frequency shift backscatter is analyzed in CHIRP sonar systems (see Figures 15.22–15.24) to discriminate between objects in close proximity.
- Simple high-frequency backscatter can characterize a target as to aspect, texture, surface features, and orientation.

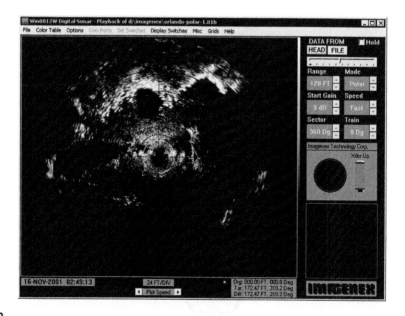

FIGURE 15.10

Combination of sand and mud bottom showing differing reflectivity based upon bottom composition (gain setting 9 dB).

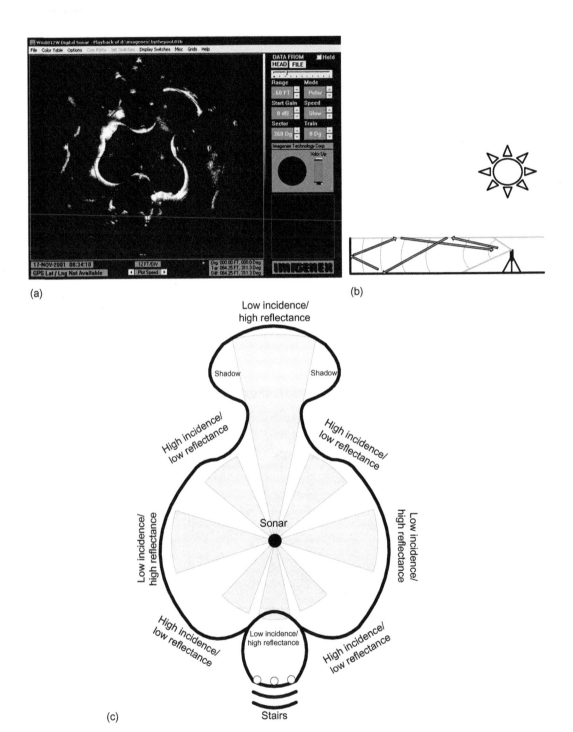

(a)

(b)

(c)

FIGURE 15.11

(a) Sonar display of a recreational swimming pool plus (b) profile view and (c) plan view containing incidence/reflectance levels per pool quadrant.

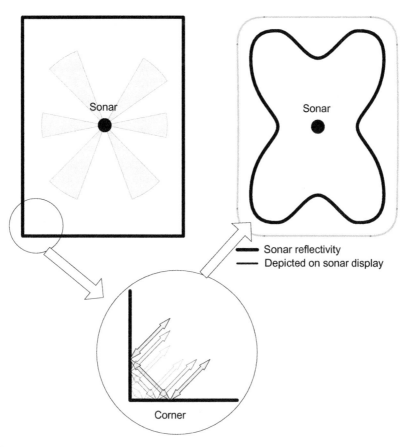

Sonar picture of an underwater rectangular room.

15.1.12 Single- versus multibeam

Active imaging or profiling sonar systems generally fall into three basic categories:

1. Multibeam
2. Mechanically or side-scanning sonar (single beam)
3. Single-beam directional sonar

Single-beam sonars are simply one pulse with one reception on a single receiving element. The single-beam mechanically scanning system comprises the lion's share of ROV-mounted sonar systems currently on the market today due to their simplicity as well as their reduced cost.

Multibeam sonar systems, on the other hand, transmit one wide pulse and receive the backscatter on a large number of receiving elements. Accurate sensing of the delays in sound arrival between elements enables the sonar to distinguish the direction of the received sound, in order to build up a detailed profile of the area being insonified.

In its simplest form, multibeam sonars operate on the principle of additive sound pulses. Recall from Figure 15.3 that sound spreads in an isotropic fashion (the same in all directions) from a sound projector known as an "isotropic source" (a sound projector projecting a sound source equally in all directions). Placing two isotropic sound sources in proximity proportional to the projector's wavelength, the additive pulses (termed "interference") can be either constructive or destructive based upon the phase of the pulses (Figure 15.13).

From this constructive interference, a highly directional pulse can be constructed along the line of equidistance with spacing $\lambda/2$. This forms the axis of a pulse in a 3D directional beam (Figure 15.14). If several projectors are aligned in a row, the beam can become more focused and highly directional.

By varying the time delay between the projector transmissions, the axis of the beam can be steered to project multiple beams. With the use of the same concept upon reception, the time variance can also be analyzed to accept sound sources only from a narrowly directive reception cone, thus forming the multiple beam transmission and reception. The reception array is perpendicular to the projector array so that reception is only allowed at the crossing points, allowing for clean multibeam transmission and reception.

Table 15.3 analyzes some of the benefits of multibeam technology. Multibeam sonar systems are an amazing recent technological development brought on by advances in digital signal processing of sonar returns as well as advances in highly directional transducer receiving elements. With the technology trickling down to the smaller ROV system through lower size costs and increased functionality, look for more multibeam sonar systems aboard these vehicles.

15.1.13 Frequency versus image quality

Determination of the optimum sonar frequency for the particular application at hand is a trade-off of desired image quality, depth of penetration into the medium as well as range of desired coverage area. As discussed above, the higher the frequency of the source, the higher the attenuation of the signal, thus lowering the effective range of the beam. The lower the frequency, the longer the

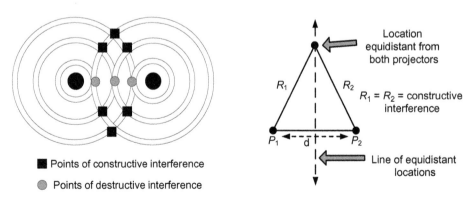

■ Points of constructive interference

● Points of destructive interference

FIGURE 15.13

Constructive and destructive interference of two point sources with spacing *d*.

propagation distance and the penetration depth into the medium being insonified. The higher the frequency, the lesser the penetration into the medium, allowing for a more detailed surface paint of the target for target shape and texture classification.

For small-item surface discrimination, the wavelength of the pulse must be no larger than the target—and much shorter for any detailed imaging of the target. For better small-object discrimination, the higher the frequency the better. What suffers at very high frequencies

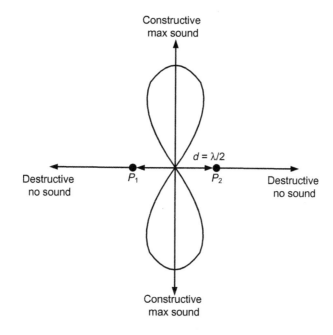

FIGURE 15.14

Constructive dipole projection of sound along a beam axis.

Table 15.3 Some Advantages and Disadvantages of ROV-Mounted Single and Multibeam Sonar Systems

Sonar System	Advantages	Disadvantages
Single beam	• Able to view 360° around vehicle	• Slow image generation
Multibeam	• Up to 10 frames per second image update rate at short range	• ROV-mounted projector allows narrow reception beam width • Higher computing power needed drives cost upward

(700 kHz and higher) is the useful range of the sonar system. For searching a wide area for small-object location, a range of only 20 m will require an extended search time to cover any appreciable area. On a more human note, medical ultrasound functions in the 2–3 MHz range (and higher) for very-small-object discrimination. At these frequencies, extremely high detail can be achieved over very short distances.

Low-frequency sonar is used for such applications as seismic surveys of geological strata (very low frequency—in the 60–100 Hz range) as well as sub-bottom profiling below the mud line (low frequency—in the 10–100 kHz range, depending upon desired mud penetration depth). These systems can gain a broad classification of the overall area, but small details of individual areas are not possible due to the wavelength being larger than any small target.

15.2 Sonar types and interpretation

15.2.1 Imaging sonar

A fan-shaped sonar beam scans surfaces at shallow angles, usually through an angle in the horizontal plane, and then displays color images or pictures. The complete echo strength information for each point is displayed primarily for visual interpretation. As depicted in Figure 15.15, with imaging sonar a fan-shaped sonar beam scans a given area by either rotating (as with ROV-mounted systems) or moving in a straight line (as with side-scan sonar systems).

A pulse of sound traveling through the water generates a backscatter intensity (or amplitude) that varies with time and is digitized to produce a time series of points. The points are assigned a color or grayscale intensity. The different colored points, representing the time (or slant range) of each echo return, plot a line on a video display screen. The image (Figure 15.16), consisting of the different colored points or pixels, depicts the various echo return strengths.

The following characteristics are necessary to produce a visual or video image of the sonar image:

- The angle through which the beam is moved is small.
- The fan-shaped beam has a narrow angle.
- The transmitted pulse is short.
- The echo return information is accurately treated.

These visual images provide the viewer with enough data to draw conclusions about the environment being scanned. The operator should be able to recognize sizes, shapes, and surface-reflecting characteristics of the chosen target. The primary purpose of the imaging sonar is to act as a viewing tool.

15.2.2 Profiling sonar

Narrow pencil-shaped sonar beams scan surfaces at a steep angle (usually on a vertical plane). The echo is then displayed as individual points or lines accurately depicting cross-sections of a surface. Echo strength for each point, higher than a set threshold, digitizes a data set for interfacing with external devices.

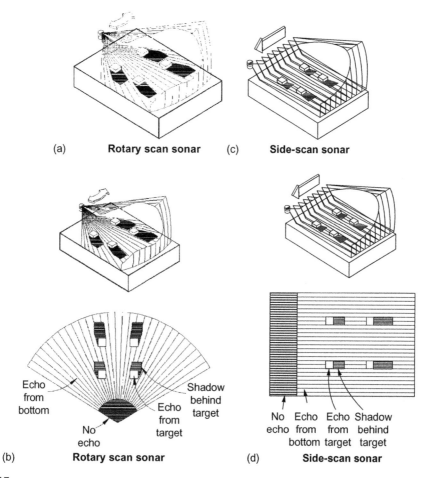

(a)   **Rotary scan sonar**   (c)   **Side-scan sonar**

(b)   **Rotary scan sonar**   (d)   **Side-scan sonar**

FIGURE 15.15

An imaging sonar builds a picture (a) and (b) via rotation of the head (i.e., mechanically scanning) or (c) and (d) with motion through the water (i.e., side scan).

The data set is small enough to be manipulated by computer (primarily a measurement tool). In profiling, a narrow pencil-shaped sonar beam scans across the surface of a given area, generating a single profile line on the display monitor (Figure 15.17). This line, consisting of a few thousand points, accurately describes the cross-section of the targeted area.

A key to the profiling process is the selection of the echo returns for plotting. The sonar selects the echo returns, typically one or two returns for each "shot," based on a given criterion for the echo return strength and the minimum profiling range. The information gathered from the selection criteria forms a data set containing the range and bearing figures. An external device, such as a

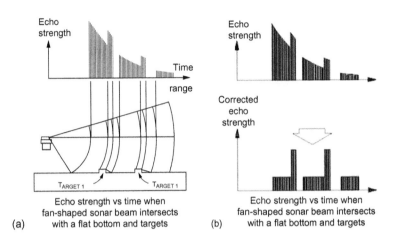

FIGURE 15.16

Echo strength versus time and range (a) plotted in raw form and (b) normalized over full range.

personal computer or data logger, accesses the data set through a serial (or other communications protocol) interface with the sonar.

The profile data is useful for making plots of bottom profiles, trench profiles, and internal and external pipeline profiles. The primary purpose of the profiling sonar is as a quantitative measuring tool such as a depth sounder or for bottom characterization. Imaging sonar can also be combined with profiling sonar to accomplish both imaging and altitude measurement (distance from bottom) on the same platform (Figure 15.18).

15.2.3 Side-scan versus mechanically/electrically scanning

The only difference between the side-scan sonar and the mechanically/electrically scanning sonar systems is the means of locomotion of the transducer head (Figure 15.19).

The side-scan sonar generates an image of the target area by the simple means of "take a shot," move the tow fish, "take a shot," move the tow fish, etc. A highly directive pulse of acoustic energy is transmitted/bounced/received as the transducer platform is either towed behind a moving platform (surface or submerged vessel) or mounted to the side of a self-propelled platform (i.e., an AUV). Problems encountered with image generation by a side-scan sonar tow fish involve the constant surge and stall of the tow fish as the tow platform (e.g., a boat) bounces and pitches over a heavy sea state. AUVs (with vehicle-mounted side-scan sonars) are perceived to be potentially superior as an image generation platform due to the decoupling of the sonar platform from the surface sea state, allowing for steady and smooth image generation without image smear.

As depicted in Figure 15.15, the mechanically or electronically scanning sonar is mounted to some other relatively stable platform (ROV, tripod, tool, dock, etc.). For the mechanically scanning sonar system, the transducer head is rotated via a stepper motor timed to move as the transmit/receive cycle completes for that shot line. An image is generated as the sonar head builds the sonar lines in either a polar fashion or in a sector scan (Figure 15.20).

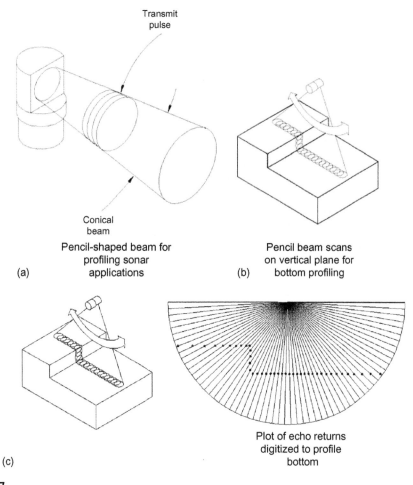

Transmit
pulse

Conical
beam

Pencil-shaped beam for
profiling sonar
(a)   applications

Pencil beam scans
on vertical plane for
(b)   bottom profiling

(c)

Plot of echo returns
digitized to profile
bottom

FIGURE 15.17

A profiling sonar (a) beam depicted, (b) beam scanning of bottom, and (c) display generation of beam scan.

The electrically scanning sonar can either be a spaced steerable array, a multibeam sonar, or a focused array:

- A steerable array involves separate point sources timed to produce a dipole beam. The time spacing of the pulses allows the beam axis to be steered due to additive pulse fronts.
- The multibeam sonar system is normally fixed with a nominal beam width of no more than 180°. The single sonar pulse is generated with multiple high directivity index receivers discriminating the backscatter return over its entire beam width to produce an image of the area under investigation.
- The focused array and acoustic lens sonar technologies are both incredible technologies that focus the acoustic beams onto a localized point at very high frequencies, generating near picture quality images with a high frame rate. More on this technology is detailed at Section 15.2.5.2 below.

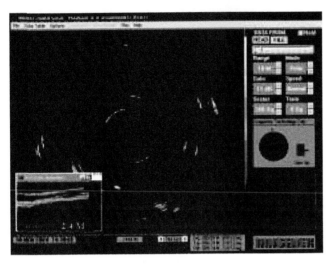

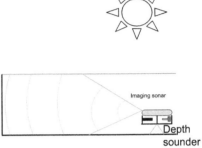

FIGURE 15.18

Here is an example of a dual imaging sonar and altimeter combination (image on screen and altitude on left).

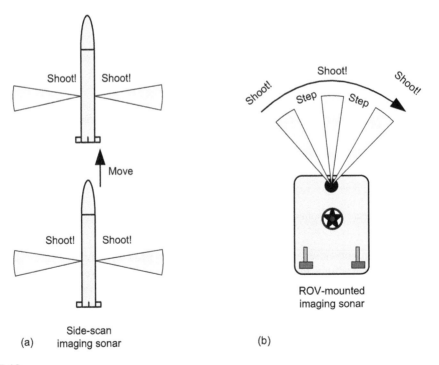

FIGURE 15.19

(a) Side-scan sonar versus (b) ROV-mounted mechanically scanning sonar.

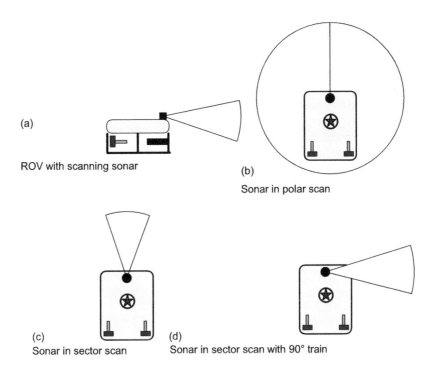

(a)
ROV with scanning sonar

(b)
Sonar in polar scan

(c)
Sonar in sector scan

(d)
Sonar in sector scan with 90° train

FIGURE 15.20

Scan modes of a mechanically scanning sonar system with (a) profile view, (b) plan view of polar mode, (c) sector scan, and (d) sector scan with train.

15.2.4 Single-/dual-/multifrequency versus tunable frequency

The single-/dual-/multifrequency, CHIRP, and tunable sonars all have their advantages and disadvantages. Definitions and discussion for each follows:

1. *Single frequency*: The single-frequency sonar system transmits and receives on one frequency and is the simplest sonar design due to its transducer selection on its nominal resonance frequency. Unfortunately, only a limited amount of target data can be ascertained from the use of a single-frequency backscatter analysis.
2. *Dual frequency*: This system arrangement allows gleaning of differing data parameters from the same target area based upon simultaneous or alternating transmissions of dual-frequency acoustic beams, generating each frequency's backscatter characteristics of the sample area. An excellent example of the advantages of dual-frequency sonar systems came recently during an oil spill in the Gulf of Mexico. A barge filled with heavy (heavier than water) heating oil struck a submerged unmarked platform destroyed by one of the storms of 2005. The oil sank to the bottom, requiring various governmental agencies to track the plume. To discriminate the submerged oil plume from the loosely consolidated mud bottom, a dual-frequency

(100 kHz/500 kHz) towed Klein 5000 sonar was used. The low-frequency band penetrated through the relatively light viscosity oil while the higher frequency bounced off the top of the oil/water interface, allowing for differentiation of the relatively rough mud bottom from the smooth oil as it migrated on the bottom (Figure 15.21). Once the search area was covered and the sonar data analyzed, an observation-class ROV visually characterized the bottom based upon sonar returns. Excellent tracking results were obtained of the oil plume, allowing for a higher incidence of recovery.

3. *CHIRP sonar*: This frequency-shifting sonar technology (along with its numerous technological advantages) is described more fully in the next section.
4. *Tunable frequency*: This type of transmitter/receiver combination uses frequency tuning to glean differing characteristics of the target area based upon the backscatter characteristics of the frequency transmitted. The strength of this technique is the ability to insonify the target and analyze backscatter at the various frequencies to gain better target characterization. The weakness is the rapid degradation in efficiency of the transducer as the frequency departs the nominal transducer design range.

15.2.5 CHIRP technology and acoustic lens systems

15.2.5.1 CHIRP sonar

CHIRP (compressed high intensity radar pulse) techniques have been used for a number of years above the water in many commercial and military radar systems. The techniques used to create an electromagnetic CHIRP pulse have now been modified and adapted to commercial acoustic imaging sonar systems.

To understand the benefits of using CHIRP acoustic techniques, one needs to analyze the limitations using conventional monotonic techniques. An acoustic pulse consists of an on/off switch modulating the amplitude of a single carrier frequency (Figure 15.22).

FIGURE 15.21

Dual-frequency sonar used during oil spill plume tracking with (a) tow fish hand launch and (b) sonar display.

The ability of the acoustic system to resolve targets is determined by the pulse length; this, however, has its drawbacks. To get enough acoustic energy into the water for good target identification and over a wide variety of ranges, the transmission pulse length has to be relatively long. The equation for determining the range resolution of a conventional monotonic acoustic system is given by:

$$Range\ resolution = velocity\ of\ sound/(bandwidth \times 2)$$

In a conventional monotonic system at moderate range, a typical pulse length is 50 μs and velocity of sound (VOS) in water 1500 m/s (typical). Therefore, the range resolution = 37.5 mm. This result effectively determines the range resolution (or ability to resolve separate targets) of the monotonic acoustic imaging system (Figure 15.23).

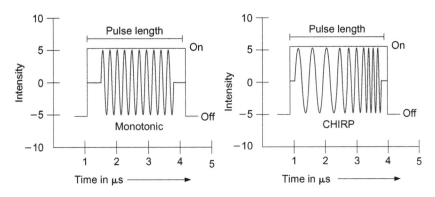

FIGURE 15.22

Comparison of monotonic (left) versus CHIRP (right) sonar techniques.

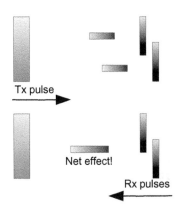

FIGURE 15.23

Example of inability of monotonic sonar to distinguish between close proximity targets.

Using the example above, if two targets are less than 37.5 mm apart, they cannot be distinguished from each other. The net effect is that the system will display a single large target, rather than multiple smaller targets.

CHIRP signal processing overcomes these limitations. Instead of using a burst of a single carrier frequency, the frequency within the burst is swept over a broad range throughout the duration of transmission pulse. This creates a "signature" acoustic pulse; the sonar knows what was transmitted and when. Using "pattern-matching" techniques, it can now look for its own unique signature being echoed back from targets.

In a CHIRP system, the critical factor determining range resolution is now the bandwidth of the CHIRP pulse. The CHIRP range resolution is given by:

$$Range\ resolution = 2 \times velocity\ of\ sound\ bandwidth$$

The bandwidth of a typical commercial CHIRP system is 100 kHz.
With VOS in water 1500 m/s (typical), our new range resolution = 7.5 mm—a theoretical improvement by a factor of 5!
This time, when two acoustic echoes overlap, the signature CHIRP pulses do not merge into a single return. The frequency at each point of the pulse is different and the sonar is able to resolve the two targets independently (Figure 15.24).
The response from the "pattern-matching" algorithms in the sonar results in the length of the acoustic pulse no longer affecting the amplitude of the echo on the sonar display. Therefore, longer transmissions (and operating ranges) can be achieved without a loss in range resolution.
Additionally, CHIRP offers improvements in background noise rejection, as the sonar is only looking for a swept frequency echo and removes random noise or out-of-band noise.

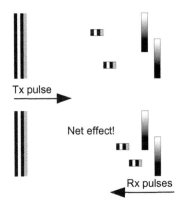

FIGURE 15.24

CHIRP sonar successfully distinguishes between close proximity targets.

In summary, CHIRP techniques provide the following advantages:

- Greatly improved range resolution compared to fixed-frequency sonars
- Larger transmission pulse lengths for increased operating ranges
- Improved discrimination between closely spaced targets
- Improved noise rejection and SNRs
- Reduced power consumption, from high-speed digital circuitry

15.2.5.2 Acoustic lens sonar

For an explanation of acoustic lens sonars, the concept will first be introduced followed by an example of a commercially available product using this technology.

From Loeser (1992) comes a general explanation of acoustic lens technology:

> *Acoustic lenses simplify the beam-forming process. The liquid lens is a spherical shell, filled with a refractive medium, that focuses sound energy in the same manner an optical lens focuses light energy. Sound waves incident on the lens are refracted to form a high-intensity focal region. The refraction is caused by a difference in acoustic wave speed in the lens media and surrounding water. The focusing ability is set by its diameter as measured in wavelengths for the frequency of interest.*
>
> *A single hydrophone located in the focal region of the lens forms a highly directive beam pattern without the necessity of auxiliary beam-forming electronics.*

The Sound Metrics name for their acoustic lens products are MIRIS and DIDSON (the latest iteration of this product series is the ARIS), referenced here as an example of the acoustic lens technology (Figure 15.25).

FIGURE 15.25

The DIDSON acoustic lens sonar.

MIRIS and DIDSON use acoustic lenses to form very narrow beams during transmission of pulses and reception of their echoes. Conventional sonars use delay lines or digital beam-forming techniques on reception and generally transmit one wide beam on transmission that covers the entire field of view. Acoustic lenses have the advantage of using no power for beam forming, resulting in a sonar that requires only 30 W to operate. A second advantage is the ease to transmit and receive from the same beam. The selective dispersal of sound and two-way beam patterns make the images cleaner due to reduced acoustic crosstalk and sharper due to higher resolution.

15.2.5.2.1 Lenses

Figure 15.26 shows a DIDSON sonar with the lens housing removed. The acoustic lenses and transducer array are shown above the electronics housing. The small cylinder under the side of the lens housing contains the focus motor and mechanism that moves the second lens forward and aft. This movement focuses the sonar on objects at ranges from 1 to 40 m. The front lens is actually a triplet made with two plastic (polymethylpentene) and one liquid (3M FC-70) component. The plastic lenses as well as the transducer array are separated by ambient water when the lens system is submerged. Optical lens design programs determined the lens curvatures. The designs were analyzed by custom software to evaluate the beam patterns over the field of view of interest.

15.2.5.2.2 Transducer array

The DIDSON and MIRIS transducer arrays are linear arrays. DIDSON has 96 elements with a pitch of 1.6 mm and a height of 46 mm. The elements are made with PZT 3:1 composite material constructed by the dice-and-fill method. The 3:1 composite provides a wide bandwidth, allowing DIDSON to operate at 1.8 or 1.0 MHz, the upper and lower ends of the transducer passband. The composite also allows the transducers to be curved in the height direction to aid in the formation of

FIGURE 15.26

DIDSON internal workings.

the elevation beam pattern. All 96 elements are used when operating at 1.8 MHz. Only the "even" 48 elements are used when operating at 1.0 MHz.

15.2.5.2.3 Beam formation

Figure 15.27 shows a ray diagram of the MIRIS system. A plane wave entering the left side through the front triplet L1 and single lenses L2 and L3 is focused to a line perpendicular to the page at the transducer T. If the normal to the plane wave is perpendicular to the front lens surface at the center, the acoustic line is formed at 0° in the diagram. If the normal is 9° off from perpendicular, the line is formed at 9° in the diagram. When a focused line of sound coincides with a long, thin transducer element, the acoustic energy is transformed into electrical energy and processed. The DIDSON beam former loses approximately 10 dB in sensitivity each way with beams 15° off-axis. Even with this reduction, DIDSON images fill the 29° field of view as shown in Figure 15.28. The average beam width in the vertical direction for both MIRIS and DIDSON is 14° (one-way). The lenses form the horizontal beam width and the curved transducer element forms the vertical beam width.

15.2.5.2.4 Image formation

The sonar transmits a short pulse and then receives its echo as it sweeps along the stripe. The echo amplitude varies in time as the reflectance varies with range along the insonified surface. Echoes from 96 adjacent lines, which together map the reflectance of the insonified sector-shaped area, are used to form a DIDSON image.

The difference between optical video and images from these sonars is more than the usage of light or sound. The sonar must be oriented to project beams with a small grazing angle to the surface of interest. The resulting image appears to be viewed from a direction perpendicular to the surface and the shadows indicate a source off to the side. Optical video could image a surface with the camera view perpendicular to the surface. If the sonar beams were perpendicular to the surface, the resulting image would show a single line perpendicular to the center beam axis. The line would be located at the range the surface is from the sonar. The sonar images are formed with "line-focused"

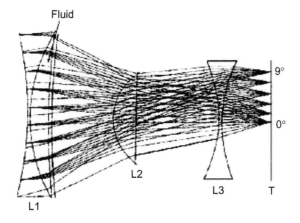

FIGURE 15.27

MIRIS internal workings.

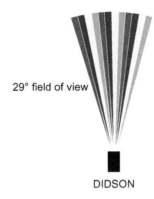

29° field of view

DIDSON

FIGURE 15.28

DIDSON beam forming.

beams that provide good images in many conditions, but not in all conditions. If objects were at the same range in the same beam but at different elevations, this type of imaging could not sort them out. An example would be trying to image an object in a pile of debris on the ocean floor. If the object were imbedded in the pile, the acoustic images from MIRIS and DIDSON would be confusing. Video using "point-focused" optics could meaningfully image the object embedded in the pile as long as it was not totally covered.

Fortunately, the great majority of imaging tasks do not have multiple objects in the same beam, at the same range, but at different elevations. In most cases, MIRIS and DIDSON provide unambiguous, near photographic-quality images. In the newest generation of lens sonar from Sound Metrics, the ARIS sonar features a higher number of beams along with better processing power and data capture/manipulation through improvements in technology.

15.3 Sonar techniques

15.3.1 Using an imaging sonar on an ROV

The imaging sonar is a useful addition to a positioning system on an ROV. Without an imaging sonar, an ROV operator must rely on flying the submersible underwater to bring new targets into view. With an imaging sonar, instead of traveling, it is more useful to spend some time with the vehicle sitting on the bottom while the sonar scans the surrounding area. Scanning a large area takes only a short time. The vehicle pilot can quickly assess the nature of the surrounding area, thereby eliminating objects that are not of interest. The ability to "see" a long distance underwater allows the pilot to use natural (or man-made) features and targets as position references (Figures 15.29 and 15.30).

If the ROV pilot is searching for a particular object, recognition can take place directly from the sonar image. In other cases, a number of potential targets may be seen. A pilot can sharpen sonar interpretation skills by viewing these targets with the vehicle's video camera in order to correctly identify/characterize them.

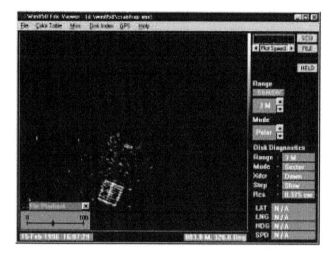

FIGURE 15.29

Sonar scan of a crab trap with imaging sonar.

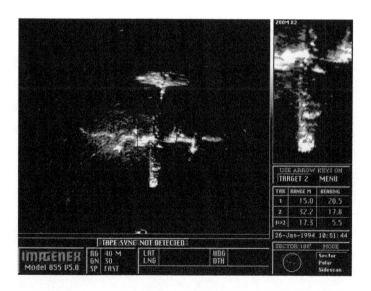

FIGURE 15.30

Convair PB4Y-2 privateer in Lake Washington, WA.

A word of caution regarding mechanically scanning sonar systems mounted on ROVs (especially OCROVs): An image is generated as the sonar transducer is rotated around its axis. If the sonar platform, that is, the submersible, is moved before the image is allowed to generate, "image smear" will occur (Figure 15.31). This phenomenon distorts the 2D display, which will not depict the correct placement of the items on the screen in *x/y* perspective.

15.3.2 Technique for locating targets with ROV-mounted scanning sonar

The accepted technique for locating the identifying items of interest insonified by scanning sonar is a four-step process:

1. Place the vehicle in a very stable position on or near the bottom to allow generation of a 360° (or wide angle) image. Depending upon the range, sector angle, and scan speed selected, the image may take up to 30 s to generate (Figure 15.32).
2. Identify the relative bearing of the item of interest.
3. Turn the vehicle to place the item at a zero bearing (in line with the bow of the vehicle), and then narrow the sector scan of the sonar system to approximately 45°. By narrowing the sector, the sonar scans only the area of the target to maintain contact.
4. Maintain contact with the item on sonar as the vehicle is driven forward toward the target until the item is in view.

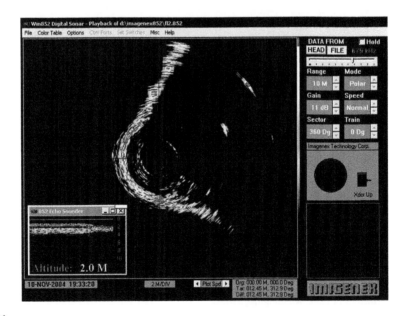

FIGURE 15.31

This image displays the sonar image of a straight wall distorted due to movement of the submersible before the image was allowed to generate.

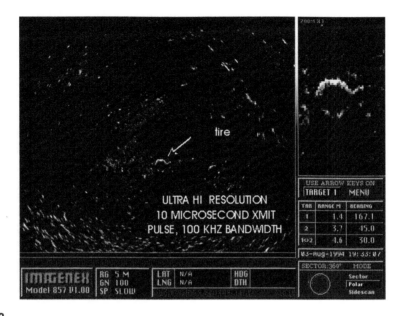

FIGURE 15.32

Ultrahigh-resolution sonar image.

15.3.3 Interpretation of sonar images

In many cases, the sonar image of a target will closely resemble an optical image of the same object. In other cases, the sonar image may be difficult to interpret and unlike the expected optical image.

The scanning process used to create a sonar image is different from the process used by the human eye or a camera to produce optical images. A sonar image will always have less resolution than an optical image, due to the nature of the ultrasonic signals used to generate it. Generally, rough objects reflect sound well in many directions and are therefore good sonar targets. Smooth angular surfaces may give a very strong reflection in one particular direction, but almost none at all in other directions (Figure 15.33). They can also act as a perfect mirror (so-called specular reflectors), reflecting the sonar pulse off in unexpected directions, never to return. This happens to people visually when they see an object reflected in a window. The human eye deals with such reflections daily, but it is unexpected to see the same thing occur with a sonar image.

As with normal vision, it is useful to scan targets from different positions to help identify them. A target unrecognizable from one direction may be easy to identify from another. It is important to note that the ranges shown to the targets on the sonar image are "slant" ranges. Usually the relative elevations of the targets are not known, only the range from the transducer. This means that two targets, which are displayed in the same location on the screen, may be at different elevations. For example, you might see a target on the bottom and a target floating on the surface in the same place (Figure 15.34).

By analyzing the shadows, an estimation of the height of objects above the bottom can be ascertained. An example of this calculation is shown in Figure 15.35.

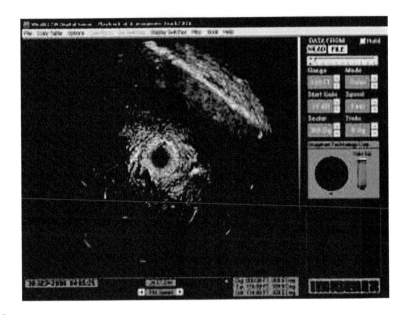

FIGURE 15.33

On the left side of the sonar display is a Ford F-150 pickup truck in 40 ft of water completely reflecting sonar signal (only shadow is recognized).

The diagrams in this chapter are examples of the sonar scanning process. Studying the diagrams will help users to better understand the images seen. A basic knowledge of this process will help users interpret what otherwise might be confusing images.

15.4 New and emerging technologies

Since the publication of the first edition of this text, a new technology—3D imaging sonar—has provided some revolutionary capabilities. This technology takes place in both hardware as well as software, allowing for full 3D rendering of *in situ* targets for various applications including metrology and modeling.

15.4.1 Image capture

The five methods for capturing 3D acoustic data are as follows:

1. Holding the platform stationary and then swinging the 2D beam projector to capture the third dimension (Figure 15.36)
2. Fixing the beam stationary relative to the platform and then moving the platform to capture the third dimension (Figure 15.37)
3. Rendering with a wider 3D beam (Figure 15.38) for full volumetric image capture
4. Separate two (or more) individual fixed projectors and render simultaneously (Figure 15.39) as the projectors are simultaneously moved (i.e., a small variation on item 2 above)
5. Physically rolling (rotating) the 2D beam/sonar housing, thus rendering the third dimension by capturing the vertical orientation (Figure 15.40)

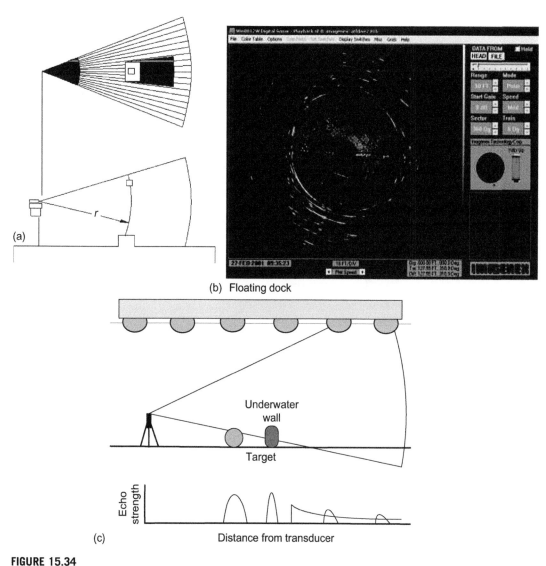

FIGURE 15.34

(a) Graphic illustration of two targets (one on surface and one on bottom) displaying colocation, (b) as depicted on the sonar display and (c) as imaged compared to signal strength.

15.4.2 Image rendering

Rendering of both 2D and 3D images from acoustic capture is done one of two ways:

1. With 2D data rendering, a bottom scene of a small physical area is scanned and imaged from several locations within the scene. Then each scan is laid on a digital raster illustration package

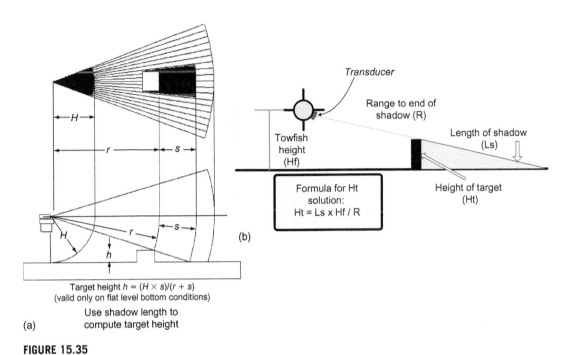

Target height $h = (H \times s)/(r + s)$
(valid only on flat level bottom conditions)
Use shadow length to
(a)   compute target height

Formula for Ht
solution:
$Ht = Ls \times Hf / R$

(b)

FIGURE 15.35

Target height calculations (a) with transducer graphically depicted and (b) shown with tow fish superimposed.

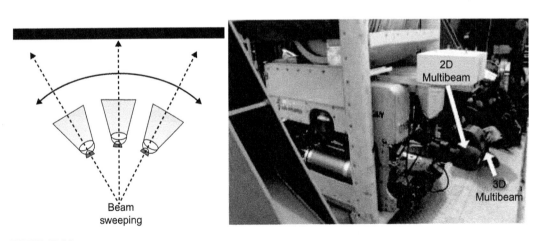

FIGURE 15.36

Concept of electronically sweeping multibeam and commercial application.

(Courtesy SeaTrepid.)

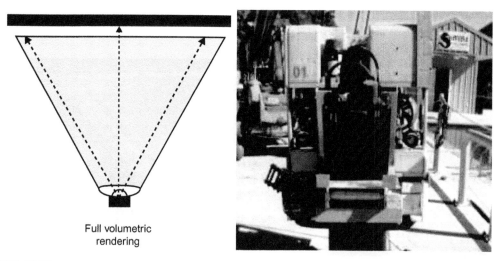

FIGURE 15.38

Concept of rendering 3D data through full volumetric coverage by sonar projector along with a Coda Octopus Echoscope integrated into an MSROV.

(Courtesy SeaTrepid.)

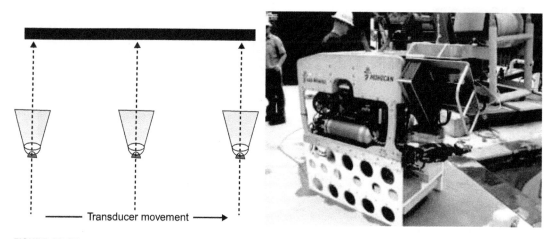

FIGURE 15.37

Concept of rendering 3D data by moving the sonar projector along with a fixed multibeam attached to an MSROV.

(Courtesy SeaTrepid.)

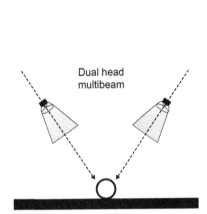

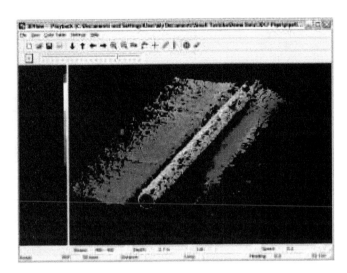

FIGURE 15.39

Concept of dual head multibeam and sample pipeline rendering.

(Courtesy Imagenex.)

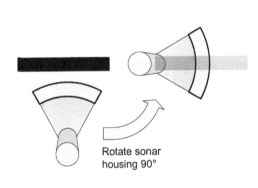

FIGURE 15.40

Concept of rotating a 2D sonar from horizontal to vertical orientation along with a Sound Metrics ARIS with X2 rotator integrated into an MSROV.

(Courtesy SeaTrepid.)

(e.g., Adobe Photoshop or similar). The individual items within the scene that are common between scans are then linked/stitched together to form a mosaic.

2. With 3D data rendering, the same series of scene capture is placed in a mosaic format. The difference between the 2D and 3D mosaic is that the 3D data is plotted into a 3D rendering software package (e.g., Leica Cyclone or similar) for the x/y/z point coordinates, thus producing a 3D "point cloud" of individual echoes (see Figure 23.8 in Chapter 23). This is particularly useful for modeling of *in situ* structures based upon real data for metrology purposes.

Acoustic Positioning

CHAPTER CONTENTS

16.1 Acoustic positioning—a technological development

The need for acoustic positioning became apparent with the loss, then difficulty, in locating the atomic attack submarine USS *Thresher*, which sank in 8400 ft (2560 m) of seawater in 1963, as well as a nuclear bomb lost at sea off the coast of Spain in 1966. The US Navy possessed the manned submersible capability to dive to the depth of the wreck site. But precision underwater navigation, through any other means but visual, was impossible with the available technology.

In the 1970s, as the search for hydrocarbons migrated into deeper water, repeatable high-accuracy bottom positioning became necessary to place the drill-string into the exact position referenced earlier through seismic instrumentation. Since radio frequency waves penetrate just a few wavelengths into water, some other form of precision navigation technology needed development. Thus was born acoustic positioning technology.

Today, water conditions in ports and harbors, as well as littoral areas around the world, are such that visual navigation below the surface is either difficult or impossible due to low visibility conditions. The need for underwater acoustic positioning remains high.

16.2 What is positioning?

According to Bowditch (2002), "positioning" is defined as "the process of determining, at a particular point in time, the precise physical location of the craft, vehicle, person or site." The position determination can vary in quality (degree of certainty as to its accuracy), in relativity (positioning relative to any number of reference frames), and point reference (versus a line of position (LOP) that is a mathematical position referenced along a given line, circle, or sphere).

A position can be derived from any number of means, including deduced (also termed "dead reckoning"), resolved (resolving bearing referenced to known fixed or moving objects—also termed "geodetic" when resolved relative to known earth-fixed objects), estimated (also termed SWAG—normally claimed by helmsman immediately before striking submerged rocks/objects), and claimed (or announced—"I claim this island in the name of King George").

16.3 Theory of positioning

All positioning is a simple matter of referencing a position relative to some other known position. From the earliest mariners, navigation was performed through "line of sight" with the coast. As explorers ventured farther from the sight of land, navigation with reference to the stars became common. Navigation with reference to the North Star for the determination of latitude was the earliest version of celestial navigation. Accurate determination of one's latitude could be gained by measuring the angular height of the North Star above the visible horizon.

The determination of position upon a known line of latitude formed a rudimentary "line of position" (LOP) in that a known position is resolved. In order to gain a higher positional resolution, a second LOP and then a third (and so forth) will be required to intersect the lines of position and resolve for two- and three-dimensional accuracy. This theory works for celestial navigation, GPS positioning, and (of course) acoustic positioning.

16.4 Basics of acoustic positioning

The basic underwater "speaker" is a transducer. This device changes electrical energy into mechanical energy to generate a sound pulse in water. For transducers used in underwater positioning, the typical transducer produces an omnidirectional sound beam capable of being picked up by other transducers in all directions from the signal source.

Acoustic positioning is a basic sound propagation and triangulation problem. The technology itself is simple, but the inherent physical errors require understanding and consideration in order to gain an accurate positional resolution.

As discussed in Chapter 2, water density is affected by water temperature, pressure, and salinity. This density also directly affects the speed of sound transmission in water. If an accurate round-trip time/speed can be calculated, the distance to a vehicle from a reference point can be ascertained. Therefore, the simple formula $R \times T = D$ (rate × time = distance) can be used. The time function is easily measurable. The rate question is dependent upon the medium through which the sound travels. The speed of sound (or "sonic" speed) through various media is listed in Table 16.1.

As shown in Table 16.1, pure water and seawater have different sound propagation speeds. For underwater port security tasks, varying degrees of water temperature and salinity conditions will be experienced. The industry-accepted default value for sound speed in water is 4921 ft/s (or 1500 m/s). If the extreme speed of pure water (4724 ft/s or 1440 m/s) to the median (4921 ft/s or 1500 m/s) is experienced, the difference of 197 ft/s (60 m/s) is approximately a 4% error (or 4 ft over a 100-foot distance). If this level of maximum error is acceptable, use the default sonic speed setting. Otherwise, consult the temperature/salinity tables for your specific conditions. Make speed adjustments within the software of the positioning system based upon those conditions.

The range to an acoustic beacon/transponder is a simple calculation:

$$R = \tfrac{1}{2}vt,$$

where R, range, is half the round-trip time, t, multiplied by the velocity, v.

If there is any latency time for a transponder/responder to process the signal, subtract that out of the time equation to produce a clean range to the beacon.

Table 16.1 Speed of Sound in Various Materials at 68°F/20°C at 1 Atmosphere

Material	Speed (ft/s)	Speed (m/s)
Air	1125	343
Air (32°F/0°C)	1086	331
Helium	3297	1005
Hydrogen	4265	1300
Pure water	4724	1440
Seawater	5118	1560
Iron and steel	16,404	5000
Glass	14,764	4500
Aluminum	16,732	5100
Hard wood	13,123	4000

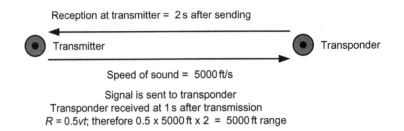

Reception at transmitter = 2 s after sending

Transmitter   Transponder

Speed of sound = 5000 ft/s

Signal is sent to transponder
Transponder received at 1 s after transmission
$R = 0.5vt$; therefore 0.5 x 5000 ft x 2 = 5000 ft range

FIGURE 16.1

Sample transmitting time calculation.

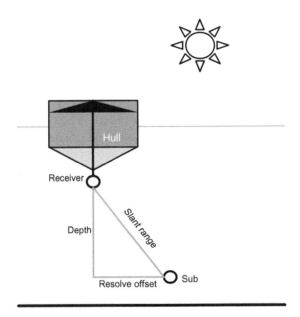

FIGURE 16.2

Computation of vehicle offset.

Using round numbers, if the speed of sound in water is 5000 ft/s and it takes 2 s (disregarding any latency time) from sending the signal to reception back at the transmitter, the beacon is 5000 ft away (Figure 16.1).

All underwater acoustical range is "slant range" for computing raw offset to the vehicle (more to point, from transducer to transducer). When tracking underwater vehicles from the surface, the depth of the vehicle is easily resolved from the vehicle's depth gauge. With the slant range and the depth producing two sides of a right-angled triangle, simple trigonometry will resolve the third side (Figure 16.2).

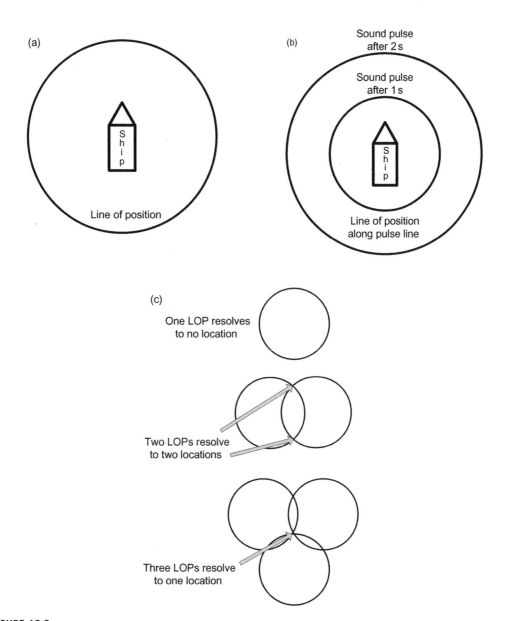

FIGURE 16.3

An omnidirectional pulse propagating outward at sonic speed. (a) Single pulse plotted as line of position, (b) multiple pulses, and (c) resolution of increasing resolved accuracy through single/dual/triple lines of position.

For determining the distance from a transducer, the resolved equidistant line of offset is known as a "line of position." This is defined as a line of known distance from a point (i.e., the transponder). The easy part of this equation is resolving the range question. The remainder of this chapter (and the difficult part of this technology) now focuses on resolving the bearing.

With two transducers, the lines of position will correctly resolve to two different locations (Figure 16.3). This has been experienced in the field, giving a resolution to an in-water position that showed the vehicle to be on land! The distance from each transducer is known, but the LOPs cannot resolve the different locations. This is termed "baseline ambiguity." A third transducer is needed in order to resolve the exact point. The software that comes with some of the positioning systems currently on the market allows the operator to select/assume a side of the baseline for operations. This allows disregarding readings on one side of the baseline, thus allowing a two-transducer arrangement.

In the scenarios shown in Figures 16.3 and 16.4, certain assumptions can be made. If it is determined that the submersible will be operated only on one side of the baseline (as shown in Figure 16.4, it is impossible to operate the system on land), all positional resolutions on one side of the baseline can be disregarded, producing accurate navigation with only two surface stations.

With underwater acoustic navigation and positioning, there are various types of underwater markers functioning as transmitters, receivers, or both. The six main classifications (Milne, 1983) are:

1. *Interrogator*: A transmitter/receiver that sends out an interrogation signal on one frequency and receives a reply on a second frequency. The channel spacing for these transmit/receive signals is often 500 Hz (0.5 kHz) apart.

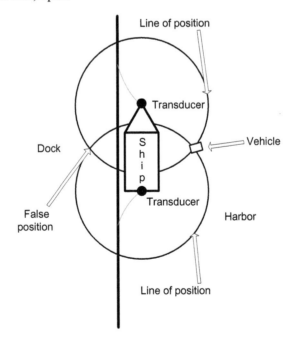

FIGURE 16.4

False position resolution.

2. *Transponder*: A receiver/transmitter installed on the seabed or a submersible (relay), which, on receipt of an interrogation signal from the *interrogator* (command) on one frequency, sends out a reply signal on a second frequency.
3. *Beacon/pinger*: A transmitter attached to the seabed or a submersible, which continually sends out a pulse on a particular frequency, that is, free running.
4. *Hydrophone*: An underwater "microphone" used to receive acoustic signals. This term also sometimes refers to a directional or omnidirectional receiver system (*hydrophone* plus receiver electronics), hull mounted, which is capable of receiving a reply from either a *transponder* or a beacon/pinger.
5. *Transducer*: A sonar transducer is the "antenna" of a transponder or interrogator. Either connected by a cable to the electronics package or hard mounted to it, a transducer can both transmit and receive acoustic signals. This is in contrast to a hydrophone, which is specified for receive operation only.
6. *Responder*: A transmitter fitted to a submersible or on the seabed, which can be triggered by a hard-wired external control signal (command) to transmit an interrogation signal for receipt by a receiver (hydrophone).

The particular arrangement of these items as well as their signal sequence determines the physical parameters of the overall positional accuracy.

16.5 Sound propagation, threshold, and multipath

Sound propagated in water as well as in air possesses many of the same characteristics. A shout in a canyon or in a theater can be echoed several times to the receiver (in this case, your ear). A concert played in an enclosed auditorium sounds much different from when it is played in an open-air location. Picking out the sound of one instrument in a full symphony is often a difficult task. Thus, trying to pick up a positioning beacon within the cacophony of underwater sounds within a busy harbor is also a challenge.

Range measurements are made by measuring the time it takes an acoustic signal (a "ping") to travel between the endpoints of the range of interest. In order for the range measurement (and hence the position determination) to be successful, the acoustic signal must be detected. A signal is "detected" if a pressure wave, of the proper frequency, has amplitude greater than a set threshold. All signal detection really means is hearing the "ping." The "ping" must be of the proper frequency and loud enough to hear (i.e., a signal-to-noise ratio of greater than 1).

Figure 16.5 shows an example of the reception of a signal from above and below the noise and reception threshold level; the signal is accepted in "1" and rejected in "2."

The performance of an acoustic positioning system can be predicted using a sonar equation that expresses the relationships between signal received and surrounding in-band noise. If a signal pulse results in a negative signal-to-noise ratio (or less than 5 dB), then most acoustic positioning systems will fail to detect the incoming signal (Wernli, 1998).

$$\text{Signal-to-noise ratio (dB)} = (E - N),$$

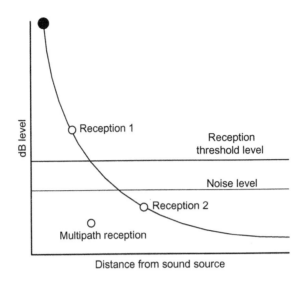

FIGURE 16.5

Reception threshold level.

where:

$$E = (SL - TL)$$
$$N = 20 \log_{10} (NT)$$
$$NT = \sqrt{(NA^2 + NS^2 + NR^2)}$$

with:

E = received signal sound pressure in dB
N = total received "in-band" noise sound pressure level
SL = source level in dB
TL = one-way transmission loss in dB
NT = total noise pressure level in μPa
NA = ambient noise pressure level (noise in the environment, both natural and man-made) in μPa
NS = self-noise pressure level (noise generated by the acoustic receiver itself) in μPa
NR = reverberation pressure level (reverberations or echoes remaining from previous pings) in μPa

The source level (SL) is the acoustic power, measured in decibels, transmitted into the water by the equipment.

There are two reasons why an in-band signal will not be detected: The signal is either too weak (too quiet) or there is too much noise.

16.5.1 Weak signals

There are many possible causes for a signal's being too weak. The most common cause by far is signal blockage. The acoustic signal cannot pass through certain objects, most notably those containing air (boat hull, fish, sea grass, etc.). Blockage will make it appear that the signal is too quiet and it may not be detected. Blockage can also be by air bubbles, kelp, rock outcroppings, mud, etc. The best way to combat signal blockage is to avoid it.

Other causes of a weak signal include damaged equipment (broken station or transducer), using the system at too great a range, the presence of thermoclines, surface effects, using a system in a liquid with a greater sound adsorption, etc. There are a large number of possible causes for a weaker than expected signal and often it will be difficult to determine the cause.

16.5.2 Noise

Noise is essentially the detection of an unwanted signal, thus drowning out the wanted signal. In order for a signal to be detected, there must be a method of separating the signal from any noise present. Essentially, the signal must be louder than the noise. Most hardware performs signal extraction through filtering and thresholding. "Filtering" is the operation of passing a signal at specific frequencies. "Thresholding" is the operation of classifying a signal based on amplitude. Basically, the system listens on a specific channel (frequency) for a signal above a certain amplitude level (threshold). If a signal of the appropriate frequency is received and it has amplitude above the threshold it is a valid signal, not noise, and will be allowed to pass.

Sound propagation in water is subject to a number of challenging environmental factors. Over the shorter distances covered by most observation-class ROV positioning systems, many of the physical factors affecting sonic speed (i.e., speed of sound in water) are insignificant. The significant physical factors over these shorter distances are sound threshold considerations and the multipath phenomenon.

16.5.3 Multipath

In Figure 16.6, all of the sounds shown originated from the same source (e.g., an acoustic beacon) but arrive at the receiver at different times. If each of these receptions appeared as a beacon on a

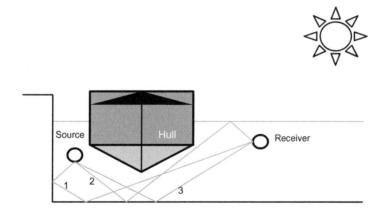

FIGURE 16.6

Multipath signals.

submersible, it would show three different distances—none of which would be correct, since none of these is line of sight. Sounds "1" and "2" have multiple reflections to get to the receiver, while sound "3" has just one reflection. If the sound source is a measured output level (units measured in dB or decibels), the receiver can be set to reject all reception below a certain dB level, thereby rejecting all multipath signals.

Acoustic positioning systems update at regular intervals. The update rate is limited by the source and reception offset. Once an acoustic sound signal has been generated, it must go to the end of its reception range and come back to the transmitter/receiver (thereby generating a range through timing difference) before the next pulse can be sent. This is the reason acoustic positioning cannot maintain real-time positioning feedback.

16.6 Types of positioning technologies

Any discussion of positioning must first begin with an explanation of the term "frame of reference."

16.6.1 Frame of reference

Whenever a position is given for anything anywhere in the universe, that position must be made within a coordinate reference frame. If a geo-referenced position is given, there are many unanswered questions simply to specify which mathematical model is used to generate that position. Some frame of reference examples are:

- "Two feet from my front door" (house reference frame)
- "The left seat of my car" (car reference frame)
- "Twenty feet off the port beam" (ship reference frame)
- "Control Yoke is located at station 45," that is, 45 inches aft of the originating datum on the engineering drawings of an example aircraft (aircraft reference frame)
- "Latitude 30° North/Longitude 90° West (WGS 84)" (earth reference frame)

In acoustic positioning, the raw positioning data is resolved from the acoustic beacon/tracking device to the transducer array (arranged in a known pattern). Your raw position will be given in an x/y/z offset from that referenced transducer array. The imaginary line forming the sides of the reference triangle(s) is known as a "baseline." The frames of reference can be fixed to the transducer array itself, a physical object (such as a dam, a ship, or a pier), or the earth. All offset coordinates may then be given with reference to that frame of reference.

The arrangement of the various acoustic sensors as well as their relative placement will determine both the accuracy of the positional resolution as well as the ease of placement and system cost. All arrangements use some form of active send/receive in order to positively measure the time of transmission and reception over the array of transducers. The known spacing and angular offset of the transducer array forms what is known as a "baseline" from which to measure angular distance (Figure 16.7). A comparison of different types of acoustic positioning arrangements is depicted in Figure 16.8.

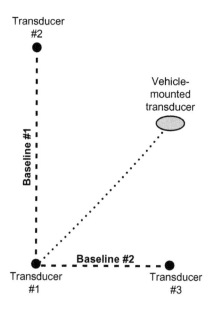

FIGURE 16.7

Basic arrangement of transducers into baselines for angular measurements.

16.6.2 **Short baseline**

A short baseline (SBL) system (Figure 16.9) is normally fitted to a vessel or fixed object with the vessel or fixed object used as the frame of reference. A number of (at least three but typically four) acoustic transducers are fitted in a triangle or a rectangle on the lower part of the vessel or fixed object. The distances between the transducers (the "baselines") are made as large as practical given physical space limitations; typically they are at least 30 ft (10 m) long. The position of each transducer within a coordinate frame fixed to the vessel or fixed object is determined by conventional survey techniques or from the "as built" survey.

The term "short" is used as a comparison with "long baseline" (LBL) techniques. If the distances from the transducers to an acoustic beacon are measured as described for LBL, then the position of the beacon, within the coordinate reference frame, can be computed. Moreover, if redundant measurements are made, a best estimate can be determined that is more accurate than the basic position calculation by averaging several fixes.

SBL systems transmit from one but receive on all transducers. The result is one distance (or range) measurement and a number of range (or time) differences.

With an SBL system, the coordinate frame is typically fixed to the vessel—which is subject to roll (change in list), pitch (change in trim), and yaw (change in heading) motion. This "disadvantage" can be overcome by using additional equipment such as a vertical reference unit (VRU) to measure roll and pitch and a gyro-compass to measure heading. The coordinates of the beacon can then be transformed mathematically to remove the effect of these rotational motions.

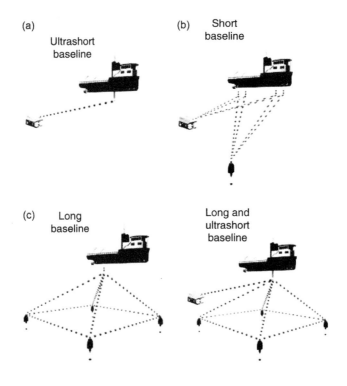

FIGURE 16.8

Comparative positioning system arrangements for (a) USBL, (b) SBL, (c) LBL and LUSBL.

(Courtesy of Sonardyne.)

At the shorter transducer spacing distances (i.e., less than 30 ft), SBL produces a higher error level than does LBL due to the greater impact of multipath and other range measurement errors. Also, any errors in transducer spacing measurement, heading indicator errors, GPS inaccuracies, and vessel instability (i.e., due to vessel motion) are propagated through to position inaccuracies. SBL is not normally considered survey quality.

16.6.3 Ultrashort baseline

Ultrashort baseline (USBL, also termed "super short baseline," or SSBL, by European operators) principles (Figure 16.10) are very similar to SBL principles (in which an array of acoustic transducers is deployed on the surface vessel) except that the transducers are all built into a single transceiver assembly—or the array of transducers is replaced by an array of transducer elements in a single transceiver assembly.

The distances or ranges are measured as they are in an SBL system but the time differences have decreased. Systems using sinusoidal signals measure the "time phase" of the signal in each element with respect to a reference in the receiver. The "time-phase differences" between transducer elements are computed by subtraction and then the system is equivalent to an SBL system.

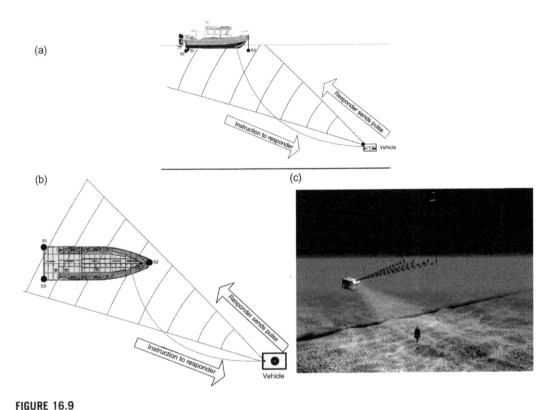

FIGURE 16.9

Typical SBL system (a) profile view, (b) plan view, and (c) graphical depiction.

Another practical difference is that the transducer elements are in a transceiver assembly that is placed somewhere in the vessel frame. The attitude of the assembly in the vessel frame must be measured during installation. It should be remembered that, intrinsically, a USBL system positions a beacon in a frame fixed to the transceiver assembly, not directly in a vessel-fixed frame as in the SBL case.

16.6.4 Long baseline

An LBL system consists of a number of acoustic transponder beacons moored in fixed locations on the seabed or mounted on fixed locations of objects such as ships or oil platforms to be surveyed (Figures 16.11–16.13). The positions of the beacons are described in a coordinate frame fixed to the seabed or the referenced object (an example of a non-seafloor LBL system is the vessel-referenced ship hull inspection system). The distances between the transponders form the "baselines" used by the system.

These transponders are interrogated by the interrogator, which is installed on the ROV, the diver, the submersible, or the tow fish to be positioned. The distance from the transducer/interrogator to a transponder beacon can be measured by causing the transducer/interrogator to transmit a short

(a)

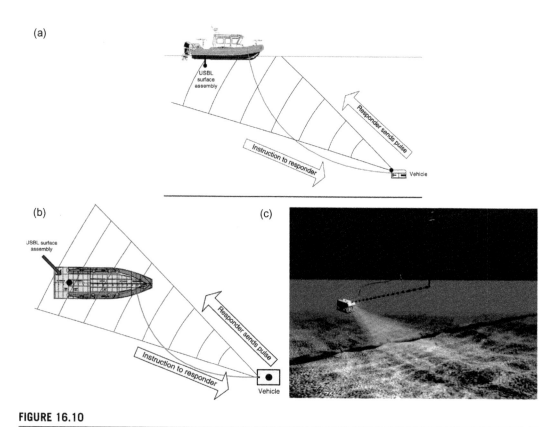

(b)

(c)

FIGURE 16.10

Typical USBL arrangement (a) profile view, (b) plan view and (c) graphical depiction.

acoustic signal, which the transponder detects, causing it to transmit an acoustic signal in response on a discrete channel. The time from the transmission of the first signal to the reception of the second is measured. As sound travels through the water at a known speed, the distance between the transducer and the beacon can be estimated. The process is repeated for the remaining beacons and the position of the vessel relative to the array of beacons is then calculated or estimated.

In principle, navigation can be achieved using just two transponder beacons, but in that case there is a possible ambiguity as to which side of the baseline (a line drawn between the beacons) the vessel may be on (the so-called baseline ambiguity—see previous section of this chapter). In addition, the depth or height of the transducer has to be assumed, unless there is an embedded depth transducer that encodes depth measurements with the transponder response. Three transponder beacons is the minimum required for unambiguous navigation in three dimensions; four is the minimum required for some degree of redundancy. This is useful for checks on the quality of navigation.

The term "long baseline" is used because, in general, the baseline distances are much greater for LBL than for SBL and certainly for USBL. Because the baselines are much larger, an LBL

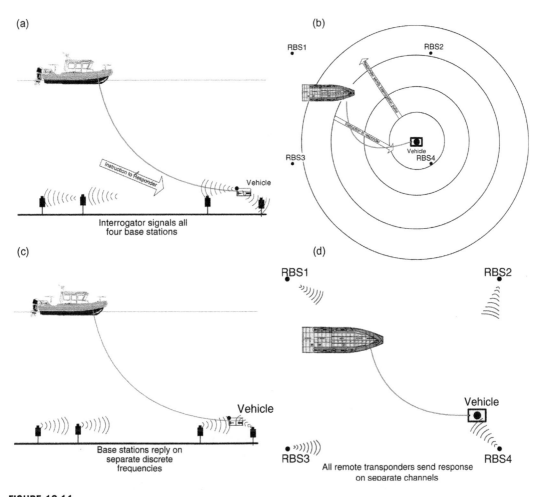

FIGURE 16.11

An LBL system first instructs the responder to interrogate ((a) profile view, (b) plan view) and transponders then reply ((c) profile view, (d) plan view).

system is very accurate and position fixes are very robust compared with the SBL and USBL versions. In addition, the transponder beacons are mounted fixed in the desired reference frame, such as on the seafloor for seafloor surveys or on a ship's hull for ship hull surveys. This removes most of the problems associated with vessel motion.

The vessel-referenced ship hull positioning system is useful in underwater port security needs (Figures 16.14 and 16.15). This system simply applies the LBL system and uses as a frame of reference the vessel to be surveyed—depicted by scaled drawings of that vessel. In ship hull inspections, certain assumptions can be made allowing positive navigation with only one transponder in "sight." If only one transponder is communicating, a deduction that the submersible is colocated in the

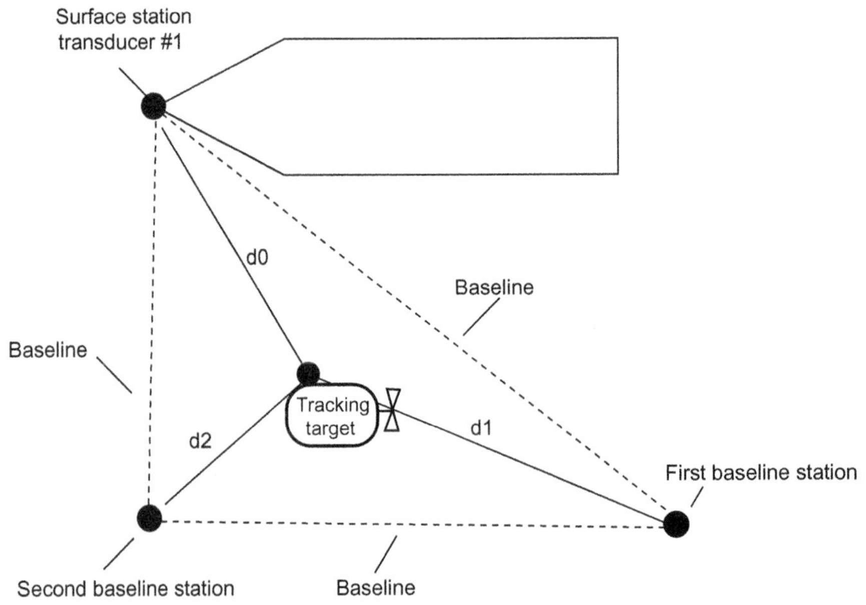

Surface station
transducer #1

d0

Baseline

Baseline

Tracking
target

d2

d1

First baseline station

Second baseline station

Baseline

FIGURE 16.12

Typical arrangement for an LBL system with transponders mounted on the seafloor.

(*Courtesy of Desert Star LLC.*)

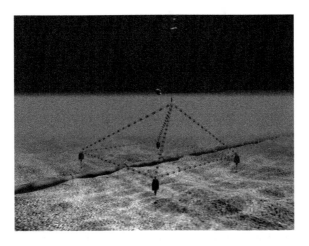

FIGURE 16.13

LBL setup on vessel.

(*Courtesy of Sonardyne.*)

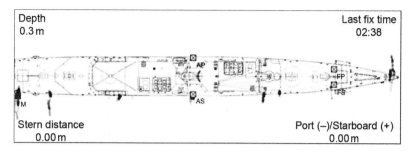

FIGURE 16.14

Depiction of vessel with ship hull inspection system transponders registered onto vessel drawing (M = mobile station, AP = aft port station, AS = aft starboard station, FP = forward port station, FS = forward starboard station).

quadrant of the ship that contains the requisite transponder can be made as long as the submersible is in visual contact with the hull.

For geo-referenced arrangements, the array of seabed beacons/transponders needs to be calibrated. There are several techniques available for achieving this. The most appropriate technique depends on the requirements of the task and the available hardware. With the continuing integration of LBL, SBL, and USBL systems, intelligent transponder beacons (that measure baselines directly), and satellite navigation systems, the calibration of seabed arrays is becoming a quick and simple operation. The operator will be free to choose the techniques appropriate to the requirements and the task based upon manufacturer-supplied specifications.

16.7 Advantages and disadvantages of positioning system types

USBL advantages:

- Low complexity
- Easy to use
- Good range accuracy
- Ship-based system

USBL disadvantages:

- Detailed calibration of vessel-based transducer assembly required (usually not performed accurately)
- Requires vessel installation of a rigid pole for mounting of the transceiver unit
- Position accuracy depends on ship's gyro and VRU
- Minimal redundancy
- Large transducer/gate valve requiring accurate/repeatable orientation

SBL advantages:

- Good range accuracy
- Redundancy

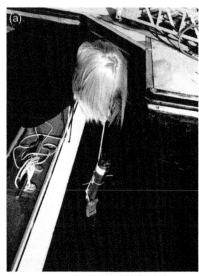

FIGURE 16.15

Ship hull inspection system being deployed with (a) and (b) hand deployment of transponder along with (c) transponder integrated into ROV.

(Courtesy Desert Star LLC.)

- Ship-based system
- Small transducers/gate valves
- No orientation requirement on deployment poles. Easy "over the side" cable-mounted transducer deployment is possible

- Good positioning accuracy is possible when operating from larger vessels or a dock (good transducer separation)

SBL disadvantages:

- Poor positioning accuracy when operating from very small vessels (limited transducer separation)
- Requires deployment of three or more surface station transducers (more wiring than USBL)

LBL advantages:

- Very good position accuracy independent of water depth
- Redundancy
- Wide area of coverage
- Single, simple deployment pole

LBL disadvantages:

- Complex systems requiring more competent operators
- Large arrays required
- Longer deployment/recovery time
- Calibration time required at each location

16.8 Capabilities and limitations of acoustic positioning

Acoustic positioning is capable of high-precision positioning in a number of frames of reference. Once the local sonic speed is computed, range is capable of being accurately measured. The larger the spacing between the transducers forming the baselines, the higher the possible bearing accuracy. The basic range/bearing resolution accuracy with LBL positioning is generally of a higher quality than the SBL and USBL techniques due to the inherent greater angular offset reception time spacing of the baselines. Range in all methods is quite accurate, assuming line of sight is maintained. Bearing accuracy of $1-3°$ for SBL and USBL is considered acceptable, while bearings of $<1°$ are possible in LBL configuration.

Acoustic positioning relies upon a line-of-sight sound signal from a transducer (generally mounted aboard the submersible) to a receiver, so that an accurate range and bearing can be resolved to give a position in some frame of reference.

Just as picking out a voice in a crowd is difficult, so is picking out an acoustic positioning beacon in a noisy harbor environment. The challenge is balancing the reception sound threshold to pick up the line-of-sight transmission and not the false signal from noise or from multipath sound reflections.

If the sound threshold level is placed too high, the possibility exists that the true line-of-sight signal from the beacon to the receiver is rejected (Figure 16.16). If the sound threshold reception level is set too low, false-positive readings from either noise or from multipath reception will show erroneous positions for your vehicle.

In Figure 16.17, sound source 1 is below the noise level and below the reception threshold—the signal is not received. In signal source 2, the source level is above the reception threshold and

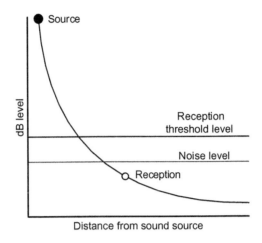

FIGURE 16.16

An example of the reception of a signal below the noise and reception threshold level—signal is rejected.

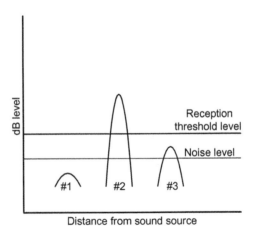

FIGURE 16.17

Sample sound source relationships to the reception threshold and noise level.

above the noise level—the signal is received and is passed. In signal source 3, the source is above the noise level but below the reception threshold—signal is received but rejected.

The two relevant underwater acoustical environmental issues are:

1. Background noise
2. Reverberation/multipath

Background noise requires the transmitting beacon to possess a higher output in order to "burn through" the noise and be recognized by the receiver. Reverberation/multipath propagation causes

Table 16.2 Environment and Typical Acoustic System Settings

Environment	Noise/Receiver Sensitivity	Reverb/Time Spacing
Lake	Low/high	Low/low
Pool	Low/high	High/high
Open ocean	High/low	Low/low
Harbor	High/low	High/high

problems by falsely recognizing signals other than the legitimate line-of-sight signal necessary to measure range to the source transducer. The two responses to these issues are to:

1. *Increase* the dB output of the transmitter.
2. *Time space* the transmission of source signals more widely to allow reverberation of sound to dissipate before sending the next signal.

Some examples of environment types are listed below with their respective characteristics:

- *Pools* are characterized by low background noise yet with a high presence of reverberation. The acoustic positioning system should be set to high sensitivity (due to the absence of background noise) with larger spacing between signals to allow reverberation to dissipate before the next signal generation.
- *Harbors, constricted waters, and surf zones* are noisy acoustical environments with a great presence of reverberation. The acoustic positioning system would be set to low sensitivity (to reject false signals) with long signal spacing.
- *Open ocean* means moderate noise and low echoes. Set the positioning system to low sensitivity and short spacing.
- *Lakes* are quiet and contain few possibilities of reverberation. Set the positioning system to high sensitivity and short spacing.

A breakdown of environment versus settings is provided in Table 16.2.

16.9 **Operational considerations**

16.9.1 **Operations in open ocean**

The open ocean can have high background noise due to broadband biological noise generation. Another factor affecting proper positional accuracy is the pitch and sway of the operation platform (i.e., the vessel). When deploying an SBL or USBL system from a rocking boat (with the frame of reference attached to the vessel), the baseline sensing station is constantly moving, disrupting a steady frame of reference (Figure 16.18).

If the transponder beacons are deployed anchored to the bottom (as in an LBL system), the frame of reference is steady, obviating the apparent bearing error from boat sway. The vessel motion errors are resolved and canceled by the VRU. The VRU determines the instantaneous platform orientation and then corrects the out-of-plane errors to produce a clean position resolution.

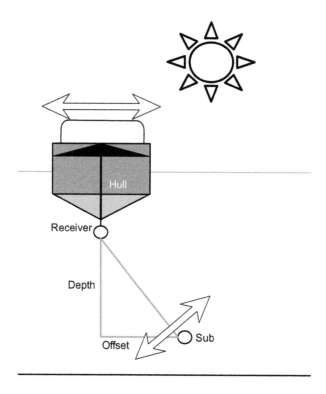

FIGURE 16.18

Effect of a moving vessel on the submersible's estimated position.

16.9.2 Operations in ports and harbors

The port and harbor acoustic environment is characterized by noisy reverberant conditions with broadband machinery noises from vessels and various sources as well as strong reverberations within the water space. Tidal current flow also affects port and harbor water sound propagation. Acoustic positioning operations require close attention to the details of background noise as well as signal spacing in order to gain accurate position readings.

16.9.3 Operations in close proximity to vessels and underwater structures

Special consideration is necessary when operating in close proximity to vessels and underwater structures such as piers, submerged structures, and anchorages. As stated previously, line-of-sight reception must be maintained to gain accurate range measurements. It is not always possible to maintain line of sight while performing a pier or hull inspection.

SBL and USBL systems require that the submersible be operated near the deployment platform in order to see all of the surface units required to gain a full position resolution. If an LBL system is used with the transponders located on the bottom, a better angle is possible to "see" around the

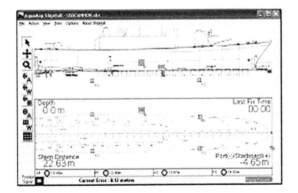

FIGURE 16.19

Vessel-referenced positioning system.

vessel being inspected. With either method, operation in and around vessels and enclosed structures may not be conducive to acoustic positioning.

One acoustic ship hull inspection system uses a series of assumptions in order to gain accurate position measurements near ship hulls despite blockage and multipath problems.

A series of acoustic transponders are hung over the side of a vessel to be inspected (Figure 16.19). The transponders are positioned below the level of the keel so that all are "in sight" once the submersible is under the vessel. Each transponder has a separate frequency corresponding to its position on the vessel (i.e., forward port, aft port, forward starboard, aft starboard). Once the submersible travels toward the surface on one side of the vessel, it loses sight of the transponders on the opposite side of the vessel (an example would be traveling up toward the surface on the starboard side, losing sight of the port side transponders). Positional accuracy is then maintained with two transponders. The "baseline ambiguity" is resolved by knowing that the submersible cannot be "inside of" the vessel being inspected, thus putting the submersible in only one of the two possible positions. If only one transponder is received, it can be assumed that the submersible is in contact with the hull in the quadrant of the receiving transponder.

16.10 Position referencing

16.10.1 Geo-referencing of position

The basic resolution of an acoustic positioning system is range and bearing. Input of the offset from the primary reference transducer to the center rotation point of the vessel will produce a reference point for the extended centerline of the vessel (Figure 16.20). From this, convert relative bearing from the vessel's axis into magnetic bearing. By comparing the relative bearing (resolved by triangulation) from the vessel to the submersible, with the magnetic heading of the vessel, a resolution is then possible for magnetic bearing to the submersible. Then add the offset of the center pivot point of the vessel to the GPS antenna, and resolution of the range and magnetic bearing to

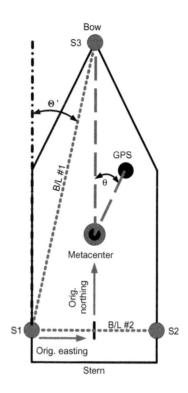

Compute:
- B/L #1 Length
- B/L #2 Length
- B/L #1 Orientation (Θ')
- B/L #1 to #2 angle
- Metacenter easting and northing
- GPS offset (range/bearing θ) from metacenter

FIGURE 16.20

Resolution of vehicle position to ship's coordinates.

place the submersible in three-dimensional space. Resolution to geographic coordinates is then possible (Figure 16.21).

Refer to the manufacturer's operations manual of the equipment to be operated to gain specific steps in setting up the equipment on your vessel. Figures 16.22 and 16.23 depict a typical screen readout of a geo-referenced positioning system.

16.10.2 Vessel referencing of position

In the geo-referenced system above, changing the submersible from relative reference (range/bearing from the vessel—more specifically to the transducer array) was complex due to the difficulty of resolving range/bearing from relative to geographical. To determine the position referenced to a ship (whose geographic coordinates are irrelevant), use a scaled drawing and place the transducer array on that scaled drawing to register position in reference to that vessel (Figure 16.24).

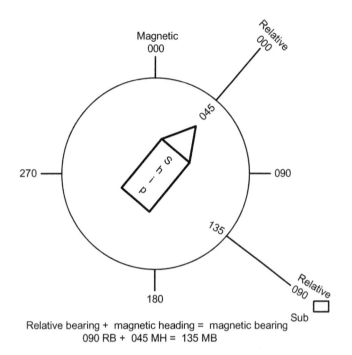

Relative bearing + magnetic heading = magnetic bearing
090 RB + 045 MH = 135 MB

FIGURE 16.21

Resolution of ship's vehicle position to geographic coordinates.

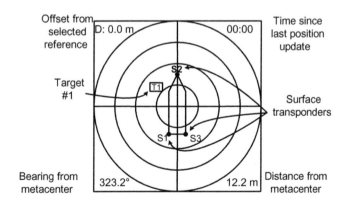

FIGURE 16.22

Typical relative bearing display of acoustical positioning to the deployment vessel.

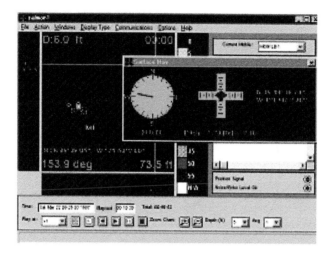

FIGURE 16.23

Readout of geo-referenced coordinates.

(Courtesy Desert Star Systems LLC.)

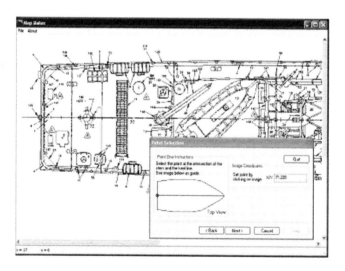

FIGURE 16.24

Position referenced to ship.

16.10.3 Relative referencing of position

It is now possible to take the ship relative reference and make any reference system that is desired. Take a nautical chart and register the coordinates for length/width to create a scaled drawing depicting the estimated surface of the earth at that location. Take a drawing of a dock and scale as

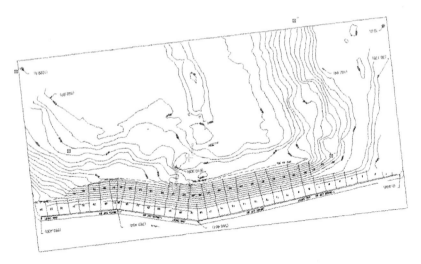

FIGURE 16.25

Relative referenced dam map with acoustic positioning.

(Courtesy Desert Star Systems LLC.)

well. This creates a scaled dock drawing where the acoustic positioning system is now relatively referenced to the drawing of the dock. Any number of combinations is possible with an understanding of scaled drawings and relative reference with regard to a frame of reference. Figure 16.25 depicts a scaled drawing of a dam with positioning beacons registered on the drawing.

16.11 General rules for use of acoustic positioning systems

The following are some guidelines to use for acoustic positioning systems:

- The best angular resolution is broadside to the longest baseline.
- Attention to detail in placement of baseline stations is paramount due to propagation of any baseline errors to the position resolution of the vehicle.
- Most acoustic positioning systems have a noise test capability. If in doubt about the background noise level, test it and adjust accordingly.
- Operate all vessel equipment to be used during performance of acoustic positioning during any noise level test.
- Carefully consider your environment type when selecting settings on your positioning system.
- If the operating platform is a heaving vessel and the frame of reference is the same vessel (without access to a VRU), expect position dithering.
- Noise levels change constantly over the course of any operation. Consider this if loss of positioning is experienced.
- If positional dithering is experienced, multipath is the most likely problem. Lower your receiver sensitivity to compensate for false (multipath) readings.

Navigational Sensors

CHAPTER CONTENTS

In this chapter, the separation of "vehicle navigational sensors" from "survey navigational sensors" will be examined along with their configurations and applications. Each of the typical survey-grade sensor's principles of operation is examined with the goal of creating a basic reader understanding of each sensor's function (vis-à-vis the task being performed). There are a wide range of sensor variations (technology, accuracy, etc.) regarding the sensors explained below; therefore, each sensor must be analyzed so as to gain a full understanding of its function and utilization.

17.1 Payload sensors versus vehicle sensors

As explained in Chapter 11, typical commercial ROV systems come equipped with a basic set of sensors for navigating the vehicle under normal conditions. These sensors are typically of a lower sensitivity as the need to position a vehicle in a basic orientation of heading/depth is simple.

For tasks demanding a higher tolerance of location (such as for most survey tasks), a much tighter tolerance for positional error is demanded. This is especially true for such tasks as pipeline survey, drilling rig/drill bit placement, or some subsea construction operations. As a result, the end-customer will require the surveyors to assure accurate and reliable vehicle positioning/orientation. Table 17.1 provides some examples of vehicle sensors versus survey sensors.

Further, the survey-grade sensor will typically have a much higher accuracy than a similar vehicle sensor. Communication with the vehicle sensors is normally performed through the vehicle's telemetry system. The survey sensor, however, is typically broken out of the vehicle's telemetry system and is isolated into a completely separate communications channel (most times with its own 1 atm pressure housing), known as a survey "pod" or "mux", which is conducted through a separate fiber (or copper) conductor in the tether. Normally, the only commonality between the vehicle's sensors and the survey "pod/mux" is common power from the vehicle's power bus.

17.1.1 Division of responsibility between ROV and survey functions

On most commercial operations, the survey function is separated from the ROV function as the two skills are unique. The survey function is responsible for accurate placement and depiction of sensor placement (i.e., geo-location) as well as the correlating of sensor readings to geo-referenced coordinates. The ROV function is simply to assure the vehicle is operational and directed to the location of the client's choosing. The survey crew typically collects and processes the data into a final deliverable.

Standard practice in the offshore oil and gas industry (as well as most other commercial applications) is for the survey team to be responsible for all items of survey and positioning above what is provided by the ROV as standard equipment. Examples of survey equipment responsibility are vehicle-mounted acoustic transponders (and conductivity, temperature, and depth (CTD) sensors for sound velocity calibration), 3D multibeam sonar, altimeter, survey-grade gyro, and other mission-specific navigational sensors.

Table 17.1 Examples of Vehicle Versus Survey Sensors

Vehicle Sensors	Survey Sensors
Flux gate magnetic compass	CTD (for Sound Velocity Profiling)
Pressure-sensitive depth gauge	RLG or FOG for orientation
Tether turn counter	DVL for speed over ground
Vehicle altimeter	Survey altimeter
Vehicle telemetry/diagnostics	Acoustic positioning system

17.1.2 Typical survey "pod/mux" configuration

The survey pod is a power and data connection point linked into a central node for powering sensors (with vehicle power) as well as for data transmission to the surface, typically through the vehicle's tether (Figure 17.1). The pod is the central gathering point where sensor communications are fed into a multiplexer for transmission to the surface. Such transmissions are performed through the tether's dedicated communications channel (typically, a separate and dedicated copper or fiber conductor) to the surface for demultiplexing and then dissemination.

The pod is powered from the junction box through the vehicle's power bus. Depending upon the power requirement of the individual sensor, the sensor may be powered through the pod or directly from the vehicle's power bus. The advantage of a survey pod is easy connection of multiple external sensors (power and data) into the vehicle power and telemetry bus without opening the vehicle's electronics can (with the possibility of electronics damage or mis-wiring).

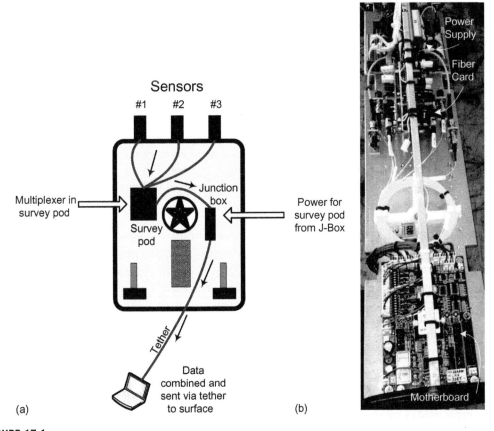

FIGURE 17.1

(a) Typical survey pod integration to vehicle, such as (b) the MacArtney Nexus IV subsea multiplexer.

17.2 Gyros

The gyroscope (referred to as simply a "gyro") is, in its simplest form, a sensor that measures rotation. If one rotates a gyro around its sensitive axis, the gyro will output a signal proportional to the rotation applied. Depending on the technology being employed, this signal can either be in the form of a force (mechanical gyros) or in the form of an electrical signal (optical gyros). An important concept with regard to gyro physics is that a gyro measures rotation in inertial space (i.e., independent of geo-orientation). This means that a gyro will not only measure any rotation applied, it will also measure the rotation of the Earth around its rotational axis. However, it will also measure the rotation of the Earth moving around the sun. In fact, it will even measure the rotation of the Milky Way! Due to most of these signals being too weak, most gyros today will, in practicality, only measure the Earth's rotation. A gyro that can measure the Earth's rotation can also be used to maintain orientation in space independent of the Earth's orientation as well as gravitational influences.

Gyros are used everywhere from intercontinental ballistic missiles to aircraft, ships, and all the way to smart phones. The principle of operation depends upon the technology. The mechanical gyro is based upon the conservation of angular momentum, whereby the spinning gyro tends to continue its orientation unless acted upon by some external force. The basic gyroscopic principle is easily demonstrated with a child's spinning top that will only stay upright as long as the top is spinning. As soon as the top stops spinning, it will lie down on its side. The spinning top needs angular momentum to stay upright. The principle is also easily understood by trying to balance on a bike when it is at a standstill. It becomes much easier once the wheels are spinning and resisting precession (i.e., falling over).

This technology evolves all the way through to portions of an inertial navigation system or to the super-accurate gyrocompasses used in ships as well as spacecraft (and, of course, in underwater vehicles). In the latest iterations of the gyro, optics has replaced mechanics as the source of precession sensing. The optical gyros have certain advantages compared to the mechanical gyro when it comes to size and repeatability due to the lack of moving parts of the optical system.

17.2.1 Mechanical gyros

The earliest mechanical gyros were invented in the classical cultures of Greece, Rome, India, and China (Figure 17.2). However, the modern mechanical gyro was first demonstrated as a useful navigational instrument during an experiment to measure the Earth's rotation by Frenchman Léon Foucault in the 1850s. Foucault named this mechanism the "gyro." The gyro gained wide acceptance in the aviation and maritime industries in the years leading up to and during World War II for accurately measuring aircraft/vessel orientation in three dimensions.

The basic gyro has no orientation with regard to Earth and only measures its orientation with regard to space itself (Figure 17.3). Further, any mechanical gyroscopic system inherently requires some type of suspension system for support of the spinning disk. The mechanical connections introduce some type/measure of friction that causes the gyro to drift over time in an error known as "precession."

For gyros used in navigation, there are two basic types depending upon their seeking capabilities:

1. The slaved gyro (typically, a gyrocompass slaved to (or "seeking") True North) and
2. The unslaved (or "rate") gyro

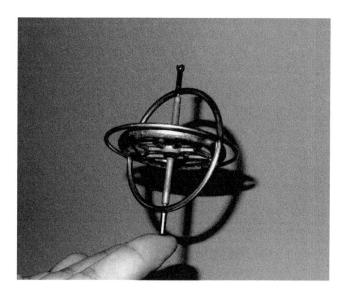

FIGURE 17.2

Mechanical gyro.

(Courtesy Graham Christ.)

17.2.1.1 Slaved gyro

The slaved gyro is arranged to seek some type of reference point (such as True North). Unlike a magnetic compass, whose only orientation is to lines of magnetic flux from the Earth's magnetic field, the gyrocompass senses its precession as it proceeds around the Earth and computes the axis of rotation. As the Earth only rotates in one direction, the pivotal point referencing to True North can then be deduced. For instance, in Figure 17.3, a gyro sitting on a table top located at Point A spins freely about its axis all day without precession as it is colocated on the North Pole (i.e., the pivotal point). A gyro at Point B maintains a constant orientation throughout the day as well since its rotation flows with the Earth's, while Point C has the gyro perpendicular to the table at noon, parallel at 2100, perpendicular again (yet in the opposite orientation) at midnight, and parallel (but flipped from 2100) at 0900. The slaving mechanism measures precession about several axes, thus computing the precession over time to derive True North.

The gyrocompass's (slaved gyro's) function is based upon the following phenomena:

1. *Gyroscopic inertia*: the gyro's conservation of angular momentum (i.e., natural resistance to change in orientation with regard to its inertial frame of reference)
2. *Gyroscopic precession*: the gyro's measured deviation in local frame of reference to the gyro's fixed frame of reference
3. *The Earth's rotation*: the gyro is fixed in space while the gyro's housing (termed the "gyro sphere") is typically fixed to a base (termed the "phantom") upon a floating vessel (which is fixed or floating with reference to Earth)
4. *The Earth's gravity*: for basic orientation to the center of the Earth (for maintaining the gyro's orientation with reference to the horizon—see Figure 17.4)

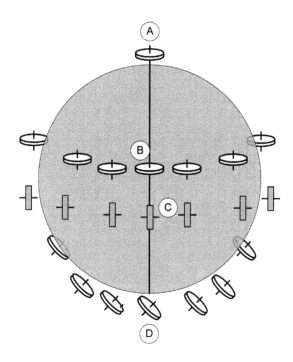

FIGURE 17.3

Gyro maintains orientation irrespective of Earth.

In order for the gyrocompass to seek True North, the gyro must be oriented and maintain the plane of the local meridian (line of longitude) by the gyro's rotor being oriented horizontally (with the help of gravity) as depicted in Figure 17.5. Next, it must remain fixed in this orientation regardless of the motion of the host platform (i.e., the ship or underwater vehicle). As the Earth rotates under the gyro's platform, the plum weight maintaining the gyro sphere's horizontal orientation tends to precess in a direction 90° to the direction of rotation/precession (due to the gyroscopic effect) thus identifying the meridian, and thus True North.

17.2.1.2 Rate gyro

For the simple rate gyro, no slaving is necessary as the only measurement is the rate of precession. Slaved gyros (e.g., gyrocompasses) are used for high-accuracy heading reference (referenced to True North), while rate gyros are used for turn and roll rate or for simple vehicle heading hold (or wing leveling in aircraft) with autopilot functions.

17.2.2 Ring laser gyros

Ring laser gyros (RLG) were developed in the 1960s and are based upon the "Sagnac Effect" (discovered in the early twentieth century by French physicist Georges Sagnac) which states that the timing difference of two light beams traveling in opposite direction around a closed path is directly

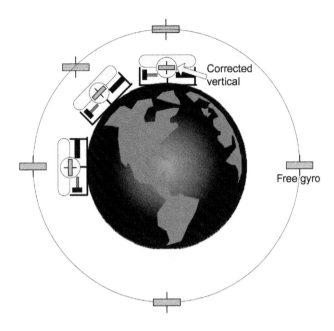

FIGURE 17.4

Free gyro versus gyro maintaining a horizontal plane.

proportional to the rotational speed of the circuit's platform. This phenomenon is the basis for all laser gyro technology.

The reasons a laser is used in RLGs are the laser's unique ability to make use of a single frequency (with small amounts of diffusion), its coherent light beam, and its ability to be focused, split, and deflected. In an RLG, two beams of laser light are projected in opposite directions around a closed circuit (Figure 17.6). The two beams are then joined upon exit of the circuit with the patterns matched in a technique called "interferometry." For a nonrotating gyro, the light patterns match (as both light beams travel the same distance). But when the gyro is rotated, the light patterns then interfere with one another as the light beam traveling in the same direction as the rotation will have traveled a longer distance than the light beam traveling against the rotation. These differences in distance/time cause an interference pattern. The degree of interference is relative to the unit's angular momentum (i.e., rate of turn) that is then measured photometrically. The laser gyros are optimally functional within the plane of the ring (i.e., along the axis of rotation).

The typical RLG has four functional elements:

1. *Excitation mechanism*: A high voltage is applied between a cathode and an anode ionizing a helium—neon gas mixture and producing two beams of light (i.e., lasing) projected in opposite directions.
2. *Gain*: The system's gain helps overcome natural losses by ionizing the low pressure gas mixture producing a fluorescent glow.

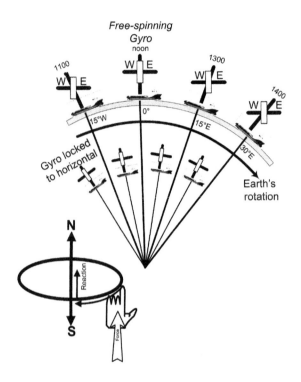

FIGURE 17.5

Depiction of free and horizontally locked gyro as Earth rotates.

3. *Feedback mechanism*: The glowing gas is reflected around corner mirrors (whether the mechanism's shape is square (Figure 17.6) or triangular (Figure 17.7)), allowing the two beams to merge and exit the circuit for pattern matching.

4. *Output coupler*: One of the corners of the unit contains a prism to allow the two beams to mix and form onto the readout detector. Photo diodes measure the light patterns on the fringe of the output, which in turn is converted into electrical pulses for output and measurement. Figure 17.7 depicts a triangular-shaped RLG.

The advantages of the RLG over a mechanical gyro include its long-term stability (mostly due to its solid-state nature with an obvious lack of moving parts), low cost, high reliability, low maintenance, small size/weight, high tolerance to vibration and acceleration, low power requirement, and minimum startup time. Its main disadvantage is its inherent problem of "frequency lock-in" during low-rate turns (i.e., the tendency for the two frequencies to couple together and indicate a zero turning rate although a low-rate turn is in progress).

17.2.3 Fiber-optic gyros

A fiber-optic gyro (FOG) is an RLG with a twist in that the light is directed around many loops of fiber-optic cable (as opposed to a simple enclosed set of mirrors with an RLG). As shown in

(a)

(b)

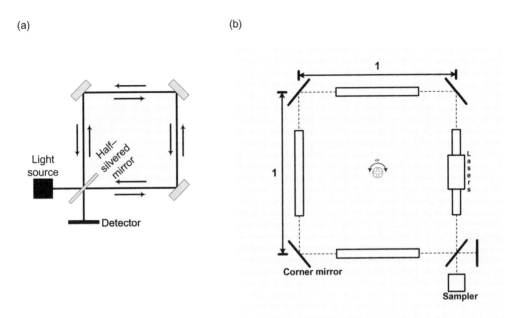

FIGURE 17.6

RLG principle of operation with (a) outside or (b) inline light source.

Figure 17.8, a light source is generated in opposing directions around a coil of fiber. As with the RLG, the beams of light are rejoined and then merged onto a sensor for pattern matching (i.e., rotational measurement). In Figure 17.8, ω is the rotational velocity while $\Delta\varphi$ is the phase difference

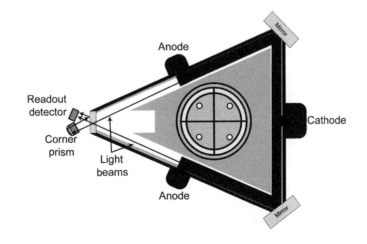

FIGURE 17.7

Triangular RLG.

(a)   (b)

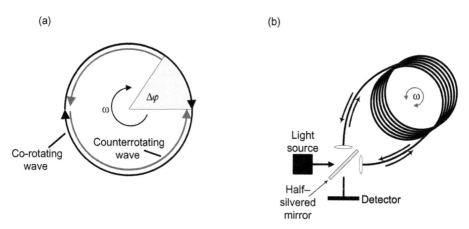

FIGURE 17.8

FOG (a) principle of operation and (b) typical arrangement.

Table 17.2 Advantages and Disadvantages of RLG and FOG

RLG		FOG	
Advantages	**Disadvantages**	**Advantages**	**Disadvantages**
1. No calibration required 2. Highly robust (high MTBF) 3. No moving parts	1. Prone to lockout at low rotational rates	1. Very precise rotational rate 2. Low sensitivity to vibration, acceleration, and shock 3. No moving parts	1. Requires calibration 2. Greater drift than RLG 3. Lower scale factor capability

(measured at the detector) from which ω can be inferred. The advantages and disadvantages of RLG versus FOG are provided in Table 17.2.

The sizes of FOG-based systems do, of course, vary in size depending upon the accuracy needed. As with all gyro systems, accuracy comes with bigger sizes. In order to get an idea of the size, a 0.5° secant latitude Attitude and Heading Reference System (AHRS) system is shown in Figure 17.9.

17.2.4 MEMS-based gyros

The march of technology has accelerated with the advent of MEMS (Micro-Electro-Mechanical-Systems) devices. MEMS is defined as miniaturized mechanical and electromechanical elements (i.e., devices and structure), combined onto a single device, produced by use of microfabrication techniques. This type of device is comprised of microminiaturized sensors, actuators, mechanical components, and electronics integrated into a single chip (which is normally made of silicon).

FIGURE 17.9

The CDL TOGS 2 is a 0.5° secant latitude AHRS system.

(Courtesy CDL.)

This class of sensors has made its way into smart phones, cars, aircraft, spacecraft, and (of course) underwater vehicles. Many MEMS devices have demonstrated performance capabilities exceeding their macroscale counterparts while newer methods of batch fabrication (as used in the integrated circuit industry) translate into a much lower per-device production cost. The main advantages of these types of devices are their ability to be mass-produced, along with their accuracy, low cost, and extremely low size profile and power requirements. However, MEMS gyros can still not compete with high-grade optical or mechanical gyros when it comes to bias stability. For this reason, there is still no True North−seeking MEMS-based system available for civilian use. However, with the current developments within the field, this breakthrough will be achieved within a few years of the writing of this book.

Look for further developments in this area as the technology evolves. This is an exciting area of sensor technology development that has (as the famous Cal Tech scientist Richard Feynman once quipped) "plenty of room at the bottom" (i.e., smaller/faster/cheaper/better).

17.3 Accelerometers

An accelerometer is, in its simplest form, a sensor measuring acceleration. It works on the principle of inertia (Force = Mass × Acceleration) by measuring the force against a known mass in order to derive the unit's acceleration. This means that if a person accelerates an accelerometer on its sensitive axis, the sensor will output a signal proportional to the acceleration applied. As with the gyros, an accelerometer measures acceleration in inertial space. This means that the accelerometer will measure not only any acceleration applied by (for instance) a person moving it around but also the Earth's gravitational acceleration. It will also measure the centripetal acceleration due to the Earth's spinning around its rotational axis.

17.3.1 Pendulum accelerometers

The operation of a pendulum accelerometer is very simple. Imagine holding a string in your hand with a stone attached to the end. If everything is stationary, then the string will hang vertically down from your hand and the angle from your hand to the string will be 90°. If someone then gave you a push on your back, you would move forward and due to the mass of the stone the angle of the string would suddenly not be 90° anymore. The new angle would be proportional to the acceleration applied.

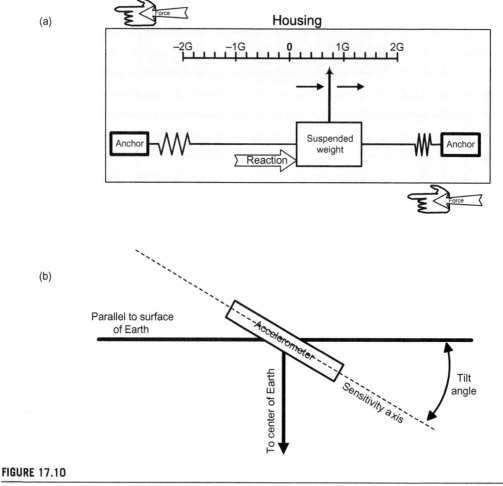

FIGURE 17.10

Accelerometers measure both (a) translational and (b) rotational moments.

In practice, it is (of course) more complicated and normally a pendulum accelerometer would use feedback control to keep the mass close to equilibrium. The amount of feedback needed to keep the mass in equilibrium would then be equivalent to the applied acceleration. Such a system is shown in Figure 17.10.

17.3.2 MEMS-based accelerometers

MEMS-based accelerometers have become very popular over the past 10 years. Due to their size, reliability, and performance, MEMS-based accelerometers are being used in many areas where more "normal" accelerometers were previously the *de facto* standard. As opposed to MEMS-based gyroscopes (which still have not achieved the performance of the high-grade optical/mechanical gyros), the MEMS-based accelerometers have achieved similar performances to the more common technologies. With this high performance, the MEMS-based accelerometers are seen more widely in use due to the obvious advantages in size and price. As with the more ordinary accelerometers, there are many different ways of engineering MEMS-based accelerometers. In order to get an idea of the scale of a MEMS-based magnetic-aided AHRS system, the CDL MiniSense3 is shown in Figure 17.11.

17.4 Inertial navigation systems

Inertial Navigation Systems (INS) are navigational systems capable of calculating position, either relative to some reference system/point or to absolute coordinates. An INS system is composed of

FIGURE 17.11

The CDL MiniSense 3 is a 2° magnetic-aided AHRS system.

(Courtesy CDL.)

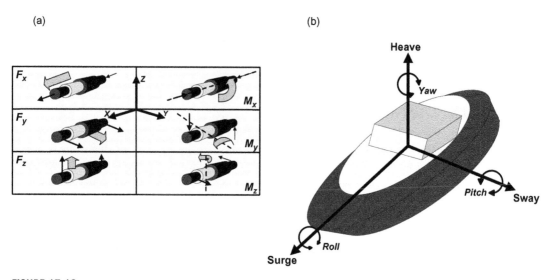

FIGURE 17.12

INS measures movement along the six degrees of freedom displayed as (a) translational (F) along with rotary (M) movement and (b) with vessel superimposed.

at least three gyros and three accelerometers enabling the system to derive a navigation solution. This navigation solution contains at least the position (normally latitude, longitude). Most INS systems today output heading, pitch, and roll. Some systems also include heave, sway, and surge.

The basic concept behind an INS system is the measurement of changes in relative motion (through the measurement of acceleration) to project a changing position in some inertial reference frame over time. The heart of an INS system is its inertial measurement unit (IMU). This mechanism is composed of three orthogonal gyros and three orthogonal accelerometers.

As depicted in Figure 17.12, a full range IMU will measure movement about six degrees of freedom including the three rotational moments M (pitch (M_y), roll (M_x), and yaw (M_z)) and the three translational moments F (heave (F_z), surge (F_x), and sway (F_y)).

The basic concept of an INS system is very simple. Imagine an IMU lying horizontally on a surface placed at the North Pole. The only forces acting on the IMU would then be the Earth's rotation acting on the vertical gyro and the Earth's gravity acting on the vertical accelerometer. If someone then gave the IMU a push forward, the accelerometer having its sensitive axis pointing forward would measure this acceleration. None of the other accelerometers would measure any change. The system would then perform basic deduced reckoning (dead reckoning) to derive an assumed position. A positional error is propagated over time due to the inherent inaccuracies of any inertial sensor. In this simple example, a bias on the accelerometer would cause the position to drift over time. This error (termed "circular error probability," or CEP) is the expected deviation from the computed position derived from the raw measurement. Bowditch (2002) describes CEP as: "(*1*) *In a circular normal distribution (the magnitudes of the two one-dimensional input errors are equal and the angle of cut is 90°), the radius of the circle containing 50 percent of the individual measurements being made or the radius of the circle inside of which there is a 50 percent probability of being located. (2) The radius of a circle inside of which there is a 50 percent probability of being located even though the actual error figure is an ellipse.*" That is, the radius of a circle of equivalent probability when the probability is specified as 50%. The CEP measurement is useful in navigation, weapons delivery computations, and search grid designations.

In order to get around this inherent positional drift in INS systems, the survey system is usually augmented by some sort of aiding device. For surface applications, an aiding device could be a global positioning system (GPS). Omega, or Loran. For subsea applications, an aiding device could be USBL, LBL, or DVL. No matter which aiding device is chosen, the purpose is the same: to decrease or even remove inherent drift in INS deduced position.

In order to remove drift in position, a nondrift position-aiding device is needed. This could be a GPS or a USBL system. If a velocity-aiding device like a DVL is used, it will only be possible to decrease the drift in position. The drift is then primarily depending on the DVL velocity accuracy. It is, of course, also possible to aid an INS system with multiple aiding devices, thus gaining redundancy and improved performance.

INS devices are becoming more accurate while being packed in continually smaller units (newer systems are based upon MEMS technology). INS is widely used in subsea vehicle applications as an integral part of the survey package. A size comparison of a FOG-based INS system and a MEMS-based INS system can be seen in Figure 17.13.

FIGURE 17.13

A size comparison between a FOG-based INS system (TOGS 2) (left) and a MEMS-based INS system (MiniSense 3) (right).

(Courtesy CDL.)

17.5 Bathymetric sensors

The science of bathymetry is the study of water depth in lakes and oceans. Early techniques for sampling water depth involved the use of a line with a weight attached. As an interesting side note, in the early days of boating on the Mississippi River the depth was called out in fathoms, often using old-fashioned words for numbers. An expression for a depth of two fathoms would be called out as "mark twain." Samuel Clemens, a former river pilot, took this expression and created his pen name: Mark Twain. This method, while certainly simple and reliable, gave only limited samples in shallow waters. In the early twentieth century, acoustics replaced the weighted line with the single beam sonar, termed a "fathometer" for this application, for measuring depth in fathoms.

Modern bathymetry systems (Figure 17.14) make use of wide-angle hull-mounted or vehicle-mounted multibeam sonar systems, arranged in a fan-like "swath" for mapping large areas of the ocean floor.

17.6 Conductivity, temperature, depth (CTD) sensors

Sound propagates through water at various speeds depending upon the density of the medium. Many factors affect the density of water (which, hence, affects the velocity of sound in water). The three main factors are:

FIGURE 17.14

Depiction of (a) hull-mounted and (b) subsea vehicle-mounted swath bathymetry systems.

(Courtesy Kongsberg Maritime.)

1. Salinity of the water (directly affecting the specific gravity of the water)
2. Temperature (also affecting density)
3. Water depth

As discussed in Chapter 2, water tends to form layers in seawater as depth changes. These layers vary depending upon local heating as well as regional and local water salinity. Cold water is typically denser (up to the 4°C cross-over point for fresh water where ice begins to form) as is higher salinity water. The higher the density of the water, of course, the higher the sound propagation speed.

As depicted in Figure 17.15, if the sound velocity profile (SVP) is not known with some degree of accuracy, large errors in distance (through measurement of inaccurate propagation rate versus time) will occur, thus foiling precise positioning. In order to have accurately derived acoustic positioning, an SVP must be plotted through the water column. The measurement of SVP is the domain of the conductivity/temperature/depth (CTD) sensor.

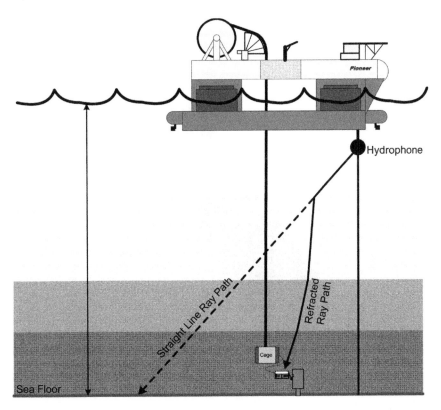

FIGURE 17.15

Differences in CTD distort SVPs.

Conductivity: Fresh water is an insulator, whereas the ever increasing level of salinity in seawater changes the water into a semiconductive medium. Through measuring the level of conductivity of the local water sample, a highly accurate measure of salinity can be derived. The traditional unit of measure for salinity has been parts per thousand (ppt) while the modern measurement unit (since salinity is typically measured electronically) is practical salinity units (PSU).

Temperature: Local water temperature is measured by a device such as a thermistor or some other temperature measurement technology.

Depth: A pressure sensor (of various technologies) is used to measure local water pressure. This measurement is then converted into an accurate measurement of depth, which is then correlated with the conductivity and temperature parameters to derive a full SVP of the water column.

Figure 17.16 and Table 17.3 provide an example of a typical CTD sensor and its operating specifications.

FIGURE 17.16

Citadel CTD-NV CTD sensor.

(Courtesy Teledyne RDI.)

Table 17.3 Operating Specifications for the Citadel CTD-NV			
Parameter	**Conductivity**	**Temperature**	**Depth (Pressure)**
Sensor	Inductive cell	Thermistor	Precision-machined silicon
Range	0–9.0 S/m (0–90 mS/cm)	− 5°C to 35°C	Customer specified
Accuracy	± 0.0009 S/m (± 0.009 mS/cm)	± 0.005°C	0.05% full scale
Stability	± 0.01 mS/cm/month	0.0005°C/month	± 0.004%
Resolution	0.00001 S/m (0.0001 mS/cm)	0.001°C	0.001% full scale

17.7 Altimeters

A variation on the hull-mounted depth sounder is the depth sounder mounted to an underwater vehicle decoupled from the surface. While a vessel-mounted depth sounder measures the distance from the hull to the seafloor, the ROV-mounted altimeter measures the distance from the ROV's frame (where the altimeter's transducer attaches) to the seafloor. This, coupled with a pressure-sensitive depth sensor, produces highly accurate local sea bottom profiles.

17.8 Doppler velocity logs

As discussed more fully in Chapter 14, a resolved velocity over bottom can be gained acoustically through use of the Doppler Velocity Log (DVL). Once the over-ground vector is determined, an accurate time/distance calculation can be gained to geo-reference the vehicle's position. In some newer applications, use of a DVL, along with Geographic Information Systems mapping applications, allows for 3D estimated tracking of a submersible with a periodic "snap to grid" updated position through some other technology (e.g., GPS or acoustic positioning). As the vehicle is maneuvered in time/distance navigation mode from a known location, the estimation error, per unit distance or time, is increased proportionally to the distance traveled or time. For instance, if there is an assumed 10% error with the DVL or the INS over distance (or time for that matter) traveled, the error will be 10% for that unit. If another unit of travel distance/time is processed, the error will be that 10% plus another 10% as the circle of possible/probable position increases. The farther the vehicle progresses on dead reckoned navigation without an updated accurate positional fix, the greater the circle of equivalent position probability (also known as CEP).

The principle of operation for a DVL is the same as for the Acoustic Doppler Current Profiler (ADCP), only instead of the upward-looking ADCP (measuring water movement), the DVL looks down (measuring bottom movement). The DVL transmits an array of normally four sonar beams in a generally downward direction toward the bottom, with each beam offset from the vertical in a measured fashion. The echo return frequency is then measured for a Doppler shift—which is proportionate to the speed over ground. In Figure 17.17, the four-transducer array of a DVL is shown along with a graphic of acoustic beam propagation.

DVLs require a bottom to be within a tolerable distance in order to obtain "bottom lock" for the transducers to be able to discern adequate frequency shifts between the beams. Maximum altitudes

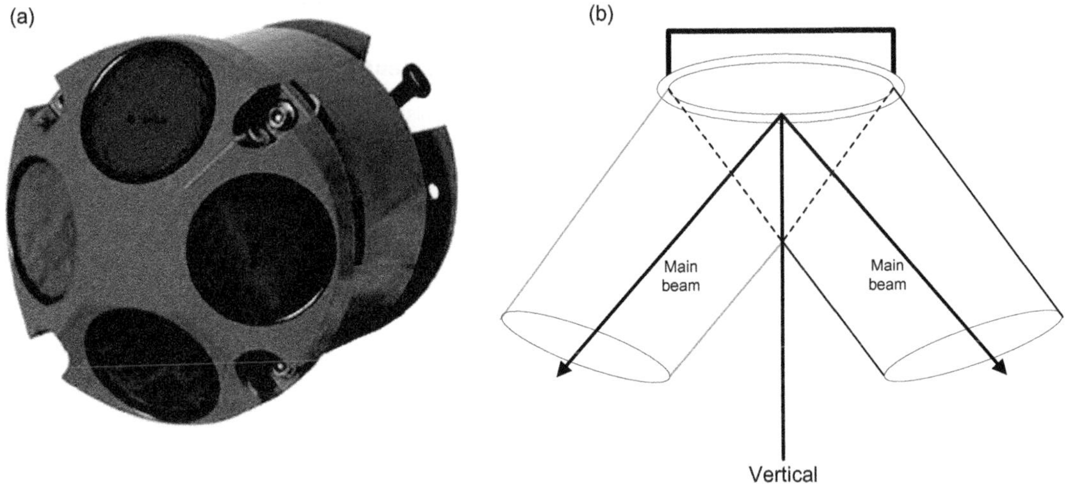

FIGURE 17.17

(a) A Workhorse Navigator DVL along with (b) beam angle illustration (Note: only two of four beams are depicted—the other two are 90° to those shown).

(Courtesy Teledyne RDI.)

are typically 100−650 ft (30−200 m), based upon the frequency used. DVLs are used extensively with subsea vehicle autopilots as well as over-ground navigation systems such as ROV dynamic positioning with relation to the bottom.

17.9 Inclinometers

An inclinometer is a senor for measuring orientation of a vehicle with reference to the gravitational field of the Earth (or other gravitational body). These types of instruments are traditionally used in aircraft and surface ships (even for ground-based measurements) but have found applications in subsea survey.

The inclinometer is critical for orientation of survey sensors within the gravitational frame of reference. The other survey sensors can place the vehicle in a "subsea box" or location, but sensors typically project away from the vehicle. The inclinometer is used to snap the sensor data from a vehicle referenced coordinate frame to an Earth- or gravity-based, referenced frame.

17.10 Long baseline arrays

To summarize the materials mentioned in Chapter 16, as it relates to combined survey instruments, a quick review of acoustic positioning follows. As GPS signals do not travel through water, thus providing the ability to achieve an absolute position fix subsea, some other way of

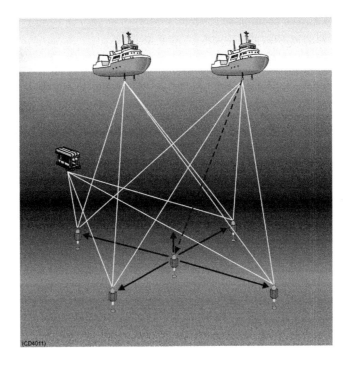

FIGURE 17.18

An LBL array with five surveyed transducers.

(Courtesy Kongsberg Maritime.)

achieving this is required. Using a long baseline (LBL) is a way of achieving a position fix within a confined area.

The basic idea is similar to that of GPS. A number of transducers are surveyed onto the seabed, thus providing an absolute position for each transducer. Any user able to receive the signals from these transducers can then use the time delays and information about each transducer to deduce an absolute position within the LBL network (Figure 17.18).

A well-surveyed high-accuracy LBL array can give positions to within centimeters. The main disadvantage of an LBL array is the cost of surveying (for high-accuracy geo-location) the LBL array. Each transducer needs to be precisely located on the seabed (which is costly). Further, the LBL only works within the area where it is installed and is not easily moved to a new location.

17.11 Ultrashort baseline arrays

An ultrashort baseline (USBL) is a similar technology to LBL. However, the transducers are moved very close together within a single unit. This unit receives a signal from a transponder/responder located on an ROV or a tow body and then calculates the relative position with reference to the ship. If the ship has a known position (e.g., from a GPS), it is easy to deduce the absolute position of the ROV or tow body (or any other client, for that matter).

The way a USBL system calculates the position of a client is to first estimate the distance by simply using the signal's travel time from the client to the USBL array. Knowledge of the speed of sound (rate) through the water column, along with the known propagation time, will give the distance to the client. Analyzing the phase delay across the USBL array will give the angle at which the signal was received. With that information, the position of the client relative to the ship can be derived.

The main disadvantage of the USBL system is the position error. The position error grows proportionally with the slant range. Although USBL systems are improving, a high-grade USBL system would have errors in the area of 0.1% of slant range. This means that if the client is operating at a depth of 10,000 ft (3000 m), the position will only be good to 10 ft (3 m) for zero slant range, with the error scaling as the slant distance increases. Less accurate systems would easily have up to 65–100 ft (20–30 m) of error (without slant range) at a 10,000 ft (3000 m) depth. One of the advantages, as opposed to the LBL array, is that it is very easy to move the USBL from site to site.

17.12 Combined instruments

The trend in subsea vehicle navigation is for manufacturers to combine several sensors into one mechanism, thereby reducing the amount of weight and power draw through sharing pressure

FIGURE 17.19

CDL TOGS-NAV 2 combines several sensors into a single package.

(Courtesy CDL.)

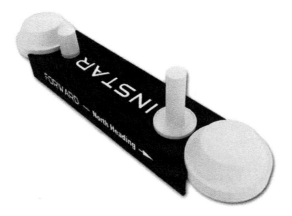

FIGURE 17.20

CDL INStar system combines several sensors and opens up for entirely new applications.

(Courtesy CDL.)

housings and electronic components. Another advantage is that a combined unit can multiplex the different sensors onto one communication line, thus reducing the number of communication lines in the umbilical.

Often an acoustics manufacturer will combine a CTD unit (measuring SVP for corrected positioning measurements) with a sonar or acoustic positioning system, while a gyro manufacturer will combine multiple gyros on differing orientations in order to produce a full accelerometer for use with an inertial navigation system.

The CDL TOGS-NAV 2 (Figure 17.19) is an example of this combination. In this system, an IMU is combined with a DVL, a depth sensor, and a sound velocity probe. This gives a single system capable of outputting heading, pitch, roll, position, velocity, depth, and speed of sound. This has a big advantage compared to having to install and integrate four different systems.

The combining of sensors also reduces installation and integration time. Combining different sensors also extends to creating completely new packages—which opens up new applications that the individual sensors would not be able to achieve as separate stand-alone systems. The CDL INStar is such a system (Figure 17.20). This package combines GPS position with an attitude sensor. It also includes communication via either satellite or radio link. This enables the system to be installed on remote devices or vehicles sending back attitude and position information either via radio link or satellite.

Ancillary Sensors

CHAPTER CONTENTS

In this chapter, further payload sensor technology is examined. These sensors are integrated *through* the ROV (by attaching to the frame with sensor power coming from the vehicle and telemetry typically routed through the vehicle tether) and fit into the category of inspection, repair, and maintenance (IRM) sensors.

18.1 Nondestructive testing definition and sensors

Nondestructive testing (NDT), by definition, is a series of techniques whereby materials are tested for structural flaws through direct and/or indirect means such that the procedures do not destroy the test subject. These types of tests include acoustic, electromagnetic field, radiographic, and other active and/or passive physical stimulations in order to verify metal integrity. The objective of this process is to identify and quantify discontinuities in the metal structure for assuring structural integrity of the test subject.

Homogeneous metals allow magnetic, sound, and radiographic waves to pass evenly through the medium. Cracks, discontinuities, and other faults in the metal interrupt the flow of these waves, producing flow interruptions that are measured for fault isolation and quantification. This is the essence of NDT.

Typically, just as a camera from a single angle in a scene will not give the full picture of the events transpiring, a single NDT sensor will not "tell the full story" on the condition of the test piece. A sensor array of varying technologies is usually deployed to test various parameters to fully characterize the item.

Technology in this field is rapidly evolving. This section will highlight some of the more common technologies available for ROV deployment. As this chapter is a high-level overview, the reader is encouraged (and cautioned) to delve deeper into each of these areas. Each of these devices has its peculiar strengths and weaknesses, requiring operator skill in gathering and interpreting the product of the sensor's measurements.

18.1.1 Magnetic particle inspection

Magnetic particle inspection (MPI) is the traditional method of measuring the magnetic flux leakage to determine metal surface discontinuity. The theory of operation involves the tracing of the lines of magnetic flux induced into a test subject and then measuring this field for anomalies.

Magnetism comes in three types (Table 18.1):

1. *Ferromagnetism*: Strongly magnetized metal exhibiting excellent magnetic characteristics
2. *Paramagnetism*: Materials exhibiting weak magnetic attraction when stimulated with a magnetic field
3. *Diamagnetism*: Materials repulsed when stimulated by a magnetic field due to the generation of an opposing magnetic field in the material

In ferromagnetic metals, the atoms form a lattice structure gathered in groups (termed "domains") with individual and isolated magnetic moments. A magnetic field can be induced into a magnetic material in any of three ways:

1. Applying a permanent magnet to the material
2. By passing a current through the material
3. By inductance through a current carrying conductor in close proximity to the material

When demagnetized, the domains have magnetic moments that are randomly distributed. These domains can be thought of as many small bar magnets arranged at random along the surface of the

Table 18.1 Sample Materials Exhibiting Magnetic Types

Ferromagnetic	Paramagnetic	Diamagnetic
Iron	Platinum	Bismuth
Nickel	Palladium	Antimony
Cobalt	Most metals	Most nonmetals
Steel	Oxygen	Concrete

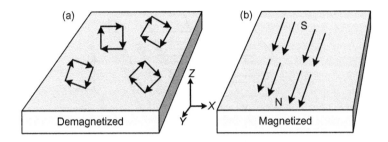

FIGURE 18.1

Lines of magnetic flux upon a (a) demagnetized surface as well as a (b) magnetized surface.

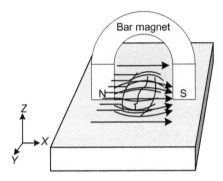

FIGURE 18.2

Lines of magnetic flux upon magnetization.

metal. Once externally magnetized, the magnets then snap to the new magnetic flux and align themselves with the magnetic field (Figure 18.1).

Once the magnetic field is stimulated (by one of the three methods above), a magnetic ink (ferromagnetic powder in liquid suspension—hence the term "magnetic particle") is introduced onto the metal, and then the lines of magnetic flux are visually observed (Figure 18.2). Any anomaly within the magnetic field becomes readily obvious, thus revealing the metal defect.

18.1.2 Alternating current field measurement

In recent years, alternating current field measurement (ACFM) has become the subsea NDT "weapon of choice" for the magnetic flux leak detection inspection sensor. ACFM is an electromagnetic technique whereby an alternating current is induced into the metal surface of a test subject (e.g., the jacket leg of an oil and gas production platform), and then the magnetic field is mapped for uniformity (or lack thereof). Once the AC magnetic field is induced, the metal produces a uniform magnetic field above the surface. Any discontinuity in the metal's surface will produce a perturbation in this field (Figure 18.3), forcing the field to flow around or under/over the fault.

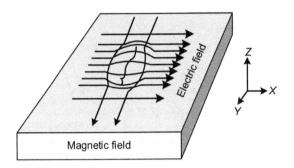

FIGURE 18.3

ACFM maps perturbations in the magnetic field around cracks.

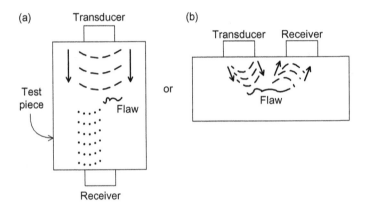

FIGURE 18.4

Ultrasonic flaw detection of metal with (a) opposing transducer/receiver arrangement and (b) corresponding transducer/receiver arrangement.

Sensors in the ACFM probe then map these field variations in order to measure the size/depth of the metal discontinuity.

18.1.3 Ultrasonic flaw detection

As stated in Chapter 15, various materials conduct sound at widely different speeds. When sound passes from steel at 16,400 ft/s (5000 m/s) to water at 4920 ft/s (1500 m/s) in a fracture of a submerged steel structure, a high level of sound attenuation is experienced. Also, discontinuities embedded within the metal further attenuate sound through the metal. The basic concept behind ultrasonic metal flaw detection is the measurement of sound attenuation through a test piece. In a homogeneous metal, longitudinal sound waves will propagate evenly through the metal, allowing sound to be measured from the transducer to the receiver. As depicted in Figure 18.4, with either breaks or internal flaws, the acoustic energy is dissipated, allowing easy measurement at a receiver and thus revealing the flaw.

18.2 Metal object detection

Metal object detection is necessary for various purposes including salvage, pipeline/cable tracking, subsea construction, mine countermeasures, and the like. There are many methods for sensing submerged metal objects through the use of electromagnetism. In this section, the three most common methods used for subsea metal object detection and their application to subsea operations will be examined—namely:

1. Active pulse inductance
2. Passive inductance sensing of toned lines
3. Sensing of anomalies in the Earth's magnetic field

18.2.1 Active versus passive

Metal detection techniques typically use some form of magnetic field detection technology. The methods vary based upon whether the magnetic field is being (i) induced or (ii) sensed passively (Figure 18.5).

Magnetic fields may be induced through active stand-off pulse inductance or through remote alternating current tone generation using a conducting line (such as a pipeline or subsea cable). Passive sensing involves varying techniques of sensing anomalies in the Earth's magnetic field generated by a magnetic field surrounding ferrous magnetic metals.

18.2.2 Active pulse inductance

Active pulse inductance (termed simply "pulse inductance") is widely used in the shore-based construction industry as a "cover meter" for determining the status and cover depth of structural rebar embedded within concrete. In the subsea industry, this technique is used to track and determine the depth of cover for buried cables or pipelines.

The basic principle of induction is quite commonly used throughout the power generation industry with applications such as transformers (Figure 18.6). A magnetic field is induced around any conductive metal coil once a current is passed through that coil. The magnitude of the field surrounding the coil is proportional to the instantaneous magnitude of the current.

With pulse inductance, an instantaneous current is generated through a coil that produces a known magnetic field surrounding the subject coil. This field decays at a constant rate over time in the absence of any other magnetic field, that is, a field induced through a nearby conducting metal. However, if other conductive materials are present in the near vicinity, a sympathetic current (termed an "eddy current") is induced within the target.

As depicted in Figure 18.7(a), a sample voltage pulse of $-15V$ amplitude with pulse width $1500\,\mu s$ drives a current into the sample search coil. At the conclusion of the drive period, the magnetic field collapses with the current falling toward a nominal zero value. The current induced into the search coil results in a magnetic field that decays at a known rate. In Figure 18.7(b), the magnetic field then induces a sympathetic current in a target (Figure 18.7(c)). The current induced into the target then generates and decays further inducing a current back into the search coil—that

FIGURE 18.5

Typical MSROV configured with (a) active pulse inductance, (b) cable tracker for toned line, and (c) passive gradiometer array.

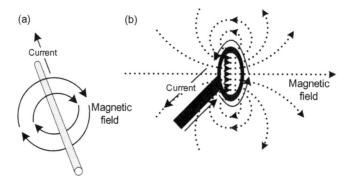

FIGURE 18.6

Magnetic field induced through both a (a) wire and through (b) a loop.

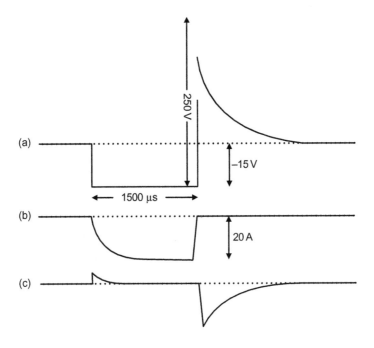

FIGURE 18.7

Pulse inductance waveform (not to scale). (a) Voltage across the search coil, (b) current pulse in the search coil, and (c) eddy currents induced in the target.

delays the full original decay within the search coil. The field decay time along with field intensity is then measured and compared to the reference decay time/intensity without target metals (i.e., with the search coil only) to derive the size and proximity of target metals.

As shown in Figure 18.8(a), ROV-mounted search coils are typically arranged in an array to sense longitudinal alignment of the target while in Figure 18.8(b) the depth of burial is derived by sensing the total distance to the target and then subtracting the distance to the bottom.

18.2.3 Passive inductance

As depicted in Figure 18.9, a magnetic field generates and collapses as an alternating current is passed through a conductor. If a target (or search) coil is passed near the reference (or target) coil, a sympathetic current will be generated in the search coil as well—which can then be measured for proximity as well as very tight tolerance frequency discrimination (for isolation of the reference cable from background noise).

In the United States, the standard power grid makes use of a 60 Hz power cycle while European nations typically make use of a 50 Hz cycle. This allows a passive inductance system to locate a typical power cable in a very noisy environment by identifying any sources of 60 Hz (or 50 Hz) alternating current inductance fields.

To isolate a single cable (Figure 18.10) from other power cables in a complicated environment, a tone can be generated in a target line from an accessible end. The tone is of a fairly wide

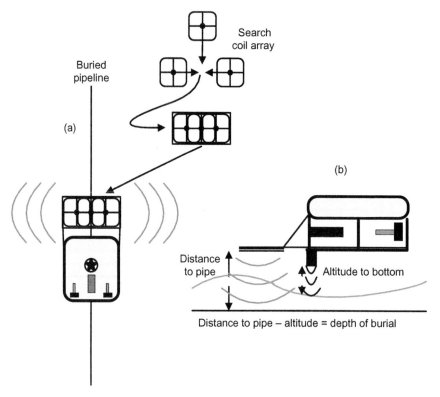

FIGURE 18.8

(a) Search coils are arrayed to sense longitudinal alignment while (b) depth of burial is derived by comparing the distance to pipe with the height above the bottom.

frequency spacing from the surrounding power cables—say, a 22 Hz tone on a length of (unpowered) power cable.

In Figure 18.11, a cable tone is defined via its frequency, peak output voltage, and peak output current. A tone generator is then placed on two of the three poles of a three-phase power cable with the third cable blocked off. If the cable terminates on a platform, the two poles with the tone must be shorted to complete the circuit—or, alternatively, if the cable is open in sea water, then water itself may serve as the conductor for the return signal.

For subsea cable trackers, orientation must be gained on three axes of alignment to derive longitudinal orientation as well as lateral offset and height (Figure 18.12). Therefore, the coils are arranged along all three translational axes of the vehicle. From this orientation, the vehicle may then sense the magnetic field of the target cable along the x, y, and z axes as the vehicle travels along the y axis.

A single towed or handheld coil can be substituted to derive cable placement (Figure 18.13). The technique involves continuously moving the sensor coil over the area of the cable and then correlating the coil readings to geographical coordinates in what is termed the "brute force" method.

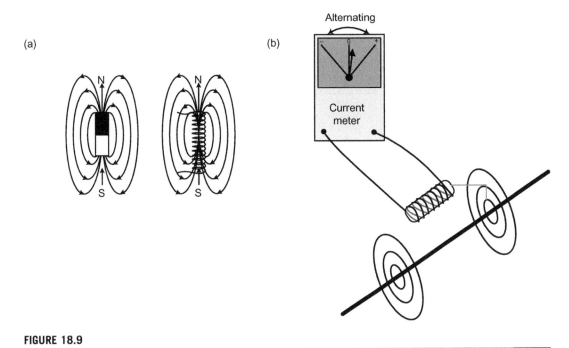

FIGURE 18.9

A coil induces a magnetic field (a) that is then measured for frequency and intensity (b).

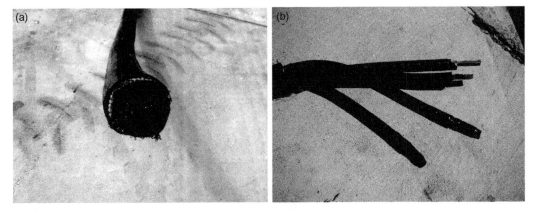

FIGURE 18.10

(a) Typical subsea cable and (b) subsea cable stripped.

(a)

(b)

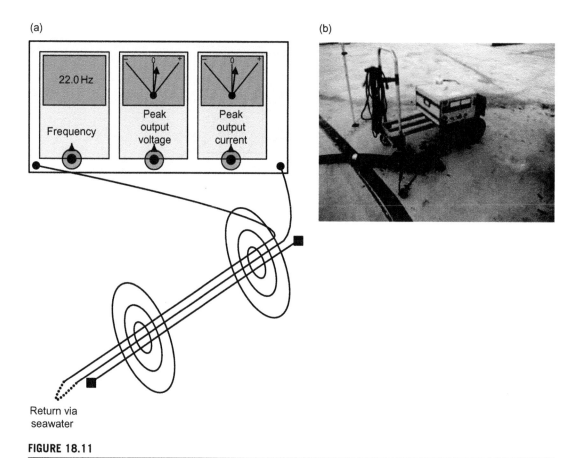

22.0 Hz

Frequency

Peak
output
voltage

Peak
output
current

Return via
seawater

FIGURE 18.11

(a) Tone is generated and returns via seawater and (b) tone generator on subsea power cable.

Once enough data points have been gained, the precise location of the cable is mapped with that area encompassing the peak readings of the inductance measurement as the exact cable location.

For single-coil tracking of a continuous cable, the search should be conducted perpendicular to the axis of the cable to isolate the location of the cable by means of the peak coil output. However, if searching for a break in the cable, the search should be conducted along the longitudinal axis of the cable as the break will be denoted by a sudden drop-off in the inductance (due to isolation of the cable downstream of the toned line).

18.2.4 Magnetometers and gradiometers

While the previous two types of magnetic field sensors sensed varying magnetism, the magnetometer is used for detection of static magnetic fields. The magnetometer was first described by Carl Friedrich Gauss in the 1830s. Since that time, many differing techniques have evolved for sensing static magnetic fields. The two most commonly used subsea search-type magnetometers, used for sensing magnetic

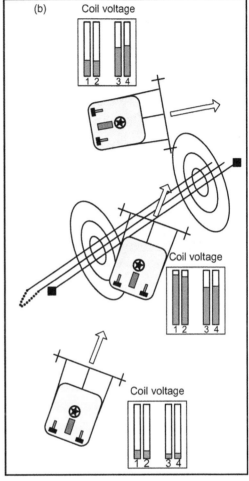

FIGURE 18.12

(a) Picture of the starboard side of the sensing coil array on cable tracker and (b) graphical depiction of a coil array output indications versus array orientation.

anomalies in the Earth's magnetic field, include the Proton Precession Magnetometer (PPM) and the Optically Pumped Alkali Vapor Magnetometer (OPM) (both of which will be covered below).

Magnetometers sense values of magnetism (vector and intensity). Since the Earth's magnetic field is (by far) the predominant magnet on Earth, magnetometers are generally used to search for anomalies in that field for various purposes. Since the application for ROV-mounted magnetometers and gradiometers is typically to determine the location of ferrous metal objects, the principles of Earth magnetism with regard to ferrous metal detection will be examined.

The Earth's magnetic field resembles a huge permanent bar magnet with the poles of the bar oriented approximately 11° off of the Earth's rotational axis (Figure 18.14). The causes of

FIGURE 18.13

(a) Handheld coil manually passed over cable and (b) submerged pole-mounted coil.

geomagnetism are not well understood, but the quantitative components are generally classed into three broad categories:

1. *Declination (also termed "variation")*: The horizontal vector difference between the Earth's geographic axis (N_g/S_g) and the Earth's magnetic axis (N_m/S_m).
2. *Inclination (or "dip")*: As further depicted in Figure 18.15, the Earth's magnetic field has a vertical component as well. Just as a bar magnet's lines of magnetic flux curve into the magnet at the poles, the flux lines curve into the Earth at the magnetic North/South poles. The lines are generally horizontal at the equator and vertical at the poles.
3. *Intensity*: The Earth's magnetic field intensity varies with its proximity to the magnetic poles with (generally) its highest intensity at the poles and its lowest intensity at the equator.

As shown in Figure 18.15, the pole on a bar magnet typically attracts to its opposite polarity; therefore, the Earth's Magnetic North Pole is considered the southern pole of the Earth's magnetic field (with the North Pole of a compass needle seeking the Earth's North Pole). The two vector quantities of the Earth's magnetic field are depicted in Figure 18.16 for (i) declination and (ii) inclination.

Magnetometers are classed in two general categories based upon their ability to sense vector quantities:

1. *Scalar*: Magnetometers able to (only) sense the magnitude of the local magnetic field
2. *Vector*: Magnetometers that sense magnitude as well as the directional component of the local magnetic field

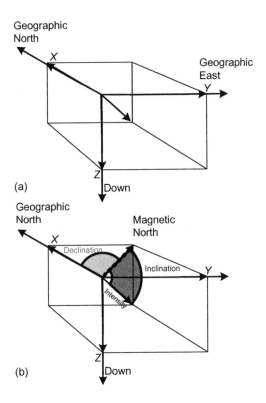

(a)

(b)

FIGURE 18.14

(a) Geographic components of Earth's magnetic field along with (b) vector components.

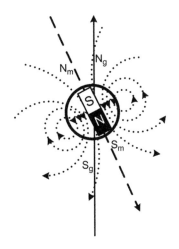

FIGURE 18.15

The Earth as a bar magnet.

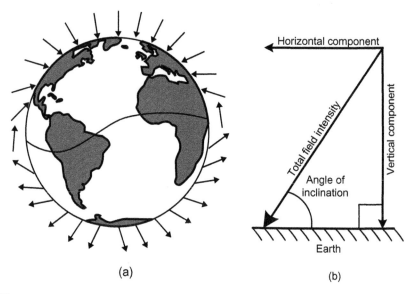

(a)   (b)

FIGURE 18.16

(a) Earth's magnetic field declination and (b) vector components.

Magnetic field intensity is measured in quantitative units based upon the flux intensity. The SI unit for magnetic flux density is the Tesla (symbol T) with the commonly accepted unit for geophysical application as the nanotesla (symbol nT). An older convention for geomagnetic intensity is the gamma (symbol γ) with 1 γ = 1 nT. The reader will notice in Figure 18.17 that the magnetic field of the Earth is very nonhomogeneous. The Earth's magnetic field varies in intensity from approximately 25,000 nT at the magnetic equator to approximately 70,000 nT at the magnetic poles with the lines of equal intensity (termed "isogonics") running generally parallel to the equator. Therefore, the general approach to anomaly detection in the Earth's magnetic field is to run the magnetometer parallel to the Earth's isogonics and note any perturbations. There you will find a magnetic field attached to some magnetized item. The intensity of the anomalous field surrounding a man-made object within the Earth's magnetic field will follow a simple rule of thumb as "1 ton of iron is 1 nT at 100 ft (33 m)." Some typical maximum magnetic anomalies for common items are shown in Table 18.2.

In Figure 18.18, a dipole magnet is inducing an anomaly in the Earth's magnetic field. The search magnetometer is run parallel to the magnetic field lines to identify the anomaly as it is approached and then crossed in a dipole manner.

All magnets are dipoles having both north and south poles. A magnet oriented to the horizon will produce a full dipole reading in the Earth's magnetic field. But if the magnet is oriented vertically, a reading in one direction (only) will result (i.e., an anomalous rise or fall (depending upon the pole's orientation as "up" or "down") in the total field intensity).

With a simple magnetometer, only field intensity can be measured without some means of deriving the vector portion of the field. A gradiometer, on the other hand, is an array of magnetometers offset

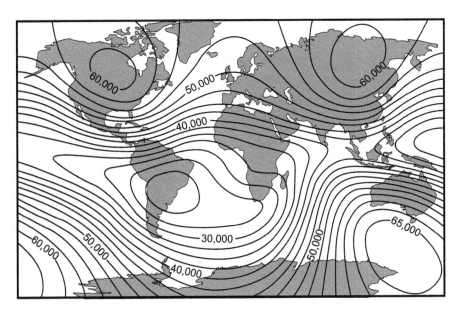

FIGURE 18.17

Earth's magnetic field intensity (in both nT and γ).

Table **18.2** Common Magnetic Anomalies		
Object	**Near Distance**	**Far Distance**
1 Ton automobile	30 ft/40 nT	100 ft/1 nT
1000 Ton ship	100 ft/300–700 nT	1000 ft/0.3–0.7 nT
Light aircraft	20 ft/10–30 nT	50 ft/0.5–2.0 nT
10 inch file	5 ft/50–100 nT	10 ft/5–10 nT
5 inch screwdriver	5 ft/5–10 nT	10 ft/0.5–1.0 nT
12 inch pipeline	25 ft/50–200 nT	50 ft/12–50 nT
Well casing and wellhead	50 ft/200–500 nT	500 ft/2–5 nT

at a known distance to derive the magnetic field "gradient" (expressed in nT/m) between the sensors, thus allowing the vector solution (Figure 18.19). The higher the gradient, the larger the anomaly for any given offset with homogeneous magnetic materials.

The Proton Precision Magnetometer (PPM) operates on the principle of the protons in certain fluids naturally aligning along the Earth's lines of magnetic flux. Within the sensor of a PPM, an electrically conducting coil surrounds the reservoir of some hydrogen-rich fluid (typically mineral spirits, diesel, or some other hydrocarbon). When the coil is nonenergized, the protons assume their natural state aligned along the lines of ambient magnetic flux. But when the coil is energized, the protons immediately snap to the local lines of flux with reference to the coil. Once the energy is

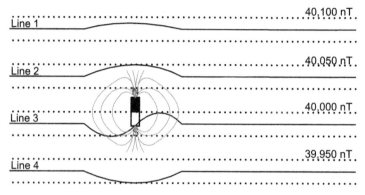

FIGURE 18.18

Sample magnetometer sensing magnet in Earth's field.

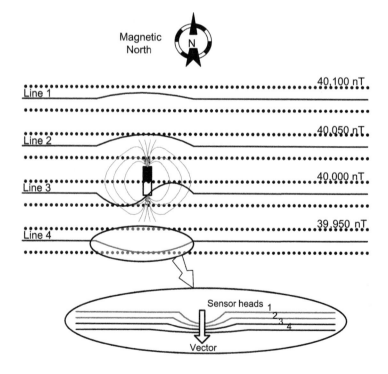

FIGURE 18.19

Gradiometer array for sensing field vector as well as intensity.

removed from the coil, the protons then "precess" back to their natural state. As the protons precess, a small yet precise electrical current is induced into the same coil that is directly proportional to the local magnetic field intensity. The rate of precession is thus measured, scaled, and outputted on a sampled basis. The typical PPM is highly accurate, but due to the time delay between the activation/deactivation of the coil for the precession process to snap between coil orientation and Earth's orientation the sampling rate (over time) is low. However, the PPM is typically simple to operate and inexpensive and requires little in the way of calibration.

The coil-based PPM has more recently been replaced by the Overhauser effect PPM with the noted advantage (over the coil-based PPM) of lower power consumption, higher sampling rates, and higher accuracy. With this method, a type of proton-rich fluid containing chemicals with free radicals is added. A constant high-frequency RF signal (approximately 60 MHz) is applied to the fluid generating the "snap" magnetic field at a much higher frequency. This allows a >2 Hz sampling rate with much higher theoretical accuracy over the solenoid method.

With the Optically Pumped Alkali Vapor Magnetometer (OPM), light is pumped through a series of gas vapor-filled chambers (typically nonradioactive cesium, but rubidium, potassium, and even helium are sometimes used), allowing the use of the electron energy and spin properties of the gas in the magnetic field measurement. A narrowly defined wavelength of light is radiated through a gas-filled cell onto a photocell. An RF signal/field is applied to the gas cell so that the properties of the light's interaction with the electrons of the excited gas will form a measurable output pattern on the photocell proportionate to the ambient magnetic field. The OPM obtains its higher performance over its PPM counterpart due to its much higher oscillating signal (between 70 and 350 KHz for the OPM versus 0.9–4.5 KHz for the PPM), allowing for much more information carrying capacity. The advantages of the OPM over the PPM is its higher sampling rate with a corresponding disadvantage of higher cost, required calibration, and lower absolute accuracy.

The field of magnetic anomaly detection is quite fascinating. Many excellent technical papers explaining the technology are freely available for further research into the subject.

18.3 Flooded Member Detection (FMD)

Fixed industrial structures, for example offshore oil and gas production platforms, attached to the seafloor are typically fabricated from welded tubular steel. Once the jacket legs are sealed at the fabrication yard, the air-filled chamber within the tubular structure remains air-filled, allowing for buoyancy offset to the often extreme weight exerted on the structure's base on the seafloor. Should the air-filled chambers of the jacket become flooded, a situation would evolve whereby some of the tubes would be flooded, destroying the buoyancy on a portion of the jacket. This would place additional stress on the structure's base in an asymmetrical load. As part of a typical IRM program, FMD is performed on selected portions of the platform to verify structural integrity of the entire platform system. This method is also used for the detection of flooded pipelines.

FMD is performed either acoustically or radiographically, depending upon several factors including member size, metal thickness, degree of marine growth, and cost.

18.3.1 Acoustic FMD

With acoustic FMD, a high-frequency transducer is placed directly on the metal of the member to be examined. Air has a much lower sound conductivity/attenuation potential than water. If the member is flooded, the sound will propagate through the metal and water and then "bounce" off of the metal on the opposite side of the flooded member. This time of flight is measured to a high degree of accuracy, thus allowing for a highly reliable test of member air integrity. If the member is air-filled, the sound will simply resonate or dissipate within the single wall of the pipe or member.

Some advantages of the acoustic method over the radiographic method are:

1. Typically, acoustic FMD is much simpler and less expensive than the radiographic technique.
2. The acoustic method does not require the HAZMAT precautions needed for the handling of the radiographic source.
3. Technician qualification is much simpler.

Disadvantages of the acoustic method over the radiographic method are:

1. The acoustic method typically requires some surface preparation on members with excessive marine growth.
2. Orientation of the transducer is critical in order to achieve sound energy propagation perpendicular to the axis of the pipe or flooded member.
3. False negative or positive readings are possible without proper data interpretation.

18.3.2 Radiographic FMD

As with acoustic FMD, the radiographic method of FMD works on the principle of differing attenuation through the medium of air versus water. With the radiological method, a low-grade gamma ray source is positioned on one side of the member or pipe while a gamma ray detector is positioned on the other side. A known metal thickness produces known gamma ray attenuation through the metal. A simple computation of the theoretical reading on the back side of the pipe will produce a very highly reliable measure of total attenuation. If the member is flooded, the total gamma ray reception will be of a much lower value than through air due to the water's attenuation of the gamma ray energy within the pipe or member.

18.4 Cathodic potential sensors

Salt water is often termed "the universal solvent." That solvent potential has an extremely high effectiveness in the corrosion of submerged metal structures in a salt water environment. To understand the principals of operation of a Cathodic Potential (CP) sensor, one must first understand the corrosion process (Figure 18.20)—specifically for structural steel in salt water.

Metal corrosion is defined as the deterioration of a metal resulting from a reaction with its environment. Three elements must be present for metal corrosion to take place:

1. Two dissimilar metals
2. An electrolyte (such as salt water)
3. A conducting path between the dissimilar metals for electrical flow

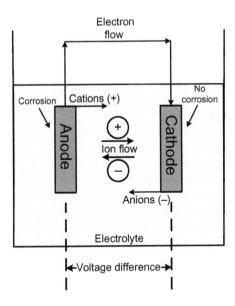

FIGURE 18.20

The metal corrosion process.

The dissimilar metals can be different elements (steel, aluminum, etc.) or simply impurities/non-homogeneous metals on a single metal sheet. Once the above conditions are satisfied, the combination forms an electrochemical cell allowing the following chemical reactions to occur:

$$2Fe \Rightarrow 2Fe^{++} + 4e^-$$

This process produces four free electrons ready to combine with water to form hydroxyl ions in the following reaction:

$$O_2 + 4e^- + 2H_2O \Rightarrow 4OH^-$$

Once the above reaction takes place, the main attraction (i.e., the really expensive heartbreak) takes place—namely, iron combining with oxygen and water to form ferrous hydroxide:

$$2Fe^{++} + 4OH^- \Rightarrow 2Fe(OH)_2$$

This process follows the second law of thermodynamics' theoretical trend toward maximum entropy, tracing the steady flow of highly organized metals toward their disorganized oxidized metal combination. The two electric potentials of the dissimilar metals form an electrical path from the anode (area of highest activity) to the cathode (area of lowest activity). The anode, therefore, is corroded away—and if that anode happens to be your steel support structure, it will eventually fail and collapse.

An unprotected metal surface of a submerged steel structure becomes an anode (active area) once the conditions for an electrochemical cell are satisfied, thus allowing the electrochemical process described above to corrode the metal.

It is often said that electricity is lazy, that is, it will take the path of least resistance, making the local electrochemical cell reactive to the area of highest potential/gradient. It is possible to change the local gradient by electrically attaching a more reactive metal to a less reactive metal, thus drawing the electrical reaction to the area of highest potential/gradient. This is known as "sacrificial anode" cathodic protection (SACP).

Sacrificial anode cathodic protection

Various types of metals are more active than others. The measure of the gradient is compared with the gradient to a "reference electrode." The definition of a reference electrode is one with a stable and precisely known potential so that other metals' potentials may be compared. Two of the more common reference electrodes used in the subsea industry are the $Cu/CuSO_4$ and the $Ag/AgCl$ combination (copper−copper sulfate electrode ($E = +0.314V$) and silver−silver chloride electrode ($E = +0.197V$ saturated)). Table 18.3 lists metal potentials with reference to the $Cu/CuSO_4$ reference electrode.

To convert the unprotected steel structure from an anode (active) to a cathode (passive), another (more reactive) metal is electrically attached to the metal surface. This allows electrolysis to react with the higher potential metal, thus sparing the steel support structure from corrosion. The Sacrificial Anode (SA) metal is "sacrificed" in order to protect the higher-value structure.

Impressed current cathodic protection

This method substitutes the passive sacrificial anode (electrically attached to the local metal surface) with an active direct current surrounding the entire facility (Figure 18.21). Impressed current

Table 18.3 Metal Potentials

Metal	Potential with Respect to a $Cu/CuSO_4$ Reference Electrode
Carbon, graphite, coke	+ 0.3
Platinum	0 to −0.1
Mill scale on steel	− 0.2
High silicon cast iron	− 0.2
Copper, brass, bronze	− 0.2
Mild steel in concrete	− 0.2
Lead	− 0.5
Cast iron (not graphitized)	− 0.5
Mild steel (rusted)	− 0.2 to −0.5
Mild steel (clean)	− 0.5 to −0.8
Commercially pure aluminum	− 0.8
Aluminum alloy (5% zinc)	− 1.05
Zinc	− 1.1
Magnesium alloy (6% Al, 3% Zn, 0.15% Mn)	− 1.6
Commercially pure magnesium	− 1.75

Source: Peabody (2001).

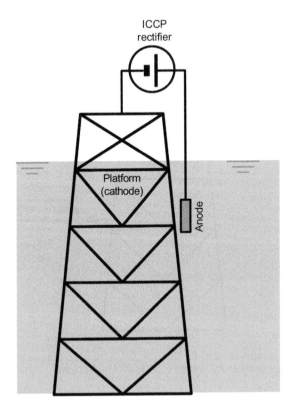

ICCP
rectifier

Platform
(cathode)

Anode

FIGURE 18.21

Diagram of an impressed current cathodic protection (ICCP) system.

cathodic protection (ICCP) uses an external electrical source (generator aboard the platform or grid power) to be "pumped," that is, AC power is rectified to DC for pumping. The current pumped into the sea thereby drives the entire platform's electrical potential to match that of the local electrolyte, for example, seawater or local soil. This inhibits the galvanization process from taking place.

The CP probe senses the local electric potential between the SACP anode attached to the steel structure and the grounded potential of the entire structure. In essence, the electrical circuit forms a huge battery with the voltage potential measured between the entire structure and the sacrificial anode (Figure 18.22).

With a higher potential between the anode and the seawater, better protection is achieved for the metal of the steel structure. With lower SA potential, the seawater ceases attacking the sacrificial anode and instead attacks the structure. By measuring the anode's potential against a standard reference electrode (typically Ag/AgCl), the flow of current onto the metal will result in a negative potential. The crossover point for stopping corrosion is in the -0.800 V versus the Ag/AgCl reference electrode. Anything above -0.800 V (i.e., -0.700, -0.600, etc.) indicates a breakdown in the protection system. The desired potential is in the -0.950 V to -1.000 V range versus the Ag/AgCl reference electrode.

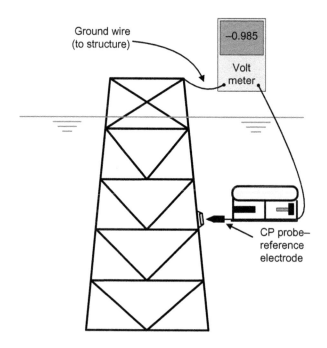

FIGURE 18.22

Voltage potential is measured between the anode and the entire structure.

Common subsea CP probes are operated in either "proximity" or "contact" modes. In proximity mode, the CP probe is placed in the near vicinity of the sacrificial anode with the electrical potential then measured. The voltage potential readings are taken throughout the platform to profile the entire platform for any local variations in potential since a nonuniform potential may indicate a breakdown in a section of the entire cathodic protection system. In contact mode, a probe is stabbed into the anode to achieve a full electrical connection with the anode, then the electrical potential is measured from the open surface lead to the platform.

There are three basic types of probes used for CP measurements (Figure 18.23):

1. Drop cell
2. Proximity probe
3. Contact probe

The drop cell is a proximity probe operated from the surface. The drop cell is normally lowered from the surface during a yearly topside inspection of an offshore platform, for example, API Recommended Practice 2A Level I Inspection. This method has a much lower accuracy than an ROV-placed probe, but the cost is much lower and this complies with regulatory requirements. The proximity probe is designed for near placement while a contact probe has a rugged stainless steel tip for stabbing into the anode and making a full electrical connection.

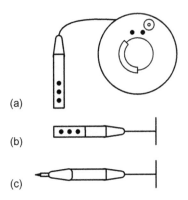

(a)

(b)

(c)

FIGURE 18.23

CP probe types. (a) Drop cell, (b) proximity probe, and (c) contact probe.

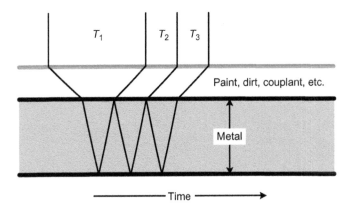

T_1 T_2 T_3

Paint, dirt, couplant, etc.

Metal

Time

FIGURE 18.24

Operation of a multiple echo metal thickness (sampling is taken at T_2 or later).

The ROV-operated CP probe has a much higher accuracy as it is placed directly adjacent to the anode being measured. This type of measurement is required by regulatory authorities to be measured every 3–5 years, for example, API Recommended Practice 2A Level II and III Inspections.

CP readings are rife with potential errors through instrument miscalibration and equipment damage. It is imperative for the technician to become completely familiar with the operation and calibration of the equipment before taking the equipment into the field.

18.5 Ultrasonic metal thickness

Further examination of metal corrosion is possible with the ultrasonic metal thickness (UT) gauge. As with the acoustic FMD sensors (above), a pulse of high-frequency sound (in the 2–5 MHz range) is generated by a transducer (either in close proximity to (or in direct contact with) a metal surface) and then transmitted through a metal. The sound will bounce off the opposite wall. The time difference is then measured to derive a very high-accuracy metal thickness measurement. These readings, however, are subject to a host of errors and interpretations.

Modern UT gauges make use of the multiple echo technique for higher accuracy reading. When sound is passed through metal at a particular frequency, the sound will resonate within the metal. This allows the taking of later echo bounces to more accurately measure the echo with regard to the metal (versus other couplants like coatings, dirt, etc.). In Figure 18.24, a diagram of the technique of measuring a subsequent echo isolates the sound timing within metal from the outside sources. This allows for accurate measurement of flat plate metal thickness along with some types of circular metal structures.

Manipulators and Tooling

Manipulators

19

CHAPTER CONTENTS

The previous version of *The ROV Manual* essentially dealt with observation-class ROVs, those with a weight of 200 lbs (91 kg) or less. Although these smaller ROVs are proliferating worldwide, most are limited in their manipulative capabilities. With the expansion of this version of the manual to include the next class of ROVs—mid-sized ROVs (MSROVs) weighing up to 2000 lbs (907 kg)—the potential to increase the manipulator and tooling capability of these vehicles grows dramatically. With that in mind, a section that deals with manipulators (this chapter) and tooling (Chapter 20) becomes an essential part of the manual. The principles and technology discussed in this section can be extended to essentially any class of ROV.

As discussed in the next section, the applications of ROVs are widespread, which means that the tasks they have to be designed to accomplish are also widespread. Without some type of manipulator/tool system, the operator is limited to observation only. Once a decision is made to incorporate a manipulator or tool system, the design decisions become more involved. The requirement can range from a simple device used to grab an object, up to a complex manipulator that requires position control and force feedback. Whether conducting oceanographic research or neutralizing a mine, the requirements of the tasks to be accomplished will drive the choice of manipulators and tools to be used by the vehicle. Those choices and their impact on the vehicles will be discussed in more detail in the following sections.

19.1 Background

As discussed in the section on the history of ROVs, the earliest commercial vehicles were essentially "flying eyeballs" (Figure 19.1) that hovered over and watched the work performed by a very capable work tool using dual-position-controlled arms with force feedback—the diver. As offshore

FIGURE 19.1

The original "Flying Eyeball"—Hydro Products RCV 225.

divers began to accept these mechanical intruders, and their reliability began to increase, they became an essential part of work operations. Whether providing topside operators with real-time feedback on the status of the underwater work or bringing down or retrieving tools, they expanded their roles offshore (Figure 19.2). As offshore operations slowly moved into waters beyond diver depth, the ROVs increased their manipulative capabilities. When the oil companies realized that remote intervention was the future, they finally began working with the ROV developers to configure the underwater equipment and tasks to be handled remotely.

The earliest ROVs had no more than a gripper attached to them, which essentially made the ROV a "flying hand." This flying hand was essentially the manipulator; it had the number of degrees of freedom (DoF) that the thrusters could provide, which was rather limited. The vehicle also had to deal with the inhospitable ocean environment that often included adverse currents, wave dynamics and poor visibility. By adding a more capable manipulator with more DoF to a free-flying ROV, the manipulative options increased, but the operator now had to deal with the manipulator's dynamics as well as the vehicle's motion. Enter a second manipulator, one with fewer DoF, that could grab onto or mate with the object to be worked on, and the operator began to get a more stabler, or more structured, work environment. Thus the standard design of work-class vehicles, from lightweight to heavy duty, includes two grabbers/manipulators.

As ROVs replaced divers in more areas, especially in environments hazardous to divers, the desire to retain the diver's manipulative capability always remained. Designers have continued to develop various versions of anthropomorphic manipulators that provide the topside operator with an ability to conduct manipulative operations as if the operator were using his/her own arms. This capability, known as telepresence, has seen wide research. As an example, research conducted by the US Navy (Figure 19.3) used such a system that also included binaural hearing and stereo vision. The operator wore a helmet that was used to move the underwater camera system as if he or she were looking around while sitting on the front of the vehicle. Whether the task at hand requires a "flying hand" or a 10,000 ft (3048 m) anthropomorphic diver replacement, the factors to be considered are similar and will be addressed in the following sections.

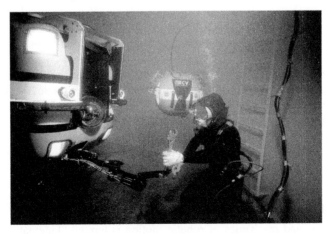

FIGURE 19.2

Hydro Products RCV 150 using manipulator to support the diver.

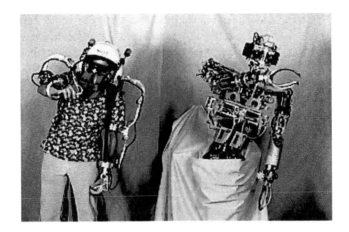

FIGURE 19.3

Anthropomorphic manipulator system.

19.1.1 **Basic robotics**

As stated in Part 1, the ROV is essentially a "ride to the work site" for the delivery of sensors and tooling in order to work within the environment. Once at the work site, the manipulators and tooling allow the vehicle to physically interact (the accepted industry term is "intervene" or "intervention") in order to accomplish this work.

The use of robotics with underwater vehicles was an outgrowth of research conducted with the "industrial robot" in other industries, predominantly during the 1960s, which was then adapted to the harsh environment of the subsea world. The robotic vehicle is consistently defined by various standards organizations, but mechanical manipulator systems are a bit murky. For instance, a typical subsea manipulator (Figure 19.4) would be considered a robotic manipulator, but a numerically controlled milling machine is typically not. If the system can be programmed or directed to perform a wide range of tasks, the device would be considered an industrial robot. Otherwise, it would be termed "fixed automation."

Likewise, the manipulator is considered robotic when either "teleoperated" or directed via logic-drive. An example of what is typically not considered a robotic manipulator is a simple knuckleboom crane. This device has all of the mechanics/kinematics ("kinematics" in this case being the science of robotic motion) of a manipulator system, but the operator directly controls the function of the valves or actuators operating the arm. If that same device was operated remotely via machine logic or telepresence, it would be considered a robotic manipulator.

19.1.2 **Manipulator mechanics and control**

The study of the range and mechanics of robotic motion is termed "kinematics" (or the science of motion). This science simply considers the ranges of motion, position, velocity, acceleration, and other position variables without regard to the forces acting upon the manipulator system.

FIGURE 19.4

Schilling "Conan" subsea manipulator system.

(Courtesy Schilling Robotics.)

In industrial manipulator systems, DoF are the possible positional variables needed in order to locate components of the manipulator system. Often in subsea manipulator systems, the term "function" is used interchangeably with DoF. The functions are easily identifiable as they contain some type of joint at each function allowing for movement of the link(s) or end effector. Each positional variable is semi-independent of the other links, although the chain is interdependent. As an example of this concept, a human can place his or her hand on a doorknob, rotate with a less than outstretched arm, and then rotate the upper and lower arms while still maintaining the position of the base (shoulder) and end effector (hand).

A manipulator system has individual components allowing it to be maneuvered to interact with the environment. The manipulator system is further broken into individual components including the "base" (or fixed pedestal upon which the manipulator mates to the operating platform—in this case, the vehicle), "links" (rigid or nearly rigid strength members, between the joints, and bearing the load), "joints" (allowing for DoF or ranges of relative motion between adjacent links—often termed "function"), and finally the end of the chain of components that actually performs the "intervention," termed the "End Effector." The end effector can be a mechanical hand (jaw), a bolt/ screw driver, a brush, a water jet, a thruster, or any mechanical device that performs work. The base of a multifunction manipulator mates to a joint (shoulder/azimuth in Figure 19.5) and then in a link-joint-link fashion to the end of the chain where we find the end effector (or "business end") that actually performs the work.

Most full-function manipulators used with ROVs are 7-function manipulator systems. The seven functions break down into six DoF (six links and six joints, arranged sequentially, with the last two links having zero length) functions (three for orientation and three for positioning) and one end effector function. A manipulator with more than six DoF is certainly possible, but it is considered kinematically redundant as the additional joints do not gain more access to the workspace (defined below).

OCROVs typically have one single- or multifunction manipulator system. Unless the vehicle is fixed to a stationary object while performing work (e.g., atop a fixed structure while thrusting down to hold the vehicle in place), coordinating the arm while the vehicle free-flies is considered difficult (if not impossible) as the vehicle frame of reference (FoR) is floating with regard to the object's FoR.

Just as a carpenter uses one hand to steady the board while the other performs the sawing/screwing/nailing function, practically all WCROVs are equipped with a dual manipulator system having one "rate" arm (for grabbing and holding onto the object, thus locking the vehicle's FoR to the object's FoR) and one dexterous "position/proportion" arm for performing the complex work task.

A base that links directly to an end effector is typically termed simply a "tool" (e.g., a brush, drill, torque tool, or other mechanical device) or a "grabber" (for end effectors that just attach to some structure). This section will consider the manipulator system, while in Chapter 20, tooling will be further examined.

19.1.2.1 The base

The manipulator base is the mechanical hard point connecting the vehicle (or pedestal) to the series of joints, links, and end effector. It is typically rigidly mounted as the forces of the entire manipulator system concentrate on the vehicle/manipulator attachment point. Depending upon the DoF of the manipulator system, the base should be braced in several orientations to account for torsion stress. The typical base for a multifunction subsea manipulator system connects to a rotary joint, allowing left/right movement, and is either fixed in the horizontal or vertical plane (depending upon desired orientation and range of motion).

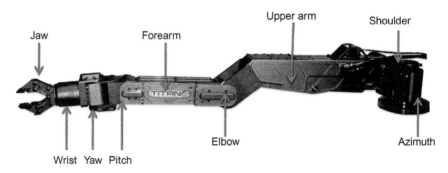

FIGURE 19.5

Basic manipulator system.

(Courtesy Schilling Robotics.)

19.1.2.2 Links

Links join to either the base, joints or end effector and form the load-bearing portions of the chain.

19.1.2.3 Joints

Joints come in two broad categories (although there are lesser type joints), "prismatic" (Figure 19.6) and "revolute" (Figure 19.7). Revolute joints translate rotary motion, and prismatic joints translate linear motion between links, base, or end effector.

Both types of joints have an axis of rotation or travel around/upon which the force applied to the connecting links translates. These axes are important as they convey the force of the manipulator's motion to the work load.

The joints, links, and end effector can be monitored visually (via remote camera) or instrumented with position measurement sensors to determine orientation and status (or both). The objective of all of this base/joint/link combination is to maneuver the end effector into a position to perform the work.

In order to interact with objects within the operating environment, the manipulator will require locating the deployment platform (the vehicle) into the three-dimensional (3D) space and then placing the end effector in an orientation to allow motion translation for the task being undertaken. The Cartesian coordinate translation between the "base Frame" (or FoR) and the end effector or "tool frame" (Figure 19.8) involves some fairly complex mathematical modeling, but most subsea manipulator systems are visually operated via telepresence (as opposed to logic driven in a controlled environment).

With consideration for the vehicle's access to the work site and the work site coordinates/orientation being within the end effector's field of motion, the tool is generally operated visually. Therefore, the coordinate translation mathematical modeling is beyond the scope of this text.

19.1.2.4 Translational versus rotational transformation

Although the typical subsea manipulator operator visually aligns the tool, it is important to determine the orientation of the axis of tool rotation for operating the tool. During the computation of

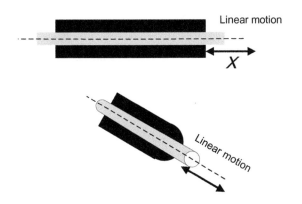

FIGURE 19.6

Prismatic (linear) joint.

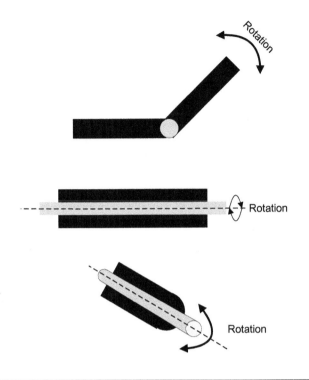

FIGURE 19.7

Revolute (rotational) joint.

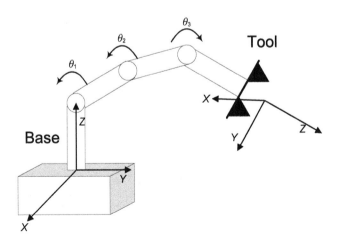

FIGURE 19.8

Base frame to tool frame kinematic coordinates.

the tooling interface, care should be given to the kinematic function of the tool/receptacle motion. The translational movement aligns the axis of rotation or movement of the end effector (Figure 19.9) while the axis of rotation must align with the tool receptacle in order to assure adequate functionality. As discussed later in this chapter, the vehicle stability with reference to the interface is quite important so as to allow seamless kinematic functioning between the end effector and the interface. Standards have evolved within the industry to establish vehicle stability while intervening with subsea valves and/or tooling, thus allowing for smooth operation of subsea systems.

19.1.2.5 Actuator
The mechanism that translates the mechanical force into motion is the actuator. For the hydraulic (or pneumatic) system(s), the pump/cylinder combination transfers fluid motion to linear or rotary motion while the electric actuator performs a similar function using an electric motor as the drive mechanism. Actuators transfer work to the manipulator in order to overcome the various forces opposing the system's movement along with configuring the geometry of the end effector for performing the work. In Figure 19.10(a), cylinder linear motion is either transferred to rotary (or linear) motion (for moving links) in a limited range while the gerotor actuator (Figure 19.10b) performs continuous rotation for rotating the wrist attached to the end effector.

19.1.2.6 Manipulator/wrist/end effector combination
The system becomes a manipulator when the series of links, joints, and structure combine with the wrist/end effector for performing work. The joint between the last link (forearm) and the end effector is the most important as it controls the final orientation of the end effector. It is thus termed the "wrist." The term "positioning" refers to placing the end effector to a point within the geometric workspace while "orienting" refers to moving it to its proper orientation once at that point. End effector positioning is for the arm, and orienting is for the wrist.

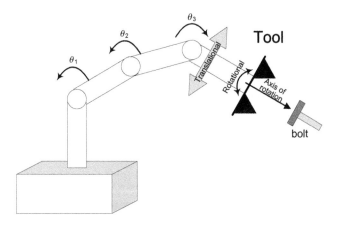

FIGURE 19.9

Translational moment versus rotational moment.

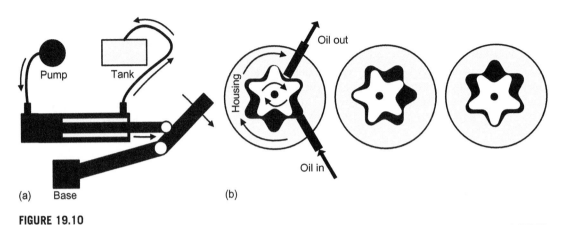

FIGURE 19.10

Hydraulic actuation for both (a) cylinder (limited rotation) and (b) gerotor (continuous rotation).

19.1.2.7 End effector

The free end of the base/joint/link chain contains the end effector. This allows for the final physical interface between the vehicle and the environment. The base/joint/link combination is present to simply maneuver the end effector into a suitable position to accomplish the work.

The simplest end effector is the gripper. It is capable of opening and closing for gripping objects. That grip can be used as a contact point for steadying the vehicle or it can be used to grip and maneuver a tool (which then becomes part of the end effector).

In standard robotics, the end effector can be any of a number of mechanical devices including a torch, grabber, grinder, electromagnet, brush, torque tool or any other mechanical device. However, in the subsea industry, end effectors other than a hand/jaw/gripper are typically classified under "tooling" as opposed to "manipulator." To correlate to anthropomorphic terms, a human would grab a wrench (spanner) from a tool chest to turn a bolt—the actual end effector is the wrench (spanner)/hand combination with the hand being a part of the end effector.

End effectors are, by convention, the working end of the manipulator, that is, the hand. But the design of the hand is highly dependent on what the required tasks are. Most manipulators have various hands that can be installed on the manipulator prior to launch of the vehicle. The hands are designed for a range of tasks, to include interfacing with and operating various tools. The various hand types and types of tooling are discussed below and in the next chapter. The manipulator or grabber hands (or jaws) are generally made up of two to four fingers/tines that operate in a parallel fashion or intermesh to grasp an object or tool. Examples of the parallel and intermeshing jaws are shown in Figure 19.11. Such grippers are well suited to grasp objects by closing the fingers around the structure or tool. Further, these parallel and intermesh hands are typically ISO 13628-8/API 17H compliant and mate readily with standardized tooling handles and wellhead interfaces. Most hydraulic manipulator jaws operate in the "fail-to-open" mode so that, should a hydraulic malfunction occur while the grabber is locked onto a platform, the manipulator will release the structure so the vehicle may be recovered and repaired.

(a)

(b)

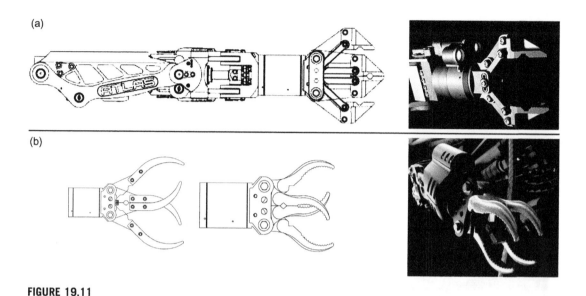

FIGURE 19.11

Parallel and intermeshing manipulator jaws. (a) 6-inch parallel jaw and (b) 11- and 7-inch intermeshing jaws.

(Courtesy Schilling Robotics.)

The large work-class manipulators are not the only ones with interchangeable hands. SeaBotix's grabber (Figure 19.12), as used on their small ROVs, not only incorporates parallel or interlocking jaws but also can be outfitted with a line cutter.

19.1.2.8 Manipulator control

As further discussed below, the two variables for arm actuation are force and slew rate. The simplest form for movement of a hydraulic arm (Figure 19.13) would be adjusting the system's pressure and flow (and pressure relief) upon the arm's actuator to either slew the arm quickly or slowly and then to apply a certain amount of pressure/force before stopping the arm's movement (and relieving the pressure back to the tank).

For electrical actuation of manipulator systems, the two types of controls are classified as either servo control (closed loop) or non-servo control (open loop). Servo manipulators are mostly used in manufacturing applications for performing redundant tasks, leaving the non-servo control for the subsea vehicle realm (the exception is with use of a position-controlled manipulator system).

The two types of controls of subsea manipulator systems are "rate control" and "position control." The simplest of these two types is the rate-controlled arm. In rate controller, each joint is singled for actuation in series, thus moving individual joints toward the end goal. In the position controller, the position of the joystick or controller arm is in a truly "master–slave" configuration, whereby the arm mimics the position of the controller. Whereas the rate controller isolates each joint for movement toward the final goal, the position controller moves several joints simultaneously to orient the arm in unison with the controller. Typically, while the position controller is more elegant, the rate controller is more robust as the position controller requires calibration periodically.

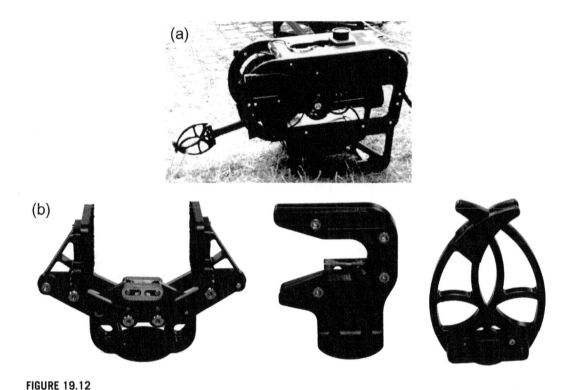

FIGURE 19.12

(a) SeaBotix vehicle with grabber and (b) various end effectors.

(Courtesy SeaBotix.)

19.1.2.9 Sensors and controller

The interface between the robot and the human is the controller (sometimes termed the "OCU" for operator control unit as used in industrial robotics). The controller can range in complexity from a simple three-position "fail-to-center" open/off/close switch to a complicated position-controlled dexterous arm controller with force feedback.

By far the biggest issue with controlling a manipulator arm is the lack of sensory feedback from the environment to the operator. An additional difficulty of operating a manipulator arm subsea is the lack of stereoscopic vision when viewing operations through a typical single-channel camera. Most manipulator operators term the simple remote operation of a manipulator system (as well as the entire vehicle for that matter) as "looking at the world through a single drinking straw." The advent of force feedback has allowed more sensory perception to circle back to the operator while stereo and 3D camera systems, along with positional indicators (and inertial sensors), assist enormously in delicate remote tasks. Technology is constantly evolving, thus allowing for more realistic telepresence. While man-in-the-environment is still far superior in sensory perception to that of telepresence (with current technology), the advent of tooling interface standards allows for complicated tasks to be performed through simple and common interfaces.

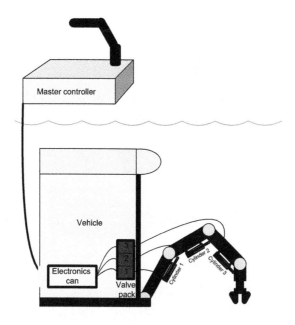

FIGURE 19.13

Surface master controller controls individual valves.

19.2 **Manipulator types**

There are a wide variety of manipulator types that range from simple grabbers to complex force feedback systems. In the past, most manipulators have been hydraulically operated; however, with the proliferation of all-electric vehicles for observation and light work tasks, small electrically operated versions are now on the market. When it comes to heavy-duty work, the strength-to-weight ratios of the hydraulically driven arms (and the fact that the work-class ROV has a massive hydraulic power supply) make them the most applicable approach.

As complexity increases so does the cost of the arms. Although today's heavy-duty manipulators are highly reliable, the more complex the arms, the more opportunities for failure exist. Therefore, the wise buyer will understand exactly what tasks he wishes the ROV to perform and outfit it accordingly. The wise buyer will also realize that even if the required manipulator/tool system is expensive, it is still only a small fraction of the cost of the entire ROV system, but if it fails, then the result is a very expensive flying eyeball. System reliability is critical to successful offshore operations.

19.2.1 **Grabbers**

Grabbers are the simplest of the manipulators available. They usually have fewer DoF, relying on the vehicle to position them. Since they are often required to hold the vehicle in position, often in a dynamic environment, they are usually more robust in design. Simple grabbers range from the

(a)

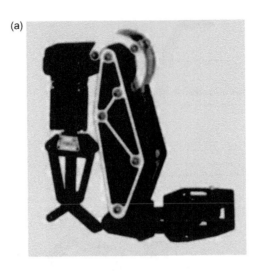

(b)

FIGURE 19.14

(a) Hydro-Lek 5-function manipulator with grabber jaw (MSROV) and (b) SeaBotix grabber (OCROV).

(*Courtesy Hydro-Lek, SeaBotix.*)

small three jaw grabber manufactured by SeaBotix Inc. and the light-duty Hydro-Lek manipulator with a wide-jaw grabber (Figure 19.14) to the heavy-duty, five-function, rate-controlled RigMaster™ heavy-lift grabber manufactured by Schilling Robotics LLC (Figure 19.15). Grabbers with more DoF are able to also support limited manipulative tasks.

Many of the grabbers come with various end effectors that can be changed out for the task at hand, such as those shown in Figure 19.12 for use with the SeaBotix grabber. Also, the Schilling RigMaster combines both the prismatic (linear motion) joint with the revolute (rotary motion) joints to enhance its range of motion and "pull-in" capacity.

When choosing a grabber, the required task should be examined, taking into consideration possible failure modes for the system. Their strength should be based on the vehicle design and the expected operating environment. If holding the vehicle in place in a cross-current is a critical task, then what happens when the environment changes due to a vehicle malfunction or another dynamic situation? Where environmental conditions are extreme, currents are increasing, or the vehicle is experiencing mechanical problems, the grabber(s) must be strong enough to hold/move the vehicle until the operation is complete or the system has "undocked" from the work object/structure. These considerations also apply to the design of the end effector: for example, if the goal is to hold onto a line in a current, then a gripper design that cannot slide off is required.

FIGURE 19.15

RigMaster WCROV grabber.

(Courtesy Schilling Robotics.)

19.2.2 **Dexterous arms**

The dexterity of the arm, that is, more DoF, essentially increases its capability from that of a grabber to the more sophisticated status, that of a "manipulator." Dexterous manipulators generally have at least six DoF plus the end effector, which could be a tool or gripper. Today's manipulators, such as Schilling's 7-function, position-controlled Titan 4 manipulator (Figure 19.16) are efficiently designed. Their hydraulic and electrical lines are located within the arm, which protects them from getting damaged when working on or in an object or structure. The wrist and hand area is also a clean design, which not only prevents snagging on an object but also does not block the view of the operator. It does no good for an operator to have the arm or tool block his vision so that he cannot see the target object. This brings up another consideration, that of the viewing system—always ensure that, whatever the tasks for which the manipulator is designed, adequate viewing options exist for the operator. Multiple perspectives are always an asset.

Most work tasks are on the seafloor or on a vertical plane in front of the ROV. Therefore, the industry standard manipulator design is the backhoe or elbow-up configuration. This design is far from anthropomorphic, or human-like, while it maximizes the work volume and minimizes the storage requirements on the ROV (Figure 19.17).

The more dexterous manipulators usually have a wrist rotate function that allows continuous gripper rotation (with use of the "gerotor" actuator). This feature is valuable for tasks like valve actuation, shackle makeup, or screw clamping. Continuous rotation allows the operator to make multiple turns of the gripper or tool without letting go of, and re-grasping, the object.

With increased dexterity comes the ability to use more sophisticated control functions. Such functions allow the operator to perform complex motions of the manipulator/tool, such as the ability to move in a straight line or to move along paths that require all arm functions to be actuated

FIGURE 19.16

Titan 4 manipulator arm.

(Courtesy Schilling Robotics.)

FIGURE 19.17

HD vehicle manipulator system.

(Courtesy Schilling Robotics.)

at the same time. The combination of the arm's dexterity and more sophisticated control types allows the operator to perform tasks such as following surfaces or inserting electrical or hydraulic connectors. With the incorporation of computer controls, additional capabilities can be added, such as stow, deploy, lock-in-position, or perform other preprogrammed routines. The manipulator's control options are discussed in Section 19.5.

19.3 **Joint design**

Manipulators are composed of a variety of joints that are operated by linear or rotary actuators. Many wrist configurations include a gerotor actuator that allows continuous wrist rotation. Smaller electric vehicles may use an all-electric design for the grabber/manipulator actuators, but as the size and capability of the arms increase, they use hydraulic actuators that take advantage of their excellent lift-to-weight ratio characteristics.

In the case of the smaller electric vehicles, the number of functions and the range of motion of the manipulator may be very small because of the ability of the vehicle to "fly" the hand into position. As the arms grow in size and capability so does the number of functions. As a minimum, the following functions are necessary to position a larger arm:

- Shoulder azimuth
- Shoulder pitch
- Wrist or forearm rotate
- Hand/gripper open/close

The above functions allow the operator to position the manipulator to grasp in both a horizontal or vertical orientation. As the manipulators grow in complexity, additional joints are added that mimic the human arm, that is, shoulder, elbow, wrist, and hand. With this in mind, a basic 7-function manipulator will include the following:

- Shoulder azimuth
- Shoulder pitch
- Elbow pitch
- Wrist pitch
- Wrist yaw or forearm rotate
- Wrist rotate (may be continuous)
- Hand/gripper open/close

Examples of 4- and 7-function arms are shown in Figures 19.18 and 19.19.

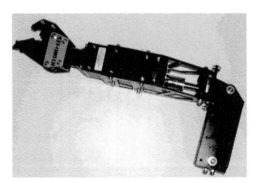

FIGURE 19.18

Hydro-Lek 4-function manipulator.

(Courtesy Hydro-Lek.)

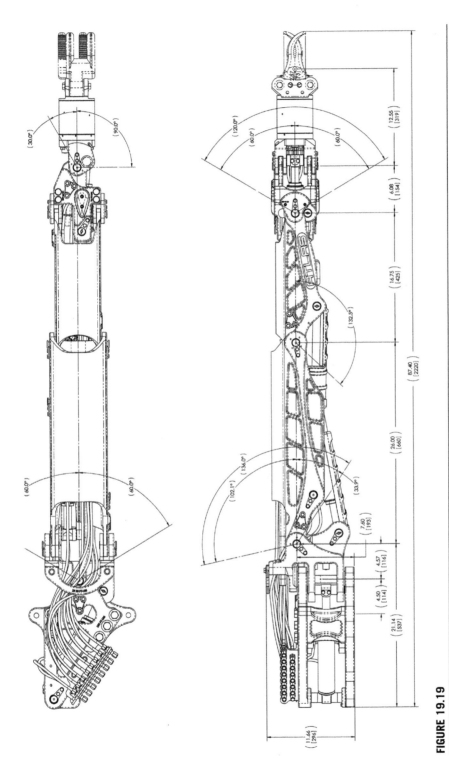

FIGURE 19.19

Atlas 7-function arm characteristics.

(Courtesy Schilling Robotics.)

19.4 **Range of motion and workspace**

Each joint added to a manipulator adds a degree of freedom (DoF). As a minimum for a work manipulator, it is desirable to be able to translate within and rotate about the x, y, z axes (six DoF) and then grasp something or hold a tool (seventh DoF). Therefore, the heavy-duty work manipulators have the seven linear (or rotary joints) DoF. Manipulator manufacturers then provide plan and elevation views of the range of motion of the subject manipulator (Figures 19.20 and 19.21).

The total volume reachable by the end effector (Figures 19.20 and 19.21) is termed the work space. The geometry of the manipulator's motion as well as the joints' mechanical constraints defines the workspace. The workspace is further broken down into "reachable" and "dexterous" workspace. The reachable workspace is the volume whereby the end effector is capable of reaching each point within the space in at least one orientation while the dexterous workspace has the end effector capable of reaching all points in all orientations.

These views provide an excellent depiction of the external shell that the tip of the manipulator can reach, that is, its range of motion. However, the warm feeling of such a magnificent range of motion must be tempered with the reality of performing work within that shell. This is not a case of Michelangelo reaching out with a paintbrush toward the ceiling of the Sistine Chapel. The tasks that manipulators are required to complete offshore are complex and may require various joint configurations to reach and/or align with the point of interest. Thus, the range of motion of a manipulator will be much larger than the practical "workspace" of the manipulator.

There are other factors that affect the real workspace of a manipulator or set of manipulators/grabbers. The first is the range of motion of the manipulator/grabber that may be used to hold the vehicle in position so that the other arm can perform the assigned task. As an example, if the grabber had a 4.5 ft (137 cm) reach and the working manipulator had a 6 ft (183 cm) reach (and was expected to work on the same vertical surface to which the grabber is attached), then the range of motion of the manipulator would be affected and the overall workspace would probably be reduced.

Now take the previous example a step further. The manipulator has to pick up a large tool and align it orthogonally to that same vertical work plane. The tip of the manipulator has now become the tip of the tool, which again reduces the maneuverability of the manipulator and thus its overall workspace. The bottom line is that the design of a complex work system that uses a variety of tools on a variety of objects is not a simple task. The manipulator(s) and/or grabber(s) specifications must be evaluated in conjunction with the working environment to be encountered and the tools to be used within that environment. The good news is that the leading offshore firms now have or use simulators of the vehicles, manipulators, and expected working environment to ensure that the chosen system and configuration can in fact perform the necessary underwater tasks.

19.5 **Types of controllers**

Manipulator and grabber controllers range from simple on/off toggle switches (rate control) to small replica controllers that match the joint configurations of the working arm (position control). Going to the next level of complexity is the force feedback controller where the forces encountered by the manipulator are scaled and fed back to the operator via the replica controller. These types of controllers are discussed in more detail in the following sections.

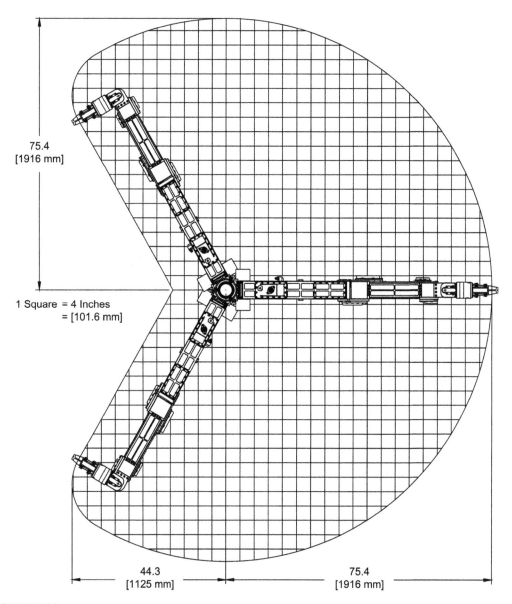

75.4
[1916 mm]

1 Square = 4 Inches
= [101.6 mm]

44.3
[1125 mm]

75.4
[1916 mm]

FIGURE 19.20

Plan view of manipulator area of reach.

(Courtesy Schilling Robotics.)

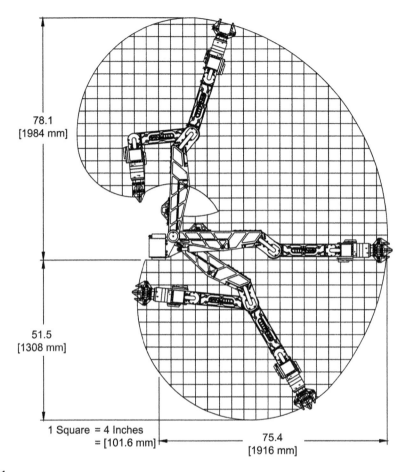

78.1
[1984 mm]

51.5
[1308 mm]

1 Square = 4 Inches
= [101.6 mm]

75.4
[1916 mm]

FIGURE 19.21

Elevation view of manipulator area of reach.

(Courtesy Schilling Robotics.)

19.5.1 Rate control

Rate control is the simplest form of control for manipulators and grabbers. The operator controls a hydraulic manifold of solenoid or servo valves through simple toggle switches or, more commonly in larger manipulators, a joystick type of controller. Through the manipulation of the joystick, or the activation of buttons or switches on the joystick, the operator activates the electrically operated valves at the remote site, opening or closing the valves to achieve the desired motion of the manipulator.

An example of an integrated rate-controlled joystick is shown in Figure 19.22. The rate hand controller, developed by Schilling Robotics, combines all manipulator functions into one joystick that can be operated using either the left or right hand. The gripper open/close switches are located

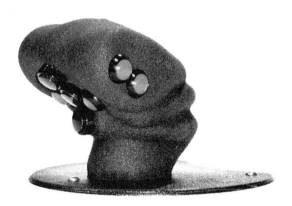 

FIGURE 19.22

Rate control joysticks.

(*Courtesy Schilling Robotics.*)

on both sides of the controller grip to allow right- or left-hand operation. Other functions are controlled by pushing buttons on the front of the grip, by twisting the grip, or by rocking it from side to side.

Rate control is an economic and robust approach to manipulator control. However, it does not provide the fluid motion of multiple joints that can be achieved with a position-controlled manipulator.

19.5.2 Position control

Position control (also called "spatially correspondent control") uses a small replica arm, or master controller, to control the manipulator (Figure 19.23). Each manipulator joint, or DoF, has a kinematic analog in the master arm. Obviously, a manipulator with position control is a more complex system because the position of the joints has to be sensed and fed back to the master controller where the positions are compared and the required signal sent to the valves to activate and adjust the position until the error is reduced to zero.

Using a master replica controller can make the movement of the manipulator simple and intuitive. In addition, multiple joints can undergo simultaneous movement, resulting in precise and fluid manipulator movements. Position-controlled manipulators also have the capability to incorporate preprogrammed operations, such as deploy or stow or move to predetermined positions. These can also limit the movement of individual joints so that the manipulator does not enter an area or configuration that could initiate a problem. Individual joint freeze, joint diagnostics, error checking, etc. are all a benefit of position-controlled manipulators.

A decade or more ago, the added complexity of such designs could have impacted their field reliability; however, with advancements in electronics, refinement of the overall designs, and the mass production of many of the commercially available manipulators, operational reliability remains extremely high. And if problems are experienced, most commercially available manipulators are repairable in the field.

FIGURE 19.23

Position controller.

(Courtesy Schilling Robotics.)

19.5.3 **Force feedback control**

The manipulators designed for position control and force feedback control are essentially the same design as both require sensing of the position of the manipulator joints. The difference is in the design of the master controller. For a position-controlled manipulator, the master controller directs the manipulator's joints to move in response to movement of the replica controller, which is known as "unilateral control." When the master controller is stationary, so is the manipulator. Movement of the manipulator, due to some external force, does not cause the replica controller to move in response.

In the case of force feedback, small electric actuators built into the joints of the master controller cause the controller to be moved in response to forces sensed by the manipulator (Figure 19.24). This bilateral control allows the operator, who is holding the master controller, to feel any forces that the manipulator is experiencing. This tactile feedback can be a valuable sensory input during complex or delicate tasks. Also, the operator has the ability to adjust the level of the force being fed back through the replica controller, thus reducing any physical fatigue that might be experienced in longer operations. As in the position-controlled manipulators, today's systems that are operating in the field are highly reliable, even if a bit more complex.

The choice of rate, position, or force feedback control is task-dependent and should be taken into consideration in the early planning stages.

19.6 **Hydraulic versus electrical power**

Since most of the heavy-duty manipulators on work-class vehicles have access to abundant hydraulic power, their choice over an all-electric system is a given. The use of hydraulic power allows the manipulator to deliver large forces in a reasonably sized package.

For the smaller OCROVs (which are nearly all electrically powered), the choice of an electrically operated manipulator is also logical since the forces needed for most tasks required of those vehicles are not as large. If electrically operated arms could be made with lift-to-weight ratios and

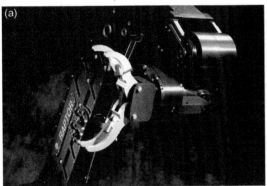

FIGURE 19.24

(a) Kraft FFB (force feedback) manipulator and (b) controller.

(Courtesy Kraft Telerobotics.)

physical envelopes the same as hydraulic manipulators, then they might be an option through not having to convert electric power to hydraulic power, which would make the system more efficient. But unfortunately, the packaging of electric motors and/or actuators does not match the packaging efficiency of hydraulic actuators. In addition, the larger work-class vehicles used in offshore commercial applications all have large hydraulic power units—some as much as 200 horsepower or more.

The choices become more complex when MSROVs are being used. As the vehicles grow in size, they soon morph into hydraulically operated systems. So when in that gray area between the small observation class and the larger work-class vehicles, the operator has a choice to make. If available electrically operated arms do not meet the operational requirements, and the vehicle is still all-electric, then there are a couple of choices. The first is to mount a hydraulic motor, valve package, and compensator system somewhere within the vehicle's frame. The second is to attach a skid below the vehicle's frame that can carry the necessary hardware (see Chapter 20). When using a skid, the manipulators could also be attached to the vehicle's main structure.

Whenever configuring a vehicle with additional equipment, whether electric or hydraulic, the stability of the vehicle must be kept in mind. The deployment of a manipulator can have a significant effect on a small vehicle's static as well as dynamic stability. This can become even more severe if using a tool or picking up a heavy object. The vehicle, and its thruster system, must maintain a stable enough configuration to allow the vehicle to maneuver properly. It will not do the operator any good if the vehicle is tilted forward at a 45° angle when trying to maneuver, especially if trying to return to the surface. However, if the goal is to grab an object and bring it back to the surface, then the vehicle can essentially turn into a shackle (Figure 19.25) between the umbilical and manipulator/grabber, as long as the strength of the umbilical and its termination with the vehicle is sufficient. Such an option may not exist when using a tether management system (see Chapter 9). Examples of the various components commercially available for configuring an ROV are provided in later sections.

FIGURE 19.25

VideoRay with grabber used for object retrieval.

(Courtesy VideoRay.)

19.7 Subsea interface standards

Current subsea vehicles are trickle-down technology from earlier military applications. Military standards, as they apply to undersea vehicles and other systems, are usually concerned with shock, vibration, and other issues that would cause the system to become inoperable due to environmental damage or wear. Until the advent of the subsea construction programs brought on by the mineral mining industry, few industry standards existed. By far, the largest current user of ROV equipment for deepwater construction applications is the offshore oil and gas industry—and thus the primary beneficiary of any standards for common interface.

As oilfield exploration and production equipment became standardized, two primary organizations took the lead on promulgating these standards—the American Petroleum Institute (API) and the International Organization for Standardization (ISO). Standards arose for common wellhead and oilfield structures so that all manufacturers could build within a framework of common interface. But as the wellheads went from human-accessible locations to the deepwater seafloor, the need for an ROV-friendly common interface designed into the beginning of the project engineering of the structure became readily apparent.

Both the API and ISO answered this call separately and then merged the standards into a mirrored international standard. As of this writing, the current standard for subsea production systems is API 17 and ISO 13628. Other standards have arisen over time. These have been adapted by various organizations including DNV, NORSOK, and IMCA as well as company-specific guidelines. As the API 17H/ISO 13628-8 standards are identical and the most commonly used (and adapted by many governmental regulatory organizations), ISO 13628 will be the controlling standard (as described herein).

API's policy on standards is to review, revise, reaffirm, or withdraw each standard at least every five years. In June 2013, the API issued the second edition to API RP 17H, adding considerably to the coverage in the previous edition. The new edition of 17H covers Remotely Operated Tooling (ROTs) while ISO breaks ROTs into a separate Part 9 to 13628. For this section, we will cover the ISO 13628 standard as it covers both ISO 13628-8 and API 17H (first edition) and leave to the reader to delve into API 17H for the second edition based upon regional preference.

ISO 13628:2002 is divided into nine parts beginning with general requirements, moving onto recommended practices of manufacturing equipment through to the common interfaces for remote intervention and remotely operated tooling:

Part 1: General requirements and recommendations
Part 2: Flexible pipe systems for subsea and marine applications
Part 3: Through flowline (TFL) systems
Part 4: Subsea wellhead and tree equipment
Part 5: Subsea umbilicals
Part 6: Subsea production control systems
Part 7: Completion/workover riser systems
Part 8: ROV interfaces on subsea production systems

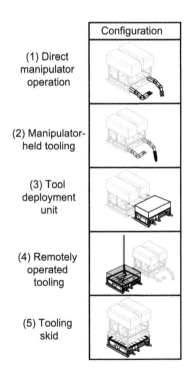

Configuration
(1) Direct manipulator operation
(2) Manipulator-held tooling
(3) Tool deployment unit
(4) Remotely operated tooling
(5) Tooling skid

FIGURE 19.26

Standardized vehicle configurations per ISO 13628-8.

Part 9: Remotely operated tool (ROT) intervention systems

A concern for ROV operators is ISO 13628 Part 8 (13628-8) and API 17H (which are identical standards) for "ROV Interfaces on Subsea Production Systems." The standard will not be detailed here but the important aspects, for later reference to the standard itself, will be highlighted.

The standard identifies several ROV intervention configurations (per 13628-8—see Figure 19.26):

1. Vehicle with manipulators for direct operation of the interface
2. Vehicle with manipulator-held tooling
3. Vehicle with tool deployment unit (TDU)
4. Dual downline method (ROTs or remotely operated tooling)
5. Vehicle with tooling skid or frame

These are discussed in more detail in the following sections.

19.7.1 Manipulator operation of the interface

This configuration is straightforward in that the typical two-manipulator operation has one grabber holding the vehicle in place while the second (more dexterous) manipulator operates the tooling interface. The interface to the subsea facility is via direct operation of the work item with the manipulator's end effector.

19.7.2 Manipulator-held tooling

As opposed to the manipulator end effector performing the work, the manipulator-held tool interfaces directly with the subsea structure. Detailed in Figure 19.11, the types of manipulator hand configurations are compatible with standard grips—the "T-bar" handle and the "fish tail" handle (see Chapter 20). The parallel hand configuration mates easily with the T-bar, and the fish tail works with the curved finger jaw.

Any number of tooling can be held with the common gripper/handle interface and be powered by the vehicle's hydraulic or electrical power system. The various types of tooling available for manipulator holding/deployment are further detailed in the next chapter.

19.7.3 Tool deployment unit

As depicted in Figure 19.27, the TDU is a separate module from the ROV frame that is attached (typically) to the front or rear of the vehicle's frame to accurately position the tool. A TDU is very similar to a skid, but typically the TDU *is* the interface module (i.e., the TDU docks with the interface as opposed to the vehicle docking to the interface with an underslung skid). The TDU is placed between the vehicle and the interface so that the tool can be specially designed for a job-specific task and then switched out once the task is completed. The TDU will typically have its own flotation (so as to minimally affect vehicle performance) and can be used in substitution of, or complimentary to, the vehicle's manipulators.

Standardized subsea production equipment interfaces will have a docking point receptacle, whereby a docking port can be positively attached to the interface for secure tool mating. Figure 19.28 depicts single and twin point docking probes located on the TDU side that locks into

FIGURE 19.27

Tooling deployment unit with dual docking probes.

(Courtesy Forum Energy Technologies.)

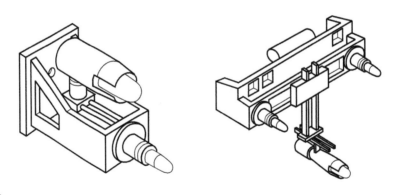

FIGURE 19.28

Tooling deployment unit with single and dual docking probes.

the corresponding receptacle on the interface side. The TDU is most useful when the intervention item is a standardized control panel.

19.7.4 Remotely operated tooling (dual downline intervention)

As detailed in the standard, the remotely operated tooling (ROT), and/or a dual downline component change out (CCO) item, is typically used to deploy or change out subsea components that exceed the payload capacity of the vehicle. An example of a CCO would be a pipeline end termination (PLET), wellhead or other item requiring retrieval and/or setting on the seafloor. However (as depicted in Figure 19.29), when performing construction or decommissioning tasks, the ROT (separate from the

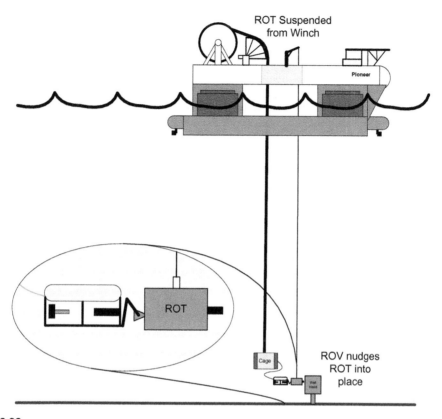

FIGURE 19.29

Dual downline ROT being nudged into position by vehicle.

vehicle) can be suspended from the deployment vessel of opportunity and surface-operated/powered, allowing the vehicle more maneuvering capability to precisely position the tool as well as provide more power to the tool (since the power is surface-supplied). This allows separation of the lift from the vehicle to a surface lift (downline), thus providing a much higher lifting capability. The CCO tool is defined within the standard along with guidelines for design and operation (including launch/catch assist design) as well as tool interface considerations plus ROV nudging techniques for the tool.

19.7.5 Tool skid

Tool skid—deployed CCO designs are also addressed within the standard, defining guidelines for subsea CCO, such as control pods and chokes. This type of intervention configuration is similar to dual downline ROT operation but requires the delicacy requirement of decoupling of the surface vessel heave from the intervention tool. As such, the CCO tool is mounted to (typically) an under-slung tooling skid (mounted to the vehicle frame beneath the vehicle).

19.7.6 Tooling stabilization techniques

With a human analogue to a subsea task, a carpenter or mechanic performing precise tasks (such as cutting a board or turning a nut) must first steady the board or bolt with one hand and perform the task with the other. This allows the manipulator with end effector (i.e., hand with saw or wrench/spanner) to steady the work item with reference to the tool (board or bolt).

The standard further provides recommendations for access to the work item along with methods of steadying the tool (locking the tool frame with the work item frame). The following stabilization methods are used:

1. Attach a flat horizontal platform to the work item (just adjacent to the interface) for the ROV to park and then steady itself onto the platform by downward thrusting onto the platform.
2. Provide a horizontal or vertical bar on the face of the interface for the ROV to grasp with a limited function grabber so that the other (more dexterous) manipulator can work while steadied by the gripper.
3. Provide docking points (such as Figure 19.27) directly on the interface.
4. Design the interface with relatively flat, smooth surfaces for using suction cup ("sticky foot") attachments, such as those depicted in Figure 20.9.

19.7.7 Standards summary

To summarize the discussion on tooling interface configurations covered within the standard, remote tooling is delivered to the work site as either integrated into the ROV or as a separate ROT in either manipulator or TDU mode. While specific differences are discussed, manufacturers of both subsea production components, as well as ROV tooling, can use this standard to design tooling interfaces compatible with the standard production system. The process from inception to final design is charted in Figure 19.30.

Within the ISO 13628 (sections other than Part 8), specific subsea production system design guidelines are discussed toward a (hopefully) seamlessly standardized ROV-friendly system.

19.7.8 Subsea facility and tooling design

Per the ISO 13628 standard, some of the facility and vehicle design philosophies that should be considered include (with section number referenced from the standard):

- ROV access to the work site (Section 4.4.8)
- Intervention system selection considering FMECA (Section 6.2.1)
- ROV load constraints while in operational mode (Section 6.2.6)
- "Fail-to-free" TDU connections (Section 4.4.3)
- Minimization of damage potential and other facility protective measures (Section 4.4.4)
- Structural design so as to minimize the potential for tether snagging (Section 4.4.6)
- Provide a positive means of stabilizing the work platform (besides thrusters) (Section 4.4.7)
- Proper identification markings (both description and function) for all components (Section 6.4.1)
- Height of interfaces above bottom (Section 6.5.5)
- Load reaction imposed on the interface (as opposed to on the vehicle) (Section 4.4.5)

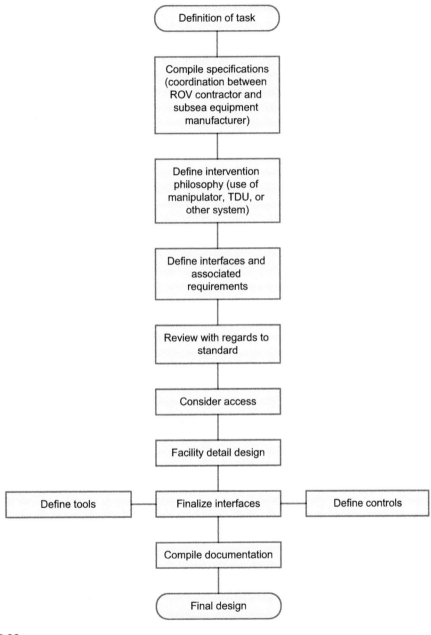

FIGURE 19.30

ROV tooling interface design process.

(Source: ISO 13628-8.)

19.7.9 Specific interface standards

While other parts of the standard discuss various interface types for the common subsea structural components (along with operational considerations, documentation standards and the like), the main concern of the ROV operator is the identification and selection of the proper tool for effectively accomplishing the subsea tasks. Detailed in Chapter 20 are descriptions of the common subsea interfaces along with standardized tooling.

Tooling and Sensor Deployment

20

CHAPTER CONTENTS

The final function of the intervention robot is to place the end effector into a point/orientation within the workspace to perform the work. As shown in Table 20.1, there are various types of possible robot kinematics along with definitive reachable workspace parameters. These robots can either be a part of the main manipulator system integrated into the vehicle or integrated within a separate module for placing the end effector/tool in order to achieve the work function.

Tools have been developed for use by remotely operated vehicles (ROVs) and manipulators to cover essentially every underwater task necessary. In the offshore oil and gas industries, because of the depth of today's intervention equipment forging beyond the reach of human intervention, industry standards have been established. The ISO 13628-8 standard is applicable to both the selection and use of ROV interfaces on subsea production equipment and provides guidance on design as well as the operational requirements for maximizing the potential of standard equipment and design principles. The information for subsea systems the standard offers will allow interfacing and actuation by ROV-operated systems, while the issues the standard identifies are those that have to be considered when designing interfaces on subsea production systems. The framework and detailed specifications set out will enable the user to select the correct interface for a specific application.

The American Petroleum Institute's standard, ANSI/API RP 17H (R2009)—*Recommended Practice for Remotely Operated Vehicle (ROV) Interfaces on Subsea Production Systems*—is identical to ISO 13628-8: 2002. While the current revision of the API 17 standard is "H," many of the older wellhead equipment still use the older version of the ROV interface standard, API 17D. There are some subtle, yet important, differences between the API 17D and 17H standards. As stated in section 19.7, API RP 17H was updated in June 2013 departing this standard from ISO 13628-8. When planning a field operation, the project manager will do well to verify the applicable local standard employed. The most prevalent region still using the older standard is the Gulf of Mexico region while most other regions have adapted the newer API 17H/ISO 13628-8 standard.

To review, the five categories of ROV tooling configurations defined within the ISO 13628-8 (discussed in Section 19.7) are:

1. Vehicle with manipulators for direct operation of the interface
2. Vehicle with manipulator-held tooling
3. Vehicle with tool deployment unit (TDU)
4. Dual down line method (ROTs or remotely operated tooling)
5. Vehicle with tooling skid or frame

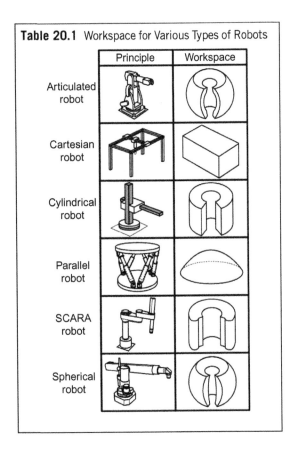

Table 20.1 Workspace for Various Types of Robots

	Principle	Workspace
Articulated robot		
Cartesian robot		
Cylindrical robot		
Parallel robot		
SCARA robot		
Spherical robot		

Examples of the tasks that are performed and examples of the tooling available are given in the following sections.

20.1 Manipulator-operated tooling

The simplest of all tooling operations is the direct manipulator-operated tooling interfacing directly with the subsea interface (*Category 1* of ISO 13628-8 ROV Intervention Configuration—Figure 20.1). From the direct operation of valves and tooling, the tool becomes less integrated with the vehicle.

The remainder of Section 20.1 is dedicated to *Category 2* of ISO 13628-8 ROV Intervention Configuration—namely, the ROV manipulator-held tooling.

- **Manipulator-operated torque tool**

The manipulator-operated torque tool (Figure 20.2) is used to operate ISO 13628 Class 1−2 (note the older standard as *API 17D*) interfaces without the need of a hydraulically operated torque tool. The stainless steel tool uses the wrist rotate function of the manipulator to generate the required torque, which is dependent on the manipulator's capability.

- **Hot stabs and receptacles**

The hot stab is described in ISO 13628-8 as an ROV-operated high-pressure subsea hydraulic connector used to power hydraulic tools, transfer fluid, perform chemical injections, and monitor pressure. Various standardized sizes, configurations, and pressure ratings are available depending upon the requirement of the task. An example of a hot stab and receptacle, with a pressure rating of 10,000 psi (690 bar) is shown in Figure 20.3. The hot stab generally consists of a flexible T-handle (mated to the standard ISO 13628-8 parallel claw—Figure 19.11), the body and removable tips that are installed to prevent damage to the receptacle in the event of an off-angle insertion.

FIGURE 20.1

Direct manipulator operation of interface.

(Courtesy Schilling Robotics.)

FIGURE 20.2

Manipulator-operated torque tool.

(Courtesy Forum Energy Technologies.)

FIGURE 20.3

Hot stab and receptacle.

(Courtesy Seanic Ocean Systems.)

- **Torque tools**

Torque tools are used to translate rotary motion to subsea structures for various purposes. The ISO 13628-8 standard, Table 20.3, describes various classes of torque tools based upon their capacity (Table 20.2).

Once the torque tool classification is selected, the shape and dimension of the actual receptacle are described in Table 20.2 by class as well as end effector shape and size (Table 20.3).

The standardized torque tool receptacle provides for tool reaction against the receptacle (as opposed to transmitting the rotary reaction to the vehicle). Figure 20.4 provides an example of a rotary torque receptacle. Note this as the interface on the *wellhead* side (not the ROV side).

The function of the torque tool, of course, is to provide a standardized interface along with rotary force for various applications. The vehicle camera can typically see the tool turning and can thus count the number of revolutions of the tool. Several manufacturers have made available a torque tool revolution counter that can be either attached as a separate module to the torque tool or integrated

Table 20.2 ISO Standard Torque Tool Classes

Class	Max. Design Torque in N-m (lbf-ft)
1	67 (50)
2	271 (200)
3	1355 (1000)
4	2711 (2000)
5	6779 (5000)
6	13,558 (10,000)
7	33,895 (25,000)

Source: From Table 3 of ISO 13628-8.

into the tool (the so-called Smart Torque Tool) for both mechanical or digital counting of the torque forces and/or turns.

Figure 20.5 provides an example of the ISO 13628 (API 17D/H) Class 1–4 torque tool. The Seanic torque tool includes a digital display that provides real-time torque and turns subsea as well as on the surface, via laptop control. The tool can be supplied as a stand-alone smart tool or with a proportional valve pack, laptop, and GUI that can limit torques, control speed, and direction as well as log values. A robust latch system securely locks the tool into the subsea bucket. In the case of an unforeseen power failure, the latches are spring loaded and will retract, allowing safe disconnection from the work site.

- **Cable cutter**

While the various types of cutters are not standardized items within the purview of ISO 13628-8, they are commonly used within the subsea construction application and are mentioned here. The open face cutting tool (Figure 20.6a) is used to cut wire rope, cables, and umbilical up to 1.5 inch (38 mm) diameter. The 49 lb (22 kg) cutter operates on 3200 psi (220 bar) maximum hydraulic pressure for chain, steel sections, and other hard materials. Also available in either vehicle-integrated or ROT deployment configurations is the gate-face cutting tool (Figure 20.6b) for cutting large diameter hard lines. For those very heavy-duty operations (best left to an ROT configuration), the 38 inch (955 mm) tall, 651 lb (295 kg) cutter can use up to 10,000 psi (690 bar) to cut wire rope and cables up to 7.5 inch (190 mm) diameter.

- **Rotary cutter**

Rotary cutters come in various sizes and types. The cutter shown in Figure 20.7 is rated for 10,000 FSW (3300 m), and with its ability to cut up to a maximum of 6 inches (152 mm), schedule 40 pipe makes it ideally suited for almost any subsea steel cutting requirements. This grinder is powered by an industry-accepted and commonly used thruster motor driving a 4 inch (102 mm) cutter with a self-contained grab. The feed and cut unit is suitable for cutting pipe, cable, or grout hose. The jaw grips the item to be cut with the cutter disk rotating and is fed onto the work piece at a controlled rate to make the cut.

- **Cable and tube grippers**

Other non-ISO tools include the cable gripper (Figure 20.8a). This tool forms the link between a recovery line from the surface vessel and a submarine cable when the cable is to be brought to the

Table 20.3 ISO Standard Receptacles

Dimension in mm (inches)	Class						
	1	2	3	4	5	6	7
A square	17.50 (0.687)	17.50 (0.687)	28.60 (1.125)	38.10 (1.50)	50.80 (2.00)	66.67 (2.625)	88.90 (3.50)
B	154.0 (6.06)	154.0 (6.06)	154.0 (6.06)	154.0 (6.06)	190.5 (7.50)	243.0 (9.56)	243.0 (9.56)
C min.	41.0 (1.62)	41.0 (1.62)	41.0 (1.62)	41.0 (1.62)	63.5 (2.50)	89.0 (3.50)	89.0 (3.50)
D	38.0 (1.50)	38.0 (1.50)	38.0 (1.50)	38.0 (1.50)	57.0 (2.25)	82.25 (3.25)	82.25 (3.25)
E	32.0 (1.25)	32.0 (1.25)	32.0 (1.25)	32.0 (1.25)	38.0 (1.50)	44.5 (1.75)	44.5 (1.75)
F	82.50 (3.25)	82.50 (3.25)	82.50 (3.25)	82.50 (3.25)	127.0 (5.00)	178.0 (7.00)	178.0 (7.00)
G min.	140.0 (5.51)	140.0 (5.51)	140.0 (5.51)	140.0 (5.51)	140.0 (5.51)	222.0 (8.75)	435.0 (17.13)
G max.	146.0 (5.75)	146.0 (5.75)	146.0 (5.75)	146.0 (5.75)	146.0 (5.75)	228.0 (9.00)	441.0 (17.38)
H	181.0 (7.12)	181.0 (7.12)	181.0 (7.12)	181.0 (7.12)	206.0 (8.12)	N/A	N/A
J	12.7 (0.50)	12.7 (0.50)	12.7 (0.50)	12.7 (0.50)	N/A	N/A	N/A
K	168.5 (6.63)	168.5 (6.63)	168.5 (6.63)	168.5 (6.63)	N/A	N/A	N/A
M	25.4 (1.00)	25.4 (1.00)	25.4 (1.00)	25.4 (1.00)	N/A	N/A	N/A
N	194.0 (7.63)	194.0 (7.63)	194.0 (7.63)	194.0 (7.63)	N/A	N/A	N/A

As an alternative to dimension A, end effector shapes as found in Annex D to 13628-8 for the appropriate torque range may be used.
All dimension tolerances are as follows:
0.x ± 0.5 (0.020)
0.xx ± 0.25 (0.010)
C: $+1.25 \left(\begin{smallmatrix} +0.05 \\ 0 \end{smallmatrix} \right)$

Note 1: Chamfer on the end of the end effector profile is 45° × 1.65 (0.06) max.
Note 2: Clearance behind anti-rotation slots (E × F × 50.8 (2)) is to allow for locking feature option provided by some tools.
Source: From Table 4 of ISO 13628-8.

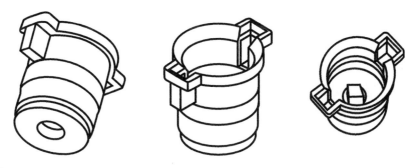

FIGURE 20.4

Example of rotary torque tool receptacle.

(Courtesy Figure 18 of ISO 13628-8.)

FIGURE 20.5

Torque tool.

(Courtesy Seanic Ocean Systems.)

surface. The cable gripper is a powerful jaw mechanism with a lifting eye. The jaw is set hydraulically and then remains mechanically locked after the hydraulics are removed. For larger objects, such as pipes and tubes, Seanic's 12.25 inch (311 mm) recovery clamp (Figure 20.8b) can lift objects up to 600 lb (273 kg). The ROV interface is via a T-handle Acme screw that closes the clamp securely.

- **Sticky foot**

Specified within the "Stabilization" section of ISO 13628-8 are various means of stabilizing the vehicle to the work platform. These include standardized dimensioned and shaped handles for use with manipulators and TDUs used in grasping the structure, standardized platforms for docking to the structure and suction foot devices (commonly called "sticky foot" docking devices).

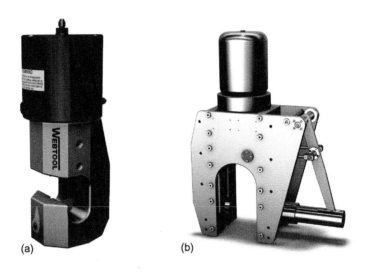

(a)   (b)

FIGURE 20.6

(a) Open face (b) and gate-face cutting tools.

(Courtesy Webtool.)

FIGURE 20.7

Self-contained rotary cutter.

(Courtesy Seanic Ocean Systems.)

The suction foot (Figure 20.9) can hold an ROV to smooth surfaces, such as ship hulls, submarines, platform tubular, and pipelines. The high level of grip makes it suitable for a wide range of ROV inspection and cleaning tasks. The attachment arm is a three-function manipulator consisting of shoulder pitch, jaw, and extend. The foot has a flexible polyurethane molded cup, tolerant of marine growth and crustaceans, and able to attach to curved surfaces down to 400 mm (16 inches) diameter. The cup is mounted on a ball swivel joint to ease self-alignment on the target surface. The swivel can be hydraulically locked once suction is achieved for greater rigidity.

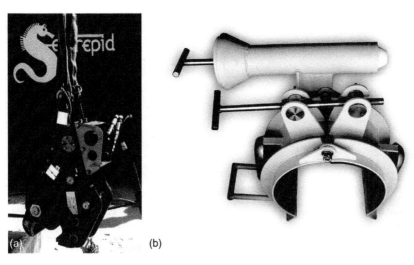

FIGURE 20.8

(a) Cable and (b) tube grippers.

(Courtesy Forum Energy Technologies and Seanic.)

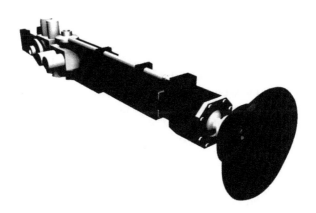

FIGURE 20.9

Sticky foot.

(Courtesy Forum Energy Technologies.)

- **Core sampling tool**

These devices are designed to obtain sediment cores in shallow or deepwater applications. The push corer consists of a Perspex tube, one-way valve, and ROV T-bar handle, which can be grasped by the manipulator as shown in Figure 20.10(a). The corers, Figure 20.10(b), are composed of two separate parts, the core tube with T-bar and the corer housing. The transparent push core is

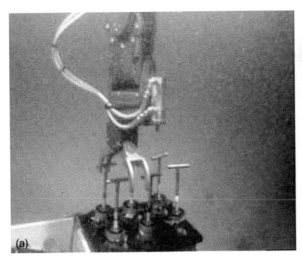

FIGURE 20.10

(a) Push corer being retrieved by manipulator and (b) typical push corers.

(Courtesy Planet Ocean.)

manipulated using the T-bar handles and pushed into the sediment of interest. A one-way valve at the top of the sample chamber allows water to escape as it is replaced by the sediment core. Upon removal from the sediment, the sample is returned to its housing. The housing is manufactured with a flared opening to allow for easy alignment of the returning core tube. At the base of the housing is a tapered rubber plug that seals the sample within the assembly. Upon recovery to the surface, the T-bar and valve are removed and the core with its rubber plug are removed from the housing by a twist fitting at the base. By taking cores with an ROV, this tool allows gathering of multiple high-quality cores from highly targeted locations with specific seabed types or habitats, for biological, physical, or chemical analysis.

- **Cleaning tool**

The cleaning tool (Figure 20.11) allows subsea cleaning and marine growth removal. The tool combines a plastic bristled brush head with an integrated injection system, delivering additional cleaning fluids when and where necessary.

- **Wellhead cleaning tool**

The motorized wellhead cleaning tool (Figure 20.12) is designed to clean the seal surface of wellheads that use AX or VX gaskets. The ROV powers the tool and holds it in position to remove debris. Flow regulation valves limit the rotation speed while still allowing high torque.

- **Gasket ring tool**

The gasket ring tool (Figure 20.13) is designed to remove and install AX and VX wellhead gaskets. The tool is aligned with the gasket and lowered onto it. Hydraulic pressure is then used to

FIGURE 20.11

Cleaning tool.

(*Courtesy Forum Energy Technologies.*)

FIGURE 20.12

Motorized wellhead cleaning tool.

(*Courtesy Forum Energy Technologies.*)

FIGURE 20.13

Gasket ring tool.

(Courtesy Forum Energy Technologies.)

expand the tool to the inner wall of the gasket. In the event of hydraulic failure after engagement, springs and small adjustable tabs located on the bottom of the tool secure the gasket.

- **Grinder**

There are a host of hydraulic power tools manufactured by traditional tooling manufacturers and marketed to the subsea industry. As an example, Stanley Hydraulic Tools manufactures a full line of tools intended for subsea use and are operated by commercial divers, underwater power units, and ROVs.

Hydraulic tools, by their very nature, operate using a closed system of hydraulic fluid. This means that they have very little susceptibility to the environment including operation fully submerged in fluids with high contamination and extreme temperatures.

Stanley's hydraulic tool design approach is to internalize its working components (such as hydraulic motors and valves), which optimizes geometry, features, and packaging to create a compact solution.

Material selection for standard subsea tools is made with the corrosive environment in mind. Typical iron and steel parts have been traded out for stainless steel and may utilize chrome plating as appropriate. This includes critical components like valves and spools, which maintains product life, but also applies to less critical components, such as fasteners, allowing maintenance or service of the product after heavy use.

This manipulator-compatible portable grinder (Figure 20.14) is primarily used for top, face, and side grinding and cutting operations; it can also be fitted with rotary wire brushes and a variety of abrasive and polishing disks.

FIGURE 20.14

Manipulator-compatible portable grinder.

(*Courtesy Forum Energy Technologies.*)

FIGURE 20.15

Manipulator-held excavation nozzle assembly.

(*Courtesy Forum Energy Technologies.*)

- **Excavation system**

The excavation system is comprised of a jetting system and a dredge pump combined into a manipulator-held nozzle assembly (Figure 20.15). The system provides sufficient power to break large solids and excavate any solid up to 3 inches (76 mm) in diameter. Dredging and jetting operations can be used alternately or simultaneously.

20.2 Remotely operated (ROV-positioned) tooling and sensors

20.2.1 Tooling and sensor skids

Tooling and sensor skids allow the operator to configure an ROV, from observation class to work class, with just about any additional manipulators, tools, hydraulics, sensors, etc., that may be necessary to perform a specific task. These skids fall under *Category 5* of the ISO 13628-8 standardized ROV configurations. The value of the skid is that the basic vehicle does not need to be rearranged to attach or install the additional hardware—that is, as long as the skid interfaces have been built into the basic system. The hydraulic power unit (HPU) can either be integrated into the skid (so that only electrical power must be provided to the skid for operation) or the skid may be powered from the vehicle's hydraulic system.

Several examples of tooling and sensor skids are provided. The skid in Figure 20.16 provides a dual manipulator system within the skid.

The Hydro-Lek tool skid (Figure 20.17), which is a turnkey package, incorporates a manipulator arm and cutter that can access a hose or cable at any angle and is controlled via Hydro-Lek Control System components, hoses, and telemetry.

Many skids are configured to take down their own "tool box" or set of deployable sensors. The skid shown in Figure 20.18, being installed on Schilling's HD ROV, has a deployable tool bin that the manipulator can interact with to use the tool or sensor required for the task at hand.

Some of the more useful skids have the power located and mounted on the skid with the actual end effector operated in manipulator-held tooling mode. The most common examples of these types of tools are cleaning skids using high-pressure water or air cavitation as the abrasive agent.

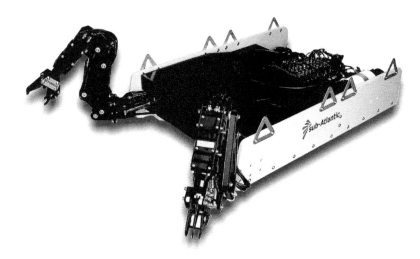

FIGURE 20.16

Dual manipulator skid.

(Courtesy Sub-Atlantic.)

FIGURE 20.17

Hydro-Lek's HLK-8100 tool skid.

(Courtesy Hydro-Lek.)

FIGURE 20.18

WCROV tooling and sensor skid.

(Courtesy Schilling Robotics.)

In Figure 20.19, both a skid-mounted 5000 psi (345 bar) water blaster pump (a) as well as a surface-supplied air cavitation pump (b) are depicted. These are used to clean sediment and marine growth from subsea structures for sealing, cathodic protection, and inspection purposes.

The traditional cleaning agent has been high-pressure water for cleaning marine growth on fragile sealing surfaces, such as wellheads and painted surfaces. However, recent advances in air cavitation cleaning technology have allowed air blasting to augment water abrasion. The advantage to the ROV

(a)

FIGURE 20.19

(a) Skid-mounted 5000 psi (345 bar) water blaster pump and (b) cavitation cleaning wand

(Courtesy (a) Seanic and (b) Cavidyne.)

user of air versus water is the substantial reduction in kinetic reaction of air versus water (i.e., the reactionary "kick" of the water jet is much more forceful than with air due to the difference in density between the two cleaning agents).

This approach, which utilizes water (salt or fresh) at low pressure and high volume, in combination with a specially designed delivery device to produce high-energy cavitation, provides a much safer underwater cleaning system. This quickly and effectively removes unwanted marine growth from underwater structures, allowing the use of much lower pressures than preexisting high-pressure water technology. Cavitation is more efficient and effective than traditional water blasters and eliminates the dangers associated with the use of high-pressure water. Cavitation units clean steel, concrete, wood, rubber, fiberglass, fabric, etc. without damaging existing surfaces or surface coatings.

Similar to skid-mounted tooling is the skid-mounted "sensor pod" or "survey skid" (Figure 20.20). Often, the sensor is not able to be mounted to the main body of the vehicle (e.g., a pipe camera that must see the sides as well as the top of a pipe while the vehicle flies over, or rides upon, the top of the pipe due to the reach angle necessary for viewing or for the proximity to electrical (or acoustic) noise. In Figure 20.20, the sensors are mounted at the end of telescoping poles (or manipulator arms) to achieve the proper offset.

20.2.2 Tooling deployment unit

Described within *Category 3* of ISO 13628-8 ROV tooling configurations is the modular TDU interfaced with the ROV's frame (very similar to an underslung skid, but mounted to the front, rear, or side of the vehicle). As shown in Section 19.7.3 of this manual, docking points are provided for subsea interface on either side of the TDU with the actual tool slung below the docking points.

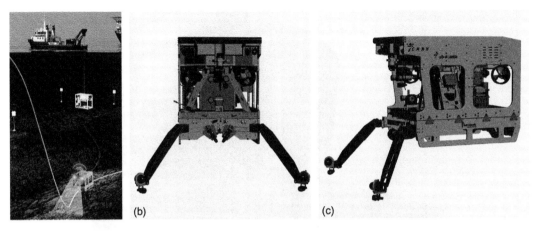

FIGURE 20.20

Pipeline survey skid with boom-mounted cameras (OCROV). (a) in operation, (b) installed skid with front and (c) side views.

(Courtesy Sub-Atlantic.)

The TDU attaches itself to the interface (as opposed to the manipulator holding the vehicle into position) so that a very precise tool positioning can be achieved.

Integrated into the TDU is a Cartesian robot (see Table 20.1 at the beginning of this chapter). Once the TDU is attached to the interface panel, this configuration allows the tool to move on railings for x, y movement along the interface plate, thus allowing individual tool placement to multiple locations (e.g., separate valves) on the interface panel. The z placement is then used to push the tool into the selected receptacle. Various tooling end effectors can be integrated into the TDU for job-specific tasking.

20.2.3 Remotely operated tooling (dual down line ROT)

- **Diamond wire saw**

One example of a surface-powered ROT is the hydraulically operated subsea diamond wire saw (Figure 20.21). This tool is designed for underwater cutting of horizontal or vertical pipe where diverless or ROV operation is required. Models allow cutting of objects from 4 to 52 inches (102–1321 mm) outside diameter. The diamond wire saw requires a high pressure and very high flow rate (2000 psi at 30 US gpm (138 bar at 114 lpm) for even the smaller saws) in order to operate the tool. Typically, this tool is operated in ROT (i.e., dual down line—*Category 4* of the ISO 13628-8 ROV tooling configurations) mode as the fluid requirement may exceed the pumping capacity of the vehicle's auxiliary hydraulic system, requiring hydraulic power from the surface.

Any number of surface-powered ROTs is possible. The tool is remotely powered from the surface and hung from a down line separate from the ROV. The vehicle is used to push/nudge the tool into position and then the surface controller is operated to power the tool, thus completing the subsea task.

FIGURE 20.21

Subsea diamond wire saw.

(Courtesy Wachs Subsea.)

20.2.4 Hydraulics

20.2.4.1 Hydraulic requirements

Most manipulators are designed to operate at 3000 psi (200 bar). While this nominal operating pressure is preferred, systems can be operated on lower pressures with reduced performance. Operation at pressures below 60% of the design pressure is not recommended due to excessive performance loss.

Flow requirements are also stated in the manufacturer's literature. The faster the arm must move, the greater the flow required. Most arms will operate satisfactorily at 30–40% of the maximum stated flow. Use on ROVs with flows below this level will cause the supply pressure to drop dramatically during arm motions and is not recommended.

20.2.4.2 Hydraulic power units

HPUs, in a variety of sizes, are available to provide hydraulic power for manipulators and tool systems. An example is the 1.3 kW HPU manufactured by Sub-Atlantic (Figure 20.22). The small HPU delivers 4 lpm (1 US gpm) at adjustable pressures up to 3000 psi (200 bar). The HPU uses a gear pump driven by a 3-phase, 440 V AC electric motor. An integral, externally adjustable pressure relief valve is used to set the system pressure and a quick-change suction filter cartridge is fitted into the body.

20.2.4.3 Hydraulic valve packages

Once the hydraulic system is energized/powered by the HPU, a valve package (valve pack) sequences and directs hydraulic fluid to/from the proper actuator for functioning of selected hydraulic equipment. Various hydraulic valve packages are available that are composed of whichever type of valve necessary. Sub-Atlantic (Figure 20.23) provides a compact, lightweight general function valve pack made up of solenoid, proportional, or mixed valve configurations controlled by a

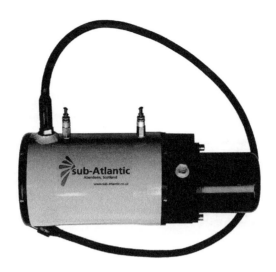

FIGURE 20.22

Sub-Atlantic HPU.

(Courtesy Sub-Atlantic.)

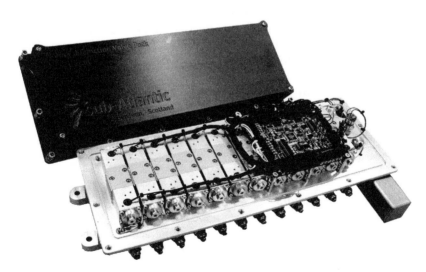

FIGURE 20.23

Sub-Atlantic general function valve pack.

(Courtesy Sub-Atlantic.)

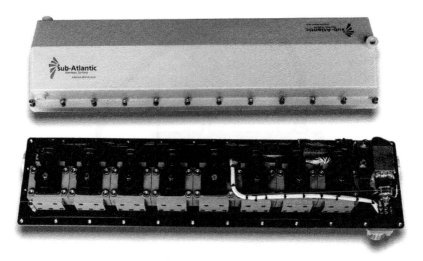

FIGURE 20.24

Sub-Atlantic servo valve pack.

(Courtesy Sub-Atlantic.)

configurable digital control system for ROV and tooling applications. The valve pack can be set up in 6, 8, 12, and 16 station versions. In industry vernacular, these types of simple valve packs are often termed "bang-bang" valves as they are typically either fully open or closed (and the valve makes a banging noise as it is opened and closed once the valve hits the stops).

A variation on the general valve pack is the servo or servo-proportional valve pack. These devices convert an analog or digital instruction set into precisely controlled valve movements for smooth operation of hydraulic equipment. This allows movement of multiple actuators simultaneously for accurate control of equipment position, velocity, and force. Tool movements can be much more smoothly controlled with the servo accelerate/stop controls along with dampening capabilities.

Sub-Atlantic's 8-station servo valve pack is shown in Figure 20.24. The compact servo valve pack weighs 33 lb (15 kg) in seawater and provides precise control of propulsion thrusters on underwater vehicles or can be used to control any tool requiring a variable speed or reversible function. The pack incorporates eight 20 US gpm (77 lpm) rated servo valves that provide low pressure drop on most standard ROV applications.

20.2.4.4 Hydraulic compensators

All oil-filled systems require some type of pressure compensation in order to avoid water intrusion into the hydraulic system from pressure-related volume contraction of entrapped air bubbles as the vehicle descends. Compensation can be provided via flexible covers or specifically designed compensators. One manufacturer provides four sizes (270, 370, 860, and 2700 cc) of rolling diaphragm, positive pressure compensators in corrosion-resistant plastic suitable for ROV and tooling applications (Figure 20.25). The range of sizes available is suitable for compensating thrusters, oil-filled junction boxes, valve packs, etc.

FIGURE 20.25

Typical pressure compensator.

(Courtesy Sub-Atlantic.)

Each and every hydraulic system *must* be compensated (Figure 20.26) lest the hydraulic system crush like a soda can once the ambient sea pressure reduces the volume of entrapped air bubbles below the volume of the entire hydraulic system (hydraulic fluid plus entrapped air). The basic compensator unit is a simple fluid reservoir with a spring attached to squeeze the reservoir, thus increasing its pressure above local ambient. As the vehicle descends, any trapped air shrinks and is immediately replaced with fluids from the compensator. Once the air bubbles compress to a point where the circuit's compensator reservoir is drained, the only thing left to fill the void is for the entire hydraulic circuit to collapse (or water to ingress)! The reverse happens as the vehicle ascends and the trapped air expands, thus driving the fluid back into the reservoir.

20.2.4.5 Compensator/reservoir

If a compensator with a larger reservoir is necessary, this 3.6 US gallon (13.5 L) positive pressure compensator is one of the available choices (Figure 20.27). It is available with an optional analogue level sensor. Mean compensation pressure of this compensator is 7 psi (0.5 bar) above ambient water pressure. An adjustable pressure relief valve, normally set at 14 psi (1 bar), protects the unit from overfill and thermal expansion of the oil.

20.2.5 Electrical actuation

Hydraulic actuators are the most common on ROVs. However, if the system is all electric and only simple actuation tasks are necessary, then the added complexity of adding an HPU and the necessary support equipment would be overkill. Electric actuators are available for use in manipulators, control surfaces, pan and tilts, tool packages, or other subsea applications, such as this DC brushless linear actuator (Figure 20.28).

The small actuator provides a 4- or 8-inch (102 or 204 mm) stroke with 250 lb (115 kg) at variable speeds to 50 mm/s. It requires only 250 W of input power and weighs 2.2 US lb (1 kg) in water. The actuator is available to operate under voltages ranging from 24 to 300 V DC. This model employs a high-efficiency ball screw that has virtually no backlash. It can be fitted with external limit switches that can be adjusted so that the available stroke length is within allowable limits for the application.

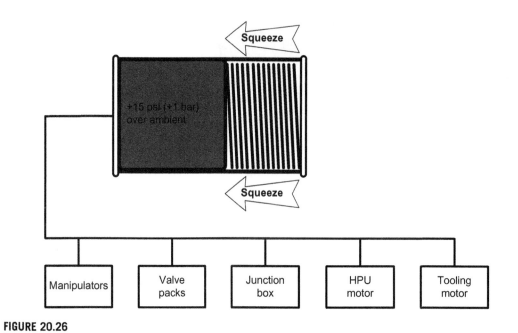

FIGURE 20.26

Compensators must be linked to all hydraulic systems.

FIGURE 20.27

Compensator/reservoir.

(Courtesy Sub-Atlantic.)

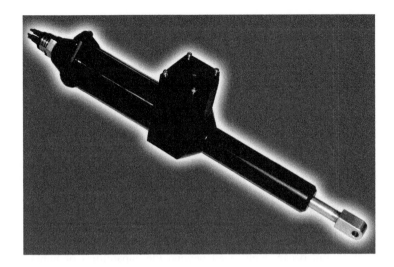

FIGURE 20.28

DC brushless linear actuator.

(Courtesy Tecnadyne.)

FIGURE 20.29

Dual manipulator system operations.

(Courtesy Schilling Robotics.)

20.3 CONCLUSION

OCROVs started out as "flying eyeballs" and soon had a grabber added. Today, the same capability exists with the smaller vehicle. However, today's vehicles are a lot more capable and reliable and cost a lot less than in previous generations of the same technology. The same thing goes for the manipulators and tooling used on today's vehicles. The goal from the beginning with vehicles incorporating manipulators has been to do what a diver can do, that is, perform dexterous work with two arms (Figure 20.29).

Work-class vehicles may do even more through the extra strength of today's manipulators—and do it faster while removing the diver from hazardous environments. Most of these missions are discussed in Chapter 21 of this manual. These new-generation manipulators and tools have become extremely capable and highly reliable. And today, those who need the services of ROVs outfitted with manipulators and tooling have recognized the need to design the deepwater infrastructure to be installed and maintained using ROVs.

Tooling completes the underwater robotic work platform with the ability to provide a full work package all the way to full ocean depth, well beyond the capabilities of humans. With industry-standard ROV-friendly tooling interfaces, provisions are made for robotic intervention from the inception of the engineering project—at the proverbial drawing board—all the way to field operations through to decommissioning. The ROV has become a fully-matured technology and an integral part of the deepwater oilfield.

In the Field

Although there will be some mines with improved capabilities, the greatest threat will be sheer numbers, rather than technological sophistication.
Lt. Gen. Rhodes and RDML Holder's doctrine paper entitled
"Concept for Future Naval Mine Countermeasures in Littoral Power Projection."

CHAPTER CONTENTS

21.1 Explosive ordnance disposal and mine countermeasures

As discussed below, the development of mine countermeasure (MCM) vehicles has radically changed direction, driven by the threat of increased numbers (versus sophistication).

21.1.1 Background

A sea mine's purpose in warfare is a basic "access denial" function for disrupting an enemy's sea navigation capability. If an amphibious landing is anticipated, a clear path to the beach requires clearing of sea mines to allow the vessel an access path. The same concept applies to shipping, since a harbor or navigation channel can be immediately shut down with any threat of sea mine presence.

Sea mining has not changed considerably since the days of the American Civil War, when Admiral Farragut declared "Damn the torpedoes" (older term for sea mines). Sea mines are still, to this day, extremely difficult to locate and neutralize. However, technology is allowing for enhanced location/identification capabilities as well as safer and more environmentally friendly means of neutralizing these sea mines.

Accordingly, the development of remotely operated vehicles (ROVs) for use in explosive ordnance disposal (EOD) and MCM missions is a military mission. This was both a blessing and a curse. The need for such systems was there, as was the funding, but the design specifications were such that the vehicles became large and expensive. The complexity of early MCM vehicles also lengthened their time of development. For example, the US Navy's mine neutralization system (MNS), AN/SLQ-48(V), took nearly two decades to move from concept to operational status (Figure 21.1). Today there are approximately 60 in the field.

The United States was not the only country building MCM vehicles. ECA of France had the workhorse of the early MCM category, the *PAP 104*, which hugged the bottom using a drag weight. Today, ECA has fielded approximately 450 PAP vehicles, 120 of which are the new PAP Mark 5 systems (Figure 21.2).

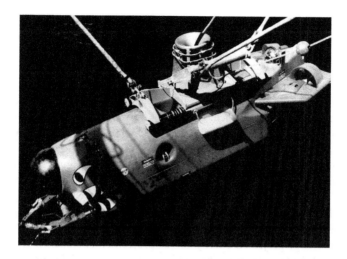

FIGURE 21.1

The US Navy's mine neutralization system.

FIGURE 21.2

ECA's PAP Mark 5.

Other heavy-duty MCM vehicles include:

- Germany's Pinguin B3 (Atlas Elektronik GmbH)
- Sweden's Double Eagle line (Saab Underwater Systems)
- Canada's Trailblazer line (International Submarine Engineering)
- Switzerland's Pluto line (Gaymarine srl)

All of the larger ROV MCM vehicles have the capability of delivering a small acoustically activated explosive charge to the target. The goal is for the MCM ship to locate the mine ahead of time using the ship's MCM sonar, send out the vehicle (while tracking it with the ship's sonar) to verify the target, and plant the charge. Once the charge is planted, the vehicle is retrieved and the charge (previously installed by the vehicle on the mine) is then acoustically detonated. This will, hopefully, destroy the mine by either sinking it (thus placing it outside the path of shipping), destroying its detonation capabilities (i.e., damaging the detonation mechanism or flooding the mine casing with seawater, thus rendering the explosive charge inert) or setting off the mine's warhead while maintaining a safe distance. Disarming a live sea mine is an EOD function.

An older technique of floating the moored mine has essentially disappeared. With this method, the ROV or diver attached an explosive charge to the cable of a moored mine to sever the mooring, thus floating the mine to the surface. The mine was then destroyed with small arms fire or with the deck gun. However, the last scenario any ship captain wants is to have a mine floating somewhere on the surface. Even if the mine could be located, the mine could be lost while floating to the surface, thus exacerbating the problem. Therefore, the best mission plan is to destroy the mine in place.

To neutralize the mine in place without "cooking off" the mine's warhead, two basic techniques are used: (1) shaped charge/projected energy or (2) projectile. The shaped charge, while probably destroying the mine, definitely destroys the vehicle. Once the vehicle (and, hopefully, the mine) has been destroyed, the mine must be reacquired to verify that it has been neutralized (a painfully tedious process after the bottom has been stirred from the explosive charge). The projectile method allows the vehicle to be reused repeatedly, thus saving the time to reacquire the target for battle damage assessment as well as the repeated transit time for the vehicle from the launch platform to the minefield.

The long chess game of mine disposal ends when the mine has been positively characterized as inert. Under the one mine/one vehicle concept of operation, the mine must be reacquired after the mine neutralization vehicle warhead discharge to conduct a battle damage assessment to verify elimination of the threat. The time and cost for the operations platform to reacquire the target is significant and may vastly outstretch the cost of the mine neutralization vehicle itself. Clearly, a reusable vehicle is preferable to an expendable vehicle from a cost and time perspective.

The change in the mine neutralization method, from explosive charge to projectile, may provide the following benefits:

- Lower environmental disruption, allowing for increased stealth
- A range of sizes available to fit the mission requirements
- Reusability of the vehicle
- Long loiter time in minefield, reducing vehicle transit time to operations area
- Immediate muzzle reload with automatic rifle mechanism
- Rapid magazine reload upon retrieval of the vehicle
- Lower risk from possible sympathetic discharge of the mine's warhead
- Environmentally friendly method of mine neutralization
- Immediate battle damage assessment after delivery of the projectile to the mine
- More operational use and training due to reuse of the vehicle
- Built to standard communications protocols for rapid platform switching
- A range of weapons attachable to the vehicle, allowing scalable response
- Lower cost structure to expendable mine neutralization vehicle
- Rapid prototyping with COTS components.

Further, a range of applications outside of MCM may exist, which are limited only by the warfighter's imagination:

- Torpedo tube swim-out from submarine for rapid ship's swimmer defense
- Harbor protection from swimmer attack allowing measured response with long loiter time (with both lethal and nonlethal payload)
- "Shot across bow" ability for a range of threats encountered in law
- Usable from either submersible-powered or surface-powered vehicle
- Tele-operated or fully autonomous modes are programmable, producing a hybrid ROV/UUV

21.1.2 EOD applications

As for EOD applications, the early ROVs didn't have the capability to counter the turbid, dynamic, near-shore environment that was the diver's domain. Little has changed in that area to date. Considerable research is being performed in this area, but if you can't see the target, you can't destroy it.

However, one area outside of the near-shore environment where EOD divers have a tough job is the inspection of ships' hulls to counter terrorist threats. Considerable research and demonstrations are being conducted to evaluate the applicability of ROVs (and autonomous underwater vehicles (AUVs)) to inspect ships' hulls. By using high-resolution sonars to view the hull, advanced navigation techniques for tracking the vehicle's location while maintaining a constant distance from the hull (from several inches to a few feet), and cameras to verify the target, the burden of the EOD diver is again being reduced through robotic technology. Other areas where the workload of the EOD diver can be considerably reduced through the use of robotic technology will be discussed in the following sections.

21.1.3 MCM today

Advances in the miniaturization and efficiency of sensors necessary for MCMs have brought the use of robotics in this mission to a new level. The MCM mission breaks down simply to "find it" and "kill it." Or, in official Navy parlance, the two phases are "Search, Classify, Map (SCM)" and "Reacquire, Identify, Neutralize (RIN)."

The reacquire aspect of this new approach is the key to how MCM is being planned for the future. Two systems are anticipated: one to perform the SCM phase and one for the RIN phase. These two systems could actually be the same vehicle but outfitted with a different payload.

The ability to find the mines while the operating platform is at a safe distance is critical. To solve this problem the US Navy is looking at AUVs. With small, capable sonars, excellent navigation, and the ability to store large quantities of data on the vehicle, AUVs are looking very promising for mapping of target areas.

The US Navy used AUVs successfully during Operation Iraqi Freedom. The Naval Special Clearance Team One used the REMUS AUV (Figure 21.3) to successfully perform MCM. The use of the AUV allowed them to map the area, which not only reduced the time to clear the harbor but also made it a much easier job for the EOD divers.

(a)

FIGURE 21.3

REMUS AUV. (a) with all standard components and (b) in operation on surface.

Other companies developing AUVs for the SCM mission include Bluefin Robotics (USA), ECA SA (France), Hafmynd Ltd (Iceland), and Kongsberg Maritime (Norway), which recently acquired Hydroid, the developer of the REMUS AUVs.

21.1.3.1 Expendable MCM vehicles

Of particular interest is the second portion of the mission—the RIN. Four companies that have developed a "killer" ROV with use of projected energy include:

- BAE System's Archerfish
- ECA SA's K-Ster
- Kongsberg's Minesniper
- Atlas Elektronik's Seafox

An example of the methods in which these systems are used is provided in Figure 21.4.

In development within the United States is a new family of vehicles known as the "K2" family of vehicles using projectile delivery as its primary means of mine neutralization.

Each of these systems will be described in more detail in the following sections. However, since all are similar in concept (i.e., they use an optical communication cable, have a speed of 6—7 knots, carry cameras, sonar, navigation systems, etc.), their pertinent physical and operational specifications are provided in Table 21.1.

21.1.3.1.1 Archerfish

The Archerfish (Figure 21.5), developed by BAE systems, is a fiber optically guided, single-shot mine disposal system (Figure 21.6) that is available in both exercise and warhead variants. Twin thrusters allow it to hover and transit. Maximum speed is 7 knots.

The US Department of Defense selected a team of Raytheon and BAE Systems for the demonstration and development of the airborne mine neutralization system (AMNS). The AMNS will be integrated into the US Navy's MH-60 helicopter.

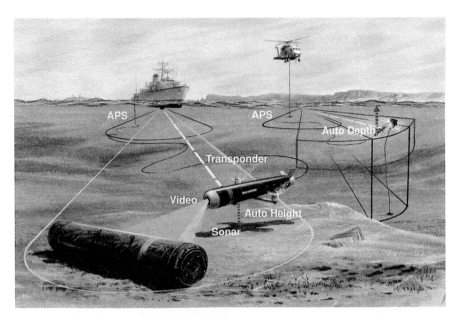

FIGURE 21.4

Archerfish operational scenario.

(Courtesy BAE Systems.)

Table 21.1 Expendable MCM Vehicle Specifications

	Archerfish	K-STER	Minesniper	Seafox
Length (m)	1.05	1.45	1.8	1.3
Width (m)	0.135	0.23	0.48	0.4
Height (m)	0.135	0.23	0.17	0.4
Weight (kg)	15	50	39	40
Depth (m)	300	300	500	300
Range (m)	Unknown	Unknown	4000 +	1200
Sonar	Yes	Dual freq.	Yes	Yes
TV	Yes	B&W, color	CCD	CCTV
Navigation	Trackpoint II	Yes	SBL	INS

21.1.3.1.2 K-STER

France's ECA SA, which has been expanding their line of undersea vehicles, has added the K-STER Mine Killer (Figure 21.7). The vehicle is an expendable mine disposal system (EMDS) that comes in both a positively buoyant, inert training variant (K-STER I) and the negatively

FIGURE 21.5

Archerfish.

(Courtesy BAE Systems.)

FIGURE 21.6

Archerfish prosecuting bottom mine.

(Courtesy BAE Systems.)

buoyant disposal vehicle (K-STER D). The system features a unique tilting head that carries the sensors and shaped charge (Figures 21.8 and 21.9).

21.1.3.1.3 Minesniper

Kongsberg Defence and Aerospace has completed the development of the Minesniper "one-shot" mine destructor vehicle. The Minesniper is controlled via a fiber-optic tether and can be fitted with either a shaped charge or semi-armor-piercing warhead to use against sea mines. The expendable vehicle is powered by twin rotatable thrusters that give the Minesniper the ability to rotate around its own axis. Minesnipers have been sold to the navies of Norway and Spain. It can be used from ships or helicopters.

FIGURE 21.7

K-STER during launch.

(Courtesy ECA SA.)

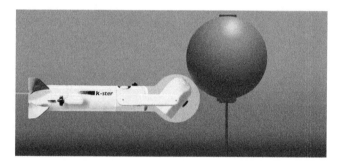

FIGURE 21.8

K-STER against tethered mine.

(Courtesy ECA SA.)

21.1.3.1.4 Seafox

The Seafox, developed by Atlas Elektronik, is a fiber-optic-guided ROV that can be used for the SCM mission (Seafox I), or it can be loaded with an explosive charge and used for the RIN phase (Seafox C, for combat). The vehicle can be used against short and long tethered mines and proud bottom mines. The vehicle system is made up of a console, a launcher, and the Seafox vehicles.

The Seafox (Figure 21.10) uses its integrated sonar to reacquire the target and a CCTV camera to identify it. Once identified, the vehicle can use four independent, reversible thrusters, and one vertical thruster for high maneuverability prior to firing the shaped charge. The system, which has

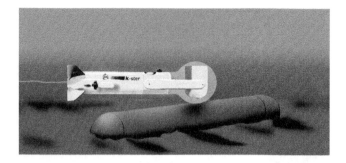

FIGURE 21.9

K-STER neutralizing bottom mine.

(Courtesy ECA SA.)

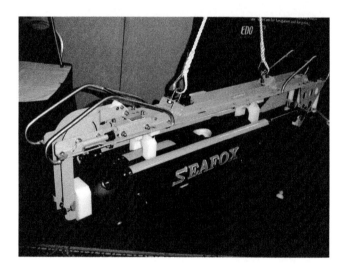

FIGURE 21.10

Seafox.

been integrated into several navies, is capable of operating from several platforms, including dedicated MCM vessels, surface combatants, craft of opportunity, and helicopters.

21.1.3.2 Nonexpendable MCM vehicles

The K2 family of "nonexpendable" vehicles (Figure 21.11), like the expendable MCM ROVs above, is a fiber-optic or copper-guided ROV that can be used for the SCM or RIN missions. The vehicle is either surface powered or can be battery operated. The vehicle, with a sensor and navigation package, is flown into the minefield, where it identifies/classifies the target and subjects the target to repeated projectiles until the target has been neutralized. Once the target has been assessed as neutralized, the vehicle proceeds to the next target.

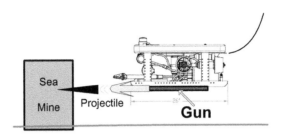

FIGURE 21.11

K2 concept drawing.

Table 21.2 Task and ROV Size Matching	
Tasking	**Best Size**
External pipeline inspection	Large OCROV, MSROV, or WCROV
External hull inspection	Medium/large OCROV
Internal wreck survey	Small OCROV
Open-water scientific transect	Medium/large OCROV or MSROV
Calm-water operations	All sizes

21.2 Commercial, scientific, and archeological operations

Since commercial, scientific, and archeological operations involve a set of how-to tactics, techniques, and procedures (TTPs) similar to those in homeland security (Section 21.4), this section will address the development of a list of specific environments and operations peculiar to these applications.

As discussed in Chapter 3, vehicle geometry does not affect the motive performance of an ROV nearly as much as the dimensions of the tether. ROV systems are a tradeoff of a number of factors including cost, size, deployment resources/platform, and operational requirement. The various sizes of observation-class ROV systems (as discussed in Chapter 3) have within them certain inherent performance capabilities. The larger systems usually have a higher payload and thruster capability, allowing better open-water functions. The smaller systems are much more agile in getting into tight places in and around underwater structures, making those more suitable for enclosed structure penetrations. The challenge is to find the right system for the job. Some examples of tasks, along with the best system size selection, are provided in Table 21.2.

Operational considerations regarding various tasks are discussed throughout this book. The remainder of this section will deal with additional considerations more applicable to commercial, scientific, and archeological tasks.

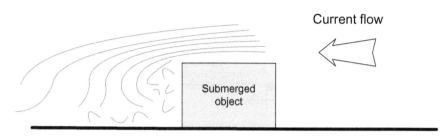

FIGURE 21.12

Currents eddy on the downstream side of a submerged structure.

21.2.1 **High current operations**

The optimum environment for operating ROV equipment is clear water, calm seas, and no current. Unfortunately, the real world intervenes with this perfect world, since one cannot wait for the weather to change in order to get the job done. It is possible to mitigate the effect of currents by doing drift work within the water column, but most commercial inspection and intervention jobs require viewing items that are anchored to the bottom or the shore.

According to Bowditch's *American Practical Navigator* (2002), horizontal movement of water is called "current." It may be either tidal or nontidal. Tidal current is the periodic horizontal flow of water accompanying the rise and fall of the tide. Nontidal currents include all other currents not due to tidal movement. Nontidal currents include the permanent currents in the general circulatory system of the oceans as well as temporary currents arising from meteorological conditions. Currents experienced during ROV operations will normally be a combination of these two types of currents.

In order to complete an underwater task successfully, work with nature to assure that the factors affecting the outcome of the operation are timed in such a way as to mitigate their effect. Planning the dive for commencement during times of slack tide (time during the reversal of tidal flows where current is at a minimum) can lessen current effects upon the operation. Also, an understanding of the dynamics of water flow over rivers, lakes, and structures will assist in taking advantage of these factors (refer to Chapter 2 for additional information on the environment).

The hydrodynamics of water flowing over an underwater structure (Figure 21.12) can have dramatic consequences on the success of the operation. If the situation allows, approach a submerged structure from the downstream side, proceeding against the current. Place the weight or managing platform (i.e., the cage or TMS) above the level of the structure (to lower the risk of fouling) and as close to the object/work site as possible.

In general, the deeper the water in the ocean environment, the lower the current. At the bottom is an area of water that is trapped by surface tension, holding the water immediately near the bottom at zero current. The boundary between the zero current water and the full current stream is an area of high gradient (and turbulent) velocity change called the "boundary layer" (Figure 21.13). If the vehicle is placed below that layer, a relatively current-free operation can occur. The boundary layer moves closer to the bottom as the current increases. In very high currents, where the layer is closer to the bottom than the height of the submersible, this positive effect may be negated.

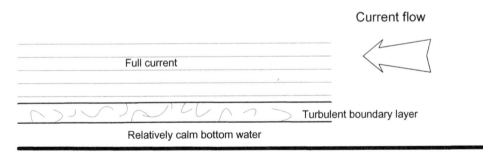

FIGURE 21.13

Bottom conditions relating to current flow.

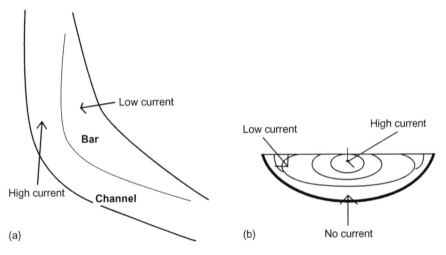

FIGURE 21.14

Examples of channeling on river bends: (a) plan view, (b) cross-section.

Once the hydrodynamic conditions are understood, some other possible solutions to operational problems become apparent. For instance, rivers channel on the outside of a bend and bar on the inside of the turn. If the approach to the object is from the lower current side on the bar, where the vehicle can sneak beneath the boundary layer, the capability is increased to successfully complete an inspection task that otherwise would be unobtainable with a more direct approach (Figure 21.14).

21.2.2 Operations on or near the bottom

By far, the most important working issue in ROV operations is tether management. In operations at or near the bottom, the objective is to reduce the amount of tether the submersible is required to pull to the work site through judicious use of the deployment cage, tether management system, or clump weight. With the weight or managing platform (i.e., the cage or TMS) forming the center

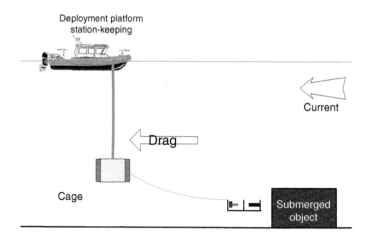

FIGURE 21.15

Position of boat with respect to current and target.

point of a circle of operation, the deployment platform (e.g., the boat, dynamically positioned vessel, or drilling rig) can be moved to within the operational radius of the work site, allowing the work to be easily performed with minimal excess tether (Figure 21.15).

The deployment platform should be placed directly over the work site with the managing platform in a position to have good angular access to all points on the work site. If direct access to a location on the job site is obstructed, the best practice, in most cases, will be to move the deployment platform, rather than risk tether entanglement upon the structure.

Perform all work operations with as little tether actually dragging the bottom as possible (to cut down on the possibility of tether hangs). Then recover the submersible before moving the deployment platform or leaving the site.

21.2.3 Enclosed structure penetrations

Probably the most exciting (and nerve wracking) application of ROV technology is the enclosed structure penetration. It is within this application that the small ROV really shines (Figure 21.16). Due to the inherent dangers to divers in underwater structure penetrations (difficulty of rescue, danger of entanglement on the structure, questionable integrity of the structure, etc.), the ROV may be the only way possible to survey the site.

Some rules of thumb gathered during many hours of enclosed structure penetration with small ROVs follow:

- Choose an entry point into the structure that has a smooth transition (to avoid tether chaffing or trapping) and allows for easy tending of the tether as close to the entry point as possible.
- Purposely choose the tether lay along the structure with a constant eye to recovery.
- First pull the entire length of tether into the structure, to the full extent possible in order to lay out the tether. Then work back toward the entry point, keeping slack in the tether to a minimum.

FIGURE 21.16

Small ROV, with a diver tending the cable, performing internal wreck penetration of the USS *Arizona*.

(Courtesy Brett Seymour, National Park Service.)

- As discussed in Section 9.3, tether traps are the vehicle's biggest danger within enclosed structures. A tether naturally channels to juncture points within the structure, possibly trapping the tether in a less-than-90° wedge point. Plan the tether lay to avoid the tether being pulled into one of these traps, possibly trapping the expensive vehicle irretrievably within the structure.
- It is best to assure a neutrally buoyant tether is available for structure penetrations since tethers can be fouled on either the ceiling or the floor of the structure.
- Enclosed structures (especially old archeological sites) normally have significant silt buildup within the structure. To assure the least stirring of silt (thus obscuring visibility), maintain the vehicle close to neutral buoyancy (cheating on the positive side so that any thrusting needed is away from the bottom silt).
- When exiting the structure, fly the vehicle out adjacent to the tether while the tether handler slowly pulls the tether out of the structure.
- Never pull the tether to free a snagged vehicle. If there is any resistance to pulling the tether from the structure, slack and then reassess.
- Plan on getting into some minor snags and plan the time budget with this in mind.

When the inevitable does happen, keep calm, continue attempting to solve the problem, and never give up.

21.2.4 **Aquaculture and scientific applications**

See Figure 21.17. ROVs are quite useful in aquaculture applications for the following reasons:

- A small ROV does not stress the stock nearly as much as a swimmer.
- Use of an ROV allows more frequent inspections of the nets, mooring, and supports for the farm without danger to divers.
- If configured properly, the ROV can recover "morts" (dead fish) remotely, thus reducing disease propagation and diver bottom time.
- Reduce excess feeding (some species—like Norwegian salmon—will only feed from the water column and will not bottom feed), thus saving feed meal when the stock is finished feeding.
- Comply with regulatory reporting through video documentation of farm conditions.
- Study feeding patterns for optimization of feed delivery.

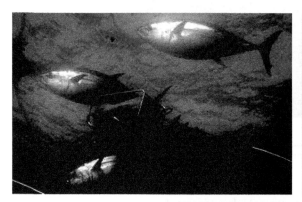
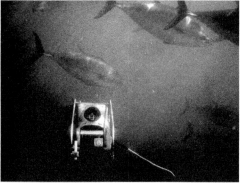

FIGURE 21.17

Small ROVs used in aquaculture applications.

- Hydrological profile (through vehicle delivery of any number of sensors to all parts of the cage) of cage for environment optimization.
- For sea cages, reduce stock loss through constant monitoring of cage integrity.

Traditionally, fish farming operations have been low-budget endeavors. With the introduction of modern management techniques, new technologies have gained wide acceptance within the industry, thus encouraging investments in robotics. In scientific applications, ROVs are replacing the mundane tasks previously performed by divers, allowing a safer and more cost-effective scientific operation (Figure 21.18).

21.3 **Public safety diving**

Many of the useful applications for this technology are in the field of public safety diving. Several public safety divers (PSDs) are killed each year in the line of duty due to the hazardous environments in which they operate. Unmanned searches, as well as electronic search techniques, have the ability to significantly increase the efficiency of the search and lower the risk to field operations

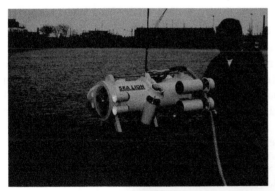

FIGURE 21.18

Small ROVs are excellent for scientific operations.

while enhancing the chance of success. With the right equipment, in properly trained hands, good results can be obtained with the ROV in the dive team's tool chest. This chapter will explore the basics of finding, documenting, and recovering items of interest to the PSD through the use of electronic search techniques.

21.3.1 Public safety diving defined

"Public safety diving" is normally defined as diving operations by or under the control of a governmental agency for the service of the general public. This covers all aspects of search, rescue, recovery, criminal investigations, and any other functions normally covered by police departments, fire departments, and other applicable governmental agencies. Examples of public safety diving would be crime work done by the federal, state, and local law enforcement agencies, search and rescue missions performed by police/fire departments, and disaster relief operations. Examples of operations excluded from this definition would be commercial inspections of public sewer outflows, inspection of bridge abutments, and other services not in the direct service of the general public.

21.3.2 Mission objectives and finding items underwater with the ROV

ROVs are a supplement to the PSD's tool chest. Under current technologies, ROVs are becoming much more useful and pertinent to the PSD team's operational mission, providing an enhancement to its capabilities. The function of the ROV in this long chess game of finding items of interest underwater is in the endgame as a final identification tool. The ROV has limited search and some recovery capabilities. The main function of the ROV is to lower the physical risk to the PSD by putting the machine at risk in situations and environments that have been traditionally borne by the PSD.

The PSD's objectives in performing an underwater search follow:

- The isolation, securing, and defining of the search area
- The clearing of the search area of possible targets
- (Upon location of the items of interest) the gleaning of what information from the site is available, to department standards of documentation
- The disposal of the item as deemed necessary by the command structure

The actual search technique for locating bottom targets with an ROV system is to run the search pattern across the bottom as high off the bottom as possible, based upon the water visibility, while maintaining visual contact with the bottom. The search is run either with a single camera display (which most OCROV systems possess) performing a zigzag pattern across the bottom along the search line (to search on either side of the submersible) or in a straight line with multiple cameras providing wider camera coverage. It is imperative to keep visual contact with the bottom at all times while transiting the coverage area, if for no other reason than to raise the possibility of a "lucky find." So-called spaghetti searches (unstructured look-around searches within the search area) are quite popular in getting the immediate result of searching and performing a task, but it has been consistently deemed that a clear, coherent, and structured clearing of the search area is the only productive means of performing an underwater search.

For an expanded explanation of public safety diving operations, refer to Teather (1994) and Linton et al. (1986) in the bibliography. There are also excellent training courses available to police and fire departments for training as a PSD in traditional scuba diving techniques.

21.3.3 When to use the diver/when to use the ROV

The ROV or the PSD should only be put into the water when a target is "trapped" by some other means (i.e., a sonar/magnetometer target is identified or a narrow search area is identified). The general search navigation procedure should be to start at a known reference point, finish at a known reference point, and maintain positive navigation throughout the search phase so that all of the search area is positively covered with the search instrument.

For the grid search to cover a defined geographical area, search lines should be spaced to the extent of the range of the equipment (with an approximate 10% overlap of coverage for error margin):

- For sonar equipment—to the effective range of the system used
- For visual equipment—to the effective range of the ROV's camera or diver's sight (i.e., clarity of the water)
- For magnetometer equipment, etc.

With a trained and proficient ROV operator on the dive team's staff, all of the mundane search/identification of possible targets should be performed by the ROV up to the final identification of the item. The diver should get wet if the physical capability of the ROV system is insufficient to handle the mass/bulk of the item or if the item is of such a fragile or sensitive nature that an ROV would be inappropriate.

21.3.4 Search theory and electronic search techniques

An ROV can be thought of as a delivery platform for a series of sensors. Since ROVs are inherently slow-speed platforms, the full search toolbox should address the needs dictated by the operational requirements. If a very small search area is defined (e.g., a possible drowning victim was seen falling off of a dock in an exact location), an ROV with just a camera may be the only requirement if the water clarity is good. If the water clarity is bad, a scanning sonar may be needed to image the victim acoustically. If a large area is defined (e.g., a drowning victim's last seen point is unknown, thus the search boundaries cover a very wide area), other sensors decoupled from the ROV would be much more effective in covering the search area.

The basic problem with any search is that until you find the item of interest, your time is spent searching where the item is not located. The objective of searching is to positively eliminate an area of interest as a target area with as high a degree of certainty as possible. The job of a PSD is to clear that area as quickly and efficiently as possible, given the resources available.

Finding items underwater is a chess game from opening moves to end game. The search area is defined and then swept to identify possible targets that will need to be prosecuted or classified.

As introduced in Section 21.4, search and identification of underwater items are accomplished through four basic steps:

1. *Research* the operation to define the area of interest for mapping and limiting the search area. Examples of items used to outline the search area follow:
   a. Intelligence
   b. Witness interview/last seen point
   c. Survey data, prior searches, etc.
2. *Wide area* search with instruments and sensors such as:
   a. Towed or watercraft/aircraft-mounted imaging and profiling acoustics (such as side scan sonar)
   b. Magnetic anomaly detection (MAD) equipment (such as a towed magnetometer)
   c. Radiological and chemical instrumentation, etc.
3. *Narrow area search* with slow-speed instruments including:
   a. Acoustics (such as a fixed location mechanically scanning imaging sonar)
   b. Fixed or towed magnetometer
   c. Optical equipment (towed/drop/ROV-mounted camera or laser line scanner), etc.
4. *Final identification*, classification, discrimination, and disposition of the item by a human—that is, through the diver's eyes or through the camera mounted on an ROV that is propelled to the inspection site and viewed by the operator.

The ROV's place within the PSD's toolkit is through the narrow area search and final identification phases of the search plan. ROV intervention as a productive and cost-effective means of final item identification, discrimination, and disposition is the focus of this section. The four basic steps are discussed in more detail in the remainder of this section.

21.3.4.1 Planning and research

The planning and research segment of a search operation is by far the most important yet least practiced segment. A PSD dive team can never obtain too much information before a search. Inexpensive time spent asking questions and poring through data could save hours of expensive, high-risk, resource-intensive time on the water.

Suppositions can be made based upon witness interviews and some basic science. Calculations of travel time and distance to impact on accidents and drowning cases help narrow the search area. For an expanded text on drowning physics and underwater crime scene investigations, Teather's work (1994) on the subject (referenced in the bibliography) is excellent.

There is a plethora of information available from many sources that contributes to the data needed to define the search area as well as to gain an understanding of the conditions and challenges faced while on the water. Examples of these public-domain files include:

- Geographic Information Systems (GIS) database for the search area maintained by the local township
- The US Coast Guard, State Environmental Agencies
- Local land survey companies
- The US Geological Survey
- US Board of Reclamation

- Previous searches within the (or other inter-agency) department
- Fishing maps
- Archeological data from local universities

Data can be obtained from any number of other sources. A competent and effective PSD will spend time locating/gathering these items of information before being faced with a time-sensitive search. Examples of the types of data acquired during two prior searches, one an archeological search and the other a murder victim search, follow.

21.3.4.1.1 Archeological search

A good example of research was during a prior job cataloging the WW2 wreck sites of torpedoed tankers in the Gulf of Mexico in support of an effort with a Southern Louisiana university. Specific steps, information gathered, and issues encountered follow:

1. Petitioned and received the then-named Defense Mapping Agency (DMA) for non-submarine contacts within their database
2. Received from the US Coast Guard's (USCG) 8th District, all submerged obstructions within their database
3. Received from the National Oceanic Survey (NOS) their database of wrecks
4. Received from Texas A&M University their "hangs" listing of all reported fishing net hangs within the geographical area in question
5. Received from the US Minerals Management Services its archeological database of historically significant items
6. Obtained copies of pipeline survey maps that displayed high-accuracy pipeline coordinates for positional referencing

The data was not ready for populating the GIS, since all of the databases were in different coordinate systems:

1. The MMS database had all information in survey block coordinates based upon North American Datum 1927 (NAD27) survey State Plane Coordinate System (since that is what the offshore oilfield leases were drawn with in the 1960s) in x, y coordinates from the reference datum point.
2. The USCG, NOS, and DMA coordinates were all in World Geodetic System 1984 (WGS84, which is the current coordinate system used by most charting systems), but the reported coordinates were sometimes as much as 70 years old, taken via dead reckoning and celestial sightings. A "reliability quotient" had to be assigned to the positioning coordinates.
3. The Texas A&M "hangs" listing was in Loran A "TDs" (i.e., timing differences), which required conversion to latitude/longitude coordinates before processing.
4. The pipeline survey maps were all in NAD27. The pipeline maps had to be merged with the production platform coordinates to accurately plot the subsea pipeline track away from the production platform. The map drawings had "shot points" where high-accuracy position readings were taken along the length of the pipeline at measured points; the remainder of the string was "connect the dots" type of dead reckoning. Therefore, all of the coordinates had to merge to the pipeline map, so they all needed to be converted to the NAD27 coordinate system.

Once all of the data had populated the GIS map, the data points should clump onto the site of the actual wreck. As an example, the *R.W. Gallagher* was a US Merchant Marine tanker sunk in July 1942 by the German submarine U-67. The DMA, USCG, NOS, and NOAA databases had that wreck in three separate locations—dives were conducted, and it was not at any of those locations. Regardless of the success of such searches, the bottom line is that without proper planning, the search team is probably destined for failure.

21.3.4.1.2 Murder victim search

A second example involved the search of a water-filled abandoned rock quarry for a murder victim who was purportedly cut in half, stuffed into two separate acid-filled sealed barrels, and rolled off a cliff into a 100 ft water depth isolated location. Although the victim was not found, the data gathered to ensure an efficient search included:

1. Satellite and aerial photos from Terraserver
2. County survey data from county records
3. Company records of excavation from landowner
4. Extensive witness interviews
5. Sheriff's investigators rolled water-filled test barrels off the cliff to test the trajectory and then marked the search area off with buoys.

21.3.4.1.3 Environmental considerations

One would prefer to swim in the warm waters of the equatorial regions on sunny cloudless days. Unfortunately, the PSD does not have the luxury of knowing the time and choosing the place of the team's deployment. Most PSDs have within their jurisdiction aquatic areas comprised of lakes, rivers, canals, and coastal waterways where most human activities prevail. These areas are normally of low visibility and/or of a high current/tidal flow that create difficult search conditions. Visual instruments are of limited use for such environments, making the intelligent use of other sensors practically a requirement.

In theory, with low visibility conditions, items of interest should be first trapped by a nonvisual sensor and then held in a steady-state condition (e.g., for a sonar target, it should be held "in sight" on the sonar screen) so the diver or ROV can positively navigate to the target. If conducting area searches, it is most convenient to use search lines of equal depth to limit the maneuvering to x, y coordinates without throwing in the additional z factor of constantly changing depth.

Smooth, flat bottom conditions are the optimum type of underwater topography for searching with sonar. Since the target can be insonified on sonar much more easily on a flat bottom, the sensors will pick up minimal "false targets." For rocky bottom conditions, more time must be allocated to clear the area, since the target may have fallen against or into some of the rocks. This would make identification visually or acoustically difficult. Additional details on environmental and bottom conditions that should be kept in mind when preparing for a search are provided in the following sections.

Types of environments

- *Rivers*: Of the search environments confronting the public safety dive team, the most difficult class is the river environment. Rivers normally drain from higher elevations and have both suspended particulate matter and high currents. Rivers also often have any number of bottom obstructions, highly concentrated surface vessel traffic, and soft, loosely consolidated or unconsolidated mud bottoms.

- *Streams*: These normally possess easy access with clear visibility and shallow depths (but possibly higher currents). Streams are normally searched visually or with shallow water, snorkeling divers.
- *Lakes*: The lake environment is probably the easiest underwater search environment due to its usual lack of currents, generally good visibility, and relative lack of concentrated surface traffic.
- *Estuaries and littoral waters*: In general, the closer the body of water is to mountainous terrain, the higher the water flow and the higher the visibility of the water (with some exceptions). In areas of generally flat terrain with field run-off, expect low visibility and difficult search conditions. The challenge of estuaries and littoral areas is the sheer size of the search area. Covering a large search area without high-speed, high-coverage search sensors (such as a towed side scan sonar or magnetometer) will, in all likelihood, be time and/or cost prohibitive.

Bottom conditions affecting searches

- *Sea grass*: Sea grass and other bottom growth near the surface is fed by photosynthesis and suspended in the water column via air-filled pockets within the structure of the leaves. As explained in Chapter 15, air is highly reflective to sonar and will cause both a sonar echo (possibly displaying a false target) and a sonar shadow on the backside of the echo. If a significant presence of sea growth is encountered in the search area, prepare for a long and difficult search experience with plenty of false targets and fouled search equipment.
- *Rocky bottom*: This type of bottom condition generally reflects sonar signals well, causing a noisy environment for searching and numerous false targets. In order to maintain a consistent sonar display, the gain on the system must be adjusted to a lower setting. However, this sometimes masks the sonar target along with the clutter. The only true straight lines in the environment are man-made, which does help in target discrimination. Rocky bottoms are also difficult for tether management of the ROV since tethers tend to snag around rocky outcroppings and get trapped in overhangs.
- *Muddy bottom*: Muddy bottoms permit easy sonar searches due to the absorption of the sonar signals, allowing a higher gain setting. Anything with a higher consistency than mud appears quickly on the sonar display. ROV systems attempting to navigate near muddy bottoms are often plagued with ineffective cameras due to low visibility. Also, when items fall into a muddy bottom condition, they are often enveloped within the mud, causing impossible visual clues and rendering them invisible to high-frequency sonar systems.
- *Sandy bottom*: Of the bottom conditions for searching, the sandy bottom is probably the easiest environment for locating and discriminating targets, both visually and with sonar. The surface tension of the sandy bottom is such that items will not normally bury in a short period of time, nor will the sonar reflectivity produce the false target data present in other less favorable conditions.

21.3.4.2 Wide area search

A wide area search is conducted with equipment and sensors mounted on or towed from a relatively high-speed platform that can cover a wide search area in a short period of time. Examples of wide area search systems include light detection and ranging (LIDAR) or MAD equipment mounted aboard an airborne platform; towed or boat-mounted side scan sonar; towed video camera and bathymetry equipment.

21.3.4.3 Narrow area search

The narrow area search is conducted with relatively immobile sensor equipment mounted aboard a fixed or slow-moving platform. Examples of narrow area search equipment include:

- Diver performing a rope-guided search
- Tripod-mounted, mechanically or electrically scanning sonar moved from location to location while covering a fixed area
- ROV performing a grid search
- Low-tech cable drags

An earlier project involved a classic narrow area search for evidence (in conjunction with the US Bureau of Alcohol, Tobacco, and Firearms) on a suspected arson case involving a boat belonging to a high-profile politician. The boat burned in the slip of a floating dock; therefore, investigators were certain that the clues rested on the muddy bottom, 62 ft (20 m) below the surface. The visibility on the bottom of the muddy lake in the southern United States was less than 2 ft (70 cm). The search area was less than 100 ft (30 m) in diameter. A tripod-mounted, mechanically scanning sonar was lowered to the bottom to image the entire search area. Two separate ROVs were used to recover all of the items of interest from the search area (over 40 items were recovered in 6 hours) with the sonar used to direct the ROVs to each target.

21.3.4.4 Some search examples

Two good examples of the effect of the environment involved separate unsuccessful drowning victim searches in 2000, both of which were in rocky bottom lake conditions. The rocky bottom caused such a large false target population that the entire search budget was expended identifying rock outcroppings. The victims were both eventually recovered, which allowed the refinement of the search techniques used.

For illustration of the search process, the efforts made on the two unsuccessful body searches are discussed below, as well as those on a successful search. These will serve as a primer to the issues involved with such an endeavor, as well as highlight the difficulties inherent in body searches.

21.3.4.4.1 Drowning considerations

Prior to conducting a mission that may involve a drowning victim, the following rules of thumb should be considered:

- For drowning victims, the body has up to a 1:1 glide ratio to its landing spot on the bottom. That glide ratio must be adjusted based upon currents for an actual track over the ground. A depiction of this phenomenon is shown in Figure 21.19.
- The average adult weighs 8–12 pounds (4–6 kg) in water (for an expanded table of weights as well as an in-depth analysis of the drowning issue, see Teather's work on the subject). If the ROV is equipped with a manipulator, it should be sufficient to bring the victim to the surface if a strong hold can be maintained to the body.
- The average temperature crossover point for a body to either remain negatively buoyant or become positively buoyant is approximately 52°F (12°C). Check the water temperature to determine the likelihood that the victim is still near the last seen point.

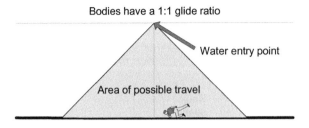

Bodies have a 1:1 glide ratio

Water entry point

Area of possible travel

FIGURE 21.19

The area of possible travel for a body from the water entry point to the bottom landing spot.

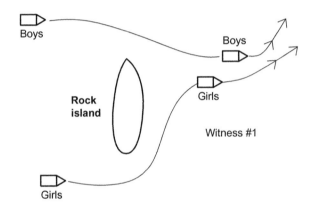

Boys

Boys

Girls

Rock
island

Witness #1

Girls

FIGURE 21.20

Witness #1 version.

21.3.4.4.2 Scenario 1

The drowning victim was a 16-year-old boy who was thrown from the back of a rapidly moving personal watercraft (PWC) and run over by another PWC in trail. There were two teenaged boys on the lead PWC and three teenaged girls on the trailing PWC. The conditions of the bottom of the rock-strewn reservoir in the north-central United States were very cold fresh water with an average bottom depth of 110 ft (33 m) and approximately 10 ft (3 m) visibility. The last seen point was very confusing due to several conflicting witness testimonies. The varying depictions of the last few seconds before the victim was thrown from the PWC follow.

Witness #1 was the middle passenger on PWC #1 (girls). She stated that PWC #2 (boys) cut in front of them after they went around the south side of the rock island and then sprayed them, throwing the victim into the path of their PWC (Figure 21.20). They immediately stopped and turned around to render assistance.

Witness #2 was the driver of PWC #2 (boys). He stated that they pulled alongside PWC #1 (girls) and then turned left to spray the girls (Figure 21.21). The victim fell off into the path of PWC #1.

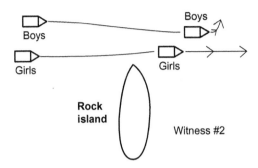

FIGURE 21.21

Witness #2 version.

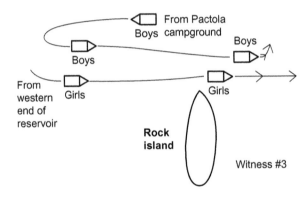

FIGURE 21.22

Witness #3 version.

Witness #3 was the rear passenger of PWC #1 (girls). She stated that the boys saw them coming and turned around to give chase (Figure 21.22). The driver of PWC #2 (boys) did a radical left turn in front of PWC #1, throwing the victim into the path of the oncoming vehicle.

Witness #4 was the driver of PWC #1 (girls). She stated that the boys saw them coming and turned around to give chase (as with witness #3). The driver of PWC #2 (boys) did a radical and complete 360° turn in front of PWC #1, throwing the victim into the path of the oncoming vehicle (Figure 21.23).

There was a previous search with side scan sonar of the search area that turned up a large number of possible targets—none of which matched the shape of the victim.

The fire chief in charge of the operation was an experienced professional in the public safety field. The second search was much more detailed due to the tedious nature of the bottom conditions. A tripod-mounted, mechanically scanning sonar system was used, deployed from a pontoon boat with a four-point mooring. The fire chief had no other choice but to assume that the search area was located in the northeastern quadrant from the rock island in the lake. He properly determined that the search lines needed to start at the island and proceed in an easterly direction until the allotted time for the search was expended. North/south search patterns were instituted proceeding in an easterly direction.

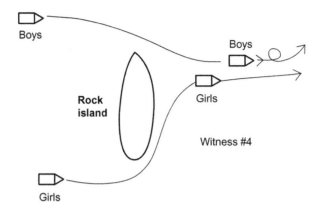

FIGURE 21.23

Witness #4 version.

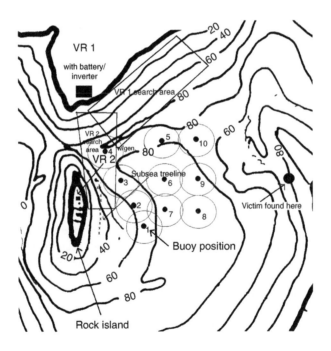

FIGURE 21.24

Composite of eyewitness data.

The grids were spaced at 100 ft intervals. The search proceeded in a systematic fashion for 3 days, at which time the search was ended without success.

The victim was eventually located approximately 800 ft (250 m) east of the rock island by recreational divers (Figure 21.24). The victim was found in an area approximately two search lines farther from the line under evaluation when the search was called off. This is a case of an impossible PSD search due to temperature and decompression issues. The search area was difficult to define due to conflicting witness testimonies. The bottom conditions were tedious and difficult, with uneven and rock-strewn conditions causing a high number of sonar targets as well as having any number of places to trap the victim in bottom holes and crevasses.

21.3.4.4.3 Scenario 2

The victim was on a field trip in Western Canada and decided to swim across a 1 mile glacial lake. The victim drowned somewhere during the swim across. A witness last saw the victim swimming with difficulty and took a sight line on the last seen point.

Several days of searching with side scan sonar produced thousands of sonar targets. Discriminating those for evaluation and elimination took more time than the search team had budgeted for the search. The bottom conditions were difficult for divers but acceptable for ROVs, with 25 ft (8 m) visibility and an average depth of 125 ft (38 m) of fresh water. The search area was simply too large due to the soft "last seen" point. The victim was eventually recovered when the lake was drained for reservoir maintenance.

21.3.4.4.4 Scenario 3

An example of a successful drowning victim recovery is the search for the bodies from an overturned public water taxi in the eastern United States. The accident happened on a cold, windy day during frontal system passage when the vessel capsized due to the turning moment as the vessel attempted to turn back into port. Rising wind conditions overturned the dual-hulled vessel, trapping the victims within the cabin. Six passengers drowned during the ensuing rescue attempt. Data gathered during the planning phase of the body recovery included:

1. Extensive witness interviews as well as interviews of the survivors and vessel operators
2. Navigational charts
3. Survey and work permit application charts for subsurface harbor construction

The initial search focused on finding a debris field before the tedious task of body search was initiated. Initial search efforts focused on the north side (upwind side on the day of the accident) of the harbor near where the vessel capsized. After several days of searching without results, the data was reanalyzed. A trajectory of the vessel track due to wind drift after the capsizing was plotted, and then the search area was redrawn based upon witness interviews of the recovery efforts. What seemed insignificant during the original search effort took on more importance—an LST ("landing ship tank" or military surplus landing vessel for loading/offloading equipment in the construction industry) had been dispatched to render assistance. Once a reestimation of the LST's position at the time of intercept of the water taxi was made, the search area was refocused on that position. Once the search was resumed in the new area, the severed canopy from the water taxi was found with side scan sonar on the first search line. The location of the debris field formed the new central datum from which the new search area could be drawn. The victims could not be (and were not) far from that point.

Within 3 days of finding the debris field, all of the remaining victims were located and recovered. They were located with side scan sonar and then reacquired with an acoustic lens sonar mounted to an ROV. An example of the detailed search capability of an acoustic lens sonar is the fact that it was able to image the stitching in the shoes of one of the victims to the shoe manufacturer's logo.

The victims were recovered with divers as the ROV held position while imaging the targets acoustically. Neither the ROV equipment nor the divers were put into the water until sonar targets were identified.

Excellent planning and research can lead to a successful search operation.

21.3.4.5 Final identification

The final identification of an item of interest requires human discrimination of an image of the item. Positive classification of the target as either the object of the search or the exclusion of the object as a possible target from the search area achieves the objective of the task.

Once the final identification is obtained, the objective of the entire field task must be concluded. In most instances, clues as to the cause of the incident under investigation may be gleaned from the scene by taking the extra time to document the area. What other evidence can be found within the area? In what position did the victim or the evidence come to rest? How can one (or should one) recover the victim or evidence?

Extreme attention to detail in the final phase of the search and recovery operation can significantly enhance the results of the PSD operation. The quality of the findings will support court proceedings as well as fact-finding tribunals where the PSD could be called to testify.

21.4 Homeland security

21.4.1 Concept of operations

ROVs provide governmental and law enforcement officials with the ability to visually inspect underwater areas of interest from a remote location. ROVs are intended to be complementary to the use of public safety and military dive teams. They can be used in conjunction with divers, or as a replacement to them, when the operation's environmental conditions allow. Factors to consider in deciding which capability to employ should include, but not be limited to, the mission objective, degree of accuracy required, current threat level, on-scene conditions, and the amount of time plus resources available.

As homeland security operations involve shallow water, light-duty operations from small platforms, this section covers usage of OCROV systems with submersibles weighing from the smallest size to 200 pounds (90 kg—excluding tether). A two-person (nominal) team is recommended to operate inspection ROVs (although single-operator functions can be performed). Experience has shown that personnel from electronics or scuba diving backgrounds are excellent candidates for assignment as the primary ROV operator. Operation and maintenance of the ROV require an average level of electronics and situational awareness skill and experience that fit well with the background of most electronics and dive team personnel. The use of volunteer or administrative personnel also helps to reduce impacts to the department's operational schedule, as many law enforcement personnel are not assigned to dedicated 24/7 security duties.

When planning ROV operations, appropriate consideration must be given to the personal protection of the ROV team. An ROV team operating in a nonsecure area under heightened threat levels may require armed escorts. Because ROV operations normally require the operators to focus on the ROV, it is strongly recommended that separate personnel carry out the task of team protection/ security.

21.4.2 Tactics, techniques, and procedures

The capabilities and limitations of the individual ROV systems determine the deployment procedures, since the more powerful systems allow for higher distance offset through greater currents than do their less powerful brothers. For the purposes of classifying the procedures, the systems are divided into smaller categories for ease of procedure assignment (as further defined below).

The tactics, techniques, and procedures (TTPs) developed for this text are intended to be manufacturer nonspecific to prevent these procedures from being tied to any single manufacturer of ROV equipment. Techniques were tested on a variety of ROV systems in the small observation class category to validate and gain confidence as to these procedures' applicability through a range of ROV sizes and capabilities. The systems tested were placed into three general observation class categories based upon their respective sizes, weight, and forward thruster output available. The size assignments were:

- *Small*: Submersible weight less than 10 lb (5 kg) with forward thruster output less than 10 lb (5 kg)
- *Medium*: Submersible weight between 10 and 70 lb (5 and 32 kg) with forward thruster output 10−20 lb (5 and 9 kg)
- *Large*: Submersible weight above 70 lb (32 kg) and/or with forward thruster output greater than 20 lb (9 kg).

21.4.3 Operating characteristics of ROV size categories

The overall size of the system partly determines the payload capacity and ability to carry larger, more powerful thrusters. Thruster output determines the vehicle's capability to deliver the submersible to a place where it can produce a useful picture while fighting currents and pulling its drag-producing tether. Generally, the larger the submersible, the more powerful the thrusters. Other control, drag, and stability considerations include the hydrodynamics of the thruster placement (affecting laminar/turbulent flow through thruster housings), vehicle stability at higher speeds and, the largest of these considerations, the diameter of the tether (circular shapes having the highest drag coefficient) being pulled behind the vehicle.

To summarize, vehicle size category determines the physical capability to maneuver into a place to accomplish the task. The viewing from that point is a function of camera coverage and distance from the item being inspected. If water clarity is 5 ft (1.5 m), the ROV can be no more than 5 ft (1.5 m) away from the target. The larger the standoff (i.e., water insertion point to work area), the larger the thrust necessary to pull the tether to that workplace. The stronger the current, the larger the thrust required to overcome the parasitic drag created by the vehicle and the tether.

Conversely, while a larger ROV may be able to overcome the above-described obstacles, the larger ROV is heavier, has a larger system footprint, and requires more electrical power (e.g., larger generator with higher fuel consumption and space requirement, louder engine with higher emissions). Although the smaller vehicles are more easily affected by the conditions described above, they offer many beneficial trade-offs (i.e., smaller footprint, lower power requirements, more mobile, capable of getting into smaller areas).

21.4.4 Port security needs

For the purposes of this manual, underwater port security tasks fall under two broad categories:

1. *Search and identification* of underwater targets or areas of interest. Examples may be limiting an inspection to a vessel's running gear or searching a particular location based on intelligence.
2. *Search and inspection* of general areas of interest for potential threats within the security area. Examples may be conducting a search/inspection of all the pier pilings within a specified zone prior to the arrival of a high-value vessel.

Search and identification of underwater targets is accomplished through four basic steps:

1. *Research* to define the area of interest
2. *Wide area search* with instruments and sensors
3. *Narrow area search* with slow-speed instruments
4. *Final identification*, visual discrimination, and disposition—that is, diver or remote camera propelled to the inspection site.

ROV intervention as a productive and cost-effective means of final identification, discrimination, and disposition is analyzed in this manual. Refer to Section 21.3.4 for a more detailed discussion of the previous four steps.

21.4.5 Underwater environment of ports

The underwater environments of ports within the United States (and the remainder of the world) vary in temperature, water clarity, operating currents, and vessel traffic. The environment in Seattle, WA (cold deep water, good visibility, moderate currents, and moderate vessel traffic), varies significantly from the Port of New Orleans (warm deep water on the Mississippi River, poor visibility, high currents, and moderate vessel traffic), the Port of Galveston (warm shallow water, poor visibility, low currents, and low vessel traffic), and New York Harbor (cold shallow water, poor visibility, high currents, high vessel traffic, medical waste, etc.).

Accordingly, the environmental factors that determine the difficulty of completing an underwater port security task with an ROV system follow:

- *Currents* determine the ability of the submersible to successfully swim to, and station-keep near, a fixed object while countering these currents.
- *Water depth* determines the offset from the deployment (water insertion) point to the bottom for bottom clearance searches as well as proximity of bottom to vessels/moorings/anchors/piers.
- *Vessel traffic* determines the relative security of the inspection operation (a more secure location produces an easier inspection task).
- *Water temperature* determines the resources and expertise necessary for a dive team to enter that environment along with loiter time.

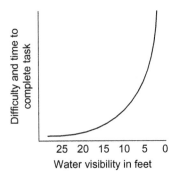

FIGURE 21.25

Degree of difficulty and time required for visual inspection versus water clarity.

- *Water clarity* determines the degree of difficulty in completing the underwater task as well as the time to fully image the items of interest (Figure 21.25).

21.4.6 Navigation accessories

There are several technologies available for achieving underwater positioning and target acquisition for further investigation. Most common ROV-mounted acoustical systems involve imaging sonar and acoustic positioning.

21.4.6.1 Imaging sonar

Imaging sonar is useful in underwater port security tasks in identifying items of interest by producing a sonar reflection or a blockage, which produces a sonar shadow. Imaging sonar manufacturers have been able to miniaturize the sonar unit to fit aboard practically all sizes of ROV systems. In order to get a high-resolution image of a small target at a nominal range, such as 50 ft (15 m), most manufacturers use high-frequency sonar with a fan beam in the 600–800 kHz range. This allows items protruding from the bottom to be imaged on sonar (Figure 21.26).

21.4.6.2 Acoustic positioning

An acoustic positioning system calculates range from a submersible-mounted transducer to other transducers at known locations with known spacing. This permits an accurate range calculation, with adjustment for water temperature/salinity/density, by computing the one-way or round-trip timing. Bearing is resolved through triangulation of the timing differences across the transducer array (i.e., merging point of the separate lines of position). In order to resolve relative bearing to magnetic bearing, a magnetic transducer array orientation needs to be determined (easily done in software with a flux-gate compass outputting standardized data streams). In order to resolve relative location to geo-referenced location, an accurate latitude/longitude position must be determined with a GPS unit (outputting standardized data streams).

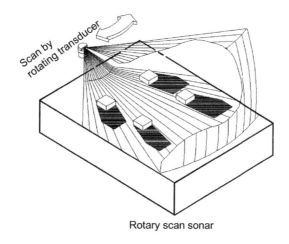

Rotary scan sonar

FIGURE 21.26

Imaging sonar.

(Courtesy Imagenex Technology Corp.)

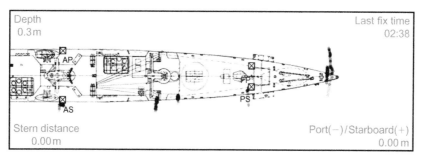

FIGURE 21.27

Vessel-referenced acoustic positioning ship hull files.

(Courtesy Desert Star LLC.)

The vessel-referenced acoustic positioning system (Figure 21.27) is similar to the geo-referenced short baseline system (Figure 21.28) except that the reference is with the ship. The transducer array is placed on a measured drawing of the target vessel with all transducer placements calculated based upon that scaled drawing. The frame of reference is relative to the ship.

21.4.6.3 Difficulties involved with sonar and acoustic positioning

The major issue involved with ROV-mounted imaging sonar is image interpretation. A basic target of interest can either be identified by acoustic reflection or by acoustic shadow. Unfortunately, image interpretation is in many cases counterintuitive and requires special instruction in sonar theory and application. Further difficulties peculiar to very small ROV systems, due to the vehicle's

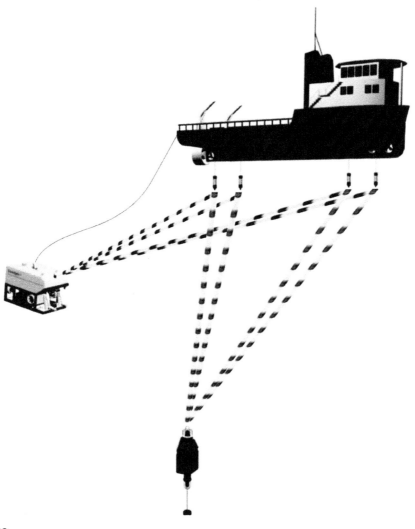

FIGURE 21.28

Short baseline positioning setup.

(Courtesy Sonardyne.)

movement during image generation, must be understood. This so-called image smear is produced from moving the vehicle/sonar platform before allowing the full image to generate.

Acoustic positioning within a port environment is problematic. A port, by its location and function, is a noisy acoustic environment. Broadband noises bounce around the water space like a ping-pong ball, causing false narrowband reception and reducing the signal-to-noise ratio of the primary positioning signal. Multipath errors also cause difficulty (a narrowband sound bouncing between a hull and a pier wall can spoil round-trip sound calculations due to false reception).

With all of the difficulties associated with these technologies, with proper training and implementation they remain powerful tools for accomplishing port security tasks.

21.4.7 Techniques for accomplishing port security tasks

ROV-based underwater port security tasks involve two broad categories of inspections:

1. Identification of targets located by other means
2. Clearing an inspection area of suspicious items

It is possible to discover an item of interest while making random searches of an area. However, it has been found to be of marginal benefit to conduct unstructured searches of suspect areas. As water visibility decreases, a high degree of certainty that an area of interest has been cleared, with all suspicious items discovered, becomes increasingly difficult, as does positive navigation through the search area. In order to construct a maximum risk/benefit model of search time covering the high-risk exposure, a combination of tools and techniques must be used to achieve best results. These are discussed below.

21.4.7.1 Hull searches

The highest risk sections of commercial and military seagoing vessel hulls are isolated to a limited set of landmarks located on the hulls of these vessels. These targets can be located and checked in a relatively short time, thereby quickly eliminating the high-risk areas. Later, a search pattern can be implemented to image a specified percentage of the hull area.

One hundred percent visual hull coverage is exceedingly difficult to achieve due to navigational considerations as well as environmental factors, with the time-requirement curve turning nearly vertical as the water visibility nears zero (Figure 21.29). Experience has shown that the 80/20 rule applies in

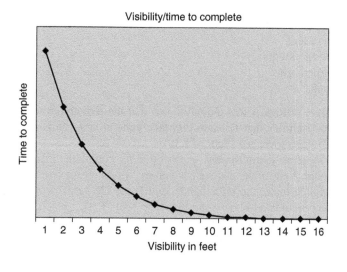

FIGURE 21.29

Visibility versus time to complete inspection task.

hull inspection missions, where 80% of the hull can be inspected in 20% of the time it would take to achieve full hull coverage. The last 20% of hull coverage is the most tedious and time-consuming.

21.4.7.2 Under-vessel bottom searches

Achieving bottom clearance beneath berthing of large vessels is time-consuming without the use of imaging sonar to identify items of interest. Bottom visibility conditions become difficult due to poor lighting as well as bottom stirring from vessel traffic, ROV submersible thrusters, and silt movement from tidal flows. Results may be improved by using technology to discriminate anomalies to isolate items of interest (with imaging sonar, magnetometer, and other instruments) and positively identify these items with an ROV-mounted camera, thus eliminating them as threats.

21.4.7.3 Pier/mooring/anchor searches

Identification of the threat could aid in reducing the time spent searching pier/mooring/anchor areas. The farther away from the high-value asset the item to be inspected resides (e.g., the placement of an explosive charge 20 pilings away from the berthing location of a large container vessel would produce minimal damage to the hull), the lower the likelihood a threat will be placed in that location. Planning the operation should take into consideration such factors and limit the time inspecting lower risk areas or ignore them altogether.

21.4.8 Development of TTPs for port security

An earlier project included the development of a list of best practices involving a series of underwater port security tasks, followed by testing the practices over a range of ROV equipment.

The tasks comprised the following major areas:

- Ship hull searches (pier-side and at anchor)
- Pier searches
- Bottom searches to include directly under vessels
- Day and night operations
- Inclement weather operations
- Launch/recovery operations
- Tether entanglements

During the procedure testing, it was quickly noted that the size and power of the individual systems translated into capabilities that fell into the three general size categories (small/medium/large as stated in Section 21.4.2 above, see Figure 21.30).

The large ROV systems demonstrated capabilities to complete difficult tasks that required power to muscle through long offsets as well as strong wind/current combinations (Figure 21.31). The smaller systems failed under the same harsh conditions. The larger systems, however, were too bulky to be easily accommodated aboard the response boat (RB), causing difficulty in movement for the crew while under way (Figure 21.32).

The tasks tested and performed included:

- Ship hull inspections (moored/anchored)
- Pier inspections
- Simulated HAZMAT spills
- Simulated potable water environment

FIGURE 21.30

Size comparison between ROV systems tested.

FIGURE 21.31

Larger ROV systems function well with distant offset but are more difficult to handle than smaller systems.

21.4.9 **Results of procedure testing by sizes**

The power of large systems countered the lack of experience with the operators. The smaller systems took a higher degree of operator proficiency, task planning, and tether handler coordination than did the larger systems to accomplish similar tasks. The smaller systems fared better in untangling their tether and in functioning in confined spaces. The results of the tests broke down into a series of what were termed "degrees of difficulty":

- E: Easy
- D: Difficult
- VD: Very difficult
- NR: Not recommended

FIGURE 21.32

Large ROV aboard RB (a) and excess green tether (b).

Table 21.3 Task Completion Versus Current and Tether Offset

Prevailing Current (kt)	0	0.5	1.0	1.5	2.0	2.5	3+
Small system	E	D	VD	NR	NR	NR	NR
Medium system	E	E	D	VD	VD	NR	NR
Large system	E	E	E	D	VD	VD	VD

Example task assumes 50 ft (15 m) of tether offset to item of interest.
Note: The current and offset decide whether the system can accomplish the task. Visibility decides the time to accomplish this task.

Table 21.3 provides a summary of the results. A detailed discussion of port security task expectations based on vehicle size is presented in Chapter 22.

As discussed in detail in Chapter 6, the ROV must produce enough thrust to overcome the drag produced by the tether and the vehicle. Although there are techniques to aid the lower-powered systems, the obvious message is that the tether drag on the vehicle is the largest factor in ROV deployment and usage. The higher the thrust-to-drag ratio and power available on the ROV system, the better the submersible will be able to pull its tether to the work site.

ROVs are being used around the world for law enforcement as well as homeland security tasks, thus protecting our ports and harbors from threats to shipping and commerce. Central to that use is the establishment of procedures and management to allow better security through cost effectiveness and more frequent inspections.

21.4.10 Inland

Inland commercial operations typically involve inspections of some type of submerged or enclosed structure (e.g., tanks, pipelines, and tunnels) or construction projects for lakes, dams, or rivers.

Each task will require an individual configuration, but the size classifications required are typically OCROV for most inspection tasks and MSROV for long-distance pipeline or tunnel penetrations. The environment is typically quite benign and the jobs are fairly straightforward.

21.4.11 Offshore

The offshore environment is by far the most taxing of all ROV assignments. Whereas the OCROV (and to a limited extent, the MSROV) rules in the inland environment, the MSROV and WCROV rule the waves for the offshore market. The extent of work scope is much too broad to cover with any justice in this short section. To ensure success, offshore operations should be properly planned with procedures agreed beforehand along with exacting scope of work and completion standards specified. Operations conducted offshore are typically expensive and risky. With the introduction of regulatory safety standards enacted in the wake of the 2010 *Deepwater Horizon* disaster, adherence to strict guidelines for personal safety and environmental considerations will become more focused as time goes by. ROV work in the offshore oil and gas industry is a rapidly expanding market. A career in this field is an exciting, profitable, and rewarding experience.

21.5 CONCLUSION

It is often said that "every operation has its own technique." Each application for ROV services involves its own peculiar nuances. This chapter has outlined a few of the lighter usages for ROVs in a variety of tasks. The usage of this technology is limited only by the operator's imagination, but all successful applications require a proficient operator employing well-maintained equipment in a structured and intelligent fashion in order to achieve proper mission results. In short, "Plan Your Dive, Then Dive Your Plan."

It's the Little Things That Matter

22

CHAPTER CONTENTS

It's the little things that matter, and are easily forgotten or ignored. This Chapter provides the information necessary to achieve success, which includes Standard Operating Procedures, Servicing and Troubleshooting, and Putting it All Together.

22.1 Standard operating procedures

This section contains detailed examples of standard operating procedures (SOPs) for remotely operated vehicle (ROV) deployments used in general underwater operations with particular focus upon port and harbor security. Every operational situation is unique, with its own peculiar set of operational requirements. These SOPs have been operationally tested and proven. Use the below procedures as a starting point for deployments. The SOPs assume an ROV operator of average proficiency under normal conditions. A very proficient operator may be able to perform items located in the "Not recommended" category, but due to equipment performance constraints, it may require an expanded timeframe. This chapter concentrates on the OCROV category as the larger vehicles typically require much more planning and expanded SOPs (which exceed the scope of this text).

22.1.1 Overall operational objectives

The professional ROV crew will approach any field assignment in a systematic fashion with constant focus on the achievement of the operation's ultimate goal. The overall objectives of the assignment will be accomplished through the following general tasks:

- An integral part of any ROV support operation includes a full understanding of the overall operation to be conducted as well as the ROV's role within that mission plan. As part of the planning for the field assignment, a full review of any customer-produced documentation is required in order to maintain full operational efficiency.

- Obtain a scope of work from the customer so that the appropriate equipment may be scheduled for the operation.
- Mobilize all equipment allocated and scheduled for the operation.
- Transport all equipment to the work site.
- Prepare all equipment for the individual tasks in proper order of use.
- Thoroughly brief the crew on the planned operation as well as the vessel crew and/or any necessary third party (Port Captain, Facility Owner/Operator, nearby vessels, etc.).
- Complete all predive functionality tests as well as checklists for deployment.
- Deploy the equipment, locate the work site, and begin the work task.
- At the completion of the work task, recover the equipment (per detailed procedures).
- Perform postdive functionality checks as well as visual condition inspection.
- Upon completion of the assignment, demobilize the equipment and transport it back to base.

22.1.2 Equipment mobilization

Equipment for deployment should be readied for travel and use as follows:

- Verify all preventative maintenance and servicing have been performed to standards established by the manufacturer.
- Verify all consumables (e.g., lights, seals, and O-rings) are serviceable and in good condition.
- Ensure an adequate spares package is available to meet the priority level of the mission as well as remoteness of the job site (adjust the spares package to compensate for the cost of downtime versus transit time for replacement parts).
- Verify all functions of the equipment before packaging it for shipment.
- Ensure the equipment is packaged adequately for travel to the work location. You would be surprised at the number of job start-up faults that are directly traceable to electronic components unseating due to over-the-road travel jarring.

The operator is responsible for the operation of his ROV system. All tests and mobilization steps should be performed before leaving base.

Standard ROV equipment needs (predeployment checklist) for field deployment fall under these broad categories:

- ROV manufacturer-provided equipment
- ROV manufacturer-provided spares
- Other than ROV manufacturer-provided equipment deemed essential to the ROV system (e.g., viewing monitor, video cables, transport cases, and power cables)
- Spares, adapters, and electrical wiring
- Test equipment for troubleshooting component parts

For the first three items above, refer to the manufacturer's operations and maintenance manual for a complete listing of items needed for deployment. For the last two items above, refer to Section 22.2.3 of this chapter for a general listing of items needed for troubleshooting, backup, and testing of ROV equipment.

22.1.3 Operational considerations

22.1.3.1 Operations on and around vessels

The following should be considered when conducting any ship hull inspection:

- The bridge of any vessel to be inspected must be informed prior to commencing the dive to determine if any conditions are present that are hazardous to ROV or vessel operations.
- When launching and recovering the vehicle, attention must be paid to any submerged obstructions in and around the deployment platform or vessel to be inspected.
- Where ROV operation is required within and around the operating structure of the inspection vessel, care must be taken to ensure that the tether or umbilical is not snagged or damaged.

22.1.3.2 Operations from the vessel of opportunity while station-keeping

The following principal points shall be noted in all ROV operations from vessels utilizing station-keeping:

- Vessel shall not be moved without the prior knowledge of the ROV operator.
- The ROV shall be launched as far away from thrusters as practical.
- The tether shall be manned at all times.
- Vehicle buoyancy shall be slightly positive.
- Good communications must be maintained between the ROV operator and the helm to ensure that the ROV operator is aware of all actions regarding the relative positions of the ROV and vessel.

22.1.3.3 ROV operations with divers in the water

Concurrent ROV/diver operations may potentially carry an increased risk to the diver for various reasons, including possible entanglement, mechanical contact between the diver and the vehicle, and various other areas of interference. Diver safety is paramount. Observe the following guidelines during all in-water concurrent operations:

- As an integral part of any industrial operation, a pre-job briefing session should be conducted with all concerned operators. Once the customer representative, vessel or facilities operator, and vendor personnel agree upon the proposed operation, the ROV supervisor shall brief his/her crew (Section 22.1.4).
- During the operational crew briefing, the diving supervisor should establish a perimeter around the diver that the ROV should not penetrate without express permission and positive approval from the diving supervisor. The ground fault interrupt (GFI) system on the ROV system should be fully operational and tested.
- Consideration should be given to thruster ingestion of items belonging to the diver. The possible foreign object damage to the thrusters and diver injury could be avoided through thruster guards or through properly stowing diver tooling.
- The sequence for diver/ROV deployment will directly affect the possibility of a diver/ROV umbilical entanglement. The ROV should be deployed to the dive site with the tether fully lain to the work site before the diver is cleared to the area. The diver should work and then clear the

area *before* the ROV is recovered. No vehicle recovery should be started unless prior approval is gained from the diving supervisor.

- Clear, open, and direct lines of communication should be maintained between the ROV operator and the diving supervisor. Emergency procedures should be fully briefed and understood before the beginning of the concurrent operation.

22.1.3.4 Precautions and limitations

The following precautions should be exercised to properly protect the ROV equipment while operating within its design limitations (Wernli, 1998):

- The ROV should only be operated by approved and appropriately trained company operators.
- Piloting ROVs in excess of 8 hours within any 12-hour period subjects the operator to excessive fatigue, thus possibly compromising personnel and equipment safety. If extended periods in the water are planned, a relief pilot should be scheduled or shifts should be staggered.
- The launch and recovery phase of the dive cycle is probably the most dangerous portion of the dive. The vehicle and launch platform are subjected to an inordinate amount of risk to damage while the vehicle is suspended in air (lift from the deck until contact with the water). Weather conditions at the time of launch should not exceed company-specified limitations. This would increase the risk of damage to the equipment to an unacceptable level.
- The ROV, as well as system components, should not be operated outside of the manufacturer-specified operating parameters.
- When operating from dynamically positioned vessels or vessels under station-keeping (as specified above), ensure that the vehicle is kept clear of thrusters and surface obstructions.
- ROV operations should not be conducted in low water visibility (less than 4 ft/1.2 m) or in uncharted and/or unstructured areas unless the vehicle is equipped with submersible-mounted imaging sonar.
- Approval from command and possibly the insurance company may be required when operating the ROV in hazardous circumstances. Examples of such circumstances include:
  1. HAZMAT spills
  2. Use of explosives
  3. Working within underwater structures
  4. Extreme dynamic environments caused by waves and/or surge
- Ensure demobilization procedures are properly followed to include postoperation maintenance and storage procedures.

22.1.3.5 Night operations and extreme operational environments

22.1.3.5.1 Night operations

As discussed in previous chapters, light is subject to scattering and absorption in water. The largest negative factor in working at night is the enhanced effect of backscattering due to the lack of ambient lighting (causing the auto iris on most CCD cameras to open further, amplifying the backscatter effect). The only significant source of lighting at night to illuminate the item of interest is the submersible's lights; therefore, the submersible may have a lower effective range of vision during night operations. The submersible should be moved closer to the item of interest while the lights are set on a slightly lower setting to counter the enhanced backscattering.

There are, however, several advantages to deploying ROV systems at night for underwater port security, including:

- Lower harbor traffic that could inhibit operations
- Easier sighting of the submersible's lights below the water, assisting in visually locating the vehicle
- Better discrimination of items by the camera due to higher lighting contrast

22.1.3.5.2 Extreme temperature operations

For operations in Arctic and desert conditions, special considerations are needed to protect the equipment from malfunctioning due to temperature damage.

High temperatures are especially hard on electronic components due to the printed circuit board's and conductor's normal heating, plus the local high temperature, exceeding the maximum temperature limits of the materials.

Low temperatures are especially hard on seals, O-rings, plastics, and other malleable components (most of which are consumables).

Shock temperature loading of components happens when the components are left in air and allowed to settle to the local extreme ambient temperature. Once the components are temperature-soaked to the local extreme ambient air temperature, the possibility exists of shock temperature damage. If the submersible is then operated in water with a large temperature difference from the local air, shock loading of the components will be experienced.

An example of this in the Arctic would be a cold-soaked submersible left out overnight in −40°F/ −40°C air and then deployed into +32°F/0°C water, causing a total shock temperature loading of 72°F/22°C. This mistake, in all likelihood, would shatter the domes and cause seal failure, with the resulting ingress of seawater into air-filled spaces of the submersible.

A similar instance happened recently in the Gulf of Mexico aboard a derrick barge during a hot, windless August day. Prior to an operation, the vehicle was left out on the steel deck without shade while the crew relaxed in the air-conditioned operations shack. When the time came to dive, the vehicle (now heated to approximately 150°F/65°C) was lowered into the 75°F/24°C water, unevenly contracting the faceplate on the lights, thus allowing saltwater incursion into the light housing. Once the salt water penetrated the light housing, there was an immediate ground fault, tripping the GFI circuit and shutting down the vehicle electronics. Upon vehicle recovery, the lighting circuit had been destroyed due to metal oxidation through the ground fault.

Practically all temperature-related issues are resolved on the submersible by gradually matching the temperature of the submersible with the temperature of the water before deploying the vehicle. Extreme temperature problems with the control console are more easily resolved since it will be with the operator, normally, in a temperature-controlled environment.

22.1.4 Predive operations and checks

22.1.4.1 Crew briefing

Any team effort requires a full understanding of the operation to be conducted by all team members. Before the start of any ROV operation, a full crew briefing is essential so that efficient tasking is accomplished in order to enhance team synergies. A proper crew briefing should include the following essential elements:

- Mission tasking and/or threat outline, detailing the specific requirements of the task
- A thorough and complete explanation of the job components
- A scope of work and specific goals to be accomplished during the task
- Ingress/egress routes to the work site
- Crew positions during the task as well as specific responsibilities
- Tactics, techniques, and procedures for accomplishing each individual task as well as completion of the mission
- Specific information needed as well as methods of documentation
- All relevant information on the object of inspection, including (but not limited to) drawings, maintenance records, damage reports, survivor statements, and any other information that would assist in the work task
- Work site coordinates
- Topographical maps, bathymetry data, tide tables, underwater obstruction analysis, prior surveys, and any other environmental information that will assist in accessing the work site
- If there was any prior work done on the work site, it is imperative to thoroughly review this information (and, if possible, interview the previous crew) and job reports to better gain an insight into the current site condition
- Schedule for completion of the task objectives (best case/worst case/most likely case) with consideration given for field delays
- A thorough detail (and, if deemed necessary, drill) on emergency procedures

The ROV supervisor should also take further steps to be fully briefed on any other conditions that would affect or interfere with his planned operations, including (but not limited to):

- Planned vessel movements
- Vessel/platform support operations (including support vessel logistics)
- Diving operations
- Drilling operations
- Pipeline lay operations
- Operation of surface and subsurface machinery
- Scheduled power generation equipment outages/changeovers

Many industrial applications for ROV operations require a job safety analysis (JSA) to be performed and condensed into writing in order to document the job-specific hazards and steps for mitigating those hazards.

22.1.4.2 Vehicle preparation

The sequence of events from crew briefing to the contact of the vehicle with the water should be conducted in a logical and fluid flow. Once the briefing is completed, the predive checklist is begun and continued until the vehicle is in the water. If there is a checklist item that is unable to be cleared, the supervisor's judgment will dictate if the items will stop the predive sequence. If a long interval is encountered between the completion of the predive checklist and the start of the actual dive, the checklist may require a restart from the top or from some intermediate step so that all checklist items are completed in a timely fashion.

Vehicle preparations are normally vehicle-specific, but should include at least the following items:

- Assure all functions of the vehicle are operational.
- Any tooling or sensor packages should be operational and tested with data and telemetry flowing in a nominal fashion.
- All video feeds from the appropriate camera packages are wired and operational.
- Verify the generator has enough fuel to support the entire operation.
- All controls are operational and functioning in the appropriate direction.
- All video systems are fed through the appropriate systems with recording devices tested and staged (e.g., tapes labeled, videotapes/DVDs logged, notes columned, test footage verified)—if the dive is not recorded, it did not happen!!!
- All documentation packages are staged and ready for action (e.g., video overlay prepared with the appropriate job number/location/date/time).
- Crew members are at their assigned stations.
- Crew communication systems are tested.
- The bridge is notified of hoist of vehicle from the deck.
- The bridge is notified of vehicle in the water.

22.1.4.3 Predive checklist

The following predive check should be carried out prior to every dive:

- Visually inspect the vehicle to ensure the propellers are not fouled, all components are secured, and there is no mechanical damage to the frame or other components.
- Check the tether for scrapes, nicks, or other visible damage. The vehicle should not be used if the tether jacket is broken through.
- Verify correct operation of thrusters. CAUTION: Most ROV manufacturers advise that the thrusters should be run for only a few seconds in air. Water forms a heat sink that cools the thrusters during normal in-water operations. Also, on vehicles with onboard hydraulic pumps, the typical on-deck operating time limit is 2 minutes due to heat buildup.
- Check that all fasteners are in place and secure.
- Ensure that the whip connectors at the electronics and tether termination cans are plugged in securely. Note: Wet connectors have a tendency to become dry over time. Use dielectric silicone grease to lubricate both sides for easy installation. See Chapter 8 for further information on cable and connector maintenance.
- Ensure all unused vehicle connectors are capped securely with dummy plugs. A forgotten or unsecured dummy plug can lead to serious electrical system damage.
- Ensure all surface cables are securely connected.
- Unexpected thruster movement can occur when powering up. To prevent this, turn any auto depth switch on the control panel to the Off position, center the manual depth control knob, and switch any auto heading selector to the Off position before powering up the system.
- Verify correct operation of manipulators. CAUTION: Remain clear of the entire range of motion of the manipulator (especially for hydraulic manipulators) during the activation of the

manipulator and/or hydraulic circuit in case a stuck control valve or control circuit is energized, thus moving the manipulator in an uncontrolled fashion.

- Test the lights, camera/manipulator/tooling functions, and thrusters. Check that adequate video is obtained and that any video recording equipment is working properly. In most cases, underwater lights are not designed for operation out of water for any length of time. After testing the function of the lights, quickly turn the lights off to prevent excessive heat buildup. (At a recent trade show booth in Paris, a sound like a gunshot made all nearby duck for cover— it was an incandescent underwater light at a nearby booth that overheated and exploded.)
- After placing the vehicle in the water, check the vehicle for ballast and trim.

22.1.5 Specific consideration for operational deployment of ROVs

22.1.5.1 General considerations

In this section, the primary focus will be on the OCROV system. As discussed in Chapter 3, there are varying levels of capability with differing sizes of OCROV systems. In the following sections, a series of operational tasks will be listed. The listing will be accompanied with suggested methods of how to approach each task, based upon the operating characteristics of different size classifications of OCROV systems.

The classification of each ROV system (small, medium, and large) is based upon the systems evaluated during OCROV procedures testing. Each is capable of performing general tasks (i.e., ship hull inspection, pier and mooring inspections, etc.) under ideal conditions. The more difficult the conditions become (e.g., higher currents, longer stand-off from work site, and higher tether drag), the less likely the system will be able to pull its tether to the work site.

22.1.5.2 Performance considerations

From Chapter 3, a series of drag curves were developed to demonstrate mathematically the crossover point where the net thrust produced by the submersible will no longer overcome the drag of the system (vehicle plus tether). This is the point where the system fails to deliver the submersible to the work site (which is an important objective). From those charts, the net thrust (total horizontal forward thrust less total drag of the vehicle plus the perpendicular drag of the tether through the water) available to take the vehicle to the work site can be determined. The approximate median "tether length in the water" crossover points for the systems listed in each category are provided in Table 22.1. Again, these figures assume the tether to be perfectly perpendicular to the water flow (Figure 22.1).

Table 22.1 Crossover Points for Net Thrust by ROV Size Category (per Figures 3.25–3.28)

Velocity (knots)	Small (ft)	Medium (ft)	Large (ft)
0.5	250	>250	>250
1.0	75	100	150
1.5	10	30	60
2.0	0	0	20

If a 50% reduction in drag due to the tether being less than perpendicular is assumed, Table 22.2 can be derived.

If a 75% tether drag reduction is assumed, Table 22.3 is derived.

With the above assumptions, the approach to an ROV underwater port security task will vary depending upon the conditions to be encountered during that task.

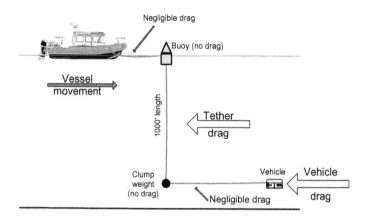

FIGURE 22.1

Vehicle and tether drag at 100% tether drag due to perpendicular water presentation.

Table 22.2 Net Thrust Crossover Points by ROV Size Category with 50% Reduction of Tether Drag Due to Less Than Perpendicular Tether Presentation to the Water Flow

Velocity (knots)	Small (ft)	Medium (ft)	Large (ft)
0.5	>250	>250	>250
1.0	125	200	250
1.5	40	75	110
2.0	5	10	50

Table 22.3 Net Thrust Crossover Points by ROV Size Category with 75% Reduction of Tether Drag Due to Less Than Perpendicular Tether Presentation to the Water Flow

Velocity (knots)	Small (ft)	Medium (ft)	Large (ft)
0.5	>250	>250	>250
1.0	200	>250	>250
1.5	75	100	175
2.0	10	15	75

22.1.5.3 Operational examples

Certain maneuvers will be possible with larger, higher performance systems that can overcome tether drag. Examples of a container ship keel run using a large and small ROV follow.

22.1.5.3.1 Large ROV system

Keel run of a large container vessel at anchor while station-keeping with a response boat (RB) is shown in Figure 22.2.

Motor-by, side hull scan of large container vessel with RB following is illustrated in Figures 22.3 and 22.4.

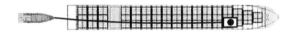

FIGURE 22.2

Keel run of large cargo vessel from stern.

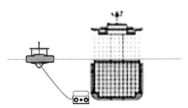

FIGURE 22.3

Side scan of hull (stern view looking forward) with submersible—RB following vehicle.

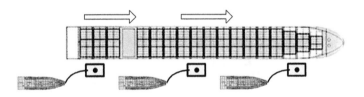

FIGURE 22.4

Side scan of hull (top view) with submersible—RB following vehicle.

22.1.5.3.2 Small vehicle procedural modifications

Procedural modifications to account for a smaller system include the following:

- Consider breaking the vessel inspection into smaller segments (perhaps half, quarter, or eighth sections), so that shorter lengths of tether are in the water.
- Consider moving the tether handling point away from the operations platform to assist in tether mechanics (in the above example, the operator would be able to control the system from the RB, while someone aboard the inspection vessel handled the tether).
- Consider changing from a longitudinal search pattern to a lateral search pattern (as depicted in Figures 22.5 and 22.6).
- Consider the use of clump weights to cut down on the cross-section drag presented to the oncoming water flow.
- Consider moving the deployment platform closer to the work site (Figure 22.7).
- Consider changing the direction of pull to the work site to a more direct pull, increasing the mechanical advantage of the vehicle's pull on the tether.

FIGURE 22.5

Example of procedural change to lateral search in order to lower tether length in water (top view).

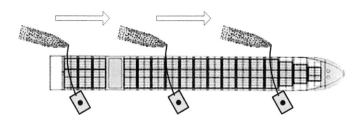

FIGURE 22.6

Lateral search pattern to lower tether length in water (side and stern views).

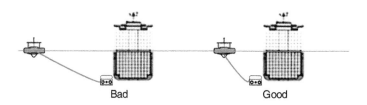

Bad   Good

FIGURE 22.7

Try to keep the vehicle as close as possible to the deployment platform.

- Consider delaying the operation until conditions are more favorable to the mission with the system.
- Consider sourcing a higher performance system.

22.1.6 Task list and guidelines

The trial-and-error method of learning new tasks, while effective, is costly and time-consuming. The learning curve for a task is greatly accelerated if one has the benefit of others' experiences with trial and error regarding the selected task. The task listing below was developed from a menu of port and harbor security requirements. The procedures were tested to determine the best method of task approach to cut down on inefficiencies during mission planning and conduct.

The task list details a method for each general task classification. The general guidelines provide for operations in the best-case scenario or under perfect conditions. Along with the general guidelines is a diagram and task description for accomplishing specific tasks as the conditions become less than ideal.

Once the task has been mastered by the ROV operator, modifications to the procedure certainly should be attempted (and, if successful, fully documented) in order to continuously update and improve these TTPs. Use the matrix below as a guideline for starting the task and modify as the situation dictates.

22.1.6.1 Table of task expectations

The following categories are provided to help determine the equipment capability and level of proficiency of the operator based on the planned task.

- P: Possible with operator of average proficiency
- D: Difficult with operator of average proficiency (possible but time-consuming)
- X: Difficult or not possible with operator of average proficiency (positive outcome questionable or doubtful)
- NA: Nonapplicable.

The following OCROV sizes, as described in Chapter 3, were considered in the task matrix:

- S: Small ROV
- M: Medium ROV
- L: Large ROV

When performing the tasks described in the matrix, all items in the "P" category are suggested methods of approaching the task with the size of equipment indicated (S, M, and L). At the beginning of each task is a descriptive note suggesting the overall approach to the task. Then select the menu of procedures based upon the "P" items within the size category.

In practically all instances, the RB is the best platform of opportunity due to the proximity of its operating platform to the water's surface, as well as its mobility and agility.

The detailed list of tasks examined in the matrix and sections to follow include:

- Ship hull inspections
- Pier inspections
- Anchor inspections
- Inspecting underwater obstructions

- HAZMAT spills
- Oil spills
- Potable water tank inspections

All ship hull inspection tasks assume the inspection target is a 600 ft (200 m) oceangoing cargo vessel. Sea conditions assume no current. As the current increases from zero, all ROV systems will trend from their nominal state toward the "X" state. Also, as the length of the vessel to be inspected decreases, all nominal states tend toward the "P" category.

22.1.6.2 Ship hull, pier, mooring, and anchor inspections

In an ideal environment (perfect water clarity, no currents, 360° viewing around the submersible, large powerful ROV system, and vessel at rest), a 100% hull or object coverage could be completed in a relatively short time. Under ideal conditions, the following approach to each inspection task should be considered.

22.1.6.2.1 Ship hull inspections

Approach the 100% coverage ship hull inspection by running a series of inspection lines horizontally/longitudinally along the hull. For side hull inspections, start at the surface and descend to the water level where the submersible can still view the surface. Place the submersible in auto depth mode to remain at that level, and run the line from stem to stern (or reverse). Reverse direction, and descend the vehicle to a level where the last line is in sight. Run that line. Continue this method until the hull has been covered to the intersection of the bilge keel (if present). For under-hull inspections, run longitudinal lines, weighting the tether to avoid entanglement of the tether with obstructions on the hull. Continue running longitudinal lines until the hull is covered.

22.1.6.2.2 Pier inspections

Approach the 100% coverage pier and structure inspection on a "voyage in/inspect out" basis. Inspect pilings on a "bay-by-bay" basis. Swim the submersible into the bay and inspect the pilings (progressing outward) as the submersible is coming back out. Start at the surface for the first piling and then run the piling toward the mud line. Proceed on the bottom to the next piling and run it to the surface. Run on the surface to the subsequent piling and repeat.

22.1.6.2.3 Anchor inspections

Approach an anchor inspection from the surface toward the bottom, exercising caution at the lay point as discussed in previous sections of this manual.

Guidelines as conditions deteriorate—for the less than ideal conditions, the following matrix of alternate approaches in varying conditions and situations is provided.

22.1.6.3 Pier inspections

Pier inspections are performed by all sizes of ROV systems. The most efficient method demonstrated for performing pier inspections is with straight pulls into the work site. One bay of pilings should be inspected, and then the submersible should be recovered (or swum out of the piling structure) before performing the next bay. Best results have been achieved by the "look-down and then swim-down" followed by the "look-up and then swim-up" piling scan method. Inspect down on one piling to the bottom, swim to the next, and scan upward toward the surface (Figure 22.8).

Task	Vehicle Size		
	S	M	L
Pier/Mooring/Anchor/Hull Search from RB			

Tied off to structure

Note: When inspecting a moored vessel (time and logistics permitting), it is best to tie off to the structure to which the vessel is moored so that the RB is stable for deployment and out of the path of other port traffic. It is best to approach a vessel from the bow for the main hull inspection and reposition to the stern for running gear inspection. This will avoid the potential tether entanglement in the running gear as the tether drags past while inspecting forward sections of the hull.

		S	M	L
a.	Run bilge keel or keel	X	D	P
b.	Inspect running gear to stuffing block	P	P	P
c.	Inspect sea chest(s)	P	P	P
d.	Inspect secondary thrusters (bow and laterals)	P	P	P
e.	Inspect through-hull fittings	P	P	P
f.	Inspect bulkhead/pilings	P	P	P
g.	Run anchor chain	D	P	P
h.	Acoustically/visually search bottom under vessel	P	P	P

Station-keeping

Note: It is possible to inspect a vessel at anchor with the ROV system operated from the RB while station-keeping. As discussed above, the larger systems are capable of some time-saving maneuvers running the submersible while following with the RB. These are very efficient maneuvers, but should be conducted with caution. The approach for smaller systems is to break the vessel inspection into segments, then run inspection lines with the minimum wetted tether following the guidelines established in the reference drag tables above. When inspecting anchors, begin on the surface and inspect down the chain toward the anchor.

(Continued)

(Continued)	Vehicle Size		
Task	**S**	**M**	**L**
a. Run bilge keel or keel	X	D	P
b. Inspect running gear to stuffing block	P	P	P
c. Inspect sea chest(s)	P	P	P
d. Inspect thrusters	P	P	P
e. Inspect through-hull fittings	P	P	P
f. Inspect bulkhead/pilings	D	P	P
g. Run anchor chain	D	P	P
h. Acoustically/visually search bottom under vessel	P	P	P

Tied off to different structure and swimming to object (100-ft/30-m offset)

Note: In some instances it may be necessary to tie off to a different structure from the vessel to be inspected with some distance offset. This method of operation complicates the inspection task and may be impossible with some smaller systems. Consider moving the deployment platform closer, handling the tether from the vessel, or delaying the task until logistical conditions improve.

	S	M	L
a. Run bilge keel or keel	X	D	P
b. Inspect running gear to stuffing block	X	D	P
c. Inspect sea chest(s)	X	D	P
d. Inspect thrusters	X	D	P
e. Inspect through-hull fittings	X	D	P
f. Inspect bulkhead/pilings	X	D	P
g. Run anchor chain	X	X	D
h. Acoustically/visually search bottom under vessel	X	D	P

Visual and Acoustic Hull Inspections from Vessel

Deployed from vessel with RB handling tether

Note: In some instances (such as with an overhanging bow or stern structure), mechanical advantage may be gained while operating the ROV system from the vessel to be inspected by having the tether tended from the RB.

	S	M	L
a. Run bilge keel or keel	X	D	P
b. Inspect running gear to stuffing block	P	P	P
c. Inspect sea chest(s)	P	P	P
d. Inspect thrusters	P	P	P
e. Inspect through-hull fittings	P	P	P

Deployed from RB with tether handled from vessel

Note: As stated above, in some cases it may be beneficial to separate the deployment platform from the location of the tether handler. If this is deemed advisable, either swim the vehicle to the inspection platform for retrieval of the tether or physically walk the vehicle to the vessel to be inspected. Then deploy the submersible over the side and commence the inspection.

	S	M	L
a. Run bilge keel or keel	X	D	P
b. Inspect running gear to stuffing block	P	P	P

(Continued)

(Continued)

Task	Vehicle Size		
	S	**M**	**L**
c. Inspect sea chest(s)	P	P	P
d. Inspect thrusters	P	P	P
e. Inspect through-hull fittings	P	P	P

Visual and Acoustic Hull Inspections from Dock

Deployed from dock with RB handling tether

Note: In some instances (such as the inspection of the seaward side of a moored vessel), some mechanical advantage may be gained by tending the tether from the RB.

	S	M	L
a. Run bilge keel or keel	X	D	P
b. Inspect running gear to stuffing block	P	P	P
c. Inspect sea chest(s)	P	P	P
d. Inspect thrusters	P	P	P
e. Inspect through-hull fittings	P	P	P

Deployed from RB with tether handled from dock

Note: In some instances (such as the inspection of the shore side of a moored vessel), some mechanical advantage may be gained by tending the tether from the dock.

	S	M	L
a. Run bilge keel or keel	X	D	P
b. Inspect running gear to stuffing block	P	P	P
c. Inspect sea chest(s)	P	P	P
d. Inspect thrusters	P	P	P
e. Inspect through-hull fittings	P	P	P

Stationary deployment from RB

Note: For some smaller moored vessels, some advantage may be gained by attempting to swim the entire hull from a single point. However, this method is, in most cases, inefficient.

	S	M	L
a. Sub swimming entire ship from single spot—bow	X	D	P
b. Sub swimming entire ship from single spot—amidships	D	P	P
c. Sub swimming entire ship from single spot—stern	X	D	P

Minimal tether in water from RB

Note: This method requires coordination and practice, but in most instances it is the most efficient.

	S	M	L
a. Station-keeping with RB moving along with sub	X	D	P
b. Move RB from stationary point to point and deploy	D	P	P

Longitudinal/lateral searches from RB

Note: Longitudinal inspections are more time-efficient than are lateral searches since it requires much less effort to reposition for each search line. However, lateral searches allow for better control of the tether due to less wetted surface on each search line. These two considerations must be balanced based upon the conditions prevalent at the time of the tasking.

	S	M	L
a. Longitudinal search from bow	X	D	D

(Continued)

(Continued)

Task	Vehicle Size		
	S	**M**	**L**
b. Longitudinal search from stern	X	D	D
c. Lateral search	P	P	P
Section searches (quarter/halve vessel) versus landmark/high-exposure searches			
Note: All hull inspections are greatly enhanced with possession of a diagram of the hull to be inspected displaying prominent features on the hull. With a full diagram of the hull, feature-based visual searches will ensure coverage of that section of the hull inspection.			
a. Obtain "as built" drawings and perform landmark/high-exposure searches	P	P	P
b. Full coverage search	NA	NA	NA
c. Convenience of breaking vessel down into sections versus full ship	NA	NA	NA
Adrift and tied to vessel hull search			
Note: During tidal flows, it may be beneficial to place the submersible in the water and then scan as the submersible drifts by. This allows the submersible to drift (and look) downstream facing down-current or attempt up-current vehicle swim while tied off to the stern of the vessel to be inspected. Caution should be exercised when doing drift-by scans due to the possibility of the submersible snagging on items in the water column or attached to the vessel to be inspected.			
a. Drift in current from bow to stern while conducting hull scan	P	P	P
b. Tie off to bow and do drift scan with sub facing down-current	D	D	D
c. Tie off to stern in current and do up-current search	X	D	P
Pier/Mooring/Anchor/Hull Search from Both Shore and Vessel			
Shore to structure/vessel			
Note: The least resource-intensive hull search scenario is with the ROV deployed from shore while conducting a hull or structure inspection. This will form a high incidence of hull inspection scenarios. In areas of low tidal flow, this is a simple and efficient method of hull and structure inspection.			
a. Run bilge keel or keel	X	D	P
b. Inspect running gear to stuffing block	P	P	P
c. Inspect sea chest(s)	P	P	P
d. Inspect thrusters	P	P	P
e. Inspect through-hull fittings	P	P	P
f. Inspect bulkhead/pilings	P	P	P
g. Run anchor chain	NA	NA	NA
h. Acoustically/visually search bottom under vessel	P	P	P
Vessel to same vessel			
Note: Inspections of hulls while positioned aboard the same vessel are convenient. It may, however, be difficult to inspect the underside of the vessel.			
a. Run bilge keel or keel	X	D	P
b. Inspect running gear to stuffing block	P	P	P
c. Inspect sea chest(s)	P	P	P

(Continued)

(Continued)	Vehicle Size		
Task	**S**	**M**	**L**
d. Inspect thrusters	P	P	P
e. Inspect through-hull fittings	P	P	P
f. Inspect bulkhead/pilings	P	P	P
g. Run anchor chain	NA	NA	NA
h. Acoustically/visually search bottom under vessel	P	P	P
From vessel swimming to other vessel (100 ft offset)			
Note: Positioning the operations platform from one vessel while inspecting another is possible. The larger the offset, the more limited is the submersible's ability to perform its inspection task. If this scenario is presented, consider moving the tether handling function to either an RB or to the vessel to be inspected.			
a. Run bilge keel or keel	X	D	P
b. Inspect running gear to stuffing block	X	D	P
c. Inspect sea chest(s)	X	D	P
d. Inspect thrusters	X	D	P
e. Inspect through-hull fittings	X	D	P
f. Inspect bulkhead/pilings	X	D	P
g. Run anchor chain	X	D	D
h. Acoustically/visually search bottom under vessel	X	D	D
Visual and Acoustic Hull Inspections from Vessel and Shore			
Deployed from vessel with shore-based handling of tether			
Note: This is a convenient and effective method of performing an inspection on the shore side of a moored vessel.			
a. Run bilge keel or keel	X	D	P
b. Inspect running gear to stuffing block	P	P	P
c. Inspect sea chest(s)	P	P	P
d. Inspect thrusters	P	P	P
e. Inspect through-hull fittings	P	P	P
f. Inspect bulkhead/pilings	P	P	P
g. Run anchor chain	NA	NA	NA
h. Acoustically/visually search bottom under vessel	P	P	P
Deployed from shore with tether handled from vessel			
Note: This method is effective for inspecting the seaward side of a moored vessel while operating from shore. Consider this method if access to the vessel to be inspected can be gained.			
a. Run bilge keel or keel	X	D	P
b. Inspect running gear to stuffing block	P	P	P
c. Inspect sea chest(s)	P	P	P
d. Inspect thrusters	P	P	P

		Vehicle Size		
Task		**S**	**M**	**L**
e.	Inspect through-hull fittings	P	P	P
f.	Inspect bulkhead/pilings	P	P	P
g.	Run anchor chain	NA	NA	NA
h.	Acoustically/visually search bottom under vessel	P	P	P
Stationary deployment on shore and from vessel				
Note: This method is not the preferred method of performing a ship hull inspection since there is no mechanical advantage gained. This may be necessary due to limited access to deployment points.				
a.	Sub swimming entire ship from single spot—bow	X	D	P
b.	Sub swimming entire ship from single spot—amidships	D	P	P
c.	Sub swimming entire ship from single spot—stern	X	D	P
Minimal tether in water				
Note: In all instances of ROV operations, the best vehicle movement will be gained with minimal tether in the water, thus minimizing the tether drag to which the vehicle is subjected. Consider this as a primary method of ROV vehicle deployment.				
a.	Tether handler moving along with sub	P	P	P
b.	Tether handler moving from stationary point to point and deploy	P	P	P
Longitudinal/lateral searches from vessel and from shore				
Note: Longitudinal pulls along the hull of any vessel involve long tether pulls with considerable tether drag for the length of the hull. Lateral searches, however, provide less drag across the hull's surface through shorter transects of the hull surface.				
a.	Longitudinal search from bow	X	D	P
b.	Longitudinal search from stern	X	D	P
c.	Lateral search	P	P	P

22.1.6.4 Inspecting underwater obstructions

Navigation to and inspection of underwater obstructions involves a peculiar set of operational problems not associated with other underwater port security tasks:

- The object may not have a positive link to the surface (such as a wreck, object dropped overboard, or other underwater obstruction).
- The identification dimensions of this obstruction may not be known.
- Navigation to this obstruction may be difficult without the use of acoustics.
- Significant unseen entanglement hazards could (and in all likelihood will) be present.
- Station-keeping above the obstruction may be difficult without dynamic positioning equipment.
- Keeping the submersible in visual sight of the object could be difficult or impossible with any measure of tidal flow.

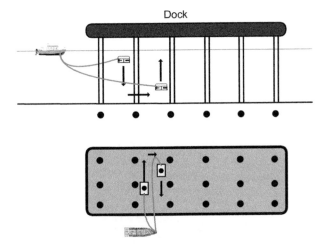

FIGURE 22.8

Pier inspections from both a profile view and plan view.

Some procedural modifications to compensate for the above factors while operating on and around underwater obstructions follow:

- Make liberal use of clump weights to keep the managed tether length to a minimum, while keeping the surface tether away from vessel thrusters. Assure that the clump weight stays above the level of the obstruction to reduce the risk of clump weight entanglement with the obstruction.
- Approach the obstruction with the vehicle from the leeward (or down-current) side, moving up-current.
- Consider use of a down-line, anchored to or near the wreck, to maintain positive navigation to the bottom.
- Avoid crossing over the obstruction to minimize risk of tether entanglement.
- The location and survey of an underwater obstruction without the use of scanning sonar may be exceedingly difficult in low-visibility conditions.

22.1.6.5 Other nonstandard operations
22.1.6.5.1 HAZMAT spills
ROV systems are effective in performing inspections of hazardous substance spills. The time and resources required to place a diver into these substances are considerable. In order to protect the vulnerable components of the submersible, a general idea of the substance into which the vehicle will dive is needed. Upon recovery of the vehicle, adequate decontamination is required (Figure 22.9).

FIGURE 22.9

ROV contaminated during an oil spill survey.

The suggested method of penetration into a HAZMAT environment is to determine the length of tether needed for the operation and then quarantine all parts of the submersible/tether combination from the insertion point to the submersible. A suggested decontamination method follows:

- Prepare the submersible for deployment into the HAZMAT substance.
- Mark the tether at the point of entry (or point of quarantine) to the submersible.
- Obtain (or create) an open container of decontaminate large enough to handle both the submersible and tether upon recovery of the vehicle.
- Perform work and recover vehicle.
- Thoroughly immerse the vehicle and tether in decontaminate until cleansed to EPA (or controlling authority) guidelines.
- Rinse all decontaminated parts and perform scheduled maintenance.

Protecting the environment at the site is critical. It is paramount for both personal safety, as well as environmental protection, to assure that all safety cleanup protocols are followed. This will allow timely and assured site recovery, preserve community health, and prevent any liability for the team.

22.1.6.5.2 Oil spills

ROV systems are effective as a quick responder to oil spills. The technique for inspecting the oil spill location is to get below the oil sheen without obstructing the domes, seals, and lights. For oil spills, the following are suggested guidelines:

- Find a location upstream of the oil slick to deploy the vehicle. If there is no current flow, place a deployment vessel (tubular object such as a garbage can or pipe), evacuate all oil from the tube, and deploy the vehicle through the clear area.
- Locate the source of the spill.
- Upon completion of the mission, recover the vehicle, clean with solvent, and perform scheduled maintenance.

22.1.6.5.3 Potable water tank inspections

Small ROV systems do quite well in the low-flow, quiet environment of a potable water tank. The following steps are suggested when performing a potable water tank inspection:

- Determine the length of tether needed to perform the inspection of the potable water tank.
- Thoroughly clean all components of the vehicle with decontaminate in accordance with the American Water Works Association (or other controlling authority) from the quarantine point on the tether to the vehicle.
- Obtain some type of container (a clean cardboard box, clean barrel, or other large container) large enough to hold the tether and the vehicle after cleaning. This will form the quarantine area for all insertion/extraction operations.
- Upon completion of the mission, perform scheduled maintenance with consideration that the potable water tank could contain chlorine (or other chemicals) affecting the seals and O-rings of the vehicle. Signs of chemical damage to O-rings and seals may include whitening, flaking, and drying of wetted areas.

22.1.7 Postdive procedures

22.1.7.1 Postdive checklist

A postdive inspection should be carried out after every dive:

- Visually inspect the vehicle following each dive to ensure no mechanical damage has occurred.
- Check the propellers for any fouling.
- Visually check through the ports to ensure that no water has entered the camera, thruster, or electronics housings.
- Inspect the tether for cuts, nicks, or kinks in the outer shell.
- Rinse the vehicle and tether in freshwater if it has been operated in salt water.
- Check all vehicle functions again before power-down.
- Store the tether and vehicle properly for the next use. Refer to specific manufacturer's instructions for system storing procedures.

22.1.7.2 Demobilization of equipment

Once the mission is completed, ensure the equipment is packaged adequately for travel from the work location. Upon return to the base, perform the following steps to demobilize the equipment for storage:

- Ensure all system functions are performing and operational.
- Verify that the system is completely free of salt water. If there is any presence of saltwater residue, rinse completely with fresh water and dry before storing.
- Perform preventative maintenance in accordance with manufacturer-specified guidelines.
- Ensure all O-rings, seals, joints, and turn points are greased and packaged for storage.
- Store in accordance with manufacturer's recommendations.

22.2 Servicing and troubleshooting

All ROV systems share the same basic operating characteristics and maintenance needs. This section contains an outline of troubleshooting and preventative maintenance that should be performed on a regular basis. These procedures are not a substitution for manufacturers' suggested operating and maintenance procedures. These are guidelines to supplement manufacturer-specific instructions. A major source for this section is Wernli (1998).

22.2.1 Maintenance

Equipment maintenance forms a vital part of a safe and efficient ROV operation. Properly maintained systems can achieve substantially reduced downtime. System schedules ranging from simple predive checklists through detailed planned maintenance procedures must therefore be used to attain and maintain the highest possible standard of operating efficiency.

All work should be undertaken in compliance with supplier's/manufacturer's recommendations. Each ROV system type is provided with a full set of vehicle manuals and vendor subsystem technical information to enable efficient maintenance and reordering of system spare parts.

ROV system maintenance is divided into the following main areas of documentation:

- Operations and maintenance manuals and drawings
- Supplier manuals and drawings
- Catalogs of equipment

The above includes the following key subsections:

- Vehicle maintenance procedures
- Subsystem maintenance procedures, that is, video cameras, sonar system, tools/motors
- Detailed repair and maintenance procedures (found in the specific ROV operations manual)

All equipment shall be suitably labeled to indicate its operational status on arrival at the deployment base in accordance with company procedures.

Every manufacturer of ROV equipment has a set of maintenance standards peculiar to their respective equipment. It is the responsibility of the ROV maintenance supervisor to assure that the individual manufacturer's maintenance schedule is meticulously followed. A sample maintenance schedule is provided below in "Operational forms" (Section 22.2.5).

22.2.2 Basics of ROV troubleshooting

22.2.2.1 Basics of an ROV system

There is not much "high technology" in designing, manufacturing, and producing an ROV system. Every year, the Marine Advanced Technology Education Center (MATE) hosts an ROV competition for high school and college students who put together their own operational ROV system. Educators in British Columbia have put forth a book on how to build an ROV in a garage out of hardware store parts (Bohm and Jensen, 1997). The difficulty is in producing a commercially viable and reliable system that can take the abuse of fieldwork and produce results.

The basic parts of a free-swimming, OCROV system follow:

- Submersible
- Tools and sensors
- Tether
- Power supply
- Controller
- Viewing device

Essentially, an ROV is a camera generating a video signal mounted in a waterproof housing with electric motors attached to a cable. Practically all of the vehicles use common consumer industry standard commercial off-the-shelf (COTS) components.

There are a few items on the system that require computer processing power, including the sonar, the acoustic positioning system, some instrument packages, controls to run the motor driver boards or manipulator, and telemetry from the submersible to the surface. Practically all of these are located on "easily changeable" printed circuit boards. The only real nonelectronic challenge for an ROV technician is working with the O-rings, seals, and tight machining tolerances needed to complete a waterproof seal on the submersible. A simplified schematic of an ROV submersible control system is shown in Figure 22.10.

From these basics (with the use of manufacturer-supplied schematics and drawings), one should be capable of performing basic troubleshooting in the field to complete the mission requirements.

22.2.2.2 The troubleshooting process

Effective and efficient troubleshooting requires gathering clues and applying deductive reasoning to isolate the problem. Once the problem is isolated, one can analyze, test, and substitute good components for suspected bad ones to find the particular part that has failed.

The use of general test equipment (such as a digital multimeter) or special test equipment (such as an NTSC pattern generator and an oscilloscope) can speed the analysis, but for many failures, deductive reasoning can suffice. Once it is determined whether the problem is electronic or mechanical, deductive analysis changes to intelligent trial-and-error replacement. Reducing the number of suspected components to just a few and then using intelligent substitution are the fastest ways to identify the faulty device.

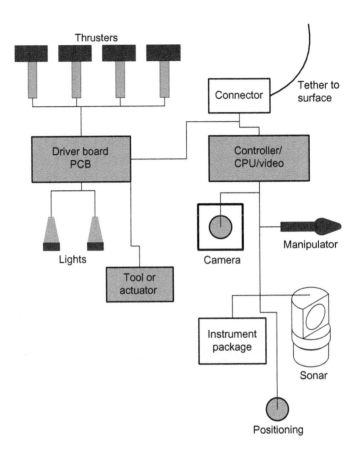

FIGURE 22.10

Basic submersible schematic.

In general, follow these steps when the ROV system fails:

1. Obtain the service manual for the ROV system in question.
2. Observe the conditions in which the symptoms appear.
3. Make note of as many of the symptoms as possible.
4. Use senses to locate the source of the problem.
5. Retry.
6. Document the finding and test results.
7. Assume one problem (when multiple symptoms are present, troubleshoot the easiest one first).
8. Diagnose to a section (fault identification).
9. Consult the troubleshooting chart of the manufacturer's maintenance manual.
10. Localize to a stage (fault localization).
11. Isolate to a failed part (fault isolation).
12. Repair.
13. Test and verify.

When something goes wrong, the first step is to determine whether the trouble results from a failure, a loose connection, or human error. Once it is confirmed that the failure has occurred (i.e., the operator lays the blame squarely on the ROV), the next step is to determine which portion of the ROV system is not operating—mechanical or electrical.

Then, step by step, partition each section into stages and track the trouble to a single component. For example, if one function is not working, the problem could be in the switch itself, the connector/connecting wires, or the electronic circuitry of the controller. For these procedures to be effective, a basic understanding of the operation and design of an observation-class ROV system is necessary.

22.2.3 Tools and spares for fieldwork

In the field, anything can happen. A good work routine with some simple spares will help solve or eliminate most field problems.

As a general rule, a clean workplace is a safe and productive workplace. If there are items unsecured, someone will naturally trip on them. If an electrical connection is frayed, it will short out and bring down the entire system (requiring extensive troubleshooting and/or trips to the hospital).

The following is a general field tools and spares listing that will assist any OCROV pilot/technician in outfitting a field pack for servicing ROV equipment (MSROV and WCROV systems require a much more robust system-specific inventory):

- A good Swiss Army–type knife or one of the newer multitools
- A good folding pocket/diving knife (although many industrial settings have gone to a "no knife" policy, for obvious reasons, in favor of high-quality scissors)
- A roll of electrical tape (black vinyl type) as well as a roll of high-quality duct tape
- Plenty and varying lengths of cable ties
- Video cable adapters:
  - BNC to RCA
  - RCA to BNC
  - BNC Tee
  - Spare adapters for as many types of connectors as anticipated
  - Spare video cables
- A pocket magnetic compass (for calibrating vehicle compass)
- A pocket magnifying glass (for the gray-haired technicians losing their eyesight)

The following electronic accessories are also recommended for the field toolkit:

- A 50 ft, 12/3 extension cord
- A minimum of one high-quality power bar (with surge protection)
- At least one A/V three-way amplifier (common Radio Shack part #15-1103)
- A four-way A/V selector (common Radio Shack #15-8250)
- An amplified audio/video selector (common Radio Shack #15-1951)
- A video RF modulator (common Radio Shack #15-1283A—for those times when the TV monitor doesn't have RCA-type video inputs)
- A video isolation transformer, an example of which is North Hills, FSCM 98821, Wideband Transformer, Video Isolation, Model 1117C, 10 Hz–5 MHz

- A small container with at least one of every cable adapter Radio Shack carries. Suggestion: If Radio Shack or some other electronics store has an adapter that is not in the toolkit, buy it. Stocking them in the kit is priceless when needed.
- A small variety of video cables, both RCA and BNC types. It is important to have a variety of the standard lengths available at any electronics retail outlet.
- Make up a 100 ft video cable extension cord. It may be handy to have one that has a male RCA plug on one end and a female RCA plug on the other so that the adapters in the kit can be used for any changes in format.
- A small digital multimeter is essential. The biggest problem with the multimeter is power. Always keep spare batteries.

In addition to the above, carry a small toolbox with a variety of small hand tools, pliers, wire cutters, 6 inch (or metric equivalent) adjustable crescent wrench, multibit screwdriver, soldering iron, solder, a small roll of wire, Allen wrenches (both standard and metric), and virtually anything else that will fit in the box.

22.2.4 Standard preventative maintenance checklist

The items in Table 22.4 are for general guidance and should not supplant the specific directives of any manufacturer-specific maintenance schedule. These maintenance items are typical of a proper

Table 22.4 Maintenance Schedule

	Check Every			
Maintenance Action	**Predive**	**Post**	**50 h**	**500 h**
Check wires, cables, and hoses for wear and damage.	X	X		
Check for loose or missing hardware. Repair or replace as necessary.	X	X		
Check sacrificial anodes for deterioration.	X	X		
Visually inspect the vehicle to ensure that the propellers are not fouled, that all components are secured, and there is no mechanical damage to the frame or other components.	X	X		
Check the tether for scrapes, nicks, kinks, or other visible damage. The vehicle should not be used if the tether jacket is broken through.	X	X	X	X
Check that all fasteners are in place and secure.	X	X		
Check ballast weights.	X	X		
Ensure the tether connection at the vehicle is plugged in all the way, and that the tether termination can is well secured by the flotation.	X	X		
Power-up test: Test all system functions.	X	X	X	X
Check for water intrusion through the ports.	X	X	X	X
Drain and check the oil in the thruster cone end.			X	X
Replace thruster seals and inspect propeller shaft. Replace when grooved.			X	X
Store properly.	X	X	X	X

manufacturer's preventative maintenance checklist that should be provided with the ROV system upon delivery from the manufacturer.

22.2.5 Operational forms

Often after the field operation is completed, field operations personnel are required to substantiate findings made while on duty. For operational and legal consideration, documentation of operations should be meticulously kept. The following pages provide sample forms that should assist in documentation of ROV operations to include:

- ROV predive operations check
- ROV postdive operations check
- ROV dive log

ROV Pre-Dive Operations Check

Dive No:	Date (Y/M/D):
Dive Location:	Start Time:
Operator:	Stop Time:

Check Conducted	Initials	Comments
Visually inspect the vehicle to ensure that the propellers are not fouled, that all components are secured, and there is no mechanical damage to the frame or other components.		
Check the tether for scrapes, nicks, or other visible damage. The vehicle should not be used if the tether jacket is broken through.		
Check that all fasteners are in place and secure.		
Check ballast and trim adjustments.		
Ensure that both whip connectors at the electronics can and tether termination can are plugged in all the way. Also check that the tether termination can is clamped securely between the lower bracket and flotation.		
Ensure all unused vehicle connectors are capped securely with dummy plugs. A forgotten or unsecured dummy plug can lead to serious electrical system damage.		
Ensure that all surface cables are properly connected.		
To prevent unexpected thruster operation, switch the Auto Depth switch to OFF, center the Manual Depth Control knob, and switch Auto Heading OFF **before powering up the system.**		
Power up the system. Test the lights, camera functions, and thrusters. Check that good video is obtained and that any video recording equipment is working properly. **Caution: refrain from running the thruster and/or lights any longer than a few seconds out of the water.**		

ROV Post-Dive Operations Check

Dive No:	Date (Y/M/D):
Dive Location:	Start Time:
Operator:	Stop Time:

Check Conducted	Initials	Comments
The vehicle should be visually inspected following each dive to ensure that no mechanical damage has occurred.		
Check the propellers for any fouling.		
Visually check through the ports to ensure that no water has entered the camera housings.		
Inspect the tether for cuts and/or nicks or kinks in the outer shell.		
A fresh water rinse is required if the vehicle has been operated in salt water.		
Re-check all vehicle functions. To prevent unexpected thruster operation, switch the Auto Depth switch to OFF, center the Manual Depth Control knob, and switch Auto Heading OFF **before storing the system.**		
Store the tether and vehicle properly for the next use. Refer to specific manufacturer's recommendations for tether storing procedures		

ROV Dive Log

Vessel: _____   No. of Sheets: _____

Location: _____

Date: _____

Dive No: _____

Operations Crew: _____

Conditions:

_____

_____

_____

Purpose of Dive:

_____

_____

_____

_____

Dive Job Summary:

_____

_____

_____

_____

_____

_____

_____

_____

Total Wet Time: _____

Dive Log Completed By: _____ Signature: _____

ROV Dive Log

Dive No: _____   Sheet: _____of: _____

TIME	OPERATION

22.3 Putting it all together

Throughout this manual, the technology applicable to ROVs was discussed, along with its application in various missions. The crowning achievement is when these newly acquired skills come together to successfully find and prosecute an underwater target. In this section, the steps to accomplish this final task will be examined in detail along with the most common tools in the underwater technician's tool chest—the side scan sonar and the ROV deployed aboard a small boat.

22.3.1 Attention to detail

In the underwater business, there is no so-called silver bullet (i.e., a single piece of equipment that can be used to solve all operational situations). Finding things underwater is more a function of gathering input from many different sources, putting them together to form the most likely conclusion, and repeatedly testing that conclusion until it is positively proven or disproved.

An oversight can cause a wasted mission through simple inattention to detail. A conversion of feet to meters can throw the projected location of the search area away from the known target. A mistake while choosing the map projection can waste an entire day of searching. Choosing the wrong operating voltage can destroy equipment. Careless wiring and arrangement of equipment can cause serious injury and even death while in the field. Attention to detail is paramount when performing any field task.

22.3.2 Training and personnel qualifications

The knowledge requirements for operators, repair technicians, and tasking personnel are:

- Operators are required to know the theory and application of ROV technology.
- Repair technicians are required to understand the operations and maintenance of the components for the ROV system.
- Tasking personnel need only concern themselves with the capabilities and limitations of an ROV inspection system in use for various missions.

This manual has provided the basics in these areas. However, a training program should be established and a record kept in the personnel file of those involved to ensure that each individual, and thus the team, is ready for the task at hand.

22.3.3 Equipment setup considerations

Some considerations while setting up equipment aboard a small boat are needed to properly and efficiently operate the equipment in a safe and productive manner. Time spent initially during the setup phase will pay dividends repeatedly while under way through a tidy, clean, and efficient working environment.

Equipment setup considerations include:

- Both side scan sonar and ROV equipment use considerable lengths of cable. Attempt to use only one piece of equipment at a time while completely stowing the second piece of equipment until needed.

- Run and stow all cables and wires away from travel spaces. A tripping hazard will repeatedly catch personnel off guard, possibly causing serious injury or death.
- The mission equipment should never interfere with the boat's operational equipment. Blocking access to engine compartments, anchor lockers, dock lines, and any boat safety equipment is a danger to both the vessel and the mission.
- Set up all equipment with consideration to the order in which it will be needed. Completely stow equipment when not in use.
- The figure-8 flaking of cables has shown repeatedly to be a very efficient method of mechanical tether management and is usually preferred to the use of a tether reel. A tidy workspace is a safe and efficient workspace.
- To satisfactorily view computer screens and video monitors, a location away from sunlight is needed, such as an enclosed cabin or a tarp over the monitors. Also, a comfortable temperature-controlled area will assist in eliminating the physical need to end the operation due to discomfort.
- A power source separate from the vessel's generator is preferred. Assure that the exhaust from the generator and/or the vessel's exhaust does not vent near the enclosed work area.

22.3.4 Division of responsibility

The captain of the boat is responsible for the vessel's safe and efficient operation. The equipment operator is normally responsible for the mission. The captain of the boat is the final authority regarding all operations aboard the vessel. It is proper protocol to gain permission before deploying any equipment that will affect the operation of the vessel. It is also proper protocol to keep the vessel operator completely informed of all of the team's intentions and planned tasks.

The mission specialist and ROV team are guests aboard a vessel of opportunity. Work along with the crew within the vessel management structure in order to get the maximum out of the equipment and to achieve the mission objectives.

Before accepting a demanding assignment aboard a vessel of opportunity, qualify the captain and crew for the mission at hand. Such tasks as station-keeping in a difficult sea state or maintaining a survey line during a side scan search are paramount to completing the mission. Many of the operational problems can be solved before leaving the dock by screening the boat crew to ensure they are properly qualified. A fishing boat crew may be the best crew for hauling in a large catch, but maintaining a tight-tolerance survey demands another skill set.

22.3.5 Boat handling

Many failed operations can be traced directly to a simple matter of boat handling. Unless the equipment can be deployed consistently onto the location of the target, the mission, in all likelihood, will fail. Also, without the deployment platform maintained in a steady and stable state over the top of the work site, the entire operation may become a complete waste of time and resources.

22.3.5.1 Side scan sonar operations

To perform a proper side scan sonar survey of an area, survey lines must be followed and tow fish altitudes must be maintained within fairly tight tolerance to achieve area coverage with a high degree of certainty. A high-quality GPS receiver with the capability for survey line input, as well as course deviation indication, is very helpful. Complete the survey of the entire area in the survey phase before attempting to switch to the identification phase of the mission (with diver or ROV). Many operations have been inefficiently run by stopping the survey to look at each suspected target (only to discover it was not the proper item). The switchover time between equipment can be considerable. Consult the equipment manufacturer's performance specifications to obtain the proper tow speed and altitude requirements. And do not be afraid to request corrective action of the boat operator if those parameters do not meet the mission requirements.

22.3.5.2 ROV operations

It is paramount to have the ROV deployed over a steady work location. If the boat drifts off location, the ROV operator will be required to repeatedly reacquire the target (once the submersible is dragged off the target), frustrating efforts to complete the mission. If an ROV operation over a target site is to be performed properly, either multipoint anchoring over the site or dynamic positioning will be necessary. In some situations, a skilled boat operator can keep the vessel steady enough to get the identification done. But as the wind and sea state worsens, the ability to keep the vessel on-station becomes increasingly difficult. For shallow-water operations, a three- or four-point anchoring system directly over the work site (or a jack-up barge) is recommended to complete the work task.

22.3.6 Marking the target(s)

To mark a target for positive identification, the following steps should be performed:

- Complete the side scan sonar survey of the entire area, and then retire to a location where a complete analysis of the data can be made. The targets can then be identified, classified, and prioritized. Once the targets are identified, further investigation can be performed.
- Fabricate a sonar reflector to place next to the target of interest. Sonar targets are highly (sonar) reflective weighted anchors connected to the surface via a line and buoy (Figure 22.11). The

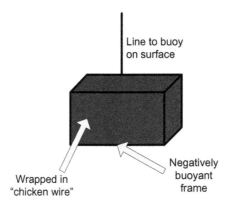

FIGURE 22.11

Sonar reflecting target.

objective of the sonar target is to give the ROV or diver a direct visual path to the target on the bottom via the buoy line. A suggestion for a low-cost sonar target is to fabricate an approximately 3 ft (1 m) cubed metal box structure wrapped in chicken wire, which reflects high-frequency sonar waves at all angles of incidence.

- Once the decision is made to positively identify a sonar target, proceed to the coordinates of the sonar target, drop the sonar reflecting target (make sure the buoy line is well secured with minimal slack to avoid accidental propeller entanglement), and rescan the area to determine the range and bearing of the sonar target to the reflector. Continue to move the sonar reflector until the reflector is as close as possible to the target (Figure 22.12).
- Anchor the boat (Figure 22.13), swim the submersible on the surface to the buoy line, and follow the line down to the sonar reflector. From there it should just be a range bearing steer to the sonar target.

22.3.7 Methods for navigating to the target

Section 22.3.6 describes the simplest method of ROV/sonar navigation to targets on the bottom. Although tedious and time-consuming, it does maintain the operational objective of a "sustained and controlled environment" in that it allows for relatively immovable visual reference points throughout the process. This method has been proven to be effective.

Other methods of navigating to a known target on the bottom are with the use of a mechanically scanning tripod-mounted sonar, in addition to the ROV-mounted sonar.

22.3.7.1 Tripod-mounted sonar/ROV interaction

Just as an aircraft can be navigated to a landing area with ground-based radar, an ROV can be navigated to a target while tracking the target and submersible from a fixed location.

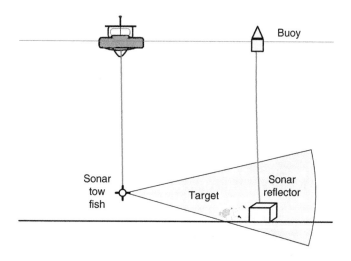

FIGURE 22.12

Positioning of the sonar reflector.

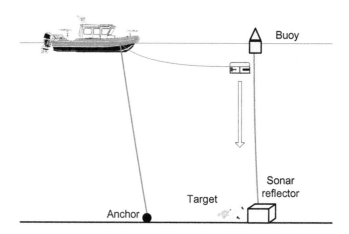

FIGURE 22.13

ROV follows the buoy line to the target.

After the tripod sonar has been lowered onto the bottom and has acquired the target, the submersible is launched; then it follows the sonar line down to the bottom. Once on the bottom, the submersible determines the orientation of the sonar head and then flies in the direction of the target. The tripod sonar operator, after locating the submersible on the sonar, can guide the submersible operator to the target (Figure 22.14).

22.3.7.2 Sonar mounted aboard the ROV

The target can also be located via the onboard scanning sonar system (Figure 22.15). But be warned—the perspective of the target from the side scan sonar will be dramatically different from

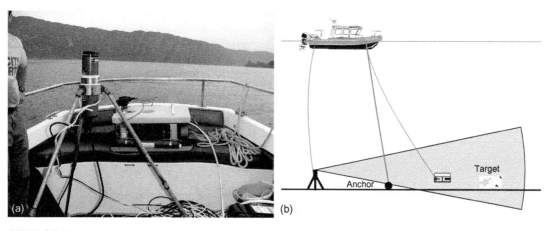

FIGURE 22.14

Fixed location tracking of ROV and target. (a) Sonar prior to deployment and (b) in operation on the bottom.

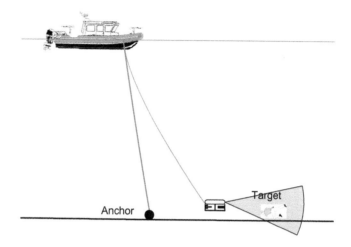

FIGURE 22.15

ROV-mounted sonar.

that viewed from the submersible. From the side scan sonar perspective (i.e., from above), a drowning victim will give the shape and appearance of a human form. From the submersible's perspective, the body may look like a log, a pipe, a rock, or any of a number of items depending upon the viewing angle.

22.3.8 Sonar/ROV interaction

A technique for gaining an orientation on the bottom is to find either a flat rock or a solid location on the bottom to plant the submersible to gain a good vantage point to image all around. A mud bottom makes this planting a bit more permanent, since it may require an excess amount of thrust to unstick the submersible.

Once a location is found, put the submersible on the bottom so that it becomes a stable platform from which to build an image of the surrounding area. The image is generated as the sonar head scans around the vehicle, isolating the target from the surrounding terrain. To better identify the target, it may be required to move the vehicle to another location to rescan the area.

The target is then brought to zero bearing (i.e., directly ahead of the vehicle) and is maintained in view while the vehicle is navigated to the target. Once the target is located and positively identified, the vehicle can remain on location for as long as necessary. If the vehicle is equipped with a manipulator or some form of physical attachment device, it is best to hook on to the target to maintain visual contact until the next step in the process is initiated.

The Future of ROV Technology

As discussed in Chapter 3, there are varying degrees of power and autonomy for remotely operated vehicles (ROVs), ranging from the basic tethered vehicle using remotely provided power and with teleoperated control coming from the surface, through onboard battery-powered vehicles with tele-operation via a fiber-optic control link and eventually to fully autonomous logic-driven vehicles with onboard power. The future of ROV technology fits into the evolution of the entire subsea control system—the main commercial application of which is the future deepwater oilfield. The oilfield-specific focus of this chapter is also applicable to any deepwater area control system.

However, before we look into our crystal ball to project the technology needs and probable vehicle developments of future deepwater oilfields, the stage should be set by taking a topside perspective of today's needs and options.

The future oilfield requires remote sensing, valve activation, chemical treatment, and all of the other aspects of in-field harvesting and separating of hydrocarbons for further transport to process facilities ashore. Currently, these tasks are performed with man-in-the-loop technology. The future is clearly further autonomy. The evolution will follow an ever-increasing removal of man/woman from the loop replacing him/her with logic circuits. The questions facing this development fold into three basic categories:

1. Nodes of the control system operated via "man-in-the-loop" or via logic-driven control
2. Control operations and inspections via internal activation/diagnostics, structurally compliant vehicle activation/inspection or free-traveling vehicle activation/inspection
3. Onboard or remote vehicle power

Logic driven

Clearly, the future is logic-driven circuits. The tasks to be performed in such a circuit are any number of items such as turning valves, pulling and installing flying leads, inspecting assets for integrity, installing new nodes, and the like. The evolution will follow the development of shore-based controls in the following areas:

1. *Placing the logic circuits onto/into the asset/node itself:* This path follows the development of shore-based controls changing the subsea wellhead from a valve panel or hydraulic control board to an intelligent control node. The wellhead will sense variations in pressures/temperatures along with gas/fluid phase, H_2S contamination, hydrate formation, sand or water mix, and the like. Based on this sensory feedback, controls will be activated for manipulating valves and restrictions, thus inhibiting hydrate formation (assuring fluid flow), circling unwanted by-products back to the formation, and protecting infrastructure from damage at the wellhead or node.
2. *Placing the logic circuit into the service vehicles:* As the saying goes in the robotics industry, "Man in the loop is so yesterday." The vehicle will communicate with the intelligent node (#1 above) to take over for operations that cannot be performed by the node itself. This involves further inspection or manipulation such as pulling leads, hot-stabbing chemicals/fluids, or other IRM and drill support tasks.
3. *Development of messaging commands for assigning goal-oriented tasking to remote nodes and vehicles:* To operate a full control system, a grand traffic cop must control the flow of the various nodes toward some higher logic, thus assuring the overall goals of the system are achieved (i.e., the entire field flows as designed). Such high-level commands typically do not tell the robot how to control itself but only to achieve a certain goal-oriented task such as *"There is a decrease in flow on subsea processing facility A; therefore, Robot 23-5, execute subroutine 35B to inspect flowline 43A from Node 23 to Node 24 for flow of hydrocarbons."* The robot will awaken, undock, fly the requisite route, and then re-dock and report its findings to the central computer (and power down per its onboard logic). The central computer will use its findings to determine the next step in the flow assurance process.

Structurally compliant or free-traveling vehicles

This discussion focuses upon the vehicles within the system; therefore, the requirements of the vehicles will depend upon the locomotive method:

1. *Make the node switchable:* As mentioned above, this method places the logic circuit on the specific asset. This does not require a vehicle in order to function, but will be able to communicate with the vehicle toward achieving a specific goal.
2. *Structurally compliant vehicles:* This method places the vehicle in semipermanent contact (much like a caterpillar or squirrel follows a branch) with the asset (e.g., a subsea oil riser), allowing the vehicle the ability to follow the assets without the excessive energy and navigational requirements of a free-traveling (i.e., autonomous) vehicle.
3. *Bottom-crawling vehicles:* This vehicle crawls the bottom on either a subsea track/rail system (much like a railroad track between two terminals) or free-traveling via vehicle-mounted tracks/wheels.
4. *Free-swimming vehicles:* These vehicles are "free-flying," neither attached to the structure nor to the bottom. These autonomous vehicles require a substantial amount of energy and logical processing for station-keeping, end-effecter manipulation, as well as navigation.

Vehicle power

The biggest question for the vehicle within the future subsea oilfield is the location of the service vehicle power source. And that is directly related to the enabling technology of battery power density. Placement of the power source folds into the following categories:

1. *Surface power:* Current ROV technology for larger vehicles has the vehicle sleeping at the surface (aboard the deck of a multiservice vessel (MSV), drilling rig, or production platform). Future vehicles will "sleep at the bottom" yet could be powered from the surface. The power is routed via a remote docking station powered from an umbilical to the controlling node or platform. This allows for unlimited power but retains some type of tether to the surface for electrical connection/power.
2. *Onboard power:* This is clearly the preferred method of power as it frees the vehicle of the parasitic drag and complexities of a tether. The problem remains one of power density to achieve the locomotion to the work site along with the tasking requirements.
3. *Hybrid:* There is a possible third solution whereby the vehicle can sleep at the bottom mated to a docking station where the battery is charged from the surface. The vehicle uses this onboard power (once undocked) to achieve the goals of its tasking.

With that preamble, the ultimate question to be addressed in this chapter is: *"Where is ROV technology today and where will it be in the future?"* This question will be answered by considering the three methods of power and control available, that is, via a powered tether mode (with copper or fiber telemetry/control), fiber-optic tether (for remote telemetry without remote power) and tetherless operating mode.

23.1 Standard ROVs

Today's standard ROVs, with power and communications down the tether or umbilical (or both), have reached the level of sophistication that the market required. When they first entered the oil patch in the 1970s and 1980s they were considered more of a nuisance to the divers (who ruled the waters) than an asset. That soon changed as the capability and reliability of the vehicles increased and the operating depth requirements went beyond diver depths. Once the oil companies understood the inevitable, they finally met the ROV manufacturers half way, actually more than half, and worked together to develop underwater equipment that could be installed, inspected, maintained, and repaired by an ROV.

There are no limits to the depths that an ROV can support. US Navy-funded vehicles broke the magic 20,000 ft (6096 m) barrier in 1990—first by the CURV III, which reached a depth of 20,106 ft (6128 m). This was followed less than a week later by the Advanced Tethered Vehicle (ATV), which reached a depth of 20,601 ft (6279 m). This record was short-lived when in 1995 Japan's KAIKO ROV reached the deepest point in the ocean—the Challenger Deep in the Mariana Trench—setting a record that can only be tied at 35,800 ft (10,911.4 m). That tie—we will call it a tie—happened when the *Nereus* vehicle reached the bottom of the Challenger Deep in 2009, recording a depth of 35,773 ft (10,903 m). The *Nereus* will be discussed more in the next section.

So, with such great acclaim for the existing state of ROV technology, where will it go from here? The greatest benefits of today's ROVs include:

- The operator can relax in a comfortable control room.
- High-definition cameras provide excellent visual feedback over high-bandwidth communication channels (fiber optics).
- The ROV can operate indefinitely (assuming no reliability problems are experienced).
- High-power hydraulics (up to 250 HP—or more!) allow for operation of tools and manipulators specifically designed to interface with the underwater equipment for conducting drilling and construction support or inspection, repair, and maintenance (IRM).
- Efficient thrusters, along with the sufficient power plants, allow the ROVs to operate in higher currents.
- These vehicles are designed to be transportable, in many cases, to vessels of opportunity to conduct work in remote locations.

That is an excellent set of attributes! However, there are also drawbacks to today's larger ROVs:

- The larger the vehicle, the larger the footprint for the cable handling as well as launch and recovery equipment.
- If not mobilized on a drilling rig of some type, the larger ROV system requires a dynamically positioned (DP) MSV.
- The umbilical/tether limits their operational footprint, especially when operating from the rig itself. Operation from an MSV extends the range of the vehicle but at day rates that may range to $100,000 or more.
- Although the manipulators and viewing systems provide adequate capabilities, they still do not give the operator an anthropomorphic-like remote telepresence at the work site.

Where will the ROVs go from here? The bottom line is still pushed by the "bottom line." Money talks. The ROV truck is there and it is powerful. So when there is a problem offshore, the ROV can be brought into play and the problem corrected. But the time it takes to fix the problem costs a lot of money. A production platform that is shut down while an ROV solves the problem may lose hundreds of thousands of dollars per day, whereas a deepwater drilling project, rig and support vessels included, could be $1 million or more per day. Just the cost of an offshore vessel and crew can reach levels of over $100,000 per day (Figure 23.1). Therefore, the working end of the vehicle has to become more efficient.

Non-anthropomorphic manipulator systems do not lend themselves to natural human motions. Our arms are not designed like a praying mantis, so our work systems should also not be designed with elbows only up or down. Today's manipulators and their respective controllers are very capable, but they can be better. If the controllers and manipulators do not feel natural to the operator, then fatigue will set in much faster.

Considerable research has been conducted in the past to give the operator the feeling of being at the work site. This remote presence included:

- Anthropomorphic manipulators with force feedback
- Stereo vision
- Binaural hearing

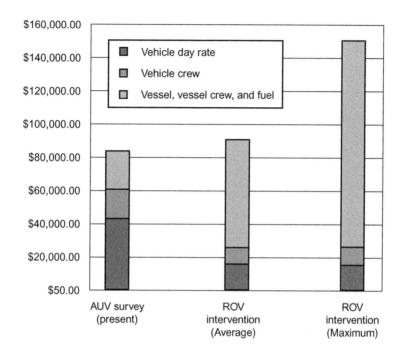

FIGURE 23.1

Offshore crew and vessel daily costs.

(Courtesy 3U Technologies.)

With today's advancements in robotics technology and the miniaturization of electronics, it should only be a matter of time until our underwater work systems evolve into something that gives the operator the feeling of telepresence during work tasks. As the operator becomes more efficient, the tasks are completed more quickly, the platform's downtime will be reduced and the company will increase its profits.

The previous comments were primarily addressing the MSROV to heavy-duty (HD) WCROVs. However, the same considerations and developments apply to the observation class ROVs, especially those that begin to include a work capability, albeit at a much lower level than the HD vehicles. The advancements in electronics and fiber-optic tethers have allowed the miniaturization of many vehicles and provided the bandwidth that allows excellent sensory feedback.

We now have the ability to build the vehicles that can get to the worksite. But who will invest into the future to develop the next generation of work systems? It seems that, as in the past, the work system manufacturers and offshore companies will have to meet in the middle so that the funds are available to solve the problem. Recent advancements in robotic technology are monumental. It seems time to exploit those advancements by applying them to the next generation of underwater vehicles.

The other drawbacks listed earlier, regarding the offshore use of ROVs, are discussed in the following sections.

23.2 Fiber-optic linked ROVs

One of the previously mentioned drawbacks to an ROV is the limitations imposed by the umbilical/tether. Yes, it provides the power and communication link necessary for the ROV to operate, but (like a leash on a dog) it constrains where it can go. And, in some (read, "Most") environments, it can become entangled—possibly causing the loss of the vehicle. One approach would be to change from a bulky and costly integrated power/telemetry cable to an inexpensive expendable (or reusable) copper-less fiber-optic tether.

The thought of using a simple fiber-optic tether for an ROV (expected to perform any level of underwater work) would have been unheard of in years past as the vehicle would be forced to carry its own energy. Today, this is commonplace on many autonomous underwater vehicles (AUVs) used in military applications for intelligence, surveillance, and reconnaissance (ISR). Much of this technology has been pushed by the world's navies for application to mine countermeasures (MCM). An example of such technology is the K-STER (Figure 23.2) developed by ECA for mine neutralization. The battery-powered vehicle can use an expendable fiber-optic tether or a reusable reinforced fiber-optic cable on a surface winch.

One of the main factors allowing for the utility of such vehicles is the advancements made in battery technology. Just as in the robotics industry (and other similar applications), the undersea industry just had to exploit the advancements made by land-based commercial firms. Today, underwater vehicle manufacturers have a range of excellent batteries for undersea applications. As an example, Bluefin Robotics has developed an easily "swappable" 1.5 kWh pressure-tolerant, lithium polymer battery (Figure 23.3).

When considering the use of a self-powered, surface-linked (via fiber-optic) vehicle, one must consider a number of the drawbacks for such an application. The first is the cost of the fiber, which

FIGURE 23.2

ECA's K-STER MCM vehicle.

runs about $2.25/yard ($2.06/m) for a 0.055 inch (1.38 mm) micro-cable spooled for use. If the fiber is expendable the costs mount up. Also, if a method is not employed to recover the fiber after it is expended, then the result is an entanglement hazard if future operations are planned for that area. If the fiber is a reinforced cable, then the vehicle can be operated like a standard ROV and the fiber recovered onto a winch, thus keeping it from becoming a hazard. Should the communication link become severed, the vehicle will require some type of fail-safe auto-homing or recovery system for the vehicle to be reliably retrieved.

For the smaller vehicles, with less onboard energy, their operational time/area will be limited. However, the inspection requirements for offshore structures should be able to keep a fleet of inspection vehicles very busy. A benefit of an expendable fiber is the freedom that the vehicle has to maneuver, which can allow it to conduct inspections in more confined spaces.

One of the best examples of the application of an expendable fiber used by an ROV is the *Nereus* vehicle (Figure 23.4) developed by Woods Hole Oceanographic Institution (WHOI). In May 2009, the *Nereus* vehicle reached a depth of 35,773 ft (10,903 m) in the Challenger Deep of the Mariana Trench. The 26 hour dive required a descent of 8.5 hours and resulted in nearly 11 hours on the bottom. The vehicle's high-bandwidth tether provided the link for video and command/control as it surveyed the ocean floor. The *Nereus* is a unique vehicle in that it can operate in either the ROV or AUV mode. When the bottom operations are completed, the vehicle severs its optical tether and returns autonomously to the surface for recovery.

The *Nereus* vehicle, with its ROV/AUV capabilities, provides the perfect segue for the book's final section that includes AUVs and the emerging hybrid vehicles.

FIGURE 23.3

Bluefin's swappable lithium polymer battery.

(Courtesy Bluefin Robotics.)

FIGURE 23.4

Nereus vehicle.

(Courtesy Advanced Imaging & Visualization Lab©, Woods Hole Oceanographic Institution (WHOI).)

23.3 Autonomous ROVs

AUVs were once confined to navy laboratories or academic institutions. Today, the technology has transitioned to industry where hundreds of vehicles are operating offshore. Companies like Bluefin Robotics (with their Bluefin line of AUVs) and Kongsberg Maritime (with their line of Hugin vehicles) have come to dominate the offshore survey market. Kongsberg also made a strategic play when Hydroid LLC, with their REMUS line of vehicles, became a wholly owned subsidiary in 2007. Such companies are keeping their military and commercial users well supplied with highly reliable systems.

23.3.1 Structurally compliant vehicles

A step toward autonomy, without using a free-swimming vehicle, is the structurally compliant vehicle. One such vehicle currently under development is the SeaTrepid Riser Crawling vehicle (Figure 23.5). With high-risk/value portions of oilfield infrastructure (e.g., risers and moorings), a structurally compliant vehicle may be semipermanently mounted to the structure for assuring asset integrity. The vehicle will regularly "crawl" the structure while using a suite of sensors for characterizing the structure.

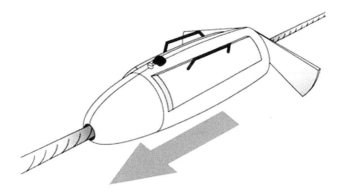

FIGURE 23.5

SeaTrepid's Riser Crawling vehicle.

With the autonomous riser crawler, the vehicle is mated to a subsea docking station (Figure 23.6(a)). The docking station is linked to the host platform via power and data communications. The vehicle communicates wirelessly with the docking station. Upon command, the vehicle departs the docking station, travels the length of the riser, and then redocks (after inspection) with the docking station for both dumping data gathered during inspection and for battery recharge (Figure 23.6(b)). This method isolates the vehicle's locomotion energy requirement to simply traveling the length of the riser (thus avoiding the considerable energy requirement for station-keeping in a subsea current).

23.3.2 AUVs

The next step in AUV incorporation is to remove them from structural support, giving them a full free-swimming capability. But to what level can such autonomous systems be incorporated into the offshore environment?

For some time now, AUVs have been used offshore for environmental and geophysical surveys. One of the more successful companies performing AUV surveys for the oil and gas industry is Lafayette, LA—based C&C Technologies. With their line of C-Surveyor vehicles, based on the Hugin AUV developed by Kongsberg, C&C has cornered the market for pipeline route and site surveys in support of deepwater exploration and development. As of this writing, C&C is approaching 125,000 miles (approx. 200,000 km) of surveys for over 60 clients (and climbing!). The question posed in the early days of AUVs entering the offshore industry was whether their use would be profitable. An AUV surveying at 4–6 knots can conduct a survey an order of magnitude faster than an ROV operating at less than 1 knot—or the now-rather-rare towed vehicles. A recent contract that C&C signed with Petrobras was for 730 days of support for a $50 million price tag!

The prior data shows that the AUV is the tool to use when conducting oil and gas exploration and pipeline surveys, however, what role would they play around the offshore rigs, platforms, and installations? Lockheed Martin believes AUVs can play a significant role in this area as they can,

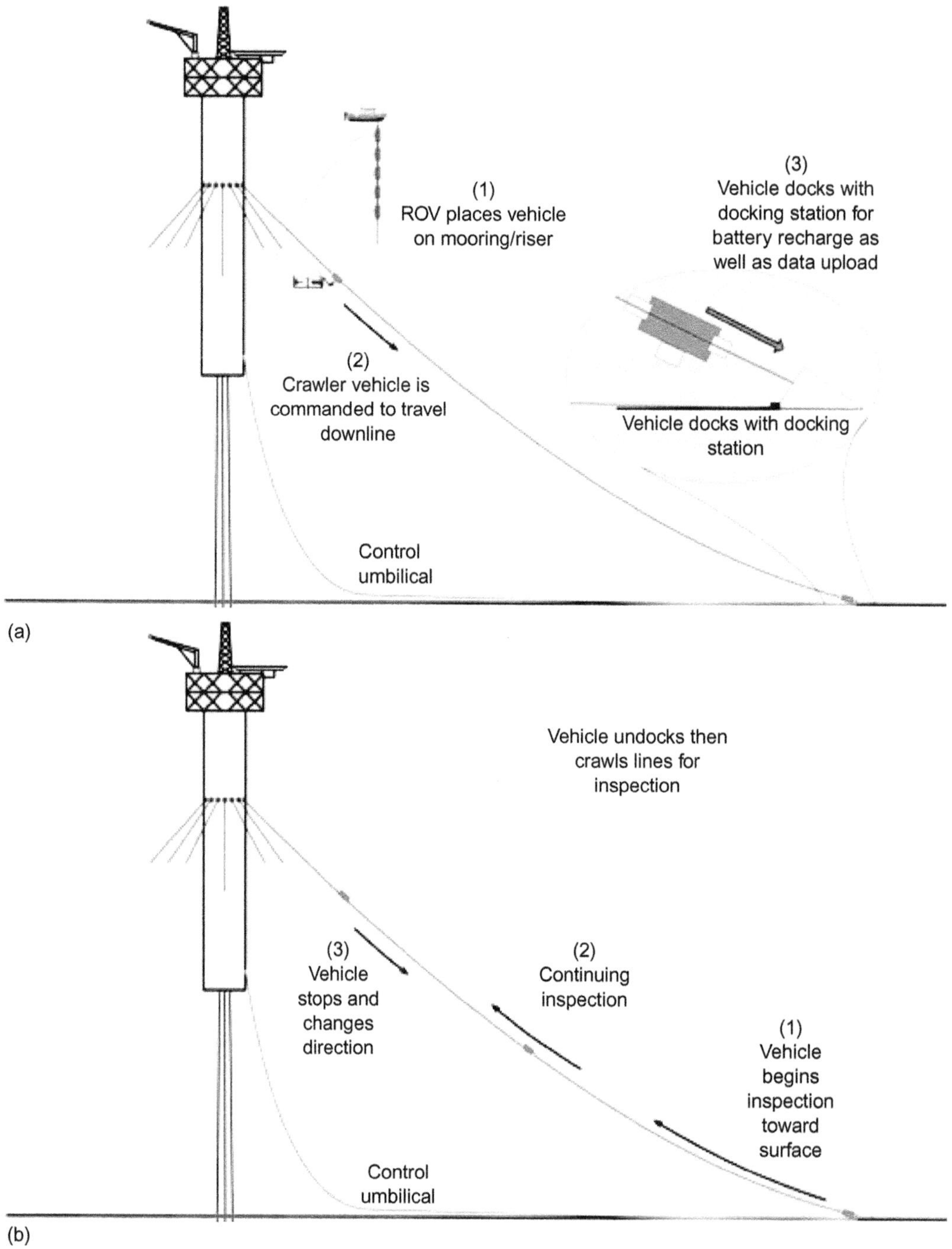

(1)
ROV places vehicle
on mooring/riser

(2)
Crawler vehicle is
commanded to travel
downline

(3)
Vehicle docks with
docking station for
battery recharge as
well as data upload

Vehicle docks with docking
station

Control
umbilical

(a)

Vehicle undocks then
crawls lines for
inspection

(3)
Vehicle
stops and
changes
direction

(2)
Continuing
inspection

(1)
Vehicle
begins
inspection
toward
surface

Control
umbilical

(b)

FIGURE 23.6

(a) The crawler is installed on the rope or riser and then is mated with the docking station. Then (b) it performs inspection.

FIGURE 23.7

Lockheed Martin's Marlin AUV.

(Courtesy Lockheed Martin.)

once the technology is properly integrated, enable faster, safer, and more efficient means of surveying offshore platforms and installations compared to traditional methods. They have developed the Marlin™ AUV (Figure 23.7), a 10 ft long (3 m) vehicle that can carry a 250 lb (113 kg) payload, travel at speeds up to 4 knots, and conduct offshore inspections in less time than ROVs (due to logic-driven operations versus teleoperations), and generate high-resolution 3D geo-tagged models of the surveyed structure within days of survey completion. The Marlin's advanced autonomy enables safer operations (by requiring fewer people at sea to perform inspections) thus reducing risk to operators. Fully autonomous operations also allow safer vehicle launch and recovery. With its hovering capability, the highly maneuverable vehicle is capable of getting into tight places for surveys and inspections. Considering the damage done by recent storms in the Gulf of Mexico (and elsewhere), there may be considerable work for such systems in the future. During recent surveys in the Gulf of Mexico, the Marlin showcased its ultrahigh-resolution Coda Octopus Echoscope™ 3D imaging sonar by conducting a structural survey of an underwater platform, allowing for a full 3D sonar point cloud rendering of the target structure (Figure 23.8) in less than 4 hours. Using AUVs equipped with the necessary sensors and instrumentation to conduct surveys of offshore platforms and installations will, once fully integrated, help operators perform more frequent surveys while at sea, receive higher fidelity information, and conduct safer operations compared to using divers or ROVs.

Research conducted by Subsea 7 and SeeByte has resulted in the development of the Autonomous Inspection Vehicle (AIV). Development of the prototype vehicle was done in conjunction with Chevron and BP as part of their AUV Strategic Research and Technology Development Program. The now operational AIV is a battery-powered inspection-class ROV used to conduct efficient underwater inspections of subsea infrastructure to include structures, risers, etc. The AIV includes a hydrodynamic design optimized for the demanding subsea environment, a 3D forward-looking sonar,

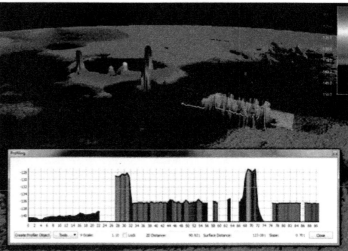

FIGURE 23.8

Marlin 3D sonar survey of offshore structure.

(Courtesy Lockheed Martin.)

color inspection video cameras, profiling sonar, and a downward-looking camera. It can use acoustic, WiFi, and satellite communications (Figure 23.9).

With today's ability to process visual and acoustic data for creating 3D rendering/modeling, there is a great potential for AUV use offshore. However, many of the platforms and installations provide a cluttered environment when it comes to underwater navigation using any type of acoustic system, especially when operating within the structure itself. An old technology that is being reassessed for use underwater is radio frequency (RF) communications. WFS Technologies (Edinburgh, Scotland) has developed underwater equipment for wireless communications, TV transmission, and power transfer. The range is limited when compared to the capabilities of acoustic modems, but the problem with acoustics in a cluttered environment is eliminated. Communications capabilities include short-range broadband at 25–156 kbps at 4–8 m up to longer range, low bandwidth of 10 bps–8 kbps at 10–30 m. Such techniques will be useful throughout the oilfield (Figure 23.10). Table 23.1 compares the various technologies for underwater communications.

Although such environments are cluttered, they are well known. And with today's ability to process optical and/or sonar images, AUVs will eventually be able to orient themselves within their working environment by referring to their knowledge base that can include the structural design of the installation being inspected. If land-based robots can maneuver within an environment using optical feedback to their onboard computer, then it is only a matter of time when the same can be performed offshore.

Conducting surveys of and around an offshore platform is only one minor part of the equation. Today's offshore oilfields are complex with wells and pipelines covering vast areas of the seafloor. Future installations will become deeper, more remote, and eventually under the Arctic ice. Once again, considering the cost of using a support vessel with an ROV to perform such routine

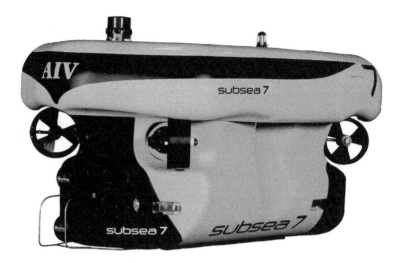

FIGURE 23.9

The Autonomous Inspection Vehicle.

(Courtesy Subsea 7.)

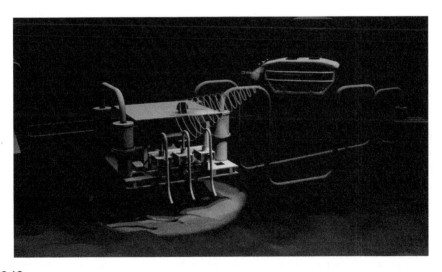

FIGURE 23.10

WiFi node communication link.

(Courtesy WFS Technologies.)

inspections, or post-hurricane inspections, the future is bright for both ROVs and AUVs. This is especially important following devastating events such as Hurricane Katrina, where many of the production installations may not be allowed to operate until the post-hurricane inspections are

Table 23.1 Pros and Cons of Various Underwater Communications Technologies

	Pros	Cons

Acoustic	• Proven
	• Up to 20 km range
	• High precision
		• Limiting conditions:
		  — Background noise
		  — Multipath
		  — Reverberation
		• Poor performance in shallow depths
		• Narrow bandwidth
		• Latency
		• May harm marine life
		• Very poor transmission at air/water interface

Optical	• Very high bandwidth • Low physical and power footprint	• Turbidity and suspended particles obstruct transmission • Alignment and focus require calibration • Short range in water • Air/water interface degrades transmission
RF	• Tolerant of water depth, turbidity, suspended particles/aeration, acoustic noise, air/water interface, etc. • Low latency • High bandwidth	• Very limited range • Low energy efficiency • Vulnerable to in-band electromagnetic interference (EMI)

Source: Courtesy WFS Technologies.

completed to assure a safe return to manned platform operations. It may be much cheaper to keep a fleet of AUVs (assigned to the facility and on standby for such an operation) than a fleet of MSVs.

And should the operator want real-time data sent back on what the AUV is finding, instead of recovering the vehicle and downloading the data, an unmanned surface vehicle (USV) could follow the AUV and provide a communication link back to the operator. Acoustic communications, especially when transmitting directly to the surface, are quite capable of providing such feedback to the operator. Also, by coupling a USV and an AUV, the AUV's navigation can be updated when required because the USV knows its exact position through GPS.

Another communications technique, as discussed in the following sections, is the provision of a fiber-optic cable, with communication nodes, that follows the pipelines back to the control station on shore or onboard the floating production, storage, and offloading (FPSO) vessel. The AUV can dock with the node, download the data, recharge its batteries, and continue the operation.

As explained throughout this text, ROVs are capable of many things. From sensor delivery to tooling operations, the ROV is quickly reaching maturity. The mundane operations of IRM will be continually thrown off to the logic-driven vehicles with onboard power (AUVs). The continued "brick wall" for operation of the subsea field will be the mechanical ability to turn valves, pull flying leads, and perform other mechanical operations for maintaining the subsea field. And the brick wall is the power density of current battery technology to muscle the vehicle to the work site and perform the mechanical operations sufficiently to complete the assigned task. For the subsea oilfield of the future, the ability to combine logic-driven operations with sufficient power delivery will determine if the ROV will be replaced by fully autonomous operations. And the energy density of future batteries will only go so far, so the systems that require the energy, the manipulators and tools, need to become highly efficient electrically driven systems.

Again, by teaming robotic systems in a cooperative manner, time (and thus money) can be saved. Such teaming leads us to the next and most advanced combinations of technology, the hybrids, where the ROV and AUV become one.

23.3.3 Hybrids

If an AUV can provide the hands-off, long-range support described in the previous section, think about the possibilities if that AUV could also perform the IRM duties that require an ROV with a work system and tools. And what if such a system could remain underwater for months at a time performing the necessary tasks or on standby for when it is needed? Such a hybrid AUV/ROV could have a series of underwater garages or docking stations where it could recharge its batteries and communicate with an operator on shore or on the production platform. It could also have the capability of reaching a remote site where it can dock with a communication node and turn over control to an operator at a remote location who could launch and operate the integral ROV.

Well, this concept is being turned into reality by CYBERNETIX. The French company is developing the Subsea Work, Inspection, and Maintenance with Minimum Environment ROV (SWIMMER). CYBERNETIX successfully demonstrated the SWIMMER prototype (Figure 23.11) with partners IFREMER and the University of Liverpool, in 2001 during full-scale sea trials. Since then, CYBERNETIX has worked with Statoil and Total to develop an operational system.

The concept of operation (Figures 23.12 and 23.13) is for the vehicle to be launched from an MSV or FPSO vessel (or from shore). The vehicle can then dock with the subsea docking station where power and data cables previously routed back to the operating consoles aboard the host vessel are embedded into the field control umbilical. This could resolve the power density question discussed previously! Further, the actual control of the vehicle can be via teleoperation performed from the host vessel or (alternatively) via satellite to an operations center ashore.

The ability to have hybrid systems such as the SWIMMER will be extremely valuable in harsh environments. In the North Sea, sea states can reach level 7 where MSVs, assuming they are even around, are not about to be launching an ROV. Having a hybrid system in place on the seafloor would eliminate that problem. One of the next hazardous frontiers is the Arctic. Within that environment complications exist from the sea ice that could form and prevent an MSV from reaching the area. With the hybrid already in place, it can be controlled remotely to perform necessary IRM operations. The system can also swim into the area below the ice to dock with the ice-covered subsea production system. The present SWIMMER vehicle with its lithium ion battery system has a projected range of 31 miles (50 km). With the use of fuel cells (which have been demonstrated on other AUVs) or auxiliary battery packs, this range can effectively be doubled.

SAAB Seaeye Ltd. (Fareham, Hampshire, UK) is also involved in advancing the AUV technology for offshore support. Their Sabertooth hybrid AUV (Figure 23.14(a)) will be capable of a

FIGURE 23.11

SWIMMER prototype.

(*Courtesy Cybernetix.*)

FIGURE 23.12

SWIMMER launches ROV from underwater docking station.

(*Courtesy Cybernetix.*)

FIGURE 23.13

SWIMMER's ROV inspects the subsea production system (SPS).

(Courtesy Cybernetix.)

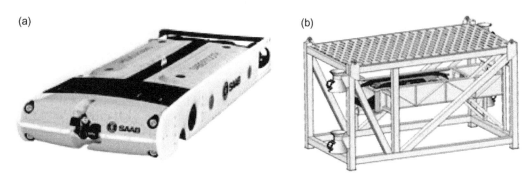

FIGURE 23.14

Sabertooth AUV (a) and docking station (b).

(Courtesy Saab Seaeye.)

12 month bottom duration. The docking station will have a 5 year lifetime (Figure 23.14(b)). It will also be capable of both RF and acoustic communications while still using a fiber-optic tether (if high-bandwidth control is necessary). Inductive coupling can provide 150 W−1 kW power for recharging the onboard batteries. An interesting capability of the Sabertooth is its ability to maneuver in any direction/orientation in order to position itself to use underwater tooling. The unique design of the vehicle allows it to position vertically, horizontally, upside-down, etc., in order to align with the object to be worked upon.

The previously described systems and research will help produce the vehicles of the future. Visionary exploration and production companies (such as Chevron) continue to investigate methods to close the technology gaps in bringing resident AUVs into the oilfield. However, as we have seen in the past (when the ROVs first tried to break into the offshore arena) it takes more than one or two companies to make this happen. Standards must be developed, preferably ahead of time, that will allow the manufacturers to develop systems for supporting all companies involved in offshore exploration and development.

Required from industry will be consensus standards allowing the enabling technologies of standardized docking stations (for disparate types of mobile robotic vehicles), communications protocols, and means by which onboard power can be recharged. Both the generation cost and acceptance of these standards must be broadly shared by stakeholders. One such initiative under way is by the Chevron-led research organization DeepStar. The 11304 subcommittee is developing recommended standards for autonomous vehicle docking stations (for power and data transfer between vehicle and host). This committee is tasked with formulating recommendations to API for adapting these enabling technologies through proposed modifications to API RP17. Another government developmental organization is the United States' Research Partnership to Secure Energy for America (RPSEA) established to foster innovation within the energy industry. It is only through broad financial participation between small innovative companies and larger industrial players that rapid technological revolution will arrive.

23.4 The crystal ball

So, what does the crystal ball say about our future in the ROV field (including AUVs and hybrids)?

In Douglas-Westwood's "World ROV Market Forecast 2011−2015," they project that annual expenditures on ROV support for underwater operations will rise from $891 million in 2010 to $1.7 billion in 2015. This is driven primarily by the offshore oil and gas sector. Oil prices are rising, deepwater activity is surging, and the number of offshore rigs and subsea construction vessels will be reaching new highs to meet the demand.

And the market also looks good for AUVs in the future. According to Douglas-Westwood's "World AUV Market Forecast 2012−2016," the market growth will continue over the next 5 years. The size of the 2012 AUV fleet was 560 active vehicles, up from 390 in 2009. The report forecasts that the number will rise to 930 active AUVs by 2016. That should keep a few manufacturers busy.

On top of the basic forecasts, one must also figure in what Mother Nature has in store for the future. One of the guarantees in life is that she will raise havoc on an annual basis somewhere in the world. Catastrophic events certainly call for assets on an immediate scale that may not be readily available. And throw into this mix the human factor. Something will eventually go wrong, producing another time-critical requirement. As an example, consider the Macondo blow-out in the US Gulf of Mexico. That disaster certainly put a considerable number of vessels, ROVs, and AUVs to work trying to assess the damage and correct the problems. Since then, oil companies have developed and staged massive capping stacks and containment systems (such as those eventually used on Macondo) around the world, at an incredible cost, in case (or for when) a similar disaster happens again. This is a great step to protect the environment in the future, but if only a small amount of those funds were put into more vehicle surveys and preventive maintenance, such extreme expenditures for containment systems may not be required.

As reliable as today's offshore systems are, with the combination of human nature, Mother Nature, Murphy's Law and the laws of physics (and statistics in general), something catastrophic will happen again in the future. The good thing is that we have the technology to address the problems as they arise, but probably not the amount of equipment and support necessary to bring things back on line in a timely fashion. That was more than proven in the aftermath of 2012's perfect storm when Hurricane Sandy hit the northeastern United States just as a winter storm came in from the west, all timed to arrive during a full-moon high tide. Who would have expected that level of devastation? The lack of assets to bring the infrastructure back to life in a timely manner only exacerbated the storm's lasting effect. The point of this discussion is that there will eventually be an overwhelming need for ROVs and support vessels due to the next disaster. But they will not be there for immediate support because the bottom line typically plans for the expected, not the unexpected.

On a more positive note, the ROVs and AUVs will become more capable and reliable as optimized production lines crank out more and more standardized vehicles. The inspection-class vehicles will also benefit as technology progresses, making them cheaper, more efficient, and having increased capability. Miniaturized systems are becoming available at a cost that may soon have them hanging near the checkout stations at the local marine supply store. An ROV for every yacht? Why not?

With all that is mind, what is missing? What must or will be developed? Here is a shopping list for the future:

- *Telepresence:* systems, both large and small, that put the operator in the environment through the use of stereo vision, audio feedback, and anthropomorphic manipulators.
- *Electric manipulators and tools:* for electric vehicle applications where the loss of energy by converting from electrical power format to hydraulics is not desired. Anthropomorphic design should also be considered when the operator is in the loop.
- *Connector-less interfaces:* for short-range data transfer and recharging of batteries. Whether inductive, optical, RF, or some other approach, they must be environmentally robust (i.e., corrosion, marine growth, hungry creatures, etc.) so they will work when the AUV shows up and needs to use them.
- *Navigation:* for operations in a cluttered environment such as an offshore structure. Whether through acoustics, RF, optics, preprogrammed knowledge, or a combination, future AUV inspections will need this technology.
- *Sensor integration:* the world is heading to 3D. The integration of 2D and 3D systems, whether optical or acoustic, will be required to efficiently present the data to the customer.
- *Teamed USV/ROV/AUV:* for operations where real-time feedback is necessary when using an AUV. A companion USV can provide the link when remote or large areas are being covered. For applications such as pipelay touchdown monitoring, having a USV deploying the ROV will remove the cost of the chase vessel and replace it with a smaller and less costly unmanned ROV deployment platform.
- *Simulators:* as we have seen in the training of ROV operators, training is critical to increase operational efficiency. And this is the case whether there is an operator directly in control or if

the vehicle's onboard pilot is running the operation. Both must be trained to do their job efficiently, correctly, and without damage to either the vehicle or subsea structure.

- *The "Inter-Sea-Net":* for future applications as discussed below.

It is time for one final prophecy. It was not that long ago that the telephone was the primary means of communication, and that was done by dropping coins in a pay phone, when people actually talked to each other. Then came the Internet, e-mail, Facebook, etc., and with the addition of cell/smart phones, iPads, etc., verbal communication was substantially curtailed (or essentially eliminated). Any information one needs can be found with the touch of an app on a digital device. Technology users can roam the world because the infrastructure is in place to allow him/her to communicate with anyone at any time from anywhere. This did not happen overnight.

So, where is this leading? When we take into consideration the extent of future offshore developments and the requirement to operate, monitor, and repair all the equipment, a requirement begins to appear—that requirement is the need for an "Inter-Sea-Net." Again, it will not happen overnight. The oil companies and commercial ROV developers did not develop their integrated approach to subsea equipment and intervention overnight either, but they did develop it.

What better way to solve future needs than to develop an integrated offshore infrastructure? One where all equipment is connected in a fashion that allows hybrid vehicles to conduct IRM operations, where advanced sensors can be installed, monitored, and replaced to sense the environment and provide a warning when something bad is about to (or has) happened. Higher level messaging-type commands are being developed on the military side, such as Joint Architecture for Unmanned Systems (JAUS), for directing individual robotic assets with goal-oriented tasking (without direct human-in-the-loop control). Or perhaps an infrastructure where AUVs can navigate, survey, recharge, and provide their feedback in real time to the operator who is on land in a warm, comfortable control room. An environment where the operator touches an app to bring up the status of all equipment he/she is responsible for.

Oceaneering and C-Innovations are taking a step in that direction with their central command centers, or war rooms, where the health and safety of their systems can be remotely monitored. This is a step in the right direction, but the path to a fully integrated oil patch is a very long road indeed.

There is a lot of communication capability already in place; however, future installations need a lot more. The future oilfield must provide for the installation of all the necessary nodes and interfaces, whether RF links on an offshore structure or acoustic beacons/modems throughout the oilfield, to allow for the future integration of a robotic infrastructure. A robotic world where robots exist that help service the other robots so they do not need to be recovered as often. Farfetched? Not really. The world of robotics is moving a lot faster than the semiviscous movement of offshore technology. If desired, and planned for, future offshore oilfields will become a fully robotic domain.

This could also include the MSVs. If a hybrid AUV/ROV can work underwater, then an MSV dedicated to the launch, recovery, and operation of an ROV, or AUV, can be developed that is operated remotely from a shore station. A much smaller, robotic MSV, without the cost of a crew might be a very economical development. If a surgeon can use robotic arms to perform an operation from a remote location, then an ROV operator could certainly run a robotic MSV (along with all of its subsystems—including the ROV!) from afar.

23.5 The bottom line

So, what's the bottom line? The bottom line is the bottom line—profits! But profits are easily lost when a disaster happens. Let us again consider the 2010 Macondo disaster in the Gulf of Mexico. Could it have been prevented with more logic-driven processes? Maybe not, since it was the result of well design and human failings. But the immediate need for robotic intervention, by both AUVs and ROVs, to solve the problem and end the environmental catastrophe could not be immediately met.

But Macondo is just the latest major offshore disaster ... not the only one to date. Could other disasters have been prevented if more inspections and maintenance had been done? What about the future? Where will the funding come from to develop the hybrid vehicles, the advanced work systems, or an "Inter-Sea-Net"–type infrastructure?

Most vehicle operators run their business on a tight budget—responding to contracts as they can get them. They cannot afford to change the system. The oil companies have realized this in the past when they went beyond diver depth and needed to use ROVs for underwater intervention, so they changed the system. Now ROVs conduct all IRM at those depths.

It will take the combined cooperation of both the oil companies and government to again change the system. Offshore operators have periodically teamed to solve specific common issues, but a wider collaboration is needed to address the ever-evolving needs of the future oilfield. As mentioned above, Chevron took the lead and teamed with several oil companies to form the DeepStar consortium to address common issues. But this organization is by no means industry-wide. The US government recognized the need for high-risk/high-reward revolutionary developmental projects to assure its future energy independence. What arose from this perceived need is the quasi-governmental Research Partnership to Secure Energy for America (RPSEA). RPSEA's goal within its mission statement is "... to identify and develop new methods and integrated systems for exploring, producing, and transporting-to-market energy or other derivative products from ultra-deepwater and unconventional natural gas and other petroleum resources ...."

What RPSEA is doing is a step in the right direction. But there must be more. In order to develop the vast array of technologies that will not only efficiently integrate the robots of the future (and hopefully prevent the next costly disaster), operators and government will be financially ahead of the game if they will cooperate for the common goal of safety and energy security. Tens of millions of dollars are lost in a flash when an oil and gas production facility goes down. Why not invest those tens of millions now to change the future ... to develop the robotic infrastructure of the future—the "Inter-Sea-Net" and the robots that operate within it.

We now look afar and see the end from the beginning. It is indeed an exciting time for the ROV industry.

Bibliography

American Petroleum Institute, 2004. Recommended Practice for Remotely Operated Vehicle (ROV) Interfaces on Subsea Production Systems (Identical to ISO 13628-8: 2002). American Petroleum Institute (Recommended Practice 17H).

AMMTIAC (A US Department of Department of Defense Analysis Center), 2006. Corrosion Prevention and Control: A Program Management Guide for Selecting Materials. (Alion Science and Technology Handbook).

Baker, D.H., 1957. Basic Principals of Unconventional Gyros. (Masters Thesis. Massachusetts Institute of Technology).

Bash, J., (Ed.), Handbook of Oceanographic Winch, Wire and Cable Technology, third ed. Available at: <http://www.unols.org/publications/winch_wire_handbook__3rd_ed/index.html>.

Benthos, Inc. StingRay Mk. II Operations and Maintenance Manual Benthos, Inc. Available at: <http://www.benthos.com/>.

Bohm, H., Jensen, V., Build Your Own Underwater Robot. West Coast Words. ISBN 0-9681610-0-6.

Bonde, L.W., Shelley, P.E., 1972. A Survey of Underwater Winches (Hydrospace Research Corporation Technical Report No. 348).

Bowditch, N., 2002. The American Practical Navigator. National Imagery and Mapping Agency, ISBN 1-57785-271-0.

Breiner, S., 1999. Applications Manual for Portable Magnetometers. (Geometrics, Inc. Technical Paper).

Burcher, R., Rydill, L., 1994. Concepts in Submarine Design. Cambridge University Press.

Busby, R.F., 1976. Manned Submersibles. Office of the Oceanographer of the Navy.

Busby, R.F., 1979. Remotely Operated Vehicles. Available at: <http://voluwww.archive.org/details/remotelyoperate00rfra>.

Cernasov, A., 2004. Digital Video Electronics. McGraw—Hill, ISBN 0-07-143715-0.

Cpl. Robert G. Teather, C.V., 1994. Royal Canadian Mounted Police Encyclopedia of Underwater Investigations. Best Publishing Company, ISBN 0-941332-26-8.

Craig, J.J., 2005. Introduction to Robotics. third ed. Pearson Prentice Hall, ISBN 0-201-54361-3.

Deep Sea Power & Light, Frequently Asked Questions. Available at: <http://www.deepsea.com/faq.html>.

Department of the Army, Corps of Engineers, 2002. Coastal Engineering Manual. Publication Number: EM 1110-2-1100.

Desert Star Systems LLC [Various Operational Manuals for the Dive Tracker Line of Acoustical Positioning Systems — Reprinted with Permission]. Available at: <http://www.desertstar.com/>.

Duxbury, A.C., Alison, B., 1997. An Introduction to the World's Oceans. fifth ed. William C. Brown Publishers, ISBN 0-697-28273-2.

Edge, M., 1999. The Underwater Photographer. second ed. Butterworth-Heinemann, ISBN 0-240-51581-1.

Everest, F.A., 1994. The Master Handbook of Acoustics. third ed. TAB Books [A Division of McGraw-Hill], ISBN 0-8306-4438-5.

Extron Electronics, 2001. UTP Technology. (Extron Electronics, Technical Paper).

Focal Technologies, Technical Brief Document Number 303. Fiber Optic Rotary Joints (2008 Moog Inc.).

Fondriest Environmental. Informational web site available at: http://www.fondriest.com/parameter.htm (reprinted with permission).

Fossen, T.I., 1994. Guidance and Control of Ocean Vehicles. John Wiley & Sons, ISBN 0-471-94113-1.

Freeman, R.L., 2002. Fiber-Optic Systems for Telecommunications. Wiley-Interscience, ISBN 0-471-41477-8.

Freeman, R.L., 2005. Fundamentals of Telecommunications. second ed. Wiley-Interscience, ISBN 0-471-71045-8.

General Corrosion Corporation, Corrosion and Cathodic Protection. Available at: <http://www.generalcorrosioncorp.com/TechInfo.html>.

Geometrics, Inc., 2000. Total Field Magnetometer Performance: Published Specifications and What They Mean. Geometrics, Inc. (Technical Report TR-120)

Gerwick Jr., B.C., 2007. Construction of Marine and Offshore Structures. third ed. CRC Press, ISBN 0-8493-3052-1.

Giancoli, D.C., 1991. Physics. third ed. Prentice-Hall, ISBN 0-13-672510-4.

Gillmore, B., et al., 2012. Field Resident AUV Systems: Chevron's Long-Term Goal for AUV Development. AUV, Southampton, UK.

Gray, R., 2004. Light sources, lamps and luminaries. Sea Technol. Mag. 45 (12), 39−43.

Gupta, P.C., 2006. Data Communications and Computer Networks. PHI Learning Private Limited, ISBN 978-81-203-2846-4.

Hedge, A., 2003. Ergonomics Considerations of LCD versus CRT Displays. (Cornell University, Technical Paper).

Heidersbach, R., 2011. Metallurgy and Corrosion Control in Oil and Gas Production. John Wiley & Sons Inc., ISBN 978-0-470-24848-5.

Hellier, C.J., 2013. Handbook of Nondestructive Evaluation. McGraw-Hill, ISBN 978-0-07-177714-8.

Huang, H.-M., 2004. Autonomy Levels for Unmanned Systems (ALFUS) Framework, vol. 1. National Institute of Standards & Technology, Special Publication 1011 (Terminology, Version 1.1).

Imagenex Technology Corporation Sonar Theory. Available at: <http://www.imagenex.com/sonar_theory.pdf> (Reprinted with Permission).

Inuktun Services, Ltd., ROV Seamor Operations Manual. Inuktun Services, Ltd. Available at: <http://www.seamor.com/>.

Jacobson, J., et al., 2013. DeepStar 11304: Laying the groundwork for AUV standards for deepwater fields. Mar. Technol. Soc. J. 47, 3.

Jazar, R.N., 2007. Theory of Applied Robotics: Kinematics, Dynamic and Control. Springer, ISBN 978-0-387-32475-3.

Joiner, J.T. (Ed.), 2001. NOAA Diving Manual. fourth ed. Best Publishing Company, ISBN 0-941332-70-5.

Jukola, H., Skogman, A., 2002. Bollard Pull. In: Paper Presented at the 17th International Tug & Salvage Convention, ITS 2002, 13−17 May 2002, Bilbao, Spain.

King, A.D., 1998. Inertial navigation—forty years of evolution. GEC Rev. 13, 3.

Kirkwood, W., 2006. AUV Technology and Application Basics. In: Notes from Tutorial given at Oceans 2006 at the Boston Hynes Convention Center, Boston, MA, USA.

Kongsberg Simrad, A.S., 2002. Introduction to Underwater Acoustics.

Lekkerkerk, H.-J., et al., 2007. Handbook of Offshore Surveying (Books One and Two). Clarkson Research Services Limited, ISBN 1-902157-73-7.

Lenk, J.D., 1997. Lenk's Video Handbook. second ed. McGraw-Hill, ISBN 0-07-037616-6.

Linton, S.J., et al., 1986. Dive Rescue Specialist Training Manual. Concept Systems, Inc. ISBN 0-943717-42-6.

Loeser, H.T., 1992. Sonar Engineering Handbook. Peninsula, ISBN 0-932146-02-3.

López-Higuera, J.M. (Ed.), 2002. Fiber Gyroscope Principals. John Wiley & Sons, ISBN 978-0471820536.

Lupini, C.A., 2004. Vehicle Multiplex Communication. SAE International, ISBN 0-7680-1218-X.

MacArtney Underwater Technology, 2008. NEXUS Mark IV Users Manual. (MacArtney Underwater Technology Revision F).

Medwin, H., Clay, C.S., 1998. Fundamentals of Acoustical Oceanography. Academic Press, ISBN 0-12-487570-X.

Milne, P.H., 1983. Underwater Acoustic Positioning Systems. E. & F. N. Spon Ltd., ISBN 0-419-12100-5.

Moore, J.E., Compton-Hall, R., 1987. Submarine Warfare: Today and Tomorrow. Adler & Adler, ISBN 091756121X.

Moore, S.W., et al., Underwater Robotics. Marine Advanced Technology Education Center, ISBN 978-0-9841737-0-9.

Murphy, R.R., 2000. Introduction to AI Robotics. MIT Press, ISBN 0-262-13383-0.

Ocean Innovations, Dummy's Guide to Marine Technology: A Useful Description of Technologies Used Beneath the Sea. Available at: <http://o-vations.com/marinetech/index.html>.

Olsson, M.S., et al., 2000. ROV Lighting with Metal Halide. White Paper on Metal Halide Technology Displayed on the Deep Sea Power & Light web site available at: http://www.deepsea.com (Reprinted with Permission) Revision 1 dated January 27, 2000.

Operational Guidelines for ROV, 1984. Marine Technology Society Subcommittee on Remotely Operated Vehicles, R. L. Wernli, Subcommittee Chairman.

Outland Technology, Inc., ROV Model Outland 1000 Operations Manual. Outland Technology, Inc. Available at: <http://www.outlandtech.com/>.

Panduit Corporation, 2003. The Evolution of Copper Cabling Systems from Cat5 to Cat5e to Cat6. (Panduit Corp. White Paper).

Parker Hannifin Corporation, 2007. Parker O-Ring Handbook ORD 5700. Parker Hannifin Corporation, Cleveland, OH, USA.

Paroscientific, Inc., Calibration of Digiquartz Instruments. Available at: <http://www.paroscientific.com/software.htm#manuals>.

Poynton, C., 2012. Digital Video and HD. Elsevier, ISBN 978-0-12-391926-7.

Remotely Operated Vehicle Subcommittee of the Marine Technology Society. Educational web site page available at: <http://www.rov.org/student/education.cfm>.

Remotely Operated Vehicles of the World, seventh ed. Clarkson Research Services Ltd., 2006/2007. ISBN 1-902157-75-3.

Seafriends.org, Underwater Photography—Water and Light. From their educational web page available at: <http://www.seafriends.org.nz/phgraph/water.htm>).

Segar, D.A., 1998. Introduction to Ocean Science. Wadsworth Publishing Company, ISBN 0-314-09705-8.

Serway, R.A., Faughn, J.S., 1995. College Physics. fourth ed. Harcourt Brace College Publishers, ISBN 0-03-003562-7.

Shamir, A., 2006. An Overview of Optical Gyroscopes Theory, Practical Aspects, Applications and Future Trends. (Adi Shamir Technical Paper).

Simpson, W., 2006. Video Over IP: A Practical Guide to Technology and Applications. Elsevier, ISBN 0-240-80557-7.

Smith, K., 1997. Cesium Optically Pumped Magnetometers: Basic Theory of Operation. Geometrics, Inc. (Technical Report M-TR91)

Sonardyne International Ltd., Acoustic Theory. Available at: <http://www.sonardyne.co.uk/theory.htm>.

Stolarz, D., 2006. Mastering Internet Video. Addison-Wesley, ISBN 0-321-12246-1.

Teledyne, T.S.S., 2003. TSS 440 Pipe and Cable Survey System Manual. VT TSS Ltd. (Document P/N 402196)

Teledyne, T.S.S., 2008. TSS 350 Survey System Manual. TSS (International) Ltd (Document P/N 402197)

Teledyne Impulse, Connector Terminology, 2009. Teledyne Impulse. Available at: <http://www.teledyneimpulse.com/PDF_FILES/96-Glossary_Connector%20Terminology_FINAL%202012/1-Glossary_Connector%20Terminology_FINAL%202012.pdf>

Tena, I., 2013. Autonomous Underwater Vehicles. Hydro International.

Texas Instruments Application Note 1031, TIA/EIA-422-B Overview. (Texas Instruments Literature Number: SNLA044A).

Texas Instruments, 2010. RS-422 and RS-485 Standards Overview and System Configurations. Texas Instruments (Literature Number: SLLA070D)

Thurman, H.V., 1994. Introductory Oceanography. seventh ed. Macmillan Publishing Company, ISBN 0-02-420811-6.

US Geological Survey. Educational web site on water science available at: <http://ga.water.usgs.gov/edu/earthhowmuch.html>.

Urick, R.J., 1975. Principals of Underwater Sound. McGraw-Hill, ISBN 0-07-066086-7.

Van Dorn, W.G., 1992. Oceanography and Seamanship. second ed. Cornell Maritime Press, ISBN 0-87033-434-4; Remotely Operated Vehicle Subcommittee of the Marine Technology Society Educational web site available at: <http://www.rov.org/student/education.cfm>.

von Karman, T., 1954. Aerodynamics. Cornell University Press, ISBN 07-067602-X.

Waite, A.D., 2002. Sonar for Practicing Engineers. third ed. John Wiley & Sons, ISBN 0-471-49750-9.

Wernli, R., 1998. Operational Effectiveness of Unmanned Underwater Systems. Marine Technology Society, ISBN 0-933957-22-X.

Wernli, R., 1999—2000. The present and future capabilities of deep ROVs. Mar. Technol. Soc. J. 33, 4.

Wilson, J.S., 2005. Sensor Technology Handbook. Elsevier, ISBN 0-7506-7729-5.

Wilson, W.D., 1960. Equation for the speed of sound in sea water. J. Acoust. Soc. Am.

Woodman, O.J., 2007. An Introduction to Inertial Navigation. (University of Cambridge, Technical Report No. 696).

Index

Note: Page numbers followed by "*f*" and "*t*" refer to figures and tables, respectively.

Printed and bound by CPI Group (UK) Ltd, Croydon, CR0 4YY

04/10/2024

01041197-0001